ИЗДАТЕЛЬСТВО
МИР

Н. С. ПИСКУНОВ

ДИФФЕРЕНЦИАЛЬНОЕ И ИНТЕГРАЛЬНОЕ ИСЧИСЛЕНИЯ

ТОМ II

ИЗДАТЕЛЬСТВО «НАУКА» МОСКВА

N. PISKUNOV

DIFFERENTIAL AND INTEGRAL CALCULUS

VOL. II

Translated from the Russian
by
George Yankovsky

MIR PUBLISHERS MOSCOW

CBS Publishers & Distributors Pvt. Ltd.

New Delhi • Bengaluru • Chennai • Kochi • Kolkata • Mumbai
Hyderabad • Uttarakhand • Nagpur • Patna • Pune • Jharkhand

ISBN: 81-239-0493-2

First Indian Reprint: 1996

This edition has been published in India by arrangement with
Mir Publishers, Moscow

Published by **Satish Kumar Jain** and produced by **Varun Jain** for
CBS Publishers & Distributors Pvt. Ltd.,
4819/XI Prahlad Street, 24 Ansari Road, Daryaganj, New Delhi - 110002
delhi@cbspd.com, cbspubs@airtelmail.in • www.cbspd.com
Ph.: 23289259, 23266861, 23266867 • Fax: 011-23243014

Corporate Office: 204 FIE, Industrial Area, Patparganj, Delhi - 110 092
Ph: 49344934 • Fax: 011-49344935
E-mail: publishing@cbspd.com • publicity@cbspd.com

Branches:

- ***Bengaluru:*** 2975, 17th Cross, K.R. Road, Bansankari 2nd Stage, Bengaluru - 70 • Ph: +91-80-26771678/79 • Fax: +91-80-26771680 E-mail: cbsbng@gmail.com, bangalore@cbspd.com
- ***Chennai:*** No. 7, Subbaraya Street, Shenoy Nagar, Chennai - 600030 Ph: +91-44-26681266, 26680620 • Fax: +91-44-42032115 E-mail: chennai@cbspd.com
- ***Kochi:*** Ashana House, 39/1904, A.M. Thomas Road, Valanjambalam, Ernakulum, Kochi • Ph: +91-484-4059061-65 Fax: +91-484-4059065 • E-mail: cochin@cbspd.com
- ***Kolkata:*** 6-B, Ground Floor, Rameshwar Shaw Road, Kolkata - 700014 Ph: +91-33-22891126/7/8 • E-mail: kolkata@cbspd.com
- ***Mumbai:*** 83-C, Dr. E. Moses Road, Worli, Mumbai - 400018 Ph: +91-9833017933, 022-24902340/41 • E-mail: mumbai@cbspd.com

Representatives:

- Hyderabad: 0-9885175004
- Patna: 0-9334159340
- Jharkhand: 0-9811541605
- Nagpur: 0-9021734563
- Pune: 0-9623451994
- Uttarakhand: 0-9716462459

Printed at:
J.S. Offset Printers, Delhi (India)

CONTENTS

CHAPTER 1 DIFFERENTIAL EQUATIONS

CHAPTER 2 MULTIPLE INTEGRALS

CHAPTER 3 LINE INTEGRALS AND SURFACE INTEGRALS

CHAPTER 4 SERIES

CHAPTER 7 OPERATIONAL CALCULUS AND CERTAIN OF ITS APPLICATIONS

CHAPTER 8 ELEMENTS OF THE THEORY OF PROBABILITY AND MATHEMATICAL STATISTICS

CHAPTER 9 MATRICES

CHAPTER 1

DIFFERENTIAL EQUATIONS

1.1 STATEMENT OF THE PROBLEM. THE EQUATION OF MOTION OF A BODY WITH RESISTANCE OF THE MEDIUM PROPORTIONAL TO THE VELOCITY. THE EQUATION OF A CATENARY

Let a function $y=f(x)$ reflect the quantitative aspect of some phenomenon. Frequently, it is not possible to establish directly the type of dependence of y on x, but it is possible to give the relation between the quantities x and y and the derivatives of y with respect to x: $y', y'', \ldots, y^{(n)}$. That is, we are able to write a **differential equation.**

From the relationship established between the variables x, y and the derivatives it is required to determine the direct dependence of y on x; that is, to find $y=f(x)$ or, as we say, **to integrate the differential equation.**

Let us consider two examples.

Example 1. A body of mass m is dropped from some height. It is required to establish the law according to which the velocity v will vary as the body falls, if, in addition to the force of gravity, the body is acted upon by the decelerating force of the air, which is proportional to the velocity (with constant of proportionality k); in other words, it is required to find $v=f(t)$.

Solution. By Newton's second law

$$m\frac{dv}{dt}=F$$

where $\frac{dv}{dt}$ is the acceleration of a moving body (the derivative of the velocity with respect to time) and F is the force acting on the body in the direction of motion. This force is the resultant of two forces: the force of gravity mg and the force of air resistance, $-kv$, which has the minus sign because it is in the opposite direction to that of the velocity. And so we have

$$m\frac{dv}{dt}=mg-kv \tag{1}$$

This relation connects the unknown function v and its derivative $\frac{dv}{dt}$; we have a *differential equation in the unknown function* v. This is the equation of motion of certain types of parachutes. To solve a differential equation means to find a function $v=f(t)$ such that identically satisfies the given differential equation. There is an infinitude of such functions. The student can easily verify that any function of the form

$$v=Ce^{-\frac{k}{m}t}+\frac{mg}{k} \tag{2}$$

satisfies equation (1) no matter what the constant C is. Which one of these functions yields the sought-for dependence of v on t? To find it we take advantage of a supplementary condition: when the body was dropped it was imparted an initial velocity v_0 (which may be zero as a particular case); we assume this initial velocity to be known. But then the unknown function $v=f(t)$ must be such that when $t=0$ (when motion begins) the condition $v=v_0$ is fulfilled. Substituting $t=0$, $v=v_0$ into formula (2), we find

$$v_0=C+\frac{mg}{k}$$

whence

$$C=v_0-\frac{mg}{k}$$

Thus, the constant C is found, and the sought-for dependence of v on t is

$$v=\left(v_0-\frac{mg}{k}\right)e^{-\frac{kt}{m}}+\frac{mg}{k} \tag{2'}$$

From this formula it follows that for sufficiently large t the velocity v depends but slightly on v_0.

It will be noted that if $k=0$ (the air resistance is absent or so small that we can disregard it), then we have a result familiar from physics: *

$$v=v_0+gt \tag{2''}$$

This function satisfies the differential equation (1) and the initial condition: $v=v_0$ when $t=0$.

Example 2. A flexible homogeneous thread is suspended at two ends. Find the equation of the curve that it describes under its own weight (it is the same as for any suspended ropes, wires, chains).

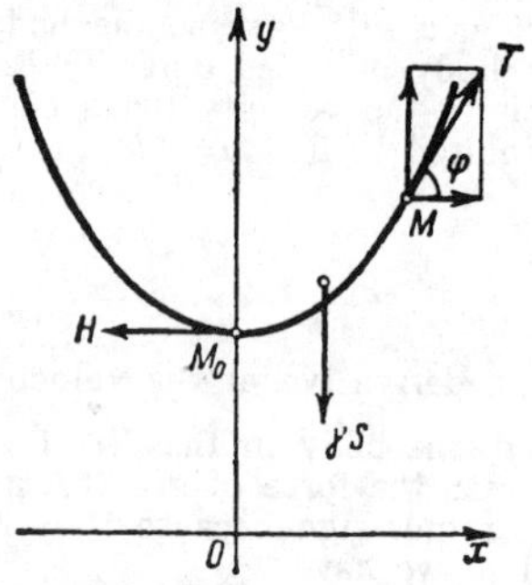

Solution. Let $M_0(0, b)$ be the lowest point of the thread, and M an arbitrary point (Fig. 1). Let us consider a part of the thread M_0M. This part is in equilibrium under the action of the resultant of three forces:

(1) the tension T, acting along the tangent at the point M and forming an angle φ with the x-axis;

(2) the tension H at M_0 acting horizontally;

(3) the weight of the thread γs acting vertically downwards, where s is the length of the arc M_0M and γ is the linear specific weight of the thread.

Fig. 1

Breaking up the tension T into horizontal and vertical components, we get the equations of equilibrium:

$$T\cos\varphi=H, \qquad T\sin\varphi=\gamma s$$

Dividing the terms of the second equation by the corresponding terms of the first, we obtain

$$\tan\varphi=\frac{\gamma}{H}s \tag{3}$$

---

* Formula (2″) can be obtained from (2′) by passing to the limit:

$$\lim_{k\to 0}\left[\left(v_0-\frac{mg}{k}\right)e^{-\frac{kt}{m}}+\frac{mg}{k}\right]=v_0+gt$$

Now suppose that the equation of the sought-for curve can be written in the form $y=f(x)$. Here, $f(x)$ is an unknown function that has to be found. It will be noted that

$$\tan\varphi=f'(x)=\frac{dy}{dx}$$

Hence,

$$\frac{dy}{dx}=\frac{1}{a}s \tag{4}$$

where the ratio $\frac{H}{\gamma}$ is denoted by a.

Differentiate both sides of (4) with respect to x:

$$\frac{d^2y}{dx^2}=\frac{1}{a}\frac{ds}{dx} \tag{5}$$

But, as we know (see Sec. 6.1, Vol. I),

$$\frac{ds}{dx}=\sqrt{1+\left(\frac{dy}{dx}\right)^2}$$

Substituting this expression into equation (5), we get the differential equation of the sought-for curve:

$$\frac{d^2y}{dx^2}=\frac{1}{a}\sqrt{1+\left(\frac{dy}{dx}\right)^2} \tag{6}$$

It expresses the relationship between the first and second derivatives of the unknown function y.

Without going into the methods of solving the equations, we note that any function of the form

$$y=a\cosh\left(\frac{x}{a}+C_1\right)+C_2 \tag{7}$$

satisfies equation (6) for any values that C_1 and C_2 may assume. This is evident if we put the first and second derivatives of the given function into (6). We shall later on indicate (Sec. 1.18), without proof, that these functions (for different C_1 and C_2) exhaust all possible solutions of equation (6).

The graphs of all the functions thus obtained are called *catenaries*.

Let us now find out how one should choose the constants C_1 and C_2 so as to obtain precisely that catenary whose lowest point M has coordinates $(0, b)$. Since for $x=0$ the point of the catenary occupies the lowest possible position, the tangent here is horizontal, $\frac{dy}{dx}=0$. Also, it is given that at this point the ordinate is equal to b $y=b$.

From (7) we find

$$y'=\sinh\left(\frac{x}{a}+C_1\right)$$

Putting $x=0$ here, we obtain $0=\sinh C_1$. Hence, $C_1=0$. If the ordinate of the point M_0 is b, then $y=b$ when $x=0$.

From equation (7) we get $b=\frac{a}{2}(1+1)+C_2$, assuming $x=0$ and $C_1=0$, whence $C_2=b-a$. Finally we have

$$y=a\cosh\left(\frac{x}{a}\right)+b-a$$

Equation (7) assumes a very simple form if we take the ordinate of M_0 equal to a. Then the equation of the catenary is

$$y = a \cosh (x/a)$$

1.2 DEFINITIONS

Definition 1. A *differential equation* is one which connects the independent variable x, unknown function $y=f(x)$, and its derivatives $y', y'', \ldots, y^{(n)}$.

Symbolically, a differential equation may be written as follows:

$$F(x, y, y', y'', \ldots, y^{(n)}) = 0$$

or

$$F\left(x, y, \frac{dy}{dx}, \frac{d^2y}{dx^2}, \ldots, \frac{d^ny}{dx^n}\right) = 0$$

If the sought-for function $y=f(x)$ is a function of **one** independent variable, then the differential equation is called *ordinary*. In this chapter we shall deal only with ordinary differential equations. *

Definition 2. The *order* of a differential equation is the order of the highest derivative which appears.

For example, the equation

$$y' - 2xy^2 + 5 = 0$$

is an equation of the first order.

The equation

$$y'' + ky' - by - \sin x = 0$$

is an equation of the second order, etc.

The equation considered in the preceding section in Example 1 is an equation of the first order, in Example 2, one of the second order.

Definition 3. The *solution* or *integral* of a differential equation is any function $y=f(x)$, which, when put into the equation, converts it into an identity.

---

* In addition to ordinary differential equations, mathematical analysis also deals with *partial differential equations*. Such an equation is a relation between an unknown function z that is dependent upon two or several variables $x, y, \ldots$, these variables $x, y, \ldots$, and the partial derivatives of z: $\frac{\partial z}{\partial x}, \frac{\partial z}{\partial y}, \frac{\partial^2 z}{\partial x^2}$, etc.

The following is an example of a partial differential equation in the unknown function $z(x, y)$:

$$x\frac{\partial z}{\partial x} = y\frac{\partial z}{\partial y}$$

It is easy to verify that this equation is satisfied by the function $z=x^2y^2$ (and also by a multitude of other functions).

In this course partial differential equations are discussed in Chapter 6.

Example 1. Suppose we have the equation

$$\frac{d^2y}{dx^2}+y=0$$

The functions $y=\sin x$, $y=2\cos x$, $y=3\sin x-\cos x$ and, in general, functions of the form $y=C_1\sin x$, $y=C_2\cos x$ or

$$y=C_1\sin x+C_2\cos x$$

are solutions of the given equation for any choice of constants C_1 and C_2; this is evident if we put these functions into the equation.

Example 2. Let us consider the equation

$$y'x-x^2-y=0$$

Its solutions are all functions of the form

$$y=x^2+Cx$$

where C is any constant. Indeed, differentiating the functions $y=x^2+Cx$, we find

$$y'=2x+C$$

Putting the expressions for y and y' into the initial equation, we get the identity

$$(2x+C)x-x^2-x^2-Cx=0$$

Each of the equations considered in Examples 1 and 2 has an infinitude of solutions.

1.3 FIRST-ORDER DIFFERENTIAL EQUATIONS (GENERAL NOTIONS)

1. A differential equation of the first order is of the form

$$F(x, y, y')=0 \tag{1}$$

If this equation can be solved for y', it can be written in the form

$$y'=f(x, y) \tag{1'}$$

In this case we say that the differential equation is solved for the derivative. For such an equation the following theorem, called the theorem of existence and uniqueness of solution of a differential equation, holds.

Theorem. *If in the equation*

$$y'=f(x, y)$$

the function $f(x, y)$ and its partial derivative with respect to y, $\frac{\partial f}{\partial y}$, are continuous in some domain D, in an xy-plane, containing some point (x_0, y_0), then there is a unique solution to this equation,

$$y=\varphi(x)$$

which satisfies the condition $y=y_0$ at $x=x_0$.

This theorem will be proved in Sec. 4.27.

The geometric meaning of the theorem consists in the fact that there exists one and only one such function $y=\varphi(x)$, the graph of which passes through the point (x_0, y_0).

It follows from this theorem that equation (1′) has an infinitude of solutions [for example, a solution the graph of which passes through (x_0, y_0); another solution whose graph passes through (x_0, y_1); through (x_0, y_2), etc., provided these points lie in the domain D].

The condition that for $x=x_0$ the function y must be equal to the given number y_0 is called the *initial condition*. It is frequently written in the form

$$y|_{x=x_0}=y_0$$

Definition 1. The *general solution* of a first-order differential equation is a function

$$y=\varphi(x, C) \tag{2}$$

which depends on a single arbitrary constant C and satisfies the following conditions:

(a) it satisfies the differential equation for any specific value of the constant C;

(b) no matter what the initial condition $y=y_0$ for $x=x_0$, that is, $(y)_{x=x_0}=y_0$, it is possible to find a value $C=C_0$ such that the function $y=\varphi(x, C_0)$ satisfies the given initial condition. It is assumed here that the values x_0 and y_0 belong to the range of the variables x and y in which the conditions of the existence and uniqueness theorem are fulfilled.

2. In searching for the general solution of a differential equation we often arrive at a relation like

$$\Phi(x, y, C)=0 \tag{2′}$$

which is not solved for y. Solving this relationship for y, we get the general solution. However, it is not always possible to express y from (2′) in terms of elementary functions; in such cases, the general solution is left in implicit form. An equation of the form $\Phi(x, y, C)=0$ which gives an implicit general solution is called the *complete integral* of the differential equation.

Definition 2. A *particular solution* is any function $y=\varphi(x, C_0)$ which is obtained from the general solution $y=\varphi(x, C)$, if in the latter we assign to the arbitrary constant C a definite value $C=C_0$. In this case, the relation $\Phi(x, y, C_0)=0$ is called a *particular integral* of the equation.

Example 1. For the first-order equation

$$\frac{dy}{dx}=-\frac{y}{x}$$

the general solution is a family of functions $y=\frac{C}{x}$; this can be checked by simple substitution in the equation.

Let us find a particular solution that will satisfy the following initial condition: $y_0=1$ when $x_0=2$.

Putting these values into the formula $y=\frac{C}{x}$, we have $1=\frac{C}{2}$ or $C=2$.

Consequently, the function $y=\frac{2}{x}$ will be the particular solution we are seeking.

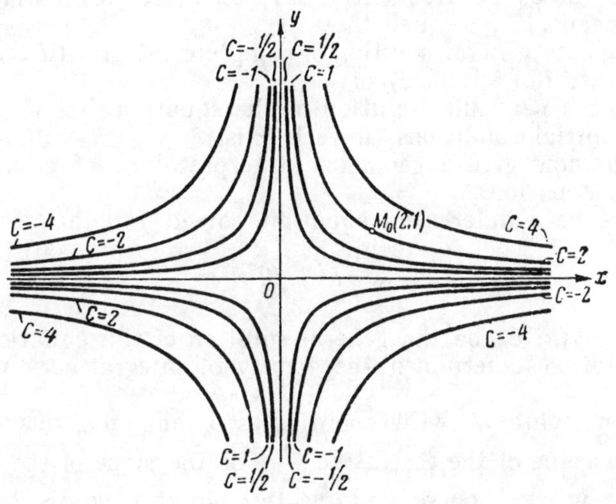

Fig. 2

From the geometric viewpoint, the **general solution (complete integral)** is a **family of curves** in a coordinate plane, which family depends on a single arbitrary constant C (or, as it is common to say, on a single parameter C). These curves are called *integral curves* of the given differential equation. A **particular** integral is associated with **one curve** of the family that passes through a certain given point of the plane.

Thus, in the latter example, the complete integral is geometrically depicted by a family of hyperbolas $y=\frac{C}{x}$, while the particular integral defined by the given initial condition is depicted by one of these hyperbolas passing through the point $M_0(2, 1)$. Fig. 2 shows the curves of a family that are associated with certain values of the parameter: $C=\frac{1}{2}$, $C=1$, $C=2$, $C=-1$, etc.

To make the reasoning still more pictorial, we shall from now on say that not only the function $y=\varphi(x, C_0)$ that satisfies the equation but also the associated **integral curve** is a **solution of the equation**. We will therefore speak of **a solution passing through the point** (x_0, y_0).

Note. The equation $\frac{dy}{dx} = -\frac{y}{x}$ has no solution passing through a point lying on the y-axis (see Fig. 2). This is because the right side of the equation is not defined for $x=0$ and consequently is not continuous.

To **solve** (or as we frequently say, to **integrate**) a differential equation means:

(a) to find its general solution or complete integral (if the initial conditions are not specified) or

(b) to find a particular solution of the equation that will satisfy the given initial conditions (if such exist).

3. Let us now give a geometric interpretation of a first-order differential equation.

Let there be a differential equation solved for the derivative:

$$\frac{dy}{dx} = f(x, y) \tag{1'}$$

and let $y = \varphi(x, C)$ be the general solution of this equation. This general solution determines the family of integral curves in the xy-plane.

For each point M with coordinates x and y, equation (1′) defines the value of the derivative $\frac{dy}{dx}$, or the slope of the tangent line to the integral curve passing through this point. Thus, the differential equation (1′) yields a collection of directions or, as we say, defines a *direction-field* in the xy-plane.

Consequently, from the geometric point of view, the problem of integrating a differential equation consists in finding curves, the directions of the tangents to which coincide with the direction of the field at the corresponding points.

For the differential equation (1) the locus of points at which the relation $\frac{dy}{dx} = C = \text{const}$ holds true is called the *isocline* of the given differential equation.

Different isoclines result from different values of C. The equation of the isocline corresponding to the value C will clearly be $f(x, y) = C$. If we have constructed a family of isoclines, it is possible to make a rough construction of the family of integral curves. We say that a knowledge of the isoclines enables one to determine qualitatively the locations of the integral curves in a plane.

Fig. 3 depicts a direction-field defined by the differential equation

$$\frac{dy}{dx} = -\frac{y}{x}$$

The isoclines of the given differential equation are

$$-\frac{y}{x}=C, \quad \text{or} \quad y=-Cx$$

This is a family of straight lines. They are constructed in Fig. 3.

4. Let us now consider the following problem.

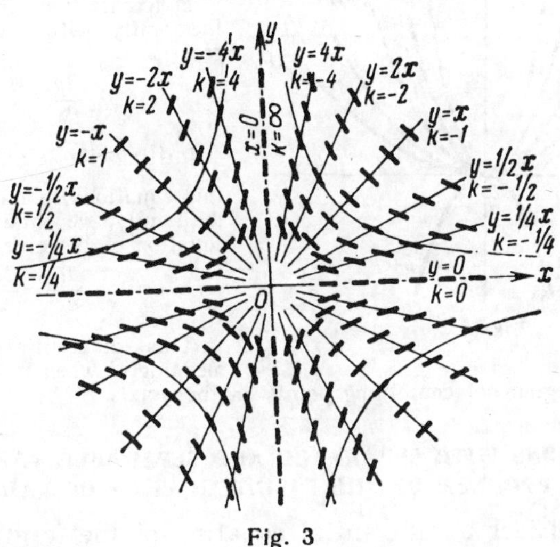

Fig. 3

Let there be given a family of functions that depends on a single parameter C:

$$y=\varphi(x, C) \tag{2}$$

and let only one curve of this family pass through each point of the plane (or some region in the plane).

For what differential equation is this family of functions a complete integral?

From relation (2), differentiating with respect to x, we find

$$\frac{dy}{dx}=\varphi'_x(x, C) \tag{3}$$

Since only one curve of the family passes through each point of the plane, for every number pair x and y, a unique value of C is determined from equation (2). Putting this value of C into (3), we find $\frac{dy}{dx}$ as a function of x and y. This is what yields the differential equation that is satisfied by every function of the family (2).

Hence, to establish a relationship between x, y and $\frac{dy}{dx}$ that is, to write a differential equation whose general solution is given by formula (2), one has to eliminate C from relations (2) and (3).

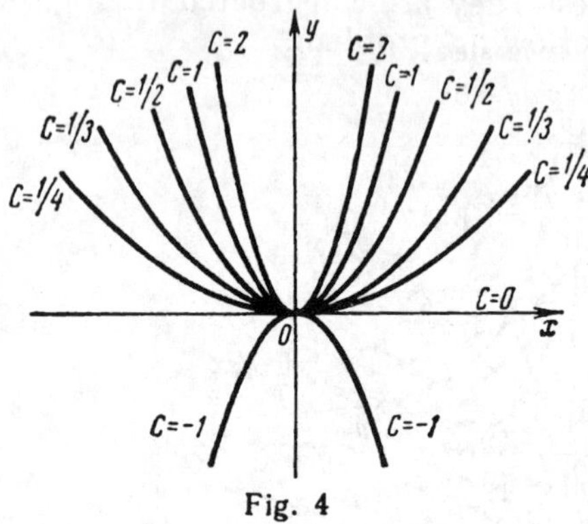

Fig. 4

Example 2. Find the differential equation of the family of parabolas $y = Cx^2$ (Fig. 4).

Differentiating the equation of the family with respect to x, we get

$$\frac{dy}{dx} = 2Cx$$

Putting the value $C = \frac{y}{x^2}$ into this equation from the equation of the family, we obtain a differential equation of the given family:

$$\frac{dy}{dx} = \frac{2y}{x}$$

This differential equation is meaningful when $x \neq 0$ which is to say, in any region not containing points on the y-axis.

1.4 EQUATIONS WITH SEPARATED AND SEPARABLE VARIABLES. THE PROBLEM OF THE DISINTEGRATION OF RADIUM

Let us consider a differential equation of the form

$$\frac{dy}{dx} = f_1(x) f_2(y) \tag{1}$$

where the right side is a product of a function dependent only on x by a function dependent only on y. We transform it in the following manner assuming that $f_2(y) \neq 0$:

$$\frac{1}{f_2(y)} dy = f_1(x)\, dx \tag{1'}$$

Considering y a known function of x, equation (1') may be regarded as an equality of two differentials, while the indefinite integrals of them will differ by a constant term. Integrating the left side with respect to y and the right with respect to x, we obtain

$$\int \frac{1}{f_2(y)} dy = \int f_1(x)\, dx + C \tag{1''}$$

which is a relationship connecting the solution of y, the independent variable x, and an arbitrary constant C; we have thus obtained a general solution (complete integral) of equation (1).

1. A type (1') differential equation

$$M(x)\, dx + N(y)\, dy = 0 \tag{2}$$

is called an equation with *separated variables*. From what has been proved, its complete integral is

$$\int M(x)\,dx+\int N(y)\,dy=C$$

Example 1. Given an equation with separated variables:

$$x\,dx+y\,dy=0$$

Integrating we get the general solution:

$$\frac{x^2}{2}+\frac{y^2}{2}=C_1$$

Since the left side of this equation is nonnegative, the right side is also nonnegative. Denoting $2C_1$ by C^2, we will have

$$x^2+y^2=C^2$$

This is the equation of a family of concentric circles (Fig. 5) with centre at the coordinate origin and radius C.

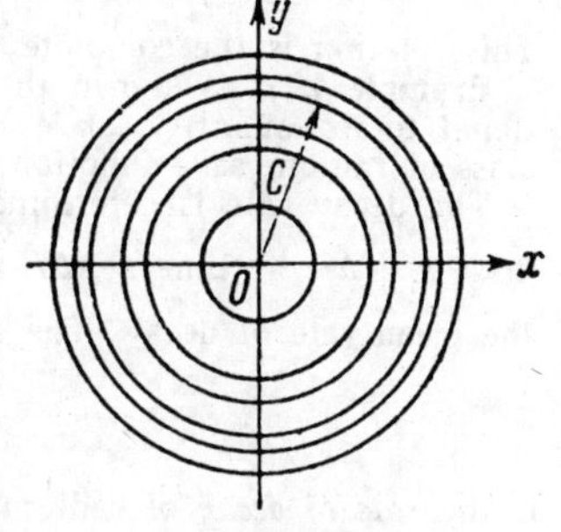

Fig. 5

2. An equation of the form

$$M_1(x)\,N_1(y)\,dx+M_2(x)\,N_2(y)\,dy=0 \quad (3)$$

is called an equation with *separable variables*. It can be reduced * to an equation with separated variables by dividing both sides by the expression $N_1(y)\,M_2(x)$:

$$\frac{M_1(x)\,N_1(y)}{N_1(y)\,M_2(x)}\,dx+\frac{M_2(x)\,N_2(y)}{N_1(y)\,M_2(x)}\,dy=0$$

or

$$\frac{M_1(x)}{M_2(x)}\,dx+\frac{N_2(y)}{N_1(y)}\,dy=0$$

that is, to an equation like (2).

Example 2. Given the equation

$$\frac{dy}{dx}=-\frac{y}{x}$$

Separating variables, we have

$$\frac{dy}{y}=-\frac{dx}{x}$$

Integrating, we find

$$\int\frac{dy}{y}=-\int\frac{dx}{x}+C$$

* These transformations are permissible only in a region where neither $N_1(y)$ nor $M_2(x)$ vanish.

which is

$$\ln|y| = -\ln|x| + \ln|C|^{*} \quad \text{or} \quad \ln|y| = \ln\left|\frac{C}{x}\right|$$

whence we get the general solution: $y = \frac{C}{x}$.

Example 3. Given the equation

$$(1+x)\,y\,dx + (1-y)\,x\,dy = 0$$

Separating variables we have

$$\frac{(1+x)}{x}dx + \frac{1-y}{y}dy = 0, \quad \left(\frac{1}{x}+1\right)dx + \left(\frac{1}{y}-1\right)dy = 0$$

Integrating, we obtain

$$\ln|x| + x + \ln|y| - y = C \quad \text{or} \quad \ln|xy| + x - y = C$$

This relation is the complete integral of the given equation.

Example 4. It is known that the decay rate of radium is directly proportional to its quantity at each given instant. Find the law of variation of a mass of radium as a function of the time if at $t=0$ the mass of radium was m_0.

The decay rate is determined as follows. Let there be mass m at time t, and mass $m+\Delta m$ at time $t+\Delta t$. During time Δt mass Δm decays. The ratio $\frac{\Delta m}{\Delta t}$ is the mean rate of decay. The limit of this ratio as $\Delta t \to 0$

$$\lim_{\Delta t \to 0} \frac{\Delta m}{\Delta t} = \frac{dm}{dt}$$

is the *rate of decay* of radium at time t.

It is given that

$$\frac{dm}{dt} = -km \tag{4}$$

where k is the constant of proportionality ($k > 0$). We use the minus sign because the mass of radium diminishes as time increases and therefore $\frac{dm}{dt} < 0$.

Equation (4) is an equation with separable variables. Let us separate the variables:

$$\frac{dm}{m} = -k\,dt$$

Solving the equation we obtain

$$\ln m = -kt + \ln C$$

whence

$$\ln\frac{m}{C} = -kt$$

$$m = Ce^{-kt} \tag{5}$$

Since at $t=0$ the mass of radium was m_0, C must satisfy the relationship

$$m_0 = Ce^{-k\cdot 0} = C$$

---

* Having in view subsequent transformations, we denoted the arbitrary constant by $\ln|C|$, which is permissible since $\ln|C|$ (when $C \neq 0$) can take on any value from $-\infty$ to $+\infty$.

Putting the value of C into (5) we get the desired mass of radium as a function of time (Fig. 6):

$$m = m_0 e^{-kt} \tag{6}$$

The constant k is determined from observations as follows. During time t_0 let $\alpha\%$ of the original mass of radium decay. Hence, the following relationship is fulfilled:

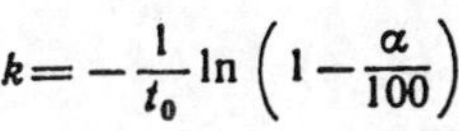

$$\left(1 - \frac{\alpha}{100}\right) m_0 = m_0 e^{-kt_0}$$

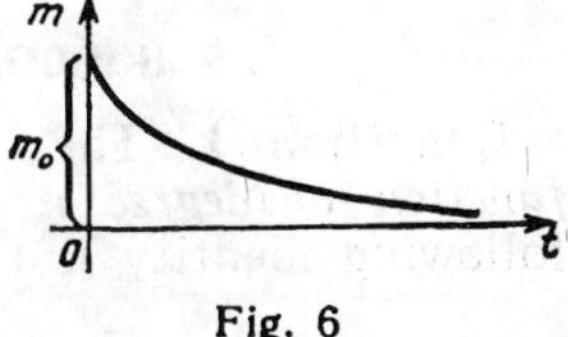

Fig. 6

whence

$$-kt_0 = \ln\left(1 - \frac{\alpha}{100}\right)$$

or

$$k = -\frac{1}{t_0}\ln\left(1 - \frac{\alpha}{100}\right)$$

Thus, it has been determined that for radium $k = 0.000436$ (the unit of time measurement is one year).

Putting this value of k into (6) we obtain

$$m = m_0 e^{-0.000436t}$$

Let us find the half-life of radium, which is the interval of time during which half of the original mass of radium decays. Putting $\frac{m_0}{2}$ in place of m in the latter formula, we get an equation for determining the half-life T:

$$\frac{m_0}{2} = m_0 e^{-0.000436T}$$

whence

$$-0.000436T = -\ln 2$$

or

$$T = \frac{\ln 2}{0.000436} = 1590 \text{ years}$$

Some other problems of physics and chemistry can be reduced to equations of the form (4).

Note 1. Let the function $f_2(y)$ of (1) have a root $y = b$, i.e., $f_2(b) = 0$. Then the function $y = b$ is clearly a solution of equation (1); this can be readily verified by direct substitution. The formula (1″) may not yield the solution $y = b$. We will not consider this case here, but will note that the uniqueness condition may fail to hold on the line $y = 0$.

By way of example, the equation $y' = 2\sqrt{y}$ has the general solution $y = (x + c)^2$ and the solution $y = 0$, which does not follow from the general solution. The uniqueness condition is violated on the line $y = 0$.

Note 2. The simplest differential equation with separated variables is one of the form

$$\frac{dy}{dx} = f(x) \quad \text{or} \quad dy = f(x)\,dx$$

Its general solution is of the form

$$y=\int f(x)\,dx+C$$

We dealt with the solution of equations of this kind in Ch. 10, Vol. I.

1.5 HOMOGENEOUS FIRST-ORDER EQUATIONS

Definition 1. The function $f(x, y)$ is called a *homogeneous function of degree n* in the variables x and y, if for any λ the following identity is true:

$$f(\lambda x, \lambda y)=\lambda^n f(x, y)$$

Example 1. The function $f(x, y)=\sqrt[3]{x^3+y^3}$ is a homogeneous function of degree one, since

$$f(\lambda x, \lambda y)=\sqrt[3]{(\lambda x)^3+(\lambda y)^3}=\lambda\sqrt[3]{x^3+y^3}=\lambda f(x, y)$$

Example 2. $f(x, y)=xy-y^2$ is a homogeneous function of degree two, since $(\lambda x)(\lambda y)-(\lambda y)^2=\lambda^2(xy-y^2)$.

Example 3. $f(x, y)=\frac{x^2-y^2}{xy}$ is a homogeneous function of degree zero since $\frac{(\lambda x)^2-(\lambda y)^2}{(\lambda x)(\lambda y)}=\frac{x^2-y^2}{xy}$, that is, $f(\lambda x, \lambda y)=f(x, y)$ or $f(\lambda x, \lambda y)=\lambda^0 f(x, y)$.

Definition 2. An equation of the first order

$$\frac{dy}{dx}=f(x, y) \tag{1}$$

is called *homogeneous* in x and y if the function $f(x, y)$ is a homogeneous function of degree zero in x and y.

Solution of a homogeneous equation. It is given that $f(\lambda x, \lambda y)=$ $=f(x, y)$. Putting $\lambda=\frac{1}{x}$ in this identity, we have

$$f(x, y)=f\left(1, \frac{y}{x}\right)$$

Thus, a homogeneous function of degree zero is dependent only on the ratio of the arguments.

In this case, equation (1) takes the form

$$\frac{dy}{dx}=f\left(1, \frac{y}{x}\right) \tag{1'}$$

Making the substitution

$$u=\frac{y}{x} \quad \text{or} \quad y=ux$$

we get

$$\frac{dy}{dx}=u+\frac{du}{dx}x$$

Putting this expression of the derivative into equation (1′), we obtain

$$u+x\frac{du}{dx}=f(1, u)$$

This is an equation with variables separable:

$$x\frac{du}{dx}=f(1, u)-u \quad \text{or} \quad \frac{du}{f(1, u)-u}=\frac{dx}{x}$$

Integrating we find

$$\int\frac{du}{f(1, u)-u}=\int\frac{dx}{x}+C$$

Putting the ratio $\frac{y}{x}$ in place of u after integration, we get the integral of equation (1′).

Example 4. Given the equation

$$\frac{dy}{dx}=\frac{xy}{x^2-y^2}$$

On the right is a zero-degree homogeneous function, which means that we have a homogeneous equation. Making the substitution $\frac{y}{x}=u$, we have

$$y=ux, \quad \frac{dy}{dx}=u+x\frac{du}{dx}$$

$$u+x\frac{du}{dx}=\frac{u}{1-u^2}, \quad x\frac{du}{dx}=\frac{u^3}{1-u^2}$$

Separating variables, we obtain

$$\frac{(1-u^2)\,du}{u^3}=\frac{dx}{x}, \quad \left(\frac{1}{u^3}-\frac{1}{u}\right)du=\frac{dx}{x}$$

Whence, integrating, we find

$$-\frac{1}{2u^2}-\ln|u|=\ln|x|+\ln|C| \quad \text{or} \quad -\frac{1}{2u^2}=\ln|uxC|.$$

Substituting $u=\frac{y}{x}$, we get the general solution of the original equation:

$$-\frac{x^2}{2y^2}=\ln|Cy|$$

It is impossible here to get y as an explicit function of x in terms of elementary functions. Incidentally, it is very easy to express x in terms of y:

$$x=y\sqrt{-2\ln|Cy|}$$

Note. An equation of the type

$$M(x, y)\,dx+N(x, y)\,dy=0$$

will be homogeneous if, and only if, $M(x, y)$ and $N(x, y)$ are homogeneous functions of the same degree. This follows from the fact that the ratio of two homogeneous functions of the same degree is a homogeneous function of degree zero.

Example 5. The equations

$$(2x+3y)\,dx+(x-2y)\,dy=0$$
$$(x^2+y^2)\,dx-2xy\,dy=0$$

are homogeneous.

1.6 EQUATIONS REDUCIBLE TO HOMOGENEOUS EQUATIONS

Equations of the following type are reducible to homogeneous equations:

$$\frac{dy}{dx}=\frac{ax+by+c}{a_1x+b_1y+c_1} \tag{1}$$

If $c_1=c=0$, then equation (1) is obviously homogeneous. Now let c and c_1 (or one of them) be different from zero. Make a change of variables:

$$x=x_1+h, \quad y=y_1+k$$

Then

$$\frac{dy}{dx}=\frac{dy_1}{dx_1} \tag{2}$$

Putting into (2) the expressions for x, y, and $\frac{dy}{dx}$, we obtain

$$\frac{dy_1}{dx_1}=\frac{ax_1+by_1+ah+bk+c}{a_1x_1+b_1y_1+a_1h+b_1k+c_1} \tag{3}$$

Choose h and k so that the following equations are fulfilled:

$$\left.\begin{aligned} ah+bk+c&=0\\ a_1h+b_1k+c_1&=0\end{aligned}\right\} \tag{4}$$

In other words, define h and k as solutions of a system of equations (4). Equation (3) then becomes homogeneous:

$$\frac{dy_1}{dx_1}=\frac{ax_1+by_1}{a_1x_1+b_1y_1}$$

Solving this equation and passing once again to x and y by formulas (2), we obtain the solution of equation (1).

The system (4) has no solution if

$$\begin{vmatrix} a & b\\ a_1 & b_1\end{vmatrix}=0$$

i.e., $ab_1=a_1b$. But in this case $\frac{a_1}{a}=\frac{b_1}{b}=\lambda$, that is, $a_1=\lambda a$, $b_1=\lambda b$, and, hence, equation (1) may be transformed to

$$\frac{dy}{dx}=\frac{(ax+by)+c}{\lambda(ax+by)+c_1} \tag{5}$$

Then by substitution

$$z = ax + by \tag{6}$$

the equation is reduced to one with variables separable.

Indeed,

$$\frac{dz}{dx} = a + b\frac{dy}{dx}$$

whence

$$\frac{dy}{dx} = \frac{1}{b}\frac{dz}{dx} - \frac{a}{b} \tag{7}$$

Putting into (5) expressions (6) and (7), we get

$$\frac{1}{b}\frac{dz}{dx} - \frac{a}{b} = \frac{z+c}{\lambda z + c_1}$$

which is an equation with variables separable.

The device applied to integrating equation (1) is also applied to the integration of the equation

$$\frac{dy}{dx} = f\left(\frac{ax+by+c}{a_1x+b_1y+c_1}\right)$$

where f is an arbitrary continuous function.

Example 1. Given the equation

$$\frac{dy}{dx} = \frac{x+y-3}{x-y-1}$$

To convert it into a homogeneous equation, make the substitution $x = x_1 + h$, $y = y_1 + k$. Then

$$\frac{dy_1}{dx_1} = \frac{x_1 + y_1 + h + k - 3}{x_1 - y_1 + h - k - 1}$$

Solving the set of two equations

$$h + k - 3 = 0, \qquad h - k - 1 = 0$$

we find

$$h = 2, \qquad k = 1$$

As a result we get the homogeneous equation

$$\frac{dy_1}{dx_1} = \frac{x_1 + y_1}{x_1 - y_1}$$

which we solve by substitution:

$$\frac{y_1}{x_1} = u$$

then

$$y_1 = ux_1, \quad \frac{dy_1}{dx_1} = u + x_1\frac{du}{dx_1}$$

$$u + x_1\frac{du}{dx_1} = \frac{1+u}{1-u}$$

and we get an equation with variables separable

$$x_1 \frac{du}{dx_1} = \frac{1+u^2}{1-u}$$

Separating the variables, we have

$$\frac{1-u}{1+u^2}\, du = \frac{dx_1}{x_1}$$

Integrating, we find

$$\arctan u - \frac{1}{2} \ln (1+u^2) = \ln |x_1| + \ln |C|$$

or

$$\arctan u = \ln |C x_1 \sqrt{1+u^2}|$$

$$Cx_1 \sqrt{1+u^2} = e^{\arctan u}$$

Putting $\frac{y_1}{x_1}$ in place of u, we obtain

$$C \sqrt{x_1^2 + y_1^2} = e^{\arctan \frac{y_1}{x_1}}$$

Passing to the variables x and y, we finally get

$$C \sqrt{(x-2)^2 + (y-1)^2} = e^{\arctan \frac{y-1}{x-2}}$$

Example 2. The equation

$$y' = \frac{2x+y-1}{4x+2y+5}$$

cannot be solved by the substitution $x = x_1 + h$, $y = y_1 + k$, since in this case the set of equations that serves to determine h and k is insolvable (here, the determinant $\begin{vmatrix} 2 & 1 \\ 4 & 2 \end{vmatrix}$ of the coefficients of the variables is equal to zero).

This equation may be reduced to one with variables separable by the substitution

$$2x + y = z$$

Then $y' = z' - 2$ and the equation is reduced to the form

$$z' - 2 = \frac{z-1}{2z+5}$$

or

$$z' = \frac{5z+9}{2z+5}$$

Solving it we find

$$\frac{2}{5} z + \frac{7}{25} \ln |5z+9| = x + C$$

Since $z = 2x + y$, we obtain the final solution of the initial equation in the form

$$\frac{2}{5}(2x+y) + \frac{7}{25} \ln |10x+5y+9| = x + C$$

or

$$10y - 5x + 7 \ln |10x+5y+9| = C_1$$

that is, as an implicit function y of x.

1.7 FIRST-ORDER LINEAR EQUATIONS

Definition. A *first-order linear equation* is an equation that is linear in the unknown function and its derivative. It is of the form

$$\frac{dy}{dx}+P(x)y=Q(x) \tag{1}$$

where $P(x)$ and $Q(x)$ are given continuous functions of x (or are constants).

Solution of linear equation (1). Let us seek the solution of equation (1) in the form of a product of two functions of x:

$$y=u(x)v(x) \tag{2}$$

One of these functions may be arbitrary, while the other will be determined from equation (1).

Differentiating both sides of (2), we find

$$\frac{dy}{dx}=u\frac{dv}{dx}+v\frac{du}{dx}$$

Putting the expression obtained of the derivative $\frac{dy}{dx}$ into (1), we have

$$u\frac{dv}{dx}+\frac{du}{dx}v+Puv=Q$$

or

$$u\left(\frac{dv}{dx}+Pv\right)+v\frac{du}{dx}=Q \tag{3}$$

Let us choose the function v such that

$$\frac{dv}{dx}+Pv=0 \tag{4}$$

Separating the variables in this differential equation with respect to the function v, we find

$$\frac{dv}{v}=-P\,dx$$

Integrating we obtain

$$-\ln|C_1|+\ln|v|=-\int P\,dx$$

or

$$v=C_1e^{-\int P\,dx}$$

Since it is sufficient for us to have some nonzero solution of equation (4), we take, for the function $v(x)$,

$$v(x)=e^{-\int P\,dx} \tag{5}$$

where $\int P\,dx$ is some antiderivative. Obviously, $v(x)\neq 0$.

Putting the value of $v(x)$ which we have found into (3), we get $\left(\text{noting that } \frac{dv}{dx}+Pv=0\right)$:

$$v(x)\frac{du}{dx}=Q(x)$$

or

$$\frac{du}{dx}=\frac{Q(x)}{v(x)}$$

whence

$$u=\int \frac{Q(x)}{v(x)}\,dx+C$$

Substituting into formula (2), we finally get

$$y=v(x)\left[\int \frac{Q(x)}{v(x)}\,dx+C\right]$$

or

$$y=v(x)\int \frac{Q(x)}{v(x)}\,dx+Cv(x) \tag{6}$$

Note. It is obvious that expression (6) will not change if in place of the function $v(x)$ defined by (5) we take some function $v_1(x)=\overline{C}v(x)$. Indeed, putting $v_1(x)$ in (6) in place of $v(x)$, we get

$$y=\overline{C}v(x)\int \frac{Q(x)}{\overline{C}v(x)}\,dx+C\overline{C}v(x)$$

The $\overline{C}$ in the first term cancel out; in the second term the product $C\overline{C}$ is an arbitrary constant, which we shall denote by C, and we again arrive at expression (6). If we denote $\int \frac{Q(x)}{v(x)}\,dx=\varphi(x)$, then expression (6) will take the form

$$y=v(x)\varphi(x)+Cv(x) \tag{6'}$$

It is obvious that this is a complete integral, since C may be chosen in such a manner that the initial condition will be satisfied:

$$\text{when } x=x_0, \quad y=y_0$$

The value of C is determined from the equation

$$y_0=v(x_0)\varphi(x_0)+Cv(x_0)$$

Example. Solve the equation

$$\frac{dy}{dx}-\frac{2}{x+1}y=(x+1)^3$$

Solution. Putting

$$y=uv$$

we have

$$\frac{dy}{dx}=u\frac{dv}{dx}+\frac{du}{dx}v$$

Putting the expression $\frac{dy}{dx}$ into the original equation, we obtain

$$u\frac{dv}{dx}+\frac{du}{dx}v-\frac{2}{x+1}uv=(x+1)^3$$

$$u\left(\frac{dv}{dx}-\frac{2}{x+1}v\right)+v\frac{du}{dx}=(x+1)^3 \qquad (7)$$

To determine v we get the equation

$$\frac{dv}{dx}-\frac{2}{x+1}v=0$$

that is,

$$\frac{dv}{v}=\frac{2dx}{x+1}$$

whence

$$\ln|v|=2\ln|x+1| \quad \text{or} \quad v=(x+1)^2$$

Putting the expression of the function v into equation (7), we get the following equation for u:

$$(x+1)^2\frac{du}{dx}=(x+1)^3 \quad \text{or} \quad \frac{du}{dx}=(x+1)$$

whence

$$u=\frac{(x+1)^2}{2}+C$$

Thus, the complete integral of the given equation will be of the form

$$y=\frac{(x+1)^4}{2}+C(x+1)^2$$

The family obtained is the **general** solution. No matter what the initial condition (x_0, y_0), where $x_0 \neq -1$, it is always possible to choose C so that the corresponding particular solution should satisfy the given initial condition. For example, the particular solution that satisfies the condition $y_0=3$ when $x_0=0$ is found as follows:

$$3=\frac{(0+1)^4}{2}+C(0+1)^2; \quad C=\frac{5}{2}$$

Consequently, the desired particular solution is

$$y=\frac{(x+1)^4}{2}+\frac{5}{2}(x+1)^2$$

However, if the initial condition (x_0, y_0) is chosen so that $x_0=-1$, we will not find the particular solution that satisfies this condition. This is due to the fact that when $x_0=-1$ the function $P(x)=-\frac{2}{x+1}$ is discontinuous and, hence, the conditions of the theorem of the existence of a solution are not observed.

Note. In applications, one frequently encounters linear equations with constant coefficients:

$$\frac{dy}{dx}+ay=b \tag{8}$$

where a and b are constants.

This equation may be solved either by the substitution (2) or by separation of variables:

$$dy=(-ay+b)\,dx, \quad \frac{dy}{-ay+b}=dx, \quad -\frac{1}{a}\ln|-ay+b|=x+C_1$$

$$\ln|-ay+b|=-(ax+C^*), \quad \text{where} \quad C^*=aC_1$$

$$-ay+b=e^{-(ax+C^*)}, \quad y=-\frac{1}{a}e^{-(ax+C^*)}+\frac{b}{a}$$

or, finally,

$$y=Ce^{-ax}+\frac{b}{a}$$

$\left(\text{where } -\frac{1}{a}e^{-C^*}=C\right)$. This is the general solution of equation (8).

1.8 BERNOULLI'S EQUATION

We consider an equation of the form*

$$\frac{dy}{dx}+P(x)\,y=Q(x)\,y^n \tag{1}$$

where $P(x)$ and $Q(x)$ are continuous functions of x (or constants), and $n\neq 0$ and $n\neq 1$ (otherwise we would have a linear equation). This equation is called *Bernoulli's equation* and reduces to a linear equation by the following transformation.

Dividing all terms of the equation by y^n, we get

$$y^{-n}\frac{dy}{dx}+Py^{-n+1}=Q \tag{2}$$

Making the substitution

$$z=y^{-n+1}$$

we have

$$\frac{dz}{dx}=(-n+1)\,y^{-n}\frac{dy}{dx}$$

---

* This equation results from the problem of the motion of a body provided the resistance of the medium F depends on the velocity: $F=\lambda_1 v+\lambda_2 v^n$. The equation of motion will then assume the form $m\frac{dv}{dt}=-\lambda_1 v-\lambda_2 v^n$ or $\frac{dv}{dt}+\frac{\lambda_1}{m}v=-\frac{\lambda_2}{m}v^n$.

Substituting into (2), we get

$$\frac{dz}{dx}+(-n+1)\,Pz=(-n+1)\,Q$$

This is a linear equation.

Finding its complete integral and substituting the expression y^{-n+1} for z, we get the complete integral of the Bernoulli equation.

Example. Solve the equation

$$\frac{dy}{dx}+xy=x^3y^3 \tag{3}$$

Solution. Dividing all terms by y^3, we have

$$y^{-3}y'+xy^{-2}=x^3 \tag{4}$$

Introducing the new function

$$z=y^{-2}$$

we get

$$\frac{dz}{dx}=-2y^{-3}\frac{dy}{dx}$$

Substituting these values into equation (4), we obtain the linear equation

$$\frac{dz}{dx}-2xz=-2x^3 \tag{5}$$

Let us find its complete integral:

$$z=uv;\quad \frac{dz}{dx}=u\frac{dv}{dx}+\frac{du}{dx}v$$

Put expressions z and $\frac{dz}{dx}$ into (5):

$$u\frac{dv}{dx}+\frac{du}{dx}v-2xuv=-2x^3$$

or

$$u\left(\frac{dv}{dx}-2xv\right)+v\frac{du}{dx}=-2x^3$$

Equate to zero the expression in the brackets:

$$\frac{dv}{dx}-2xv=0,\quad \frac{dv}{v}=2x\,dx$$

$$\ln|v|=x^2,\quad v=e^{x^2}$$

For u we get the equation

$$e^{x^2}\frac{du}{dx}=-2x^3$$

Separating variables, we have

$$du=-2e^{-x^2}x^3\,dx,\quad u=-2\int e^{-x^2}x^3\,dx+C$$

Integrating by parts, we find

$$u=x^2e^{-x^2}+e^{-x^2}+C,$$
$$z=uv=x^2+1+Ce^{x^2}$$

Consequently, the complete integral of the given equation is

$$y^{-2}=x^2+1+Ce^{x^2} \quad \text{or} \quad y=\frac{1}{\sqrt{x^2+1+Ce^{x^2}}}$$

Note. Just as was done for linear equations, it may be shown that the solution of the Bernoulli equation may be sought in the form of a product of two functions:

$$y=u(x)\,v(x)$$

where $v(x)$ is some nonzero function that satisfies the equation $v'+Pv=0$.

1.9 EXACT DIFFERENTIAL EQUATIONS

Definition. The equation

$$M(x, y)\,dx+N(x, y)\,dy=0 \tag{1}$$

is called an *exact differential equation* if $M(x, y)$ and $N(x, y)$ are continuous differentiable functions for which the following relationship is fulfilled

$$\frac{\partial M}{\partial y}=\frac{\partial N}{\partial x} \tag{2}$$

and $\frac{\partial M}{\partial y}$ and $\frac{\partial N}{\partial x}$ are continuous in some region.

Integrating exact differential equations. We shall prove that if the left side of equation (1) is an exact differential, then condition (2) is fulfilled, and, conversely, if condition (2) is fulfilled the left side of equation (1) is an exact differential of some function $u(x, y)$. That is, equation (1) is an equation of the form

$$du(x, y)=0 \tag{3}$$

and, consequently, its complete integral is

$$u(x, y)=C$$

Let us first assume that the left side of (1) is an exact differential of some function $u(x, y)$; that is,

$$M(x, y)\,dx+N(x, y)\,dy=du=\frac{\partial u}{\partial x}dx+\frac{\partial u}{\partial y}dy$$

then

$$M=\frac{\partial u}{\partial x}, \quad N=\frac{\partial u}{\partial y} \tag{4}$$

Differentiating the first relation with respect to y, and the second with respect to x, we obtain

$$\frac{\partial M}{\partial y}=\frac{\partial^2 u}{\partial x\,\partial y}, \quad \frac{\partial N}{\partial x}=\frac{\partial^2 u}{\partial y\,\partial x}$$

Assuming continuity of the second derivatives, we have

$$\frac{\partial M}{\partial y}=\frac{\partial N}{\partial x}$$

that is, (2) is a **necessary** condition for the left side of (1) to be an exact differential of some function $u(x, y)$. We shall show that this condition is also **sufficient:** if (2) is fulfilled then the left side of (1) is an exact differential of some function $u(x, y)$.

From the relation

$$\frac{\partial u}{\partial x}=M(x, y)$$

we find

$$u=\int_{x_0}^{x} M(x, y)\,dx+\varphi(y)$$

where x_0 is the abscissa of any point of the domain of existence of the solution.

When integrating with respect to x we consider y constant, and therefore the arbitrary constant of integration may be dependent on y. Let us choose a function $\varphi(y)$ so that the second of the relations (4) is fulfilled. To do this, we differentiate* both sides of the latter equation with respect to y and equate the result to $N(x, y)$:

$$\frac{\partial u}{\partial y}=\int_{x_0}^{x}\frac{\partial M}{\partial y}\,dx+\varphi'(y)=N(x, y)$$

but since $\frac{\partial M}{\partial y}=\frac{\partial N}{\partial x}$, we can write

$$\int_{x_0}^{x}\frac{\partial N}{\partial x}\,dx+\varphi'(y)=N \quad \text{that is,} \quad N(x, y)\Big|_{x_0}^{x}+\varphi'(y)=N(x, y)$$

or

$$N(x, y)-N(x_0, y)+\varphi'(y)=N(x, y)$$

Hence,

$$\varphi'(y)=N(x_0, y)$$

---

* The integral $\int_{x_0}^{x} M(x, y)\,dx$ is dependent on y. To find the derivative of this integral with respect to y, differentiate the integrand with respect to y: $\frac{\partial}{\partial y}\int_{x_0}^{x} M(x, y)\,dx=\int_{x_0}^{x}\frac{\partial M}{\partial y}\,dx$. This follows from Leibniz' theorem for differentiating a definite integral with respect to a parameter (see Sec. 11.10. Vol. I).

or

$$\varphi(y)=\int_{y_0}^{y} N(x_0,\ y)\,dy+C_1$$

Thus, the function $u(x,\ y)$ will have the form

$$u=\int_{x_0}^{x} M(x,\ y)\,dx+\int_{y_0}^{y} N(x_0,\ y)\,dy+C_1$$

Here $P(x_0,\ y_0)$ is a point in the neighbourhood of which there is a solution of the differential equation (1).

Equating this expression to an arbitrary constant C, we get the complete integral of equation (1):

$$\int_{x_0}^{x} M(x,\ y)\,dx+\int_{y_0}^{y} N(x_0,\ y)\,dy=C \tag{5}$$

Example. Given the equation

$$\frac{2x}{y^3}\,dx+\frac{y^2-3x^2}{y^4}\,dy=0$$

Let us check to see whether this is an exact differential equation.

Denoting

$$M=\frac{2x}{y^3};\ N=\frac{y^2-3x^2}{y^4}$$

we have

$$\frac{\partial M}{\partial y}=-\frac{6x}{y^4},\quad \frac{\partial N}{\partial x}=-\frac{6x}{y^4}$$

For $y \neq 0$, condition (2) is fulfilled. Hence, the left side of this equation is an exact differential of some unknown function $u(x,\ y)$. Let us find this function.

Since $\frac{\partial u}{\partial x}=\frac{2x}{y^3}$, it follows that

$$u=\int \frac{2x}{y^3}\,dx+\varphi(y)=\frac{x^2}{y^3}+\varphi(y)$$

where $\varphi(y)$ is an as yet undetermined function of y.

Differentiating this relation with respect to y and noting that

$$\frac{\partial u}{\partial y}=N=\frac{y^2-3x^2}{y^4}$$

we find

$$-\frac{3x^2}{y^4}+\varphi'(y)=\frac{y^2-3x^2}{y^4}$$

hence

$$\varphi'(y)=\frac{1}{y^2},\qquad \varphi(y)=-\frac{1}{y}+C_1$$

$$u(x,\ y)=\frac{x^2}{y^3}-\frac{1}{y}+C_1$$

Thus the complete integral of the original equation is

$$\frac{x^2}{y^3}-\frac{1}{y}=C$$

1.10 INTEGRATING FACTOR

Let the left side of the equation

$$M(x, y)\,dx + N(x, y)\,dy = 0 \tag{1}$$

not be an exact differential. It is sometimes possible to choose a function $\mu(x, y)$ such that after multiplying all terms of the equation by it the left side of the equation is converted into an exact differential. The general solution of the equation thus obtained coincides with the general solution of the original equation; the function $\mu(x, y)$ is called the *integrating factor* of equation (1).

In order to find the integrating factor μ, do as follows. Multiply both sides of the given equation by the as yet unknown integrating factor μ:

$$\mu M\,dx + \mu N\,dy = 0$$

For this equation to be an exact differential equation, it is necessary and sufficient that the following relation be fulfilled:

$$\frac{\partial(\mu M)}{\partial y} = \frac{\partial(\mu N)}{\partial x}$$

that is,

$$\mu\frac{\partial M}{\partial y} + M\frac{\partial \mu}{\partial y} = \mu\frac{\partial N}{\partial x} + N\frac{\partial \mu}{\partial x}$$

or

$$M\frac{\partial \mu}{\partial y} - N\frac{\partial \mu}{\partial x} = \mu\left(\frac{\partial N}{\partial x} - \frac{\partial M}{\partial y}\right)$$

After dividing both sides of the latter equation by μ, we get

$$M\frac{\partial \ln \mu}{\partial y} - N\frac{\partial \ln \mu}{\partial x} = \frac{\partial N}{\partial x} - \frac{\partial M}{\partial y} \tag{2}$$

It is obvious that any function $\mu(x, y)$ that satisfies this equation is the integrating factor of equation (1). Equation (2) is a partial differential equation in the unknown function μ dependent on the two variables x and y. It can be proved that under definite conditions it has an infinitude of solutions and that, consequently, equation (1) has an integrating factor. But in the general case, the problem of finding $\mu(x, y)$ from equation (2) is harder than the original problem of integrating equation (1). Only in certain particular cases does one manage to find the function $\mu(x, y)$.

For instance, let equation (1) admit an integrating factor **dependent only on y.** Then

$$\frac{\partial \ln \mu}{\partial x} = 0$$

and to find μ we obtain an **ordinary** differential equation,

$$\frac{\partial \ln \mu}{\partial y}=\frac{\frac{\partial N}{\partial x}-\frac{\partial M}{\partial y}}{M}$$

from which we determine (by a single quadrature) $\ln \mu$, and, hence, μ as well. It is clear that this can be done only if the expression $\frac{\frac{\partial N}{\partial x}-\frac{\partial M}{\partial y}}{M}$ is not dependent on x.

Similarly, if the expression $\frac{\frac{\partial N}{\partial x}-\frac{\partial M}{\partial y}}{N}$ is not dependent on y but only on x, then it is easy to find an integrating factor that **depends only on x.**

Example. Solve the equation

$$(y+xy^2)\,dx-x\,dy=0$$

Solution. Here, $M=y+xy^2$, $N=-x$

$$\frac{\partial M}{\partial y}=1+2xy, \quad \frac{\partial N}{\partial x}=-1, \quad \frac{\partial M}{\partial y}\neq\frac{\partial N}{\partial x}$$

Thus, the left side of the equation **is not** an exact differential. Let us see whether this equation allows for an integrating factor dependent only on y or not. Noting that

$$\frac{\frac{\partial N}{\partial x}-\frac{\partial M}{\partial y}}{M}=\frac{-1-1-2xy}{y+xy^2}=-\frac{2}{y}$$

we conclude that the equation permits of an integrating factor dependent only on y. We find it:

$$\frac{d\ln \mu}{dy}=-\frac{2}{y}$$

whence

$$\ln \mu=-2\ln y, \text{ i.e., } \mu=\frac{1}{y^2}$$

After multiplying through by the integrating factor μ, we obtain the equation

$$\left(\frac{1}{y}+x\right)dx-\frac{x}{y^2}\,dy=0$$

as an exact differential equation $\left(\frac{\partial M}{\partial y}=\frac{\partial N}{\partial x}=-\frac{1}{y^2}\right)$. Solving this equation, we find its complete integral:

$$\frac{x}{y}+\frac{x^2}{2}+C=0$$

or

$$y=-\frac{2x}{x^2+2C}$$

1.11 THE ENVELOPE OF A FAMILY OF CURVES

Let there be an equation of the form

$$\Phi(x, y, C)=0 \tag{1}$$

where x and y are variable Cartesian coordinates and C is a parameter that can take on a variety of fixed values.

For each given value of the parameter C, equation (1) defines some curve in the xy-plane. Assigning to C all possible values, we obtain a **family of curves** dependent on a single parameter, or using the more common term, a **one-parameter family of curves**. Thus, equation (1) is the equation of a one-parameter family of curves (because it contains only one arbitrary constant).

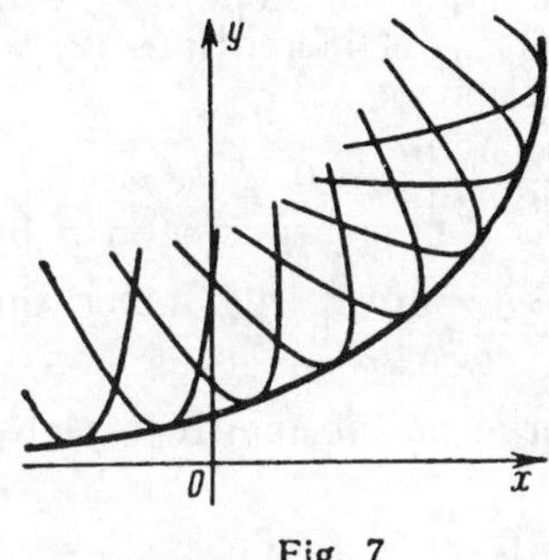

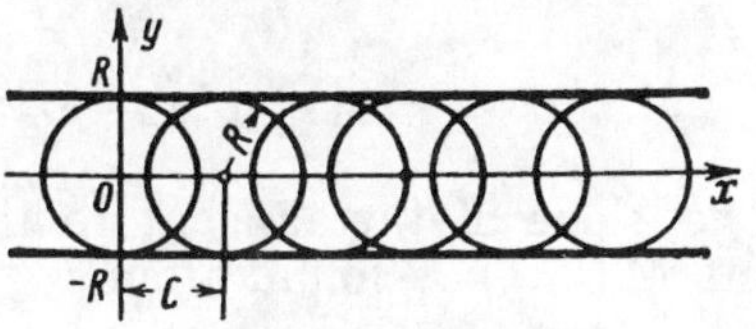

Fig. 7

Fig. 8

Definition. A curve L is called the *envelope* of a one-parameter family of curves if at each point it touches a curve of the family, and different curves of the given family touch the curve L at different points (Fig. 7).

Example 1. Consider the family of curves

$$(x-C)^2+y^2=R^2$$

where R is a constant and C is a parameter.

This is a family of circles of radius R with centres on the x-axis. This family will obviously have as envelopes the straight lines $y=R$ and $y=-R$ (Fig. 8).

Finding the equation of the envelope of a given family. Let there be given a family of curves,

$$\Phi(x, y, C)=0 \tag{1}$$

that depend on the parameter C.

Let us assume that this family has an envelope whose equation may be written in the form $y=\varphi(x)$, where $\varphi(x)$ is a continuous and differentiable function of x. We consider some point $M(x, y)$ lying on the envelope. This point also lies on some curve of the family (1). To this curve there corresponds a definite value of the parameter C, which value is determined from equation (1), for given (x, y): $C=C(x, y)$ Thus, for all points of the envelope the following equation holds:

$$\Phi[x, y, C(x, y)]=0 \tag{2}$$

Suppose that $C(x, y)$ is a differentiable function that is not constant in any interval of the values of x and y under consideration. From equation (2) of the envelope we find the slope of the tangent to the envelope at the point $M(x, y)$. Differentiate (2) with respect to x assuming that y is a function of x:

$$\frac{\partial \Phi}{\partial x}+\frac{\partial \Phi}{\partial C}\frac{\partial C}{\partial x}+\left[\frac{\partial \Phi}{\partial y}+\frac{\partial \Phi}{\partial C}\frac{\partial C}{\partial y}\right]y'=0$$

or

$$\Phi'_x+\Phi'_y y'+\Phi'_C\left[\frac{\partial C}{\partial x}+\frac{\partial C}{\partial y}y'\right]=0 \tag{3}$$

The slope of the tangent to the curve of the family (1) at the point $M(x, y)$ is found from

$$\Phi'_x+\Phi'_y y'=0 \tag{4}$$

(on this curve, C is constant).

We assume that $\Phi'_y \neq 0$, otherwise we would consider x as the function and y as the argument. Since the slope k of the envelope is equal to the slope k of the curve of the family, from (3) and (4) we obtain

$$\Phi'_C\left[\frac{\partial C}{\partial x}+\frac{\partial C}{\partial y}y'\right]=0$$

But since on the envelope $C(x, y) \neq \text{const}$, it follows that

$$\frac{\partial C}{\partial x}+\frac{\partial C}{\partial y}y' \neq 0$$

and so for its points the following equation holds:

$$\Phi'_C(x, y, C)=0 \tag{5}$$

Thus, the following two equations serve to determine the envelope:

$$\left.\begin{aligned}\Phi(x, y, C)=0\\ \Phi'_C(x, y, C)=0\end{aligned}\right\} \tag{6}$$

Conversely, if, by eliminating C from these equations, we get an equation $y=\varphi(x)$, where $\varphi(x)$ is a differentiable function and $C \neq \text{const}$ on this curve, then $y=\varphi(x)$ is the equation of the envelope.

Note 1. If for the family (1) a certain function $y=\varphi(x)$ is the equation of the locus of *singular points*, that is, of points where $\Phi'_x=0$ and $\Phi'_y=0$, then the coordinates of these points also satisfy equations (6).

Indeed, the coordinates of singular points may be expressed in terms of the parameter C that enters into equation (1):

$$x=\lambda(C), \quad y=\mu(C) \tag{7}$$

If these expressions are substituted in equation (1), we get an identity in C:

$$\Phi[\lambda(C), \mu(C), C]=0$$

Differentiating this identity with respect to C, we obtain

$$\Phi'_x \frac{d\lambda}{dC}+\Phi'_y \frac{d\mu}{dC}+\Phi'_C=0$$

Since for any points the equalities $\Phi'_x=0$, $\Phi'_y=0$ are fulfilled, it follows that for them the equality $\Phi'_C=0$ is also satisfied.

We have thus proved that the coordinates of singular points satisfy equations (6).

Summarizing, equations (6) define either the envelope or the locus of singular points of the curves of the family (1), or a combination of both. Thus, after obtaining a curve that satisfies equations (6), one has further to find out whether it is an envelope or the locus of singular points.

Example 2. Find the envelope of the family of circles

$$(x-C)^2+y^2-R^2=0$$

that are dependent on the single parameter C.

Solution. Differentiating the equation of the family with respect to C, we get

$$2(x-C)=0$$

Eliminating C from these two equations, we obtain the equation

$$y^2-R^2=0 \quad \text{or} \quad y=\pm R$$

It is clear, by geometric reasoning, that the pair of straight lines is the **envelope** (and not the locus of singular points, since the circles of a family do not have singular points).

Example 3. Find the envelope of the family of straight lines

$$x\cos\alpha+y\sin\alpha-p=0 \tag{a}$$

where α is a parameter.

Solution. Differentiating the given equation of the family with respect to α, we have

$$-x\sin\alpha+y\cos\alpha=0 \tag{b}$$

To eliminate the parameter α from equations (a) and (b), multiply the terms of the first by $\cos\alpha$, and of the second, by $\sin\alpha$, and then subtract the second from the first; we then have

$$x=p\cos\alpha$$

Putting this expression into (b), we find

$$y=p\sin\alpha$$

Squaring the terms of the last two equations and adding termwise, we get

$$x^2+y^2=p^2$$

This is a circle. It is the **envelope** of the family (and not the locus of singular points, since straight lines do not have singular points) (Fig. 9).

Example 4. Find the envelope of the trajectories of projectiles fired from a gun with velocity v_0 at different angles of inclination of the barrel to the horizon. We shall consider that the gun is located at the coordinate origin and that the trajectories of projectiles lie in the xy-plane (air resistance is disregarded).

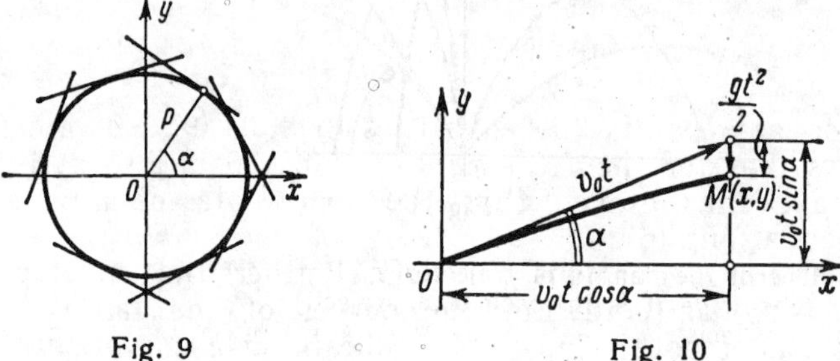

Fig. 9 Fig. 10

Solution. First find the equation of the trajectory for the case when the barrel makes an angle α with the positive x-axis. In flight, the projectile participates simultaneously in two motions: a uniform motion with velocity v_0 in the direction of the barrel and a falling motion due to the force of gravity. Therefore, at each instant of time t the position of the projectile M (Fig. 10) will be defined by the equations

$$x=v_0t\cos\alpha$$

$$y=v_0t\sin\alpha-\frac{gt^2}{2}$$

These are parametric equations of the trajectory (the parameter is the time t). Eliminating t, we get the equation of the trajectory in the form

$$y=x\tan\alpha-\frac{gx^2}{2v_0^2\cos^2\alpha}$$

Finally, introducing the notation $\tan\alpha=k$, $\frac{g}{2v_0^2}=a$, we get

$$y=kx-ax^2(1+k^2) \tag{8}$$

This equation defines a parabola with vertical axis passing through the origin and concave down. We obtain a variety of trajectories for the different values of k. Consequently, equation (8) is the equation of a one-parameter family of parabolas, which are the trajectories of a projectile for different angles α and for a given initial velocity v_0 (Fig. 11).

Let us find the envelope of this family of parabolas. Differentiating both sides of (8) with respect to k, we have

$$x-2akx^2=0 \tag{9}$$

Eliminating k from equations (8) and (9), we get

$$y=\frac{1}{4a}-ax^2$$

This is the equation of a parabola with vertex at the point $\left(0, \frac{1}{4a}\right)$, the axis of which coincides with the y-axis. It is not a locus of singular points [since parabolas (8) do not have singular points]. Thus, the parabola

$$y=\frac{1}{4a}-ax^2$$

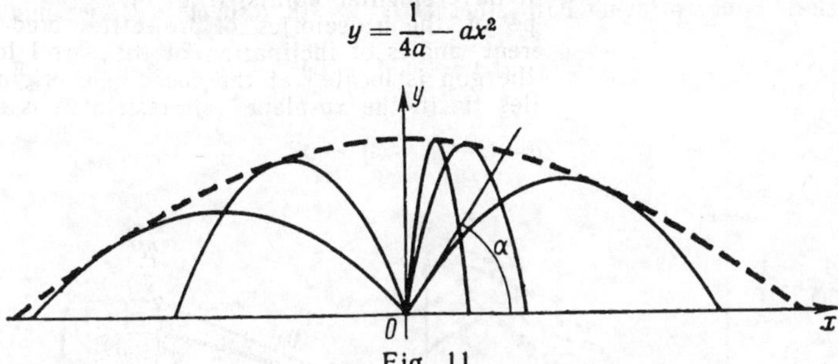

Fig. 11

is the envelope of the family of trajectories. It is called a **safety parabola** because no point outside it is in reach of a projectile fired from a given gun with a given initial velocity v_0.

Example 5. Find the envelope of the family of semicubical parabolas

$$y^3-(x-C)^2=0$$

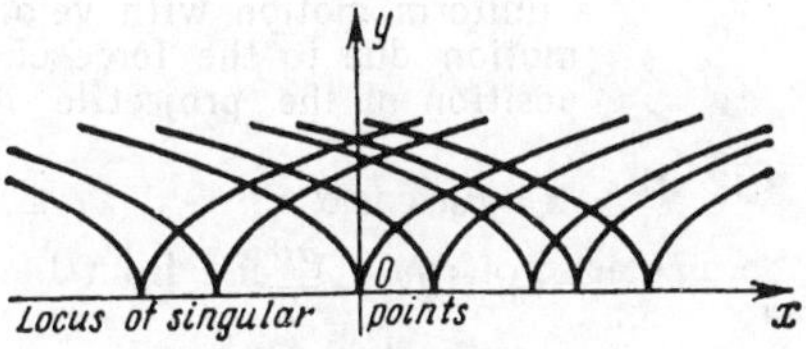

Locus of singular points

Fig. 12

Solution. Differentiate the given equation of the family with respect to the parameter C:

$$2(x-C)=0$$

Eliminating the parameter C from the two equations, we get

$$y=0$$

The x-axis is a locus of singular points—a cusp of the first kind (Fig. 12). Indeed, let us find the singular points of the curve

$$y^3-(x-C)^2=0$$

for a fixed value of C. Differentiating with respect to x and y, we find

$$F'_x=-2(x-C)=0$$

$$F'_y=3y^2=0$$

Solving the three foregoing equations simultaneously, we find the coordinates of the singular point: $x=C$, $y=0$; thus, each curve of the given family has a singular point on the x-axis.

For continuous variation of the parameter C, the singular points will fill the entire x-axis.

Example 6. Find the envelope and locus of singular points of the family

$$(y-C)^2-\frac{2}{3}(x-C)^3=0 \qquad (10)$$

Solution. Differentiating both sides of (10) with respect to C, we find

$$-2(y-C)+\frac{2}{3}3(x-C)^2=0$$

or

$$y-C-(x-C)^2=0 \qquad (11)$$

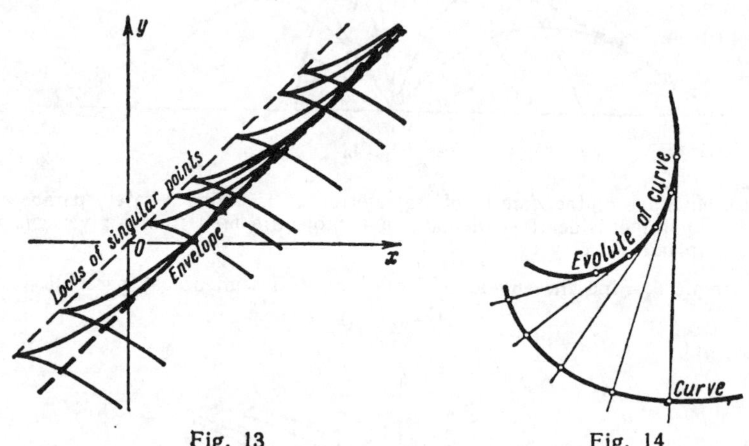

Fig. 13 Fig. 14

Now eliminate the parameter C from (11) and from the equation (10) of the family. Putting the expression

$$y-C=(x-C)^2$$

into the equation of the family, we get

$$(x-C)^4-\frac{2}{3}(x-C)^3=0$$

or

$$(x-C)^3\left[(x-C)-\frac{2}{3}\right]=0$$

whence we obtain two possible values of C and two solutions of the problem corresponding to them.

First solution:

$$C=x$$

and so from (11) we find

$$y-x-(x-x)^2=0$$

or

$$y=x$$

Second solution.

$$C=x-\frac{2}{3}$$

and so from (11) we find

$$y-x+\frac{2}{3}-\left[x-x+\frac{2}{3}\right]^2=0$$

or

$$y=x-\frac{2}{9}$$

We have obtained two straight lines: $y=x$ and $y=x-\frac{2}{9}$. The first is the locus of singular points, the second is the envelope (Fig. 13).

Note 2. In Sec. 6.7, Vol. I it was proved that the normal to a curve serves as a tangent to its evolute. Hence, the family of normals to a given curve is at the same time a family of tangents to its evolute. Thus, *the evolute of the curve is the envelope of the family of normals of this curve* (Fig. 14).

This remark enables us to point out another method for finding evolutes: to obtain the equation of an evolute, first find the family of all normals of the given curve and then find the envelope of this family.

1.12 SINGULAR SOLUTIONS OF A FIRST-ORDER DIFFERENTIAL EQUATION

Let the differential equation

$$F\left(x, y, \frac{dy}{dx}\right)=0 \tag{1}$$

have a complete integral

$$\Phi(x, y, C)=0 \tag{2}$$

Let us assume that the family of integral curves that corresponds to equation (2) has an envelope. We shall prove that this envelope is also an integral curve of the differential equation (1).

Indeed, at each point the envelope touches some curve of the family; that is, it has a common tangent with it. Thus, at each common point the envelope and the curve of the family have the same values of x, y, y'.

But for a curve of the family, the numbers x, y, and y' satisfy equation (1). Consequently, the very same equation is satisfied by the abscissa, the ordinate and the slope of each point of the envelope. But this means that the envelope is an integral curve and its equation is a solution of the given differential equation.

Since, generally speaking, the envelope is not a curve of the family, its equation cannot be obtained from the complete integral (2) for any particular value of C. The solution of the differential equation which is not obtained from the complete integral for any value of C and which has as its graph the envelope of the family of integral curves entering into the general solution, is called a *singular solution* of the differential equation.

Let the complete integral be known:

$$\Phi(x, y, C)=0$$

eliminating C from this equation and from the equation $\Phi'_C(x, y, C)=0$, we get $\psi(x, y)=0$. If this function satisfies the differential equation [and does not belong to the family (2)], then it is a *singular integral.*

It should be noted that at least two integral curves pass through each point of the curve that describes a singular solution; that is, **uniqueness of solution is violated at each point of a singular solution.**

We note that a point at which uniqueness of solution of a differential equation has been violated, that is, a point through which at least two integral curves pass, is called a *singular point* *. Thus, a singular solution consists of singular points.

Example. Find a singular solution of the equation

$$y^2(1+y'^2)=R^2 \tag{*}$$

Solution. Let us find its complete integral. We solve the equation for y':

$$\frac{dy}{dx}=\pm\frac{\sqrt{R^2-y^2}}{y}$$

Separating variables, we obtain

$$\frac{y\,dy}{\pm\sqrt{R^2-y^2}}=dx$$

Whence, integrating, we find the complete integral:

$$(x-C)^2+y^2=R^2$$

It is easy to see that the family of integral lines is a family of circles of radius R with centres on the x-axis. The pair of straight lines $y=\pm R$ will be the envelope of the family of curves.

The functions $y=\pm R$ satisfy the differential equation (*). This, consequently, is a singular integral.

1.13 CLAIRAUT'S EQUATION

Let us consider the so-called *Clairaut equation:*

$$y=x\frac{dy}{dx}+\psi\left(\frac{dy}{dx}\right) \tag{1}$$

It is integrated by introducing an auxiliary parameter. Put $\frac{dy}{dx}=p$; then equation (1) will take the form

$$y=xp+\psi(p) \tag{1'}$$

Differentiate all the terms of this equation with respect to x, bearing in mind that $p=\frac{dy}{dx}$ is a function of x:

$$p=x\frac{dp}{dx}+p+\psi'(p)\frac{dp}{dx}$$

---

* The boundary points of the domain of existence of a solution are also termed *singular points*. An interior point through which a unique integral curve of a differential equation passes is called an *ordinary point*.

or

$$[x+\psi'(p)]\frac{dp}{dx}=0$$

Equating each factor to zero, we get

$$\frac{dp}{dx}=0 \tag{2}$$

and

$$x+\psi'(p)=0 \tag{3}$$

(1) Integrating (2) we obtain $p=C\,(C=\text{const})$. Putting this value of p into (1′), we find its complete integral:

$$y=xC+\psi(C) \tag{4}$$

which, geometrically, is a **family of straight lines.**

(2) If from (3) we find p as a function of x and put it into (1′), we obtain the function

$$y=xp(x)+\psi[p(x)] \tag{1″}$$

which may be readily shown to be the solution of equation (1).

Indeed, by virtue of (3), we have

$$\frac{dy}{dx}=p+[x+\psi'(p)]\frac{dp}{dx}, \text{ i.e. } \frac{dy}{dx}=p$$

And so, by substituting the function (1″) into equation (1) we get the identity

$$xp+\psi(p)=xp+\psi(p)$$

The solution (1″) is not obtained from the complete integral (4) for any value of C. This is a **singular solution;** it is obtained by elimination of the parameter p from the equations

$$\left.\begin{aligned} y&=xp+\psi(p)\\ x&+\psi'(p)=0\end{aligned}\right\}$$

or, which is the same thing, by eliminating C from the equations

$$y=xC+\psi(C)$$
$$x+\psi'_C(C)=0$$

Thus, **the singular solution of Clairaut's equation defines the envelope of a family of straight lines represented by the complete integral (4).**

Example. Find the general and singular solutions of the equation

$$y=x\frac{dy}{dx}+\frac{a\frac{dy}{dx}}{\sqrt{1+\left(\frac{dy}{dx}\right)^2}}$$

Solution. The general solution is obtained by substituting C for $\frac{dy}{dx}$:

$$y = xC + \frac{aC}{\sqrt{1+C^2}}$$

To obtain the **singular** solution, differentiate the latter equation with respect to C:

$$x + \frac{a}{(1+C^2)^{\frac{3}{2}}} = 0$$

The singular solution (the equation of the envelope) is obtained in parametric form (where the parameter is C):

$$\begin{cases} x = -\dfrac{a}{(1+C^2)^{\frac{3}{2}}} \\ y = \dfrac{aC^3}{(1+C^2)^{\frac{3}{2}}} \end{cases}$$

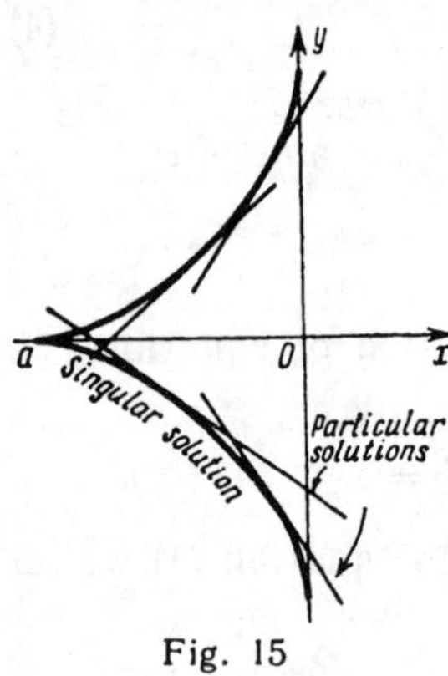

Fig. 15

Eliminating C, we get a direct relationship between x and y. Raising both sides of each equation to the power $\frac{2}{3}$ and adding the resultant equations termwise, we get the singular solution in the following form:

$$x^{\frac{2}{3}} + y^{\frac{2}{3}} = a^{\frac{2}{3}}$$

This is an astroid. However, the envelope of the family (and, hence, the singular solution) is not the entire astroid, but only its left half (since it is evident from the parametric equations that $x \leqslant 0$) (Fig. 15).

1.14 LAGRANGE'S EQUATION

The *Lagrange equation* is an equation of the form

$$y = x\varphi(y') + \psi(y') \qquad (1)$$

where φ and ψ are known functions of $\frac{dy}{dx}$.

This equation is linear in y and x. Clairaut's equation, which was considered in the preceding section, is a particular case of the Lagrange equation when $\varphi(y') \equiv y'$. The Lagrange equation, like Clairaut's, is integrated by means of introducing an auxiliary parameter p. Put

$$y' = p$$

then the original equation is written in the form

$$y = x\varphi(p) + \psi(p) \qquad (1')$$

Differentiating with respect to x, we obtain

$$p=\varphi(p)+[x\varphi'(p)+\psi'(p)]\frac{dp}{dx}$$

or

$$p-\varphi(p)=[x\varphi'(p)+\psi'(p)]\frac{dp}{dx} \tag{1''}$$

From this equation we can straightway find certain solutions: namely, it becomes an identity for any constant value $p=p_0$ that satisfies the condition

$$p_0-\varphi(p_0)=0$$

Indeed, for a constant value p the derivative $\frac{dp}{dx}\equiv 0$, and both sides of equation (1″) vanish.

The solution corresponding to each value $p=p_0$, that is, $\frac{dy}{dx}=p_0$ is a **linear** function of x $\left(\text{since the derivative } \frac{dy}{dx} \text{ is constant only in the case of linear functions}\right)$. To find this function it is sufficient to put into (1′) the value $p=p_0$:

$$y=x\varphi(p_0)+\psi(p_0)$$

If it turns out that this solution is not obtainable from the general solution for any value of the arbitrary constant, it will be a **singular solution.**

Let us now find the **general solution**. Write (1″) in the form

$$\frac{dx}{dp}-x\frac{\varphi'(p)}{p-\varphi(p)}=\frac{\psi'(p)}{p-\varphi(p)}$$

and regard x as a function of p. Then the equation obtained will be a linear differential equation in the function x of p.

Solving it, we find

$$x=\omega(p, C) \tag{2}$$

Eliminating the parameter p from equations (1′) and (2), we get the **complete integral** of (1) in the form $\Phi(x, y, C)=0$.

Example. Given the equation

$$y=xy'^2+y'^2 \tag{I}$$

Putting $y'=p$ we have

$$y=xp^2+p^2 \tag{I'}$$

Differentiating with respect to x, we get

$$p=p^2+[2xp+2p]\frac{dp}{dx} \tag{I''}$$

Let us find the **singular solutions.** Since $p=p^2$ for $p_0=0$ and $p_1=1$, the solutions will be linear functions [see (I′)]:

$$y=x\cdot 0^2+0^2,\quad \text{that is, } y=0$$

and

$$y=x+1$$

When we find the complete integral, we will see whether these functions are particular or singular solutions. To do so, write equation (I″) in the form

$$\frac{dx}{dp}-x\frac{2}{1-p}=\frac{2}{1-p}$$

and regard x as a function of the independent variable p. Integrating this linear (in x) equation, we find

$$x=-1+\frac{C^2}{(p-1)^2} \tag{II}$$

Eliminating p from equations (I′) and (II), we get the **complete integral**

$$y=(C+\sqrt{x+1})^2$$

The **singular integral** of the original equation is

$$y=0$$

since this solution is not obtainable from the general solution for any value of C.

However, the function $y=x+1$ is not a singular but a particular solution; it is obtained from the general solution when $C=0$.

1.15 ORTHOGONAL AND ISOGONAL TRAJECTORIES

Suppose we have a one-parameter family of curves

$$\Phi(x, y, C)=0 \tag{1}$$

Lines intersecting all the curves of the given family (1) at a constant angle are called *isogonal trajectories*. If this angle is a right angle, they are *orthogonal trajectories*.

Orthogonal trajectories. Let us find the equation of orthogonal trajectories. Write the differential equation of the given family of curves, eliminating the parameter C from the equations

$$\Phi(x, y, C)=0$$

and

$$\frac{\partial\Phi}{\partial x}+\frac{\partial\Phi}{\partial y}\frac{dy}{dx}=0$$

Let this differential equation be

$$F\left(x, y, \frac{dy}{dx}\right)=0 \tag{1′}$$

Here, $\frac{dy}{dx}$ is the slope of the tangent to some member of the family at the point $M(x, y)$. Since an orthogonal trajectory passing through the point $M(x, y)$ is perpendicular to the corresponding curve of the family, the slope of the tangent to it, $\frac{dy_T}{dx}$, is connected with $\frac{dy}{dx}$ by the relation (Fig. 16)

$$\frac{dy}{dx} = -\frac{1}{\frac{dy_T}{dx}} \tag{2}$$

Putting this expression into equation (1′) and dropping the subscript T, we get a relationship between the coordinates of an arbitrary point (x, y) and the slope of the orthogonal trajectory at this point, that is, a **differential equation of orthogonal trajectories:**

$$F\left(x, y, -\frac{1}{\frac{dy}{dx}}\right) = 0 \tag{3}$$

The complete integral of this equation

$$\Phi_1 (x, y, C) = 0$$

yields a **family of orthogonal trajectories.**

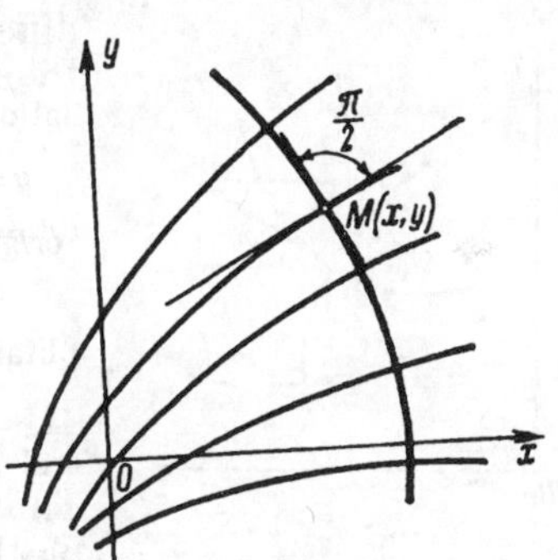

Fig 16.

A consideration of the plane flow of a fluid involves orthogonal trajectories.

Let us suppose that the flow of fluids in a plane takes place in such a manner that at each point of the xy-plane the velocity vector $\boldsymbol{v}(x, y)$ is defined. If this vector depends solely on the position of the point in the plane, but is independent of the time, the motion is called stationary or steady-state. We shall consider such motion. In addition, we shall assume that there exists a potential of velocities, that is, a function $u(x, y)$ such that the projections of the vector $\boldsymbol{v}(x, y)$ on the coordinate axes, $v_x(x, y)$ and $v_y(x, y)$, are its partial derivatives with respect to x and y:

$$\frac{\partial u}{\partial x} = v_x \quad \frac{\partial u}{\partial y} = v_y \tag{4}$$

The lines of the family

$$u(x, y) = C \tag{5}$$

are called *equipotential lines* (lines of equal potential).

The lines, whose tangents at all points coincide with the vector $\boldsymbol{v}(x, y)$, are called *flow lines* and yield the trajectories of moving points.

We shall show that the flow lines are the orthogonal trajectories of a family of equipotential lines (Fig. 17).

Let φ be an angle formed by the velocity vector $\boldsymbol{v}$ with the x-axis. Then, by relation (4),

$$\frac{\partial u(x, y)}{\partial x}=|\boldsymbol{v}|\cos\varphi, \quad \frac{\partial u(x, y)}{\partial y}=|\boldsymbol{v}|\sin\varphi$$

whence we find the slope of the tangent to the flow line

$$\tan\varphi=\frac{\dfrac{\partial u(x, y)}{\partial y}}{\dfrac{\partial u(x, y)}{\partial x}} \tag{6}$$

We obtain the slope of the tangent to the equipotential line by differentiating relation (5) with respect to x:

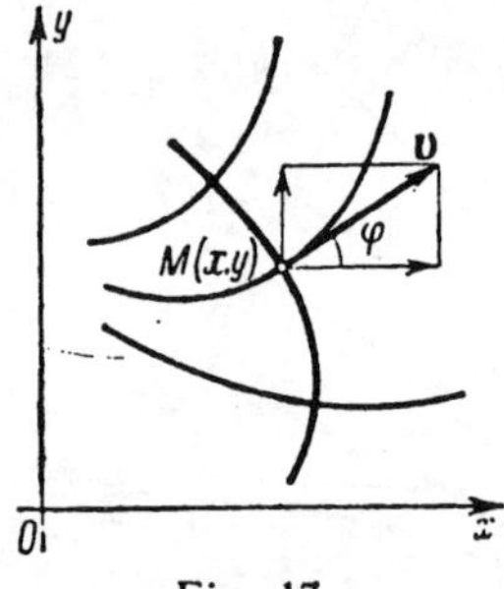

Fig. 17

$$\frac{\partial u}{\partial x}+\frac{\partial u}{\partial y}\frac{dy}{dx}=0$$

whence

$$\frac{dy}{dx}=-\frac{\dfrac{\partial u}{\partial x}}{\dfrac{\partial u}{\partial y}} \tag{7}$$

Thus, in magnitude and sign, the slope of the tangent to the equipotential line is the inverse of the slope of the tangent to the flow line. Whence it follows that equipotential lines and flow lines are mutually orthogonal.

In the case of an electric or magnetic field, the lines of force of the field serve as the orthogonal trajectories of the family of equipotential lines.

Example 1. Find the orthogonal trajectories of the family of parabolas

$$y=Cx^2$$

Solution. We write the differential equation of the family:

$$y'=2Cx$$

Eliminating C, we get

$$\frac{y'}{y}=\frac{2}{x}$$

Substituting $-\frac{1}{y'}$ for y', we obtain a differential equation of the family of orthogonal trajectories

$$-\frac{1}{yy'}=\frac{2}{x}$$

or

$$y\,dy=-\frac{x\,dx}{2}$$

Its complete integral is

$$\frac{x^2}{4}+\frac{y^2}{2}=C^2$$

Hence, the orthogonal trajectories of the given family of parabolas will be represented by a certain family of ellipses with semi-axes $a=2C$, $b=C\sqrt{2}$ (Fig. 18).

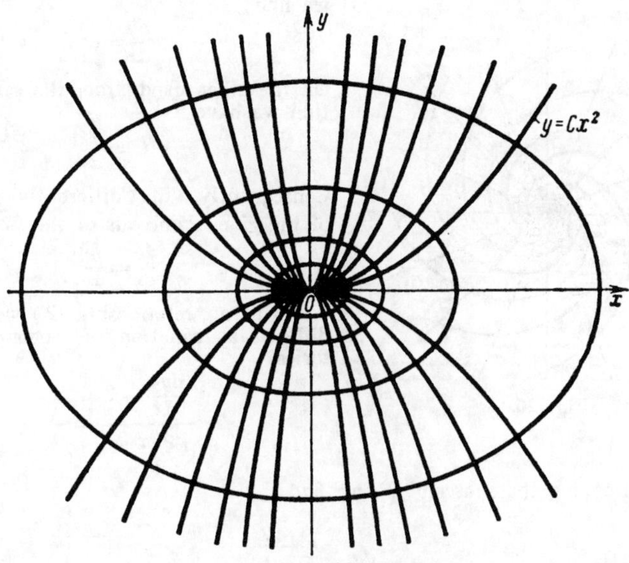

Fig. 18

Isogonal trajectories. Let the trajectories cut the curve of a given family at an angle α, where $\tan\alpha=k$.

The slope $\frac{dy}{dx}=\tan\varphi$ (Fig. 19) of the tangent to a member of the family and the slope $\frac{dy_T}{dx}=\tan\psi$ to the isogonal trajectory are connected by the relationship

$$\tan\varphi=\tan(\psi-\alpha)=\frac{\tan\psi-\tan\alpha}{1+\tan\alpha\tan\psi}$$

that is,

$$\frac{dy}{dx}=\frac{\frac{dy_T}{dx}-k}{k\frac{dy_T}{dx}+1} \qquad (2')$$

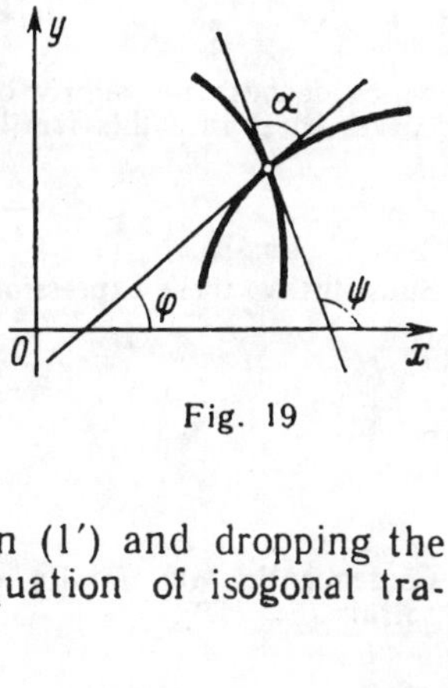

Fig. 19

Substituting this expression into equation (1′) and dropping the subscript T, we obtain the differential equation of isogonal trajectories.

Example 2. Find the isogonal trajectories of a family of straight lines,

$$y=Cx \tag{8}$$

that cut the lines of the given family at an angle α, the tangent of which equals k.

Solution. Let us write the differential equation of the given family. Differentiating equation (8) with respect to x we find

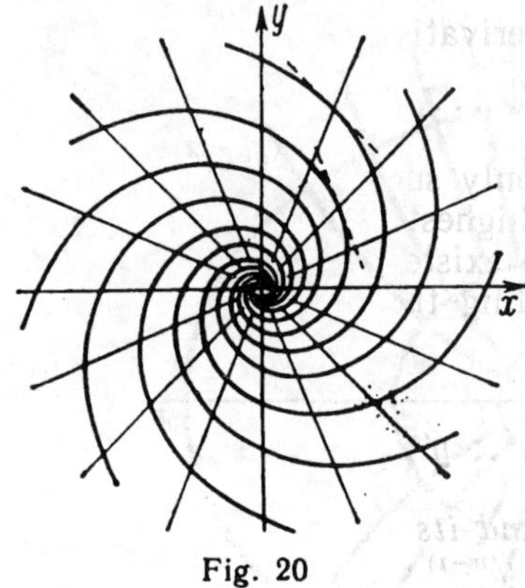

Fig. 20

$$\frac{dy}{dx}=C$$

On the other hand, from the same equation we have

$$C=\frac{y}{x}$$

Consequently, the differential equation of the given family is of the form

$$\frac{dy}{dx}=\frac{y}{x}$$

Utilizing relationship (2′) we get the differential equation of isogonal trajectories

$$\frac{\frac{dy_T}{dx}-k}{k\frac{dy_T}{dx}+1}=\frac{y}{x}$$

whence, dropping the subscript T, we find

$$\frac{dy}{dx}=\frac{k+\frac{y}{x}}{1-k\frac{y}{x}}$$

Integrating this homogeneous equation, we get the complete integral:

$$\ln\sqrt{x^2+y^2}=\frac{1}{k}\arctan\frac{y}{x}+\ln C \tag{9}$$

which defines the family of isogonal trajectories. To find out precisely which curves enter into this family, let us change to polar coordinates:

$$\frac{y}{x}=\tan\varphi, \qquad \sqrt{x^2+y^2}=\rho$$

Substituting these expressions into (9), we obtain

$$\ln\rho=\frac{1}{k}\varphi+\ln C$$

or

$$\rho=Ce^{\frac{\varphi}{k}}$$

Consequently, the family of isogonal trajectories is a family of logarithmic spirals (Fig. 20).

1.16 HIGHER-ORDER DIFFERENTIAL EQUATIONS (FUNDAMENTALS)

As has already been indicated above (see Sec. 1.2), a differential equation of the nth order may be written symbolically in the form

$$F(x, y, y', y'', \ldots, y^{(n)}) = 0 \tag{1}$$

or, if it can be solved for the nth derivative,

$$y^{(n)} = f(x, y, y', y'', \ldots, y^{(n-1)}) \tag{1'}$$

In this chapter we shall consider only such equations of higher order that may be solved for the highest derivative. For these equations we have a theorem on the existence and uniqueness of a solution, similar to the corresponding theorem on the solution of first-order equations.

Theorem. *If in the equation*

$$y^{(n)} = f(x, y, y', \ldots, y^{(n-1)})$$

the function $f(x, y, y', \ldots, y^{(n-1)})$ *and its partial derivatives with respect to the arguments* $y, y', \ldots, y^{(n-1)}$ *are continuous in some region containing the values* $x = x_0$, $y = y_0$, $y' = y'_0$, $\ldots$, $y^{(n-1)} = y_0^{(n-1)}$, *then there is one and only one solution,* $y = y(x)$, *of the equation that satisfies the conditions*

$$\left.\begin{aligned} y_{x=x_0} &= y_0 \\ y'_{x=x_0} &= y'_0 \\ &\cdots\cdots \\ y^{(n-1)}_{x=x_0} &= y_0^{(n-1)} \end{aligned}\right\} \tag{2}$$

These conditions are called *initial conditions*. The proof is beyond the scope of this book.

If we consider a second-order equation $y'' = f(x, y, y')$, then the initial conditions for the solution, when $x = x_0$, will be

$$y = y_0, \; y' = y'_0$$

where x_0, y_0, y'_0 are given numbers, these conditions have the following geometric meaning: only one curve passes through a given point (x_0, y_0) of the plane with given tangent of the angle of inclination of the tangent line y'_0. From this it follows that if we want to assign different values of y'_0 for constant x_0 and y_0, we get an infinitude of integral curves with different angles of inclination passing through the given point.

We now introduce the concept of a general solution of an equation of the nth order.

Definition. The *general solution* of a differential equation of the nth order is a function

$$y=\varphi(x, C_1, C_2, \ldots, C_n)$$

which is dependent on n arbitrary constants $C_1, C_2, \ldots, C_n$ and such that:

(a) it satisfies the equation for any values of the constants $C_1, C_2, \ldots, C_n$;

(b) for specified initial conditions

$$y_{x=x_0}=y_0$$
$$y'_{x=x_0}=y'_0$$
$$\cdot\ \cdot\ \cdot\ \cdot\ \cdot\ \cdot\ \cdot$$
$$y^{(n-1)}_{x=x_0}=y_0^{(n-1)}$$

the constants $C_1, C_2, \ldots, C_n$ may be chosen so that the function $y=\varphi(x, C_1, C_2, \ldots, C_n)$ will satisfy these conditions (on the assumption that the initial values $x_0, y_0, y'_0, \ldots, y_0^{(n-1)}$ belong to the region where the conditions of the existence of a solution are fulfilled).

A relationship of the form $\Phi(x, y, C_1, C_2, \ldots, C_n)=0$, which implicitly defines the general solution, is called the *complete integral* of the differential equation.

Any function obtained from the general solution for specific values of the constants $C_1, C_2, \ldots, C_n$ is called a *particular solution*. The graph of a particular solution is called an *integral curve* of the given differential equation.

To solve (integrate) a differential equation of the nth order means:

(1) to find its general solution (if the initial conditions are not given) or

(2) to find a particular solution of the equation that satisfies the given initial conditions (if there are such).

In the following sections we shall present methods of solving various equations of the nth order.

1.17 AN EQUATION OF THE FORM $y^{(n)}=f(x)$

The simplest type of equation of the nth order is of the form

$$y^{(n)}=f(x) \tag{1}$$

Let us find the complete integral of this equation.

Integrating the left and right sides with respect to x and taking into account that $y^{(n)}=(y^{(n-1)})'$, we obtain

$$y^{(n-1)}=\int_{x_0}^{x} f(x)\,dx+C_1$$

where x_0 is any fixed value of x, and C_1 is the constant of integration.

Integrating once more, we get

$$y^{(n-2)}=\int_{x_0}^{x}\left(\int_{x_0}^{x} f(x)\,dx\right)dx+C_1(x-x_0)+C_2$$

Continuing, we finally get (after n integrations) the expression of the complete integral:

$$y=\int_{x_0}^{x}\dots\int_{x_0}^{x} f(x)\,dx\dots dx+\frac{C_1(x-x_0)^{n-1}}{(n-1)!}+C_2\,\frac{(x-x_0)^{n-2}}{(n-2)!}+\dots+C_n$$

In order to find a particular solution satisfying the initial conditions

$$y_{x=x_0}=y_0,\quad y'_{x=x_0}=y'_0,\ \dots,\ y^{(n-1)}_{x=x_0}=y_0^{(n-1)}$$

it is sufficient to put

$$C_n=y_0,\quad C_{n-1}=y'_0,\ \dots,\ C_1=y_0^{n-1}$$

Example 1. Find the complete integral of the equation

$$y''=\sin(kx)$$

and a particular solution satisfying the initial conditions

$$y_{x=0}=0,\quad y'_{x=0}=1$$

Solution.

$$y'=\int_0^x \sin kx\,dx+C_1=-\frac{\cos kx-1}{k}+C_1$$

$$y=-\int_0^x\left(\frac{\cos kx-1}{k}\right)dx+\int_0^x C_1\,dx+C_2$$

or

$$y=-\frac{\sin kx}{k^2}+\frac{x}{k}+C_1x+C_2$$

This is the complete integral. To find a particular solution satisfying the given initial conditions, it is sufficient to determine the corresponding values of C_1 and C_2.

From the condition $y_{x=0}=0$, we find $C_2=0$.

From the condition $y'_{x=0}=1$, we find $C_1=1$.

Thus, the desired particular solution is of the form

$$y=-\frac{\sin kx}{k^2}+x\left(\frac{1}{k}+1\right)$$

Differential equations of this kind are encountered in the theory of the bending of girders.

Example 2. Let us consider an elastic prismatic girder bending under the action of external forces both continuously distributed (weight, load) and concentrated. Let the x-axis be horizontal along the axis of the girder in its underformed state and let the y-axis be directed vertically downwards (Fig. 21).

Each force acting on the girder (the load of the girder, and the reaction of the supports, for instance) has a moment, relative to some cross section of the girder, equal to the product of the force by the distance of the point of application of the force from the given cross section. The sum, $M(x)$, of the moments of all the forces applied to that part of the girder situated to one side of the given cross section with abscissa x is called the bending moment of the girder relative to the given cross section. In courses of strength of materials, it is proved that the bending moment of the girder is

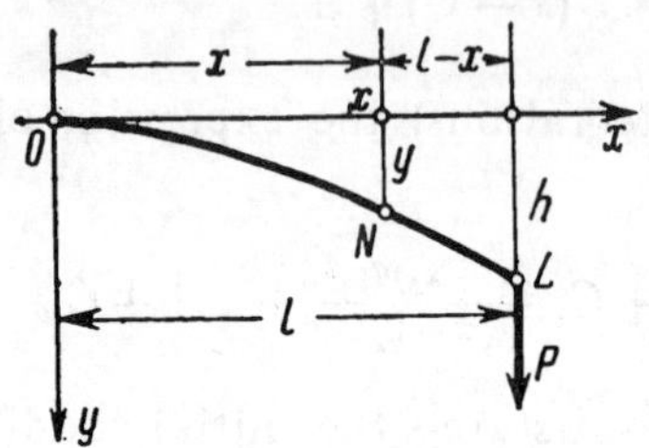

Fig. 21

$$\frac{EJ}{R}$$

where E is the so-called modulus of elasticity which depends on the material of the girder, J is the moment of inertia of the cross-sectional area of the girder relative to the horizontal line passing through the centre of gravity of the cross-sectional area, and R is the radius of curvature of the axis of the bent girder, which radius is expressed by the formula (Sec. 6.6, Vol. I)

$$R=\frac{(1+y'^2)^{3/2}}{|y''|}$$

Thus, the differential equation of the bent axis of a girder has the form

$$\frac{y''}{(1+y'^2)^{3/2}}=\frac{M(x)}{EJ} \tag{2}$$

If we consider that the deformations are small and that the tangents to the axis of the girder, when bent, form a small angle with the x-axis, we can disregard the square of the small quantity y'^2 and consider

$$R=\frac{1}{y''}$$

Then the differential equation of the bent girder will have the form

$$y''=\frac{M(x)}{EJ} \tag{2'}$$

but this equation is of the form of (1).

Example 3. A girder is fixed in place at the extremity O and is subjected to the action of a concentrated vertical force P applied to the end of the girder L at a distance l from O (Fig. 21). The weight of the girder is ignored.

We consider a cross section at the point $N(x)$. The bending moment relative to section N is, in the given case, equal to

$$M(x)=(l-x)P$$

The differential equation (2′) has the form

$$y''=\frac{P}{EJ}(l-x)$$

The initial conditions are: for $x=0$ the deflection y is equal to zero and the tangent to the bent axis of the girder coincides with the x-axis; that is,

$$y_{x=0}=0, \quad y'_{x=0}=0$$

Integrating the equation, we find

$$y'=\frac{P}{EJ}\int_0^x (l-x)\,dx=\frac{P}{EJ}\left(lx-\frac{x^2}{2}\right)$$

$$y=\frac{P}{2EJ}\left(lx^2-\frac{x^3}{3}\right) \quad (3)$$

In particular, from formula (3) we determine the deflection h at the extremity of the girder L:

$$h=y_{x=l}=\frac{Pl^3}{3EJ}$$

1.18 SOME TYPES OF SECOND-ORDER DIFFERENTIAL EQUATIONS REDUCIBLE TO FIRST-ORDER EQUATIONS. ESCAPE-VELOCITY PROBLEM

I. *An equation of the type*

$$\frac{d^2y}{dx^2}=f\left(x, \frac{dy}{dx}\right) \quad (1)$$

does not explicitly contain the unknown function y.

Solution. Let us denote the derivative $\frac{dy}{dx}$ by p, that is, we set $\frac{dy}{dx}=p$. Then $\frac{d^2y}{dx^2}=\frac{dp}{dx}$.

Putting these expressions of the derivatives into equation (1), we get a first-order equation,

$$\frac{dp}{dx}=f(x, p)$$

in the unknown function p of x. Integrating this equation, we find its general solution:

$$p=p(x, C_1)$$

and then from the relation $\frac{dy}{dx}=p$ we get the complete integral of equation (1):

$$y=\int p(x, C_1)\,dx+C_2$$

Example 1. Let us consider the differential equation of a catenary (see Sec. 1.1):

$$\frac{d^2y}{dx^2}=\frac{1}{a}\sqrt{1+\left(\frac{dy}{dx}\right)^2}$$

Set

$$\frac{dy}{dx}=p$$

then

$$\frac{d^2y}{dx^2}=\frac{dp}{dx}$$

and we get a first-order differential equation in the auxiliary function p of x:

$$\frac{dp}{dx}=\frac{1}{a}\sqrt{1+p^2}$$

Separating variables, we have

$$\frac{dp}{\sqrt{1+p^2}}=\frac{dx}{a}$$

whence

$$\ln\left(p+\sqrt{1+p^2}\right)=\frac{x}{a}+C_1$$

$$p=\sin h\left(\frac{x}{a}+C_1\right)$$

But since $p=\frac{dy}{dx}$, the latter relation is a differential equation in the sought-for function y. Integrating it, we obtain the equation of a catenary (see Sec. 1.1):

$$y=a\cos h\left(\frac{x}{a}+C_1\right)+C_2$$

Let us find the particular solution that satisfies the following initial conditions:

$$y_{x=0}=a$$

$$y'_{x=0}=0$$

The first condition yields $C_2=0$, the second, $C_1=0$.
We finally obtain

$$y=a\cos h\left(\frac{x}{a}\right)$$

Note. We can similarly integrate the equation

$$y^{(n)}=f(x,\ y^{(n-1)})$$

Setting $y^{(n-1)}=p$, we get for a determination of p the first-order equation

$$\frac{dp}{dx}=f(x,\ p)$$

From here we get p as a function of x, and from the relation $y^{(n-1)}=p$ we find y (see Sec. 1.17).

II. *An equation of the type*

$$\frac{d^2y}{dx^2}=f\left(y,\ \frac{dy}{dx}\right) \tag{2}$$

does not contain the independent variable x explicitly.

To solve it, we again set

$$\frac{dy}{dx}=p \tag{3}$$

but now **we shall consider p as a function of y** (and not of x, as before). Then

$$\frac{d^2y}{dx^2}=\frac{dp}{dx}=\frac{dp}{dy}\frac{dy}{dx}=\frac{dp}{dy}p$$

Putting into (2) the expressions $\frac{dy}{dx}$ and $\frac{d^2y}{dx^2}$, we get a first-order equation in the auxiliary function p:

$$p\frac{dp}{dy}=f(y, p) \tag{4}$$

Integrating it, we find p as a function of y and the arbitrary constant C_1:

$$p=p(y, C_1)$$

Substituting this value in (3), we get a first-order differential equation for the function y of x:

$$\frac{dy}{dx}=p(y, C_1)$$

Separating variables, we have

$$\frac{dy}{p(y, C_1)}=dx$$

Integrating this equation, we get the complete integral of the initial equation:

$$\Phi(x, y, C_1, C_2)=0$$

Example 2. Find the complete integral of the equation

$$3y''=y^{-\frac{5}{3}}$$

Solution. Put $p=\frac{dy}{dx}$ and consider p as a function of y. Then $y''=p\frac{dp}{dy}$ and we get a first-order equation for the auxiliary function p:

$$3p\frac{dp}{dy}=y^{-\frac{5}{3}}$$

Integrating this equation, we find

$$p^2=C_1-y^{-\frac{2}{3}} \quad \text{or} \quad p=\pm\sqrt{C_1-y^{-2/3}}$$

But $p=\frac{dy}{dx}$; consequently, for a determination of y we get the equation

$$\pm\frac{dy}{\sqrt{C_1-y^{-2/3}}}=dx, \quad \text{or} \quad \frac{y^{1/3}\,dy}{\pm\sqrt{C_1y^{2/3}-1}}=dx$$

whence

$$x+C_2=\pm\int\frac{y^{1/3}\,dy}{\sqrt{C_1y^{2/3}-1}}$$

To compute the latter integral we make the substitution

$$C_1y^{2/3}-1=t^2$$

Then

$$y^{1/3}=(t^2+1)^{\frac{1}{2}}\frac{1}{C_1^{1/2}}$$

$$dy=3t\,(t^2+1)^{1/2}\frac{1}{C_1^{3/2}}\,dt$$

Consequently,

$$\int\frac{y^{1/3}\,dy}{\sqrt{C_1y^{2/3}-1}}=\frac{1}{C_1^2}\int\frac{3t\,(t^2+1)}{t}\,dt=\frac{3}{C_1^2}\left(\frac{t^3}{3}+t\right)=\frac{1}{C_1^2}\sqrt{C_1y^{2/3}-1}\,(C_1y^{2/3}+2)$$

Finally we get

$$x+C_2=\pm\frac{1}{C_1^2}\sqrt{C_1y^{2/3}-1}\,(C_1y^{2/3}+2)$$

Example 3. Let a point move along the x-axis under the action of a force that depends solely on the position of the point. The differential equation of motion will be

$$m\frac{d^2x}{dt^2}=F(x)$$

At $t=0$ let $x=x_0$, $\frac{dx}{dt}=v_0$.

Multiplying both sides of the equation by $\frac{dx}{dt}dt$ and integrating from 0 to t, we have

$$\frac{1}{2}m\left(\frac{dx}{dt}\right)^2-\frac{1}{2}mv_0^2=\int_{x_0}^{x}F(x)\,dx$$

or

$$\frac{1}{2}m\left(\frac{dx}{dt}\right)^2+\left[-\int_{x_0}^{x}F(x)\,dx\right]=\frac{1}{2}mv_0^2=\text{const}$$

The first term of this equation is the kinetic energy, the second term, the potential energy of the moving point. From this equation it follows that the sum of the kinetic and potential energies remains constant throughout the time of motion.

The problem of a simple pendulum. Let there be a material point of mass m, which is in motion (due to the force of gravity) along a circle L lying in the vertical plane. Let us find the equation of motion of the point neglecting resistance forces (friction, air resistance, etc.).

Putting the origin at the lowest point of the circle, we put the x-axis along the tangent to the circle (Fig. 22).

Denote by l the radius of the circle, by s the arc length from the origin O to the variable point M where the mass m is located; this length is taken with the appropriate sign ($s > 0$, if the point M is on the right of O; $s < 0$ if M is on the left of O).

Our problem consists in establishing s as a function of the time t.

Let us decompose the force of gravity mg into tangential and normal components. The former, equal to $-mg\sin\varphi$, produces motion, the latter is cancelled by the reaction of the curve along which the mass m is moving.

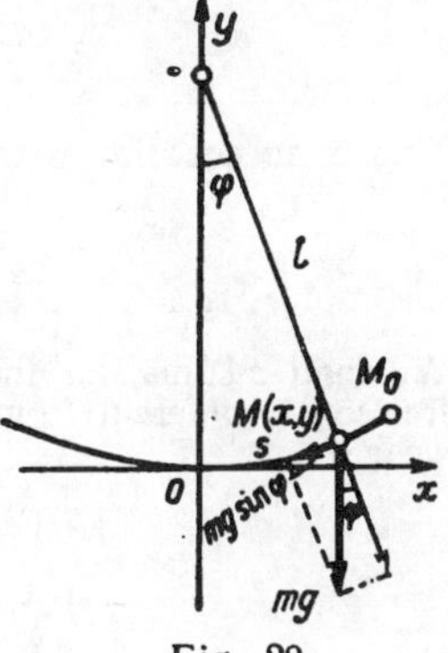

Fig. 22

Thus, the equation of motion is of the form

$$m\frac{d^2s}{dt^2}=-mg\sin\varphi$$

Since the angle $\varphi=\frac{s}{l}$ for a circle, we get the equation

$$\frac{d^2s}{dt^2}=-g\sin\frac{s}{l}$$

This is a Type II differential equation (since it does not contain the independent variable t explicitly).

Let us integrate it in the appropriate fashion:

$$\frac{ds}{dt}=p,\quad \frac{d^2s}{dt^2}=\frac{dp}{ds}p$$

Hence,

$$p\frac{dp}{ds}=-g\sin\frac{s}{l}$$

or

$$p\,dp=-g\sin\frac{s}{l}\,ds$$

whence

$$p^2=2gl\cos\frac{s}{l}+C_1$$

Let us denote by s_0 the greatest arc length to which the point M swings. For $s=s_0$ the velocity of the point is zero:

$$\left.\frac{ds}{dt}\right|_{s=s_0}=\left.p\right|_{s=s_0}=0$$

This enables us to determine C_1:

$$0=2gl\cos\frac{s_0}{l}+C_1$$

whence

$$C_1=-2gl\cos\frac{s_0}{l}$$

Therefore,

$$p^2=\left(\frac{ds}{dt}\right)^2=2gl\left(\cos\frac{s}{l}-\cos\frac{s_0}{l}\right)$$

or, applying to the last expression the formula for the difference of cosines,

$$\left(\frac{ds}{dt}\right)^2=4gl\sin\frac{s+s_0}{2l}\sin\frac{s_0-s}{2l} \tag{5}$$

or *

$$\frac{ds}{dt}=2\sqrt{gl}\sqrt{\sin\frac{s+s_0}{2l}\sin\frac{s_0-s}{2l}} \tag{6}$$

This is an equation with variables separable. Separating the variables, we get

$$\frac{ds}{\sqrt{\sin\frac{s+s_0}{2l}\sin\frac{s_0-s}{2l}}}=2\sqrt{gl}\,dt \tag{7}$$

We shall assume, for the time being, that $s\neq s_0$, then the denominator of the fraction is different from zero. If we consider that $s=0$ for $t=0$, then from (7) we get

$$\int_0^s\frac{ds}{\sqrt{\sin\frac{s+s_0}{2l}\sin\frac{s_0-s}{2l}}}=2\sqrt{gl}\,t \tag{8}$$

This is the equation that yields s as a function of t. The integral on the left cannot be expressed in terms of elementary functions; neither can the function s of t. Let us consider this problem approximately. We shall assume that the angles $\frac{s_0}{l}$ and $\frac{s}{l}$ are small. The angles $\frac{s+s_0}{2l}$ and $\frac{s_0-s}{2l}$ will not exceed $\frac{s_0}{l}$. In (6) let us replace, approximately, the sines of the angles by the angles

$$\frac{ds}{dt}=2\sqrt{gl}\sqrt{\frac{s+s_0}{2l}\,\frac{s_0-s}{2l}}$$

or

$$\frac{ds}{dt}=\sqrt{\frac{g}{l}}\sqrt{(s_0^2-s^2)} \tag{6'}$$

Separating variables, we get (assuming, for the time being, that $s\neq s_0$)

$$\frac{ds}{\sqrt{s_0^2-s^2}}=\sqrt{\frac{g}{l}}\,dt \tag{7'}$$

Again we consider that $s=0$ when $t=0$. Integrating this equation, we get

$$\int_0^s\frac{ds}{\sqrt{s_0^2-s^2}}=\sqrt{\frac{g}{l}}\,t \tag{8'}$$

or

$$\arcsin\frac{s}{s_0}=\sqrt{\frac{g}{l}}\,t$$

whence

$$s=s_0\sin\sqrt{\frac{g}{l}}\,t \tag{9}$$

---

* We take the plus sign in front of the root. From the note at the end of the solution it follows that there is no need to consider the case with the minus sign.

Note. When solving, we assumed that $s \neq s_0$. But it is clear, by direct substitution, that the function (9) is the solution of equation (6′) for any value of t.

Let it be recalled that the solution (9) is an approximate solution of equation (5), since equation (6) was replaced by the approximate equation (6′).

Equation (9) shows that the point M (which may be regarded as the extremity of the pendulum) performs harmonic oscillations with a period $T = 2\pi\sqrt{\frac{l}{g}}$. This period is independent of the amplitude s_0.

Example 4. Escape-velocity problem.

Determine the smallest velocity with which a body must be thrown vertically upwards so that it will not return to the earth. Air resistance is neglected.

Solution. Denote the mass of the earth and the mass of the body by M and m respectively. By Newton's law of gravitation, the force of attraction f acting on the body m is

$$f = k\frac{M \cdot m}{r^2}$$

where r is the distance between the centre of the earth and the centre of gravity of the body, and k is the gravitational constant.

The differential equation of motion of this body with mass m will be

$$m\frac{d^2r}{dt^2} = -k\frac{M \cdot m}{r^2}$$

or

$$\frac{d^2r}{dt^2} = -k\frac{M}{r^2} \tag{10}$$

The minus sign indicates that the acceleration is negative. The differential equation (10) is an equation of type (2). We shall solve it for the following initial conditions:

$$\text{for } t = 0 \quad r = R, \quad \frac{dr}{dt} = v_0$$

Here, R is the radius of the earth and v_0 is the launching velocity. We denote

$$\frac{dr}{dt} = v, \qquad \frac{d^2r}{dt^2} = \frac{dv}{dt} = \frac{dv}{dr} \cdot \frac{dr}{dt} = v\frac{dv}{dr}$$

where v is the velocity of motion. Putting this into (10), we get

$$v\frac{dv}{dr} = -k\frac{M}{r^2}$$

Separating variables, we obtain

$$v\,dv = -kM\frac{dr}{r^2}$$

Integrating this equation, we find

$$\frac{v^2}{2} = kM\frac{1}{r} + C_1 \tag{11}$$

From the condition that $v = v_0$ at the earth's surface (for $r = R$), we determine C_1:

$$\frac{v_0^2}{2} = kM\frac{1}{R} + C_1$$

or

$$C_1 = -\frac{kM}{R} + \frac{v_0^2}{2}$$

We put the value of C_1 into (11):

$$\frac{v^2}{2}=kM\frac{1}{r}-\frac{kM}{R}+\frac{v_0^2}{2}$$

or

$$\frac{v^2}{2}=kM\frac{1}{r}+\left(\frac{v_0^2}{2}-\frac{kM}{R}\right) \tag{12}$$

It is given that the body should move so that the velocity is always positive; hence, $\frac{v^2}{2}>0$. Since for a boundless increase of r the quantity $\frac{kM}{r}$ becomes arbitrarily small, the condition $\frac{v^2}{2}>0$ will be fulfilled for any r only for the case

$$\frac{v_0^2}{2}-\frac{kM}{R}\geqslant 0 \tag{13}$$

or

$$v_0\geqslant\sqrt{\frac{2kM}{R}}$$

Hence, the lowest velocity will be determined by the equation

$$v_0=\sqrt{\frac{2kM}{R}} \tag{14}$$

where

$$k=6.66\cdot 10^{-8}\ \text{cm}^3/\text{g}\cdot\text{sec}^2$$
$$R=63\cdot 10^7\ \text{cm}$$

At the earth's surface, for $r=R$, the acceleration of gravity is g ($g=981$ cm/sec²). For this reason, from (10) we obtain

$$g=k\frac{M}{R^2}$$

or

$$M=\frac{gR^2}{k}$$

Putting this value of M into (14), we obtain

$$v_0=\sqrt{2gR}=\sqrt{2\cdot 981\cdot 63\cdot 10^7}\approx 11.2\cdot 10^5\ \text{cm/sec}=11.2\ \text{km/sec}$$

1.19 GRAPHICAL METHOD OF INTEGRATING SECOND-ORDER DIFFERENTIAL EQUATIONS

Let us find out the geometric meaning of a second-order differential equation. Suppose we have an equation

$$y''=f(x,\ y,\ y') \tag{1}$$

Denote by φ the angle formed by the positive x-axis and the tangent to the curve; then

$$\frac{dy}{dx}=\tan\varphi \tag{2}$$

To find the geometric meaning of the second derivative, recall the formula that determines the radius of curvature of a curve at a given point: *

$$R=\frac{(1+y'^2)^{3/2}}{y''}$$

whence

$$y''=\frac{(1+y'^2)^{3/2}}{R}$$

But

$$y'=\tan\varphi,\quad 1+y'^2=1+\tan^2\varphi=\sec^2\varphi,$$
$$(1+y'^2)^{3/2}=|\sec^3\varphi|=\frac{1}{|\cos^3\varphi|}$$

therefore

$$y''=\frac{1}{R\,|\cos^3\varphi|} \tag{3}$$

Now putting into (1) the expressions obtained for y and y'', we have

$$\frac{1}{R\,|\cos^3\varphi|}=f(x,\ y,\ \tan\varphi)$$

or

$$R=\frac{1}{|\cos^3\varphi|\cdot f(x,\ y,\ \tan\varphi)} \tag{4}$$

It is thus evident that a second-order differential equation determines the magnitude of the radius of curvature of an integral curve if the coordinates of the point and the direction of the tangent to this point are specified.

From the foregoing there follows a method of approximate construction of an integral curve by means of a smooth curve composed of arcs of circles. **

To illustrate, let it be required to find the solution of equation (1) that satisfies the following initial conditions:

$$y_{x=x_0}=y_0,\quad y'_{x=x_0}=y'_0$$

Through the point $M_0(x_0,\ y_0)$ draw a ray M_0T_0 with slope $y'=\tan\varphi_0=y'_0$ (Fig. 23). From equation (4) we find the value $R=R_0$. Lay off a segment M_0C_0, equal to R_0, perpendicular to M_0T_0, and from the point C_0 (as centre) strike an arc $\overset{\frown}{M_0M_1}$

* Up till now we have always considered the radius of curvature **positive**; in this section we shall consider it a number that can take on both positive and negative values: if the curve is convex ($y''<0$), we consider the radius of curvature negative ($R<0$); if the curve is concave ($y''>0$), it is positive ($R>0$).

** A curve is called *smooth* if it has tangents at all points and the angle of inclination of the tangent is a continuous function of the arc length s.

with radius R_0. It should be noted that if $R_0 < 0$, then the segment M_0C_0 must be drawn in that direction so that the arc of the circle is convex upwards, and for $R_0 > 0$, convex down (see footnote on page 67).

Then let x_1, y_1 be the coordinates of the point M_1 which lies on the constructed arc and is sufficiently close to the point M_0 while $\tan \varphi_1$ is the slope of the tangent M_1T_1 to the circle drawn at M_1. From equation (4) we find the value $R = R_1$ that corresponds to M_1. Draw the segment M_1C_1, perpendicular to M_1T_1, equal to R_1, and from C_1 (as centre) strike an arc $\overset{\frown}{M_1M_2}$ with radius R_1. Then on this arc take a point $M_2(x_2, y_2)$ close to M_1 and continue construction as before until we get a sufficiently large piece of the curve consisting of the arcs of circles. From the foregoing it is clear that this curve is approximately an integral curve that passes through the point M_0. Obviously, the smaller the arcs $\overset{\frown}{M_0M_1}$, $\overset{\frown}{M_1M_2}$, ..., the closer the constructed curve will be to the integral curve.

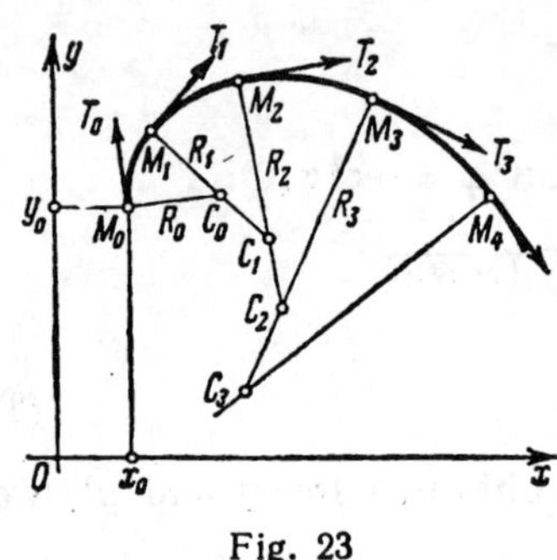

Fig. 23

1.20 HOMOGENEOUS LINEAR EQUATIONS. DEFINITIONS AND GENERAL PROPERTIES

Definition 1. An nth-order differential equation is called *linear* if it is of the first degree in the unknown function y and its derivatives y', ..., $y^{(n-1)}$, $y^{(n)}$; that is, if it is of the form

$$a_0y^{(n)} + a_1y^{(n-1)} + \ldots + a_ny = f(x) \qquad (1)$$

where a_0, a_1, a_2, ..., a_n and $f(x)$ are given functions of x or constants, and $a_0 \neq 0$ for all values of x in the domain in which we consider equation (1). From now on we shall presume that the functions a_0, a_1, ..., a_n and $f(x)$ are continuous for all values of x and that the coefficient $a_0 = 1$ (if it is not equal to 1 we can divide all terms of the equation by it). The function $f(x)$ on the right side of the equation is called the *right-hand member of the equation*.

If $f(x) \not\equiv 0$, then the equation is called a *nonhomogeneous* linear equation. But if $f(x) \equiv 0$, then the equation has the form

$$y^{(n)} + a_1y^{(n-1)} + \ldots + a_ny = 0 \qquad (2)$$

and is called a *homogeneous* linear equation (the left side of this equation is a homogeneous function of the first degree in y, y', y'', $\ldots$, $y^{(n)}$).

Let us determine some of the basic properties of homogeneous linear equations, confining our proof to second-order equations.

Theorem 1. *If y_1 and y_2 are two particular solutions of a homogeneous linear equation of the second order*

$$y''+a_1y'+a_2y=0 \tag{3}$$

then y_1+y_2 is also a solution of this equation.

Proof. Since y_1 and y_2 are solutions of the equation, we have

$$\left.\begin{aligned} &y_1''+a_1y_1'+a_2y_1=0 \\ \text{and}\quad &\\ &y_2''+a_1y_2'+a_2y_2=0 \end{aligned}\right\} \tag{4}$$

Putting into equation (3) the sum y_1+y_2 and taking into account the identities (4), we will have

$$\begin{aligned}&(y_1+y_2)''+a_1(y_1+y_2)'+a_2(y_1+y_2)\\ &=(y_1''+a_1y_1'+a_2y_1)+(y_2''+a_1y_2'+a_2y_2)=0+0=0\end{aligned}$$

Thus, y_1+y_2 is a solution of the equation.

Theorem 2. *If y_1 is a solution of equation* (3) *and C is a constant, then Cy_1 is also a solution of* (3).

Proof. Substituting into (3) the expression Cy_1, we get

$$(Cy_1)''+a_1(Cy_1)'+a_2(Cy_1)=C[y_1''+a_1y_1'+a_2y_1]=C\cdot 0=0$$

and the theorem is thus proved.

Definition 2. The two solutions of equation (3), y_1 and y_2, are called *linearly independent on an interval* $[a, b]$ if their ratio on this interval is not a constant; that is, if

$$\frac{y_1}{y_2}\neq \text{const}$$

Otherwise the solutions are called *linearly dependent*. In other words, two solutions, y_1 and y_2, are called *linearly dependent* on an interval $[a, b]$ if there exists a **constant** number λ such that $\frac{y_1}{y_2}=\lambda$ when $a\leqslant x\leqslant b$. In this case, $y_1=\lambda y_2$.

Example 1. Let there be an equation $y''-y=0$. It is easy to verify that the functions e^x, e^{-x}, $3e^x$, $5e^{-x}$ are solutions of this equation. Here, the functions e^x and e^{-x} are linearly independent on any interval because the ratio $\frac{e^x}{e^{-x}}=e^{2x}$ does not remain constant as x varies. But the functions e^x and $3e^x$ are linearly dependent, since $\frac{3e^x}{e^x}=3=\text{const}$.

Definition 3. If y_1 and y_2 are functions of x, the determinant

$$W(y_1, y_2) = \begin{vmatrix} y_1 & y_2 \\ y_1' & y_2' \end{vmatrix} = y_1 y_2' - y_1' y_2$$

is called the *Wronskian* of the given functions.

Theorem 3. *If the functions y_1 and y_2 are linearly dependent on an interval $[a, b]$, then the Wronskian on this interval is identically zero.*

Indeed, if $y_2 = \lambda y_1$ where $\lambda = \text{const}$, then $y_2' = \lambda y_1'$ and

$$W(y_1, y_2) = \begin{vmatrix} y_1 & y_2 \\ y_1' & y_2' \end{vmatrix} = \begin{vmatrix} y_1 & \lambda y_1 \\ y_1' & \lambda y_1' \end{vmatrix} = \lambda \begin{vmatrix} y_1 & y_1 \\ y_1' & y_1' \end{vmatrix} = 0$$

Theorem 4. *If the Wronskian $W(y_1, y_2)$, formed for the solutions y_1 and y_2 of the homogeneous linear equation (3), is not zero for some value $x = x_0$ on an interval $[a, b]$ where the coefficients of the equation are continuous, then it does not vanish for any value of x whatsoever on that interval.*

Proof. Since y_1 and y_2 are two solutions of equation (3), we have

$$y_2'' + a_1 y_2' + a_2 y_2 = 0$$
$$y_1'' + a_1 y_1' + a_2 y_1 = 0$$

Multiplying the terms of the first equation by y_1, the terms of the second equation by y_2 and subtracting the latter from the former, we get

$$(y_1 y_2'' - y_1'' y_2) + a_1 (y_1 y_2' - y_1' y_2) = 0 \tag{5}$$

The difference in the second parenthesis is the Wronskian $W(y_1, y_2)$, namely, $W(y_1, y_2) = (y_1 y_2' - y_1' y_2)$. The difference in the first parenthesis is the derivative of the Wronskian:

$$W_x'(y_1, y_2) = (y_1 y_2' - y_1' y_2)' = y_1 y_2'' - y_1'' y_2$$

Hence, equation (5) takes the form

$$W' + a_1 W = 0 \tag{6}$$

Let us find the solution to the last equation that satisfies the initial condition $W|_{x=x_0} = W_0$. We first find the general solution to equation (6) on the assumption that $W \neq 0$. Separating variables in (6), we get $\frac{dW}{W} = -a_1\, dx$.

Integrating we find

$$\ln W = -\int_{x_0}^{x} a_1\, dx + \ln C$$

or

$$\ln \frac{W}{C} = -\int_{x_0}^{x} a_1\,dx$$

whence

$$W = Ce^{-\int_{x_0}^{x} a_1\,dx} \tag{7}$$

It will be noted that we can write the function (7) and say that this function satisfies equation (6), which is clear if we substitute this function directly into (6). The assumption $W \neq 0$ is not required.

Formula (7) is called *Liouville's formula.*

We define C so that the initial condition is satisfied. Substituting $x = x_0$ into the left and right sides of (7), we get

$$W_0 = C$$

Consequently, the solution that satisfies the initial conditions takes the form

$$W = W_0 e^{-\int_{x_0}^{x} a_1\,dx} \tag{7'}$$

By hypothesis, $W_0 \neq 0$. But then from (7′) it follows that $W \neq 0$ for any value of x, because the exponential function does not vanish for any finite value of the argument. The proof is complete.

Note 1. If the Wronskian is zero for some value $x = x_0$, then it is also zero for any value x in the interval under consideration. This follows directly from (7): if $W = 0$ when $x = x_0$, then

$$(W)_{x=x_0} = C = 0$$

consequently, $W \equiv 0$, no matter what the value of the upper limit of x in formula (7).

Theorem 5. *If the solutions y_1 and y_2 of equation* (3) *are linearly independent on an interval* $[a, b]$, *then the Wronskian W, formed for these solutions, does not vanish at any point of the given interval.*

Proof. To start with, note the following. The function $y \equiv 0$ is a solution of equation (3) on the interval $[a, b]$ that satisfies the initial conditions

$$y_{x=x_0} = 0, \quad y'_{x=x_0} = 0$$

where x_0 is any point in the interval $[a, b]$. From the existence and uniqueness theorem (see Sec. 1.16), which is applicable to equation (3), it follows that there is no other solution to equation (3) that satisfies the initial conditions

$$y_{x=x_0}=0, \quad y'_{x=x_0}=0$$

From this same theorem, it also follows that if the solution to equation (3) is identically zero on some closed interval or the open interval (α, β) belonging to the interval $[a, b]$, then this solution is identically zero over the whole interval $[a, b]$. Indeed, at the point $x=\beta$ (and at the point $x=\alpha$) the solution satisfies the initial conditions

$$y_{x=\beta}=0, \quad y'_{x=\beta}=0$$

Hence, by the uniqueness theorem, it is also zero on some interval $\beta-d<x<\beta+d$, where d is determined by the coefficients of equation (3). Thus, extending the interval each time by the amount d, where $y\equiv 0$, we can prove that $y\equiv 0$ over the entire interval $[a, b]$.

Let us now begin the proof of Theorem 5. Assume that $W(y_1, y_2)=0$ at some point of the interval $[a, b]$. Then, by Theorem 3, the Wronskian $W(y_1, y_2)$ will be zero at all points of $[a, b]$:

$$W=0 \quad \text{or} \quad y_1y'_2-y'_1y_2=0$$

Suppose that $y_1\neq 0$ on the interval $[a, b]$. Then, by the last equation, we can write

$$\frac{y_1y'_2-y'_1y_2}{y_1^2}=0 \quad \text{or} \quad \left(\frac{y_2}{y_1}\right)'=0$$

whence it follows that

$$\frac{y_2}{y_1}=\lambda=\text{const}$$

that is, the solutions y_1 and y_2 are linearly dependent, which runs counter to the assumption that they are linearly independent.

Suppose, furthermore, that $y_1=0$ at the points $x_1, x_2, \ldots, x_k$ in $[a, b]$. We consider the interval (a, x_1). On this interval, $y_1\neq 0$. Hence, by what has just been proved, it follows that on the interval (a, x_1)

$$\frac{y_2}{y_1}=\lambda=\text{const} \quad \text{or} \quad y_2=\lambda y_1$$

We consider the function $y=y_2-\lambda y_1$. Since y_2 and y_1 are solutions of equation (3), then $y=y_2-\lambda y_1$ is a solution of (3) and $y\equiv 0$ over the interval (a, x_1). Thus, by the remark made at the beginning of the proof it follows that $y=y_2-\lambda y_1\equiv 0$ on the interval $[a, b]$ or

$$\frac{y_2}{y_1}=\lambda$$

on the interval $[a, b]$, which is to say y_2 and y_1 are linearly dependent.

But this contradicts the assumption that the solutions y_2 and y_1 are linearly independent. We have thus proved that the Wronskian does not vanish at any point of the interval $[a, b]$.

Theorem 6. *If y_1 and y_2 are two linearly independent solutions of equation* (3), *then*

$$y=C_1y_1+C_2y_2 \tag{8}$$

where C_1 and C_2 are arbitrary constants, is its general solution.

Proof. From Theorems 1 and 2 it follows that the function

$$C_1y_1+C_2y_2$$

is a solution of equation (3) for any values of C_1 and C_2.

We shall now prove that no matter what the initial conditions $y_{x=x_0}=y_0$, $y'_{x=x_0}=y'_0$, it is possible to choose the values of the arbitrary constants C_1 and C_2 so that the corresponding particular solution $C_1y_1+C_2y_2$ should satisfy the given initial conditions.

Substituting the initial conditions into (8), we have

$$\left.\begin{aligned} y_0&=C_1y_{10}+C_2y_{20}\\ y'_0&=C_1y'_{10}+C_2y'_{20}\end{aligned}\right\} \tag{9}$$

where we put

$$(y_1)_{x=x_0}=y_{10};\quad (y_2)_{x=x_0}=y_{20};\quad (y'_1)_{x=x_0}=y'_{10};\quad (y'_2)_{x=x_0}=y'_{20}$$

From system (9) we can determine C_1 and C_2, since the determinant of this system

$$\begin{vmatrix} y_{10} & y_{20}\\ y'_{10} & y'_{20}\end{vmatrix}=y_{10}y'_{20}-y'_{10}y_{20}$$

is the Wronskian for $x=x_0$ and, hence, is not equal to 0 (by virtue of the linear independence of the solutions y_1 and y_2). The particular solution obtained from the family (8) for the found values of C_1 and C_2 satisfies the given initial conditions. Thus, the theorem is proved.

Example 2. The equation

$$y''+\frac{1}{x}y'-\frac{1}{x^2}y=0$$

whose coefficients $a_1=\frac{1}{x}$ and $a_2=-\frac{1}{x^2}$ are continuous on any interval that does not contain the point $x=0$, admits the particular solutions

$$y_1=x,\quad y_2=\frac{1}{x}$$

(this is readily verified by substitution). Hence, its general solution is of the form

$$y=C_1x+C_2\frac{1}{x}$$

Note 2. There are no general methods for finding (in closed form) the general solution of a linear equation with variable coefficients. However, such a method exists for an equation with constant coefficients. It will be given in the next section. For the case of equations with variable coefficients, certain devices will be given in Chapter 4 (Series) that will enable us to find approximate solutions satisfying definite initial conditions.

Here we shall prove a theorem that will enable us to find the general solution of a second-order differential equation with variable coefficients if one of its particular solutions is known. Since it is sometimes possible to find or guess one particular solution directly, this theorem will prove useful in many cases.

Theorem 7. *If we know one particular solution of a second-order homogeneous linear equation, the finding of the general solution reduces to integrating the functions.*

Proof. Let y_1 be some known particular solution of the equation

$$y'' + a_1 y' + a_2 y = 0$$

We find another particular solution of the given equation so that y_1 and y_2 are linearly independent. Then the general solution will be expressed by the formula $y = C_1 y_1 + C_2 y_2$, where C_1 and C_2 are arbitrary constants. By virtue of formula (7) (see proof of Theorem 4), we can write

$$y_2' y_1 - y_2 y_1' = Ce^{-\int a_1\,dx}$$

Thus, for a determination of y_2 we obtain a first-order linear equation. Integrate it as follows. Divide all terms by y_1^2:

$$\frac{y_2' y_1 - y_2 y_1'}{y_1^2} = \frac{1}{y_1^2} Ce^{-\int a_1 dx}$$

or

$$\frac{d}{dx}\left(\frac{y_2}{y_1}\right) = \frac{1}{y_1^2} Ce^{-\int a_1 dx}$$

whence

$$\frac{y_2}{y_1} = \int \frac{Ce^{-\int a_1 dx}}{y_1^2} dx + C'$$

Since we are seeking a particular solution, we get (by putting $C' = 0$ and $C = 1$)

$$y_2 = y_1 \int \frac{e^{-\int a_1 dx}}{y_1^2} dx \tag{10}$$

It is obvious that y_1 and y_2 are linearly independent solutions since $\frac{y_2}{y_1} \neq \text{const}$.

Thus, the general solution of the original equation is of the form

$$y = C_1 y_1 + C_2 y_1 \int \frac{e^{-\int a_1 dx}}{y_1^2} dx \tag{11}$$

Example 3. Find the general solution of the equation

$$(1-x^2) y'' - 2xy' + 2y = 0$$

Solution. It is evident, by direct verification, that this equation has a particular solution $y_1 = x$. Let us find a second particular solution y_2, so that y_1 and y_2 should be linearly independent.

Noting that in our case $a_1 = \frac{-2x}{1-x^2}$, we have, by (10),

$$y = x \int \frac{e^{\int \frac{2x\,dx}{1-x^2}}}{x^2} dx = x \int \frac{e^{-\ln|1-x^2|}}{x^2} dx = x \int \frac{dx}{x^2 |1-x^2|}$$

$$= x \int \left(\pm \frac{1}{x^2} + \frac{1}{2(1-x)} + \frac{1}{2(1+x)} \right) dx = x \left[\mp \frac{1}{x} + \frac{1}{2} \ln \left| \frac{1+x}{1-x} \right| \right]$$

Consequently, the general solution is of the form

$$y = C_1 x + C_2 \left(\frac{1}{2} x \ln \left| \frac{1+x}{1-x} \right| \mp 1 \right)$$

1.21 SECOND-ORDER HOMOGENEOUS LINEAR EQUATIONS WITH CONSTANT COEFFICIENTS

We have a second-order homogeneous linear equation

$$y'' + py' + qy = 0 \tag{1}$$

where p and q are real constants. To find the complete integral of this equation, it is sufficient (as has already been proved) to find two linearly independent particular solutions.

Let us look for the particular solutions in the form

$$y = e^{kx}, \text{ where } k = \text{const} \tag{2}$$

then

$$y' = ke^{kx}, \quad y'' = k^2 e^{kx}$$

Substituting the expressions of the derivatives into equation (1), we find

$$e^{kx}(k^2 + pk + q) = 0$$

Since $e^{kx} \neq 0$, this means that

$$k^2 + pk + q = 0 \tag{3}$$

Thus, if k satisfies equation (3), then e^{kx} will be a solution of (1). Equation (3) is called an *auxiliary equation* with respect to equation (1).

The auxiliary equation is a quadratic equation with two roots; let us denote them by k_1 and k_2. Then

$$k_1=-\frac{p}{2}+\sqrt{\frac{p^2}{4}-q},\quad k_2=-\frac{p}{2}-\sqrt{\frac{p^2}{4}-q}$$

The following cases are possible:

I. k_1 and k_2 are real numbers and not equal ($k_1 \neq k_2$),
II. k_1 and k_2 are complex numbers,
III. k_1 and k_2 are real and equal numbers ($k_1 = k_2$).

Let us consider each case separately.

I. The roots of the auxiliary equation are real and distinct, $k_1 \neq k_2$. Here, the particular solutions are the functions

$$y_1=e^{k_1x},\quad y_2=e^{k_2x}$$

These solutions are linearly independent because

$$\frac{y_2}{y_1}=\frac{e^{k_2x}}{e^{k_1x}}=e^{(k_2-k_1)x}\neq \text{const}$$

Hence, the complete integral has the form

$$y=C_1e^{k_1x}+C_2e^{k_2x}$$

Example 1. Given the equation

$$y''+y'-2y=0$$

The auxiliary equation is of the form

$$k^2+k-2=0$$

We find the roots of the auxiliary equation:

$$k_{1,2}=-\frac{1}{2}\pm\sqrt{\frac{1}{4}+2}$$

$$k_1=1,\quad k_2=-2$$

The complete integral is

$$y=C_1e^x+C_2e^{-2x}$$

II. The roots of the auxiliary equation are complex. Since complex roots are conjugate in pairs, we write

$$k_1=\alpha+i\beta,\quad k_2=\alpha-i\beta$$

where

$$\alpha=-\frac{p}{2},\quad \beta=\sqrt{q-\frac{p^2}{4}}$$

The particular solutions may be written in the form

$$y_1=e^{(\alpha+i\beta)x},\quad y_2=e^{(\alpha-i\beta)x} \tag{4}$$

These are complex functions of a real argument that satisfy the differential equation (1) (see Sec. 7.4, Vol. I).

It is obvious that if some complex function of a real argument

$$y = u(x) + iv(x) \tag{5}$$

satisfies (1), then this equation is satisfied by the functions $u(x)$ and $v(x)$.

Indeed, putting expression (5) into (1), we have

$$[u(x) + iv(x)]'' + p[u(x) + iv(x)]' + q[u(x) + iv(x)] \equiv 0$$

or

$$(u'' + pu' + qu) + i(v'' + pv' + qv) \equiv 0$$

But a complex function is equal to zero if, and only if, the real part and the imaginary part are equal to zero; that is,

$$u'' + pu' + qu = 0$$
$$v'' + pv' + qv = 0$$

Thus we have proved that $u(x)$ and $v(x)$ are solutions of the equation.

Let us rewrite the complex solutions (4) in the form of a sum of the real part and the imaginary part:

$$y_1 = e^{\alpha x} \cos \beta x + ie^{\alpha x} \sin \beta x$$
$$y_2 = e^{\alpha x} \cos \beta x - ie^{\alpha x} \sin \beta x$$

From what has been proved, the particular solutions of (1) are the real functions

$$\tilde{y}_1 = e^{\alpha x} \cos \beta x \tag{6'}$$
$$\tilde{y}_2 = e^{\alpha x} \sin \beta x \tag{6''}$$

The functions $\tilde{y}_1$ and $\tilde{y}_2$ are linearly independent, since

$$\frac{\tilde{y}_1}{\tilde{y}_2} = \frac{e^{\alpha x} \cos \beta x}{e^{\alpha x} \sin \beta x} = \cot \beta x \neq \text{const}$$

Consequently, the general solution of equation (1) in the case of complex roots of the auxiliary equation is of the form

$$y = C_1 \tilde{y}_1 + C_2 \tilde{y}_2 = C_1 e^{\alpha x} \cos \beta x + C_2 e^{\alpha x} \sin \beta x$$

or

$$y = e^{\alpha x} (C_1 \cos \beta x + C_2 \sin \beta x) \tag{7}$$

where C_1 and C_2 are arbitrary constants.

An important particular case of the solution (7) is the case where the roots of the auxiliary equation are *pure imaginary*.

This occurs when $p = 0$ in equation (1) and it has the form

$$y'' + qy = 0$$

The auxiliary equation (3) assumes the form

$$k^2+q=0, \quad q>0$$

The roots of the auxiliary equation are

$$k_{1,2}=\pm i\sqrt{q}=\pm i\beta, \quad \alpha=0$$

The solution (7) becomes

$$y=C_1\cos\beta x+C_2\sin\beta x$$

Example 2. Given the equation

$$y''+2y'+5y=0$$

Find the complete integral and a particular solution that satisfies the initial conditions $y_{x=0}=0$, $y'_{x=0}=1$. Construct the graph.

Solution. (1) We write the auxiliary equation

$$k^2+2k+5=0$$

and find its roots:

$$k_1=-1+2i, \quad k_2=-1-2i$$

Thus, the complete integral is

$$y=e^{-x}(C_1\cos 2x+C_2\sin 2x)$$

(2) We find a particular solution that satisfies the given initial conditions and determine the corresponding values of C_1 and C_2.

From the first condition we find

$$0=e^{-0}(C_1\cos 2\cdot 0+C_2\sin 2\cdot 0), \text{ whence } C_1=0$$

Noting that

$$y'=e^{-x}2C_2\cos 2x-e^{-x}C_2\sin 2x$$

we obtain from the second condition

$$1=2C_2, \quad \text{i. e.,} \quad C_2=\frac{1}{2}$$

Thus, the desired particular solution is

$$y=\frac{1}{2}e^{-x}\sin 2x$$

Its graph is shown in Fig. 24.

Example 3. Given the equation

$$y''+9y=0$$

Find the general solution and a particular solution satisfying the initial conditions

$$y_{x=0}=0, \quad y'_{x=0}=3$$

Solution. Write the auxiliary equation

$$k^2+9=0$$

We find its roots to be

$$k_1=3i, \quad k_2=-3i$$

The general solution is

$$y=C_1\cos 3x+C_2\sin 3x$$

Let us find a particular solution. First we find

$$y' = -3C_1 \sin 3x + 3C_2 \cos 3x$$

The constants C_1 and C_2 are defined from the initial conditions

$$0 = C_1 \cos 0 + C_2 \sin 0$$
$$3 = -3C_1 \sin 0 + 3C_2 \cos 0$$

They are

$$C_1 = 0, \qquad C_2 = 1$$

The particular solution is

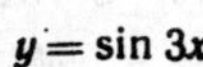

$$y = \sin 3x$$

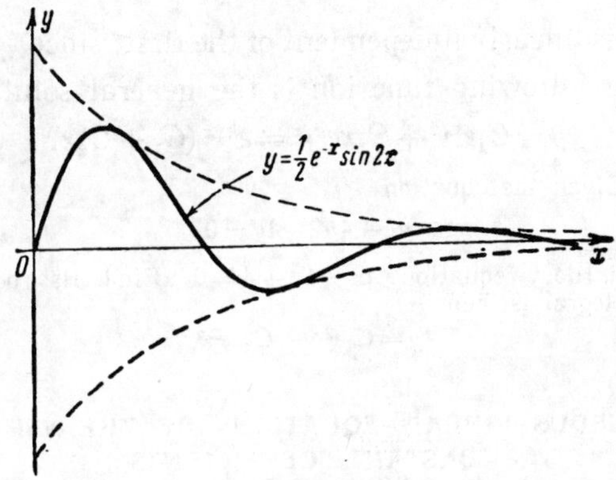

Fig. 24

III. The roots of the auxiliary equation are real and equal. Here, $k_1 = k_2$.

One particular solution, $y_1 = e^{k_1 x}$, is obtained from earlier reasoning. We must find a second particular solution, which is linearly independent of the first (the function $e^{k_2 x}$ is identically equal to $e^{k_1 x}$ and therefore cannot be regarded as a second particular solution).

We shall seek the second particular solution in the form

$$y_2 = u(x)\, e^{k_1 x}$$

where $u(x)$ is the unknown function to be determined.

Differentiating, we find

$$y_2' = u' e^{k_1 x} + k_1 u e^{k_1 x} = e^{k_1 x}(u' + k_1 u)$$
$$y_2'' = u'' e^{k_1 x} + 2k_1 u' e^{k_1 x} + k_1^2 u e^{k_1 x} = e^{k_1 x}(u'' + 2k_1 u' + k_1^2 u)$$

Putting the expressions of the derivatives into (1), we obtain

$$e^{k_1 x}[u'' + (2k_1 + p)u' + (k_1^2 + pk_1 + q)u] = 0$$

Since k_1 is a multiple root of the auxiliary equation, we have

$$k_1^2 + pk_1 + q = 0$$

Besides, $k_1 = k_2 = -\frac{p}{2}$ or $2k_1 = -p$, $2k_1 + p = 0$.

Hence, in order to find $u(x)$ we must solve the equation $e^{k_1x}u'' = 0$ or $u'' = 0$. Integrating, we get $u = Ax + B$. In particular, we can set $A = 1$ and $B = 0$; then

$$u = x$$

Thus, for the second particular solution we can take

$$y_2 = xe^{k_1x}$$

This solution is linearly independent of the first, since $\frac{y_2}{y_1} = x \not\equiv \text{const}$. Therefore, the following function is the general solution:

$$y = C_1e^{k_1x} + C_2xe^{k_1x} = e^{k_1x}(C_1 + C_2x)$$

Example 4. Given the equation

$$y'' - 4y' + 4y = 0$$

Write the auxiliary equation $k^2 - 4k + 4 = 0$. Find its roots: $k_1 = k_2 = 2$. The complete integral is then

$$y = C_1e^{2x} + C_2xe^{2x}$$

1.22 HOMOGENEOUS LINEAR EQUATIONS OF THE *N*TH ORDER WITH CONSTANT COEFFICIENTS

Let us consider a homogeneous linear equation of the *n*th order:

$$y^{(n)} + a_1y^{(n-1)} + \ldots + a_ny = 0 \quad (1)$$

We shall assume that $a_1, a_2, \ldots, a_n$ are constants. Before giving a method for solving equation (1), we introduce a definition that will be needed later on.

Definition 1. If for all x of the interval $[a, b]$ we have the equality

$$\varphi_n(x) = A_1\varphi_1(x) + A_2\varphi_2(x) + \ldots + A_{n-1}\varphi_{n-1}(x)$$

where $A_1, A_2, \ldots, A_n$ are constants, not all equal to zero, then we say that $\varphi_n(x)$ *is expressed linearly in terms of the functions* $\varphi_1(x), \varphi_2(x), \ldots, \varphi_{n-1}(x)$.

Definition 2. n functions $\varphi_1(x), \varphi_2(x), \ldots, \varphi_{n-1}(x), \varphi_n(x)$ *are called linearly independent* if not one of the functions is expressed linearly in terms of the rest.

Note 1. From the definitions it follows that if the functions $\varphi_1(x), \varphi_2(x), \ldots, \varphi_n(x)$ *are linearly dependent*, there will be constants $C_1, C_2, \ldots, C_n$, not all zeroes, such that for all x in the interval $[a, b]$ the following identity will hold true:

$$C_1\varphi_1(x) + C_2\varphi_2(x) + \ldots + C_n\varphi_n(x) \equiv 0$$

Examples:

1. The functions $y_1=e^x$, $y_2=e^{2x}$, $y_3=3e^x$ are linearly dependent, since for $C_1=1$, $C_2=0$, $C_3=-\frac{1}{3}$ we have the identity

$$C_1e^x+C_2e^{2x}+C_3 3e^x\equiv 0$$

2. The functions $y_1=1$, $y_2=x$, $y_3=x^2$ are linearly independent, since the expression

$$C_1 1+C_2x+C_3x^2$$

will not be identically zero for any C_1, C_2, C_3 that are not simultaneously equal to zero.

3. The functions $y_1=e^{k_1x}$, $y_2=e^{k_2x}$, ..., $y_n=e^{k_nx}$, ..., where k_1, k_2, ..., k_n, ... are different numbers, are linearly independent. (This assertion is given without proof.)

Let us now solve equation (1). For this equation, the following theorem holds.

Theorem. *If the functions y_1, y_2, ..., y_n are linearly independent solutions of equation* (1), *then its general solution is*

$$y=C_1y_1+C_2y_2+\ldots+C_ny_n \tag{2}$$

where C_1, ..., C_n are arbitrary constants.

If the coefficients of equation (1) are constant, the general solution is found in the same way as in the case of second-order equations.

(1) Form the auxiliary equation

$$k^n+a_1k^{n-1}+a_2k^{n-2}+\ldots+a_n=0.$$

(2) Find the roots of the auxiliary equation

$$k_1,\ k_2,\ \ldots,\ k_n$$

(3) From the character of the roots write out the particular linearly independent solutions, taking note of the fact that:

(a) to every real root k of order one there corresponds a particular solution e^{kx};

(b) to every pair of complex conjugate roots $k^{(1)}=\alpha+i\beta$ and $k^{(2)}=\alpha-i\beta$ there correspond two particular solutions $e^{\alpha x}\cos\beta x$ and $e^{\alpha x}\sin\beta x$;

(c) to every real root k of multiplicity r there correspond r linearly independent particular solutions

$$e^{kx},\ xe^{kx},\ \ldots,\ x^{r-1}e^{kx};$$

(d) to each pair of complex conjugate roots $k^{(1)}=\alpha+i\beta$, $k^{(2)}=\alpha-i\beta$ of multiplicity μ there correspond 2μ particular solutions:

$$\begin{array}{llll} e^{\alpha x}\cos\beta x, & xe^{\alpha x}\cos\beta x, & \ldots, & x^{\mu-1}e^{\alpha x}\cos\beta x, \\ e^{\alpha x}\sin\beta x, & xe^{\alpha x}\sin\beta x, & \ldots, & x^{\mu-1}e^{\alpha x}\sin\beta x \end{array}$$

The number of these particular solutions is exactly equal to the degree of the auxiliary equation (that is, to the order of the given linear differential equation). It may be proved that these solutions are linearly independent.

(4) After finding n linearly independent particular solutions $y_1, y_2, \ldots, y_n$ we construct the general solution of the given linear equation:

$$y = C_1y_1 + C_2y_2 + \ldots + C_ny_n$$

where $C_1, C_2, \ldots, C_n$ are arbitrary constants.

Example 4. Find the general solution of the equation

$$y^{\mathrm{IV}} - y = 0$$

Solution. Form the auxiliary equation

$$k^4 - 1 = 0$$

Find the roots of the auxiliary equation:

$$k_1 = 1, \quad k_2 = -1, \quad k_3 = i, \quad k_4 = -i$$

Write the complete integral

$$y = C_1e^x + C_2e^{-x} + C_3 \cos x + C_4 \sin x$$

where C_1, C_2, C_3, C_4 are arbitrary constants.

Note 2. From the foregoing it follows that the whole difficulty in solving homogeneous linear differential equations with constant coefficients lies in the solution of the auxiliary equation.

1.23 NONHOMOGENEOUS SECOND-ORDER LINEAR EQUATIONS

Let there be a nonhomogeneous second-order linear equation

$$y'' + a_1y' + a_2y = f(x) \tag{1}$$

The structure of the general solution of such an equation is determined by the following theorem.

Theorem 1. *The general solution of the nonhomogeneous equation* (1) *is represented as the sum of some particular solution of the equation* y^* *and the general solution* $\bar{y}$ *of the corresponding homogeneous equation*

$$\bar{y}'' + a_1\bar{y}' + a_2\bar{y} = 0 \tag{2}$$

Proof. We need to prove that the sum

$$y = \bar{y} + y^* \tag{3}$$

is the **general solution** of equation (1). Let us first prove that the function (3) is a **solution** of (1).

Substituting the sum $\bar{y}+y^*$ into (1) in place of y, we get

$$(\bar{y}+y^*)''+a_1(\bar{y}+y^*)'+a_2(\bar{y}+y^*)=f(x)$$

or

$$(\bar{y}''+a_1\bar{y}'+a_2\bar{y})+(y^{*''}+a_1y^{*'}+a_2y^*)=f(x) \tag{4}$$

Since $\bar{y}$ is a solution of (2), the expression in the first brackets is identically zero. Since y^* is a solution of (1), the expression in the second brackets is equal to $f(x)$. Consequently, (4) is an identity. Thus, the first part of the theorem is proved.

We shall now prove that expression (3) is the **general** solution of equation (1); in other words, we shall prove that the arbitrary constants that enter into the expression may be chosen so that the following initial conditions are satisfied:

$$\left.\begin{aligned} y_{x=x_0}&=y_0\\ y'_{x=x_0}&=y'_0 \end{aligned}\right\} \tag{5}$$

no matter what the numbers x_0, y_0 and y'_0 [provided that x_0 is taken from the region where the functions a_1, a_2 and $f(x)$ are continuous].

Noting that $\bar{y}$ may be given in the form

$$\bar{y}=C_1y_1+C_2y_2$$

where y_1 and y_2 are linearly independent solutions of equation (2), and C_1 and C_2 are arbitrary constants, we can rewrite (3) in the form

$$y=C_1y_1+C_2y_2+y^* \tag{3'}$$

Then, by the conditions (5), we will have *

$$C_1y_{10}+C_2y_{20}+y_0^*=y_0$$

$$C_1y'_{10}+C_2y'_{20}+y_0^{*'}=y'_0$$

From this system of equations we have to determine C_1 and C_2. Rewriting the system in the form

$$\left.\begin{aligned} C_1y_{10}+C_2y_{20}&=y_0-y_0^*\\ C_1y'_{10}+C_2y'_{20}&=y'_0-y_0^{*'} \end{aligned}\right\} \tag{6}$$

we note that the determinant of this system is the Wronskian for the functions y_1 and y_2 at the point $x=x_0$. Since it is given that these functions are linearly independent, the Wronskian is not zero; consequently, system (6) has a definite solution, C_1

* Here, y_{10}, y_{20}, y_0^*, y'_{10}, y'_{20}, $y_0^{*'}$ denote the numerical values of the functions y_1, y_2, y^*, y'_1, y'_2, $y^{*'}$ when $x=x_0$.

and C_2; in other words, there exist values C_1 and C_2 such that formula (3) defines the solution of equation (1) which satisfies the given initial conditions. The theorem is completely proved.

Thus, if we know the general solution $\bar{y}$ of the homogeneous equation (2), the basic difficulty, when integrating the nonhomogeneous equation (1), lies in finding some particular solution y^*.

We shall give a general method for finding the particular solutions of a nonhomogeneous equation.

The method of variation of arbitrary constants (parameters). We write the general solution of the homogeneous equation (2):

$$y = C_1 y_1 + C_2 y_2 \tag{7}$$

We shall seek a particular solution of the nonhomogeneous equation (1) in the form (7), considering C_1 and C_2 as some (as yet) undetermined **functions** of x.

Differentiate (7):

$$y' = C_1 y_1' + C_2 y_2' + C_1' y_1 + C_2' y_2$$

Now choose the needed functions C_1 and C_2 so that the following equation holds:

$$C_1' y_1 + C_2' y_2 = 0 \tag{8}$$

If we take note of this additional condition, the first derivative y' will take the form

$$y' = C_1 y_1' + C_2 y_2'$$

Differentiating this expression, we find y'':

$$y'' = C_1 y_1'' + C_2 y_2'' + C_1' y_1' + C_2' y_2'$$

Putting y, y' and y'' into (1), we get

$$C_1 y_1'' + C_2 y_2'' + C_1' y_1' + C_2' y_2' + a_1 (C_1 y_1' + C_2 y_2') + a_2 (C_1 y_1 + C_2 y_2) = f(x)$$

or

$$C_1 (y_1'' + a_1 y_1' + a_2 y_1) + C_2 (y_2'' + a_1 y_2' + a_2 y_2) + C_1' y_1' + C_2' y_2' = f(x)$$

The expressions in the first two brackets vanish, since y_1 and y_2 are solutions of the homogeneous equation. Hence, the latter equation takes on the form

$$C_1' y_1' + C_2' y_2' = f(x) \tag{9}$$

Thus, the function (7) will be a solution of the nonhomogeneous equation (1) provided the functions C_1 and C_2 satisfy the system of equations (8) and (9); that is, if

$$C_1' y_1 + C_2' y_2 = 0, \qquad C_1' y_1' + C_2' y_2' = f(x)$$

Since the determinant of this system is the Wronskian for the linearly independent solutions y_1 and y_2 of equation (2), it is not equal to zero. Hence, in solving the system we will find C_1' and C_2' as definite functions of x:

$$C_1' = \varphi_1(x), \quad C_2' = \varphi_2(x)$$

Integrating, we obtain

$$C_1 = \int \varphi_1(x)\,dx + \overline{C}_1, \quad C_2 = \int \varphi_2(x)\,dx + \overline{C}_2$$

where $\overline{C}_1$ and $\overline{C}_2$ are constants of integration.

Substituting the expressions obtained for C_1 and C_2 into (7), we find an integral that is dependent on the two arbitrary constants $\overline{C}_1$ and $\overline{C}_2$; that is, we find the general solution of the nonhomogeneous equation.*

Example. Find the general solution of the equation

$$y'' - \frac{y'}{x} = x$$

Solution. Let us find the general solution of the homogeneous equation

$$y'' - \frac{y'}{x} = 0$$

Since

$$\frac{y''}{y'} = \frac{1}{x} \text{ we have } \ln y' = \ln x + \ln C, \ y' = Cx$$

and so

$$y = C_1x^2 + C_2$$

For the latter expression to be a solution of the given equation, we have to define C_1 and C_2 as functions of x from the system

$$C_1'x^2 + C_2' \cdot 1 = 0, \quad 2C_1'x + C_2' \cdot 0 = x$$

Solving this system, we find

$$C_1' = \frac{1}{2}, \quad C_2' = -\frac{1}{2}x^2$$

whence, after integration, we get

$$C_1 = \frac{x}{2} + \overline{C}_1, \quad C_2 = -\frac{x^3}{6} + \overline{C}_2$$

Putting the functions obtained into the formula $y = C_1x^2 + C_2$, we get the general solution of the nonhomogeneous equation:

$$y = \overline{C}_1x^2 + \overline{C}_2 + \frac{x^3}{2} - \frac{x^3}{6}$$

or $y = \overline{C}_1x^2 + \overline{C}_2 + \frac{x^3}{3}$, where $\overline{C}_1$ and $\overline{C}_2$ are arbitrary constants.

When seeking particular solutions, it is useful to take advantage of the results of the following theorem.

---

* If we put $\overline{C}_1 = \overline{C}_2 = 0$, we get a particular solution of equation (1).

Theorem 2. *The solution y^* of the equation*

$$y'' + a_1 y' + a_2 y = f_1(x) + f_2(x) \qquad (10)$$

where the right member is a sum of two functions $f_1(x)$ and $f_2(x)$, may be represented as the sum $y^* = y_1^* + y_2^*$, where y_1^* and y_2^* are, respectively, the solutions of the equations*

$$y_1^{*''} + a_1 y_1^{*'} + a_2 y_1^* = f_1(x) \qquad (11)$$

$$y_2^{*''} + a_1 y_2^{*'} + a_2 y_2^* = f_2(x) \qquad (12)$$

Proof. Adding the right and left members of (11) and (12), we get

$$(y_1^* + y_2^*)'' + a_1 (y_1^* + y_2^*)' + a_2 (y_1^* + y_2^*) = f_1(x) + f_2(x) \qquad (13)$$

From this equation it follows that the sum

$$y_1^* + y_2^* = y^*$$

is a solution of equation (10).

Example. Find the particular solution y^* of the equation

$$y'' - 4y = x + 3e^x$$

Solution. The particular solution to the equation

$$y_1^{*''} + 4y_1^* = x$$

is

$$y_1^* = \frac{1}{4} x$$

The particular solution to the equation

$$y_2^{*''} + 4y_2^* = 3e^x$$

is

$$y_2^* = \frac{3}{5} e^x$$

The particular solution y^* of this equation is

$$y^* = \frac{1}{4} x + \frac{3}{5} e^x$$

1.24 NONHOMOGENEOUS SECOND-ORDER LINEAR EQUATIONS WITH CONSTANT COEFFICIENTS

Suppose we have the equation

$$y'' + py' + qy = f(x) \qquad (1)$$

where p and q are real numbers.

---

* It is clear that the theorem holds true for any number of terms in the right member of the equation.

A general method for finding the solution of a nonhomogeneous equation was given in the preceding section. In the case of an equation with constant coefficients, it is sometimes easier to find a particular solution without resorting to integration. Let us consider several such possibilities for equation (1).

I. Let the right side of (1) be the product of an exponential function by a polynomial; that is, of the form

$$f(x) = P_n(x)e^{\alpha x} \tag{2}$$

where $P_n(x)$ is a polynomial of degree n. Then the following particular cases are possible:

(a) The number α *is not a root* of the auxiliary equation

$$k^2 + pk + q = 0$$

In this case, the particular solution must be sought for in the form

$$y^* = (A_0x^n + A_1x^{n-1} + \ldots + A_n)e^{\alpha x} = Q_n(x)e^{\alpha x} \tag{3}$$

Indeed, substituting y^* into equation (1) and cancelling $e^{\alpha x}$ out of all terms, we will have

$$Q''_n(x) + (2\alpha + p)Q'_n(x) + (\alpha^2 + p\alpha + q)Q_n(x) = P_n(x) \tag{4}$$

$Q_n(x)$ is a polynomial of degree n, $Q'_n(x)$ is a polynomial of degree $n-1$, and $Q''_n(x)$ is a polynomial of degree $n-2$. Thus, n-degree polynomials are found on the left and right of the equality sign. Equating the coefficients of the same degrees of x (the number of unknown coefficients is $n+1$), we get a system of $n+1$ equations for determining the unknown coefficients A_0, A_1, A_2, ..., A_n.

(b) The number α is a **simple (single) root** of the auxiliary equation.

If in this case we should seek the particular solution in the form (3), then on the left side of (4) we would have a polynomial of degree $n-1$, since the coefficient of $Q_n(x)$, that is, $\alpha^2+p\alpha+q$, is equal to zero, and the polynomials $Q'_n(x)$ and $Q''_n(x)$ have degrees less than n. Hence, (4) would not be an identity, no matter what A_0, A_1, ..., A_n. For this reason, the particular solution in this case has to be taken in the form of a polynomial of degree $n+1$, but without the absolute term (since the absolute term of this polynomial vanishes upon differentiation): *

$$y^* = xQ_n(x)e^{\alpha x}$$

* We remark that all the results given above also hold for the case where α is a complex number (this follows from the rules for differentiating the function e^{mx}, where m is any complex number; see Sec. 7.4, Vol. I).

(c) The number α is a **double root** of the auxiliary equation. Then, as a result of the substitution of the function $Q_n(x)e^{\alpha x}$ into the differential equation, the degree of the polynomial is diminished by two units. Indeed, if α is a root of the auxiliary equation, then $\alpha^2+p\alpha+q=0$; moreover, since α is a double root, it follows that $2\alpha=-p$ (since by a familiar theorem of elementary algebra, the sum of the roots of a reduced quadratic equation is equal to the coefficient of the unknown to the first power with sign reversed). And so $2\alpha+p=0$.

Consequently, on the left side of (4) there remains $Q''_n(x)$, that is, a polynomial of degree $n-2$. To obtain a polynomial of degree n as a result of substitution, one should seek the particular solution in the form of a product of $e^{\alpha x}$ by the $(n+2)$ *th* degree polynomial. Then the absolute term of this polynomial and the first-degree term will vanish upon differentiation; for this reason, they need not be included in the particular solution.

Thus, when α is a double root of the auxiliary equation, the particular solution may be taken in the form

$$y^* = x^2Q_n(x)e^{\alpha x}$$

Example 1. Find the general solution of the equation

$$y''+4y'+3y=x$$

Solution. The general solution of the corresponding homogeneous equation is

$$\bar{y}=C_1e^{-x}+C_2e^{-3x}$$

Since the right-hand side of the given nonhomogeneous equation is of the form xe^{0x} [that is, of the form $P_1(x)e^{0x}$], and 0 is not a root of the auxiliary equation $k^2+4k+3=0$, it follows that we should seek the particular solution in the form $y^*=Q_1(x)e^{0x}$; in other words, we put

$$y^*=A_0x+A_1$$

Substituting this expression into the given equation, we will have

$$4A_0+3(A_0x+A_1)=x$$

Equating the coefficients of identical powers of x, we get

$$3A_0=1,\ 4A_0+3A_1=0$$

whence

$$A_0=\frac{1}{3},\ A_1=-\frac{4}{9}$$

Consequently,

$$y^*=\frac{1}{3}x-\frac{4}{9}$$

The general solution of $y=\bar{y}+y^*$ will be

$$y=C_1e^{-x}+C_2e^{-3x}+\frac{1}{3}x-\frac{4}{9}$$

Example 2. Find the general solution of the equation

$$y''+9y=(x^2+1)e^{3x}$$

Solution. The general solution of the homogeneous equation is readily found:

$$\bar{y}=C_1\cos 3x+C_2\sin 3x$$

The right side of the given equation $(x^2+1)e^{3x}$ has the form

$$P_2(x)e^{3x}$$

Since the coefficient 3 in the exponent is not a root of the auxiliary equation, we seek the particular solution in the form

$$y^*=Q_2(x)e^{3x} \text{ or } y^*=(Ax^2+Bx+C)e^{3x}$$

Substituting this expression in the differential equation, we have

$$[9(Ax^2+Bx+C)+6(2Ax+B)+2A+9(Ax^2+Bx+C)]e^{3x}=(x^2+1)e^{3x}$$

Cancelling out e^{3x} and equating the coefficients of identical powers of x, we obtain

$$18A=1,\ 12A+18B=0,\ 2A+6B+18C=1$$

whence $A=\frac{1}{18}$, $B=-\frac{1}{27}$, $C=\frac{5}{81}$. Consequently the particular solution is

$$y^*=\left(\frac{1}{18}x^2-\frac{1}{27}x+\frac{5}{81}\right)e^{3x}$$

and the general solution is

$$y=C_1\cos 3x+C_2\sin 3x+\left(\frac{1}{18}x^2-\frac{1}{27}x+\frac{5}{81}\right)e^{3x}$$

Example 3. To solve the equation

$$y''-7y'+6y=(x-2)e^x$$

Solution. Here, the right side is of the form $P_1(x)e^{1x}$ and the coefficient 1 in the exponent is a simple root of the auxiliary polynomial. Hence, we seek the particular solution in the form $y^*=xQ_1(x)e^x$ or

$$y^*=x(Ax+B)e^x$$

Putting this expression in the equation, we get

$$[(Ax^2+Bx)+(4Ax+2B)+2A-7(Ax^2+Bx)-7(2Ax+B)+6(Ax^2+Bx)]e^x=(x-2)e^x$$

or

$$(-10Ax-5B+2A)e^x=(x-2)e^x$$

Equating the coefficients of identical powers of x, we get

$$-10A=1,\ -5B+2A=-2$$

whence $A=-\frac{1}{10}$, $B=\frac{9}{25}$. Consequently, the particular solution is

$$y^*=x\left(-\frac{1}{10}x+\frac{9}{25}\right)e^x$$

and the general solution is

$$y=C_1e^{6x}+C_2e^x+x\left(-\frac{1}{10}x+\frac{9}{25}\right)e^x$$

II. Let the right side have the form

$$f(x) = P(x)e^{\alpha x}\cos\beta x + Q(x)e^{\alpha x}\sin\beta x \qquad (5)$$

where $P(x)$ and $Q(x)$ are polynomials.

We may handle this case by the technique used in I, if we pass from trigonometric functions to exponential functions. Replacing $\cos\beta x$ and $\sin\beta x$ by exponential functions via Euler's formulas (see Sec. 7.5, Vol. I), we obtain

$$f(x) = P(x)e^{\alpha x}\frac{e^{i\beta x}+e^{-i\beta x}}{2} + Q(x)e^{\alpha x}\frac{e^{i\beta x}-e^{-i\beta x}}{2i}$$

or

$$f(x) = \left[\frac{1}{2}P(x) + \frac{1}{2i}Q(x)\right]e^{(\alpha+i\beta)x} + \left[\frac{1}{2}P(x) - \frac{1}{2i}Q(x)\right]e^{(\alpha-i\beta)x} \qquad (6)$$

Here, the square brackets contain polynomials whose degrees are equal to the highest degree of the polynomials $P(x)$ and $Q(x)$. We have thus obtained the right side of the form considered in Case I.

It can be proved (we omit the proof) that it is possible to find particular solutions which do not contain complex numbers.

Thus, if the right side of equation (1) is of the form

$$f(x) = P(x)e^{\alpha x}\cos\beta x + Q(x)e^{\alpha x}\sin\beta x \qquad (7)$$

where $P(x)$ and $Q(x)$ are polynomials in x, then the form of the particular solution is determined as follows:

(a) if the number $\alpha + i\beta$ is not a root of the auxiliary equation, then the particular solution of equation (1) should be sought in the form

$$y^* = U(x)e^{\alpha x}\cos\beta x + V(x)e^{\alpha x}\sin\beta x \qquad (8)$$

where $U(x)$ and $V(x)$ are polynomials of degree equal to the highest degree of the polynomials $P(x)$ and $Q(x)$;

(b) if the number $\alpha + i\beta$ is a root of the auxiliary equation, we then write the particular solution in the form

$$y^* = x\left[U(x)e^{\alpha x}\cos\beta x + V(x)e^{\alpha x}\sin\beta x\right] \qquad (9)$$

Here, in order to avoid mistakes we must note that these forms of particular solutions, (8) and (9), are obviously retained when one of the polynomials $P(x)$ and $Q(x)$ on the right side of equation (1) is identically zero; that is, when the right side is of the form

$$P(x)e^{\alpha x}\cos\beta x \quad \text{or} \quad Q(x)e^{\alpha x}\sin\beta x$$

Let us further consider an important special case. Let the right side of a second-order linear equation have the form

$$f(x) = M\cos\beta x + N\sin\beta x \qquad (7')$$

where M and N are constants.

(a) If βi is not a root of the auxiliary equation, the particular solution should be sought in the form

$$y^* = A\cos\beta x + B\sin\beta x \tag{8'}$$

(b) If βi is a root of the auxiliary equation, then the particular solution should be sought in the form

$$y^* = x(A\cos\beta x + B\sin\beta x) \tag{9'}$$

We remark that the function (7′) is a special case of the function (7) $[P(x) = M,\ Q(x) = N,\ \alpha = 0]$; the functions (8′) and (9′) are special cases of the functions (8) and (9).

Example 4. Find the general solution of the nonhomogeneous linear equation

$$y'' + 2y' + 5y = 2\cos x$$

Solution. The auxiliary equation $k^2 + 2k + 5 = 0$ has roots $k_1 = -1 + 2i$; $k_2 = -1 - 2i$. Therefore, the general solution of the corresponding homogeneous equation is

$$\bar{y} = e^{-x}(C_1\cos 2x + C_2\sin 2x)$$

We seek the particular solution of the nonhomogeneous equation in the form

$$y^* = A\cos x + B\sin x$$

where A and B are constant coefficients to be determined.

Putting y^* into the given equation, we have

$$-A\cos x - B\sin x + 2(-A\sin x + B\cos x) + 5(A\cos x + B\sin x) = 2\cos x$$

Equating the coefficients of $\cos x$ and $\sin x$, we get two equations for determining A and B:

$$-A + 2B + 5A = 2,\quad -B - 2A + 5B = 0$$

whence $A = \frac{2}{5}$, $B = \frac{1}{5}$. The general solution of the given equation is $\bar{y} = y + y^*$, that is,

$$y = e^{-x}(C_1\cos 2x + C_2\sin 2x) + \frac{2}{5}\cos x + \frac{1}{5}\sin x$$

Example 5. Solve the equation

$$y'' + 4y = \cos 2x$$

Solution. The auxiliary equation has roots $k_1 = 2i$, $k_2 = -2i$; therefore, the general solution of the homogeneous equation is of the form

$$\bar{y} = C_1\cos 2x + C_2\sin 2x$$

We seek the particular solution of the nonhomogeneous equation in the form

$$y^* = x(A\cos 2x + B\sin 2x)$$

Then

$$y^{*\prime} = 2x(-A\sin 2x + B\cos 2x) + (A\cos 2x + B\sin 2x)$$
$$y^{*\prime\prime} = -4x(-A\cos 2x - B\sin 2x) + 4(-A\sin 2x + B\cos 2x)$$

Putting these expressions of the derivatives into the given equation and equating the coefficients of $\cos 2x$ and $\sin 2x$, we get a system of equations for determining A and B:

$$4B=1, \quad -4A=0$$

whence $A=0$ and $B=\frac{1}{4}$. Thus, the general solution of the given equation is

$$y=C_1 \cos 2x+C_2 \sin 2x+\frac{1}{4} x \sin 2x$$

Example 6. Solve the equation

$$y''-y=3e^{2x} \cos x$$

Solution. The right side of the equation has the form

$$f(x)=e^{2x}(M \cos x+N \sin x)$$

and $M=3$, $N=0$. The auxiliary equation $k^2-1=0$ has roots $k_1=1$, $k_2=-1$. The general solution of the homogeneous equation is

$$\bar{y}=C_1e^x+C_2e^{-x}$$

Since the number $\alpha+i\beta=2+i\cdot 1$ is not a root of the auxiliary equation, we seek the particular solution in the form

$$y^*=e^{2x}(A \cos x+B \sin x)$$

Putting this expression into the equation, we get (after collecting like terms)

$$(2A+4B)e^{2x} \cos x+(-4A+2B)e^{2x} \sin x=3e^{2x} \cos x$$

Equating the coefficients of $\cos x$ and $\sin x$, we obtain

$$2A+4B=3, \qquad -4A+2B=0$$

whence $A=\frac{3}{10}$ and $B=\frac{3}{5}$. Consequently, the particular solution is

$$y^*=e^{2x}\left(\frac{3}{10} \cos x+\frac{3}{5} \sin x\right)$$

and the general solution is

$$y=C_1e^x+C_2e^{-x}+e^{2x}\left(\frac{3}{10} \cos x+\frac{3}{5} \sin x\right)$$

Note. All the arguments of this section hold true also for a linear equation of the first order. To illustrate, let us consider a first-order equation with constant coefficients (this equation is frequently encountered in engineering applications):

$$\frac{dy}{dx}+ay=b \tag{10}$$

where a and b are constants. We find the general solution of the homogeneous equation

$$\frac{dy}{dx}+ay=0$$

Form the auxiliary equation

$$k+a=0, \quad k=-a$$

The general solution of the homogeneous equation is

$$\bar{y}=Ce^{-ax}$$

We seek a particular solution y^* of the nonhomogeneous equation in the form

$$y^*=B$$

Substituting into equation (10), we get

$$aB=b, \quad B=b/a$$

Thus

$$y^*=b/a$$

The general solution of (10) will then be

$$y=\bar{y}+y^* \quad \text{or} \quad y=Ce^{-ax}+b/a \tag{11}$$

1.25 HIGHER-ORDER NONHOMOGENEOUS LINEAR EQUATIONS

Let us consider the equation

$$y^{(n)}+a_1y^{(n-1)}+\ldots+a_ny=f(x) \tag{1}$$

where $a_1, a_2, \ldots, a_n, f(x)$ are continuous functions of x (or constants).

Suppose we know the general solution

$$\bar{y}=C_1y_1+C_2y_2+\ldots+C_ny_n \tag{2}$$

of the corresponding homogeneous equation

$$y^{(n)}+a_1y^{(n-1)}+a_2y^{(n-2)}+\ldots+a_ny=0 \tag{3}$$

As in the case of a second-order equation, the following assertion holds for equation (1).

Theorem. *If $\bar{y}$ is the general solution of the homogeneous equation* (3) *and y^* is a particular solution of the nonhomogeneous equation* (1), *then*

$$Y=\bar{y}+y^*$$

is the general solution of the nonhomogeneous equation.

Thus, the problem of integrating equation (1), as in the case of a second-order equation, reduces to finding a particular solution of the nonhomogeneous equation.

As in the case of a second-order equation, the particular solution of equation (1) may be found by the method of variation of parameters, considering $C_1, C_2, \ldots, C_n$ in expression (2) as functions of x.

We form the system of equations (cf. Sec. 1.23):

$$\left.\begin{array}{l} C_1'y_1+C_2'y_2+\ldots+C_n'y_n=0 \\ C_1'y_1'+C_2'y_2'+\ldots+C_n'y_n'=0 \\ \ldots\ldots\ldots\ldots\ldots\ldots \\ C_1'y_1^{(n-2)}+C_2'y_2^{(n-2)}+\ldots+C_n'y_n^{(n-2)}=0 \\ C_1'y_1^{(n-1)}+C_2'y_2^{(n-1)}+\ldots+C_n'y_n^{(n-1)}=f(x) \end{array}\right\} \quad (4)$$

This system of equations in the unknown functions C_1', C_2', ..., C_n' has very definite solutions. (The determinant of the coefficients of C_1', C_2', ..., C_n' is the Wronskian formed for the particular solutions y_1, y_2, ..., y_n of a homogeneous equation, and since these particular solutions are, by definition, linearly independent, the Wronskian is not zero.)

Thus, the system (4) may be solved for the functions C_1', C_2', ..., C_n'. Finding them and integrating, we obtain

$$C_1=\int C_1'dx+\overline{C}_1,\quad C_2=\int C_2'dx+\overline{C}_2,\quad \ldots,\quad C_n=\int C_n'dx+\overline{C}_n$$

where $\overline{C}_1$, $\overline{C}_2$, ..., $\overline{C}_n$ are the constants of integration.

We shall prove that in such a case the expression

$$y^*=C_1y_1+C_2y_2+\ldots+C_ny_n \quad (5)$$

is the general solution of the nonhomogeneous equation (1).

Differentiate expression (5) n times, each time taking into account equations (4); this yields

$$\begin{array}{l} y^*=C_1y_1+C_2y_2+\ldots+C_ny_n \\ y^{*\prime}=C_1y_1'+C_2y_2'+\ldots+C_ny_n' \\ \ldots\ldots\ldots\ldots\ldots\ldots \\ y^{*(n-1)}=C_1y_1^{(n-1)}+C_2y_2^{(n-1)}+\ldots+C_ny_n^{(n-1)} \\ y^{*(n)}=C_1y_1^{(n)}+C_2y_2^{(n)}+\ldots+C_ny_n^{(n)}+f(x) \end{array}$$

Multiplying the terms of the first, second, ... and, finally, last equation by a_n, a_{n-1}, ..., a_1, and 1 respectively, and adding, we get

$$y^{*(n)}+a_1y^{*(n-1)}+\ldots+a_ny^*=f(x)$$

since y_1, y_2, ..., y_n are particular solutions of the homogeneous equation; for this reason, the sums of the terms obtained in adding vertical columns are equal to zero.

Hence, the function $y^*=C_1y_1+\ldots+C_ny_n$ [where C_1, ..., C_n are functions of x determined from equations (4)] is a solution of the nonhomogeneous equation (1), and since this solution depends on the n arbitrary constants $\overline{C}_1$, $\overline{C}_2$, ..., $\overline{C}_n$, it is the general solution. This can be proved, as in the case of the second-order equation.

The proposition is thus proved.

For the case of a higher-order nonhomogeneous equation with constant coefficients (cf. Sec. 1.24), the particular solutions are found more easily, namely:

I. Let there be a function on the right side of the differential equation: $f(x) = P(x)e^{\alpha x}$, where $P(x)$ is a polynomial in x; then we have to distinguish two cases:

(a) if α is not a root of the auxiliary equation, then the particular solution may be sought in the form

$$y^* = Q(x)e^{\alpha x}$$

where $Q(x)$ is a polynomial of the same degree as $P(x)$, but with undetermined coefficients;

(b) if α is a root of multiplicity μ of the auxiliary equation, then the particular solution of the nonhomogeneous equation may be sought in the form

$$y^* = x^{\mu} Q(x) e^{\alpha x}$$

where $Q(x)$ is a polynomial of the same degree as $P(x)$.

II. Let the right side of the equation have the form

$$f(x) = M\cos\beta x + N\sin\beta x$$

where M and N are constants. Then the form of the particular solution will be determined as follows:

(a) if the number βi is not a root of the auxiliary equation, then the particular solution has the form

$$y^* = A\cos\beta x + B\sin\beta x$$

where A and B are constant undetermined coefficients;

(b) if the number βi is a root of the auxiliary equation of multiplicity μ, then

$$y^* = x^{\mu}(A\cos\beta x + B\sin\beta x)$$

III. Let

$$f(x) = P(x)e^{\alpha x}\cos\beta x + Q(x)e^{\alpha x}\sin\beta x$$

where $P(x)$ and $Q(x)$ are polynomials in x. Then:

(a) if the number $\alpha + \beta i$ is not a root of the auxiliary polynomial, then we seek the particular solution in the form

$$y^* = U(x)e^{\alpha x}\cos\beta x + V(x)e^{\alpha x}\sin\beta x$$

where $U(x)$ and $V(x)$ are polynomials of degree equal to the highest degree of the polynomials $P(x)$ and $Q(x)$;

(b) if the number $\alpha + \beta i$ is a root of multiplicity μ of the auxiliary polynomial, then we seek the particular solution in the form

$$y^* = x^{\mu}[U(x)e^{\alpha x}\cos\beta x + V(x)e^{\alpha x}\sin\beta x]$$

where $U(x)$ and $V(x)$ have the same meaning as in Case (a).

General remarks on Cases II and III. Even when the right side of the equation contains an expression with only $\cos\beta x$ or only $\sin\beta x$, we must seek the solution in the form indicated, that is, with sine and cosine. In other words, from the fact that the right side does not contain $\cos\beta x$ or $\sin\beta x$, it does not in the least follow that the particular solution of the equation does not contain these functions. This was evident when we considered Examples 4, 5, 6 of the preceding section, and also Example 2 of the present section.

Example 1. Find the general solution of the equation

$$y^{IV} - y = x^3 + 1$$

Solution. The auxiliary equation $k^4 - 1 = 0$ has the roots

$$k_1 = 1, \quad k_2 = -1, \quad k_3 = i, \quad k_4 = -i$$

We find the general solution of the homogeneous equation (see Example 4, Sec. 1.22):

$$\bar{y} = C_1 e^x + C_2 e^{-x} + C_3 \cos x + C_4 \sin x$$

We seek the particular solution of the nonhomogeneous equation in the form

$$y^* = A_0 x^3 + A_1 x^2 + A_2 x + A_3$$

Differentiating y^* four times and substituting the expressions obtained into the given equation, we get

$$-A_0 x^3 - A_1 x^2 - A_2 x - A_3 = x^3 + 1$$

Equating the coefficients of identical powers of x, we have

$$-A_0 = 1, \quad -A_1 = 0, \quad -A_2 = 0, \quad -A_3 = 1$$

Hence

$$y^* = -x^3 - 1$$

The general solution of the nonhomogeneous equation is found from the formula $y = \bar{y} + y^*$:

$$y = C_1 e^x + C_2 e^{-x} + C_3 \cos x + C_4 \sin x - x^3 - 1$$

Example 2. Solve the equation

$$y^{IV} - y = 5 \cos x$$

Solution. The auxiliary equation $k^4 - 1 = 0$ has the roots $k_1 = 1$, $k_2 = -1$, $k_3 = i$, $k_4 = -i$. Hence, the general solution of the corresponding homogeneous equation is

$$\bar{y} = C_1 e^x + C_2 e^{-x} + C_3 \cos x + C_4 \sin x$$

Further, the right side of the given nonhomogeneous equation has the form

$$f(x) = M \cos x + N \sin x$$

where $M = 5$ and $N = 0$.

Since i is a simple root of the auxiliary equation, we seek the particular solution in the form

$$y^* = x(A \cos x + B \sin x)$$

Putting this expression into the equation, we find

$$4A \sin x - 4B \cos x = 5 \cos x$$

whence

$$4A=0, \quad -4B=5$$

or $A=0$, $B=-\frac{5}{4}$. Consequently, the particular solution of the differential equation is

$$y^*=-\frac{5}{4}x\sin x$$

and the general solution is

$$y=C_1e^x+C_2e^{-x}+C_3\cos x+C_4\sin x-\frac{5}{4}x\sin x$$

1.26 THE DIFFERENTIAL EQUATION OF MECHANICAL VIBRATIONS

In this and the following sections we will consider a problem in applied mechanics and will investigate and solve it by means of linear differential equations.

Let a load of mass Q be at rest on an elastic spring (Fig. 25). We denote by y the deviation of the load from the equilibrium position. We shall consider deviation downwards as positive, upwards as negative. In the equilibrium position, the weight is balanced by the elasticity of the spring. Let us suppose that the force that tends to return the load to equilibrium (the so-called restoring force) is proportional to the deflection, that is, equal to $-ky$, where k is some constant for the given spring (the so-called "spring rigidity"). *

Let us suppose that the motion of the load Q is restricted by a resistance force operating in a direction opposite to that of motion and proportional to th‸ velocity of the load relative to the lower point of the spring; that is, a force $-\lambda v=-\lambda\frac{dy}{dt}$, where $\lambda=\text{const}>0$ (shock absorber). Write the differential equation of the motion of the load on the spring. By Newton's second law we have

$$Q\frac{d^2y}{dt^2}=-ky-\lambda\frac{dy}{dt} \qquad (1)$$

(here, k and λ are positive numbers). We thus have a homogeneous linear differential equation of the second order with constant coefficients.

This equation may be rewritten as follows:

$$\frac{d^2y}{dt^2}+p\frac{dy}{dt}+qy=0 \qquad (1')$$

* Springs whose restoring force is proportional to the deflection are called springs with a "linear characteristic".

where

$$p=\frac{\lambda}{Q}, \quad q=\frac{k}{Q}$$

Let it further be assumed that the lower point of the spring A executes vertical motions under the law $z=\varphi(t)$. This will occur, for instance, if the lower end of the spring is attached to a roller, which moves over an uneven spot together with the spring and the load (Fig. 26).

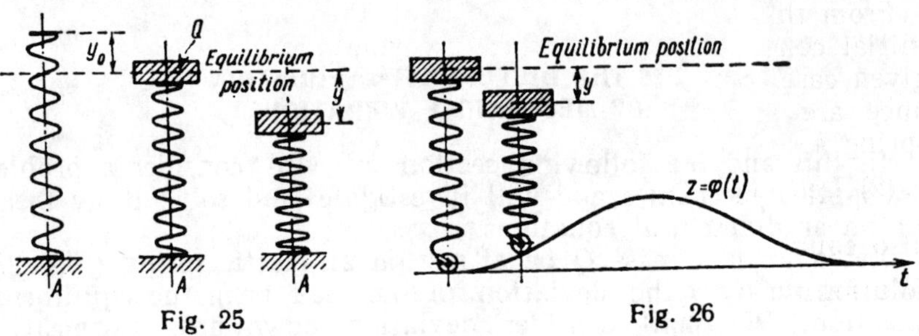

Fig. 25

Fig. 26

In this case the restoring force will be equal not to $-ky$, but to $-k[y+\varphi(t)]$, the force of resistance will be $-\lambda[y'+\varphi'(t)]$, and in place of equation (1) we will have the equation

$$Q\frac{d^2y}{dt^2}+\lambda\frac{dy}{dt}+ky=-k\varphi(t)-\lambda\varphi'(t) \tag{2}$$

or

$$\frac{d^2y}{dt^2}+p\frac{dy}{dt}+qy=f(t) \tag{2'}$$

where

$$f(t)=-\frac{k\varphi(t)+\lambda\varphi'(t)}{Q}$$

We thus have a nonhomogeneous second-order differential equation.

Equation (1′) is called an equation of *free oscillations*, equation (2′) is an equation of *forced oscillations*.

1.27 FREE OSCILLATIONS

Let us first consider the equation of free oscillations

$$y''+py'+qy=0 \quad (p>0,\ q>0; \text{ see Sec. 1.26}) \tag{1}$$

We write the corresponding auxiliary equation

$$k^2+pk+q=0$$

and find its roots:

$$k_1=-\frac{p}{2}+\sqrt{\frac{p^2}{4}-q},\quad k_2=-\frac{p}{2}-\sqrt{\frac{p^2}{4}-q}$$

(1) Let $\frac{p^2}{4}>q$. Then the roots k_1 and k_2 are real negative numbers. The general solution is expressed in terms of exponential functions:

$$y=C_1e^{k_1t}+C_2e^{k_2t}\quad (k_1<0,\ k_2<0) \tag{2}$$

From this formula it follows that the deviation of y for any initial conditions approaches zero asymptotically if $t\to\infty$. In the given case, there will be no oscillations, since the forces of resistance are great compared to the coefficient of rigidity of the spring k.

(2) Let $\frac{p^2}{4}=q$; then the roots k_1 and k_2 are equal $\left(\text{and are also equal to the negative number } -\frac{p}{2}\right)$. Therefore, the general solution will be

$$y=C_1e^{-\frac{p}{2}t}+C_2te^{-\frac{p}{2}t}=(C_1+C_2t)\,e^{-\frac{pt}{2}} \tag{3}$$

Here the deviation also approaches zero as $t\to\infty$, but not so rapidly as in the preceding case (due to the factor C_1+C_2t).

(3) Let $p=0$ (no resistance). Equation (1) is of the form

$$y''+qy=0 \tag{4}$$

The auxiliary equation is $k^2+q=0$ and its roots are $k_1=\beta i$, $k_2=-\beta i$, where $\beta=\sqrt{q}$. The general solution is

$$y=C_1\cos\beta t+C_2\sin\beta t \tag{5}$$

In the latter formula, we replace the arbitrary constants C_1 and C_2 with others. We introduce the constants A and φ_0, which are connected with C_1 and C_2 by the relations

$$C_1=A\sin\varphi_0,\qquad C_2=A\cos\varphi_0$$

A and φ_0 are defined as follows in terms of C_1 and C_2:

$$A=\sqrt{C_1^2+C_2^2},\quad \varphi_0=\arctan\frac{C_1}{C_2}$$

Substituting the values of C_1 and C_2 into formula (5), we get

$$y=A\sin\varphi_0\cos\beta t+A\cos\varphi_0\sin\beta t$$

or

$$y=A\sin(\beta t+\varphi_0) \tag{6}$$

These oscillations are called *harmonic*. The integral curves are sine curves. The time interval T, during which the argument of the sine varies by 2π, is called the *period* of oscillation; here, $T=\frac{2\pi}{\beta}$. The *frequency* is the number of oscillations during time 2π; here, the frequency is β; A is the greatest deviation from equilibrium and is called the *amplitude*; φ_0 is the *initial phase*. The graph of the function (6) is shown in Fig. 27.

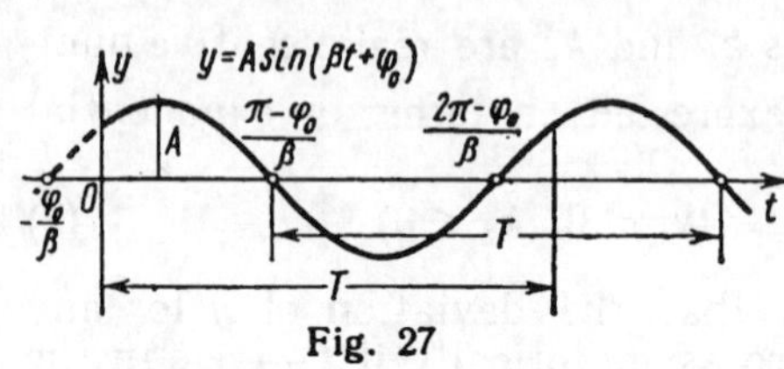

Fig. 27

In electrical engineering and elsewhere, wide use is made of complex and vectorial representations of harmonic oscillations.

Let us take the complex plane xOy and consider a radius vector $\boldsymbol{A}=\boldsymbol{A}(t)$ of constant length $|\boldsymbol{A}|=A=\text{const}$.

As the parameter t varies (here, t is the time), the terminus of the vector $\boldsymbol{A}$ describes a circle of radius A centred at the coordinate origin (Fig. 28). Suppose the angle ψ formed by the vector $\boldsymbol{A}$ and the x-axis is expressed thus: $\psi=\beta t+\varphi_0$. Then β is called the *angular velocity of the vector* $\boldsymbol{A}$. The projections of $\boldsymbol{A}$ on the y- and x-axes are

$$\left.\begin{aligned} y &= A\sin(\beta t+\varphi_0) \\ x &= A\cos(\beta t+\varphi_0) \end{aligned}\right\} \quad (7)$$

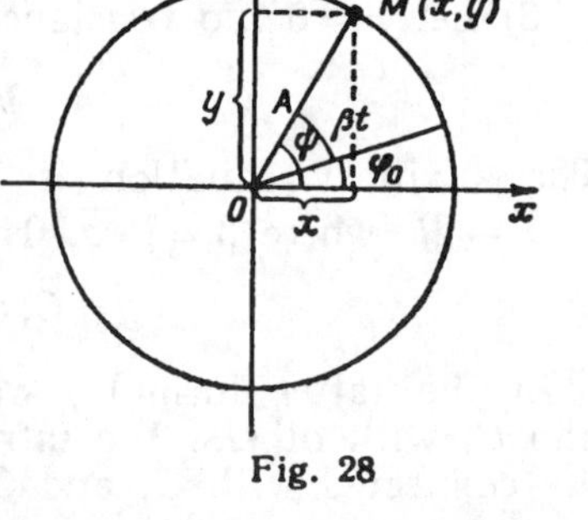

Fig. 28

The expressions (7) are the solutions of equation (4).

We consider the complex quantity

$$z=x+iy=$$
$$=A\cos(\beta t+\varphi_0)+iA\sin(\beta t+\varphi_0)$$

or

$$z=A[\cos(\beta t+\varphi_0)+i\sin(\beta t+\varphi_0)] \quad (8)$$

As was pointed out in Sec. 7.1, Vol. I, the complex quantity z in (8) is represented by the vector $\boldsymbol{A}$.

Thus, the solutions to the equation of harmonic oscillations (4) may be regarded as the *projections of the vector* $\boldsymbol{A}$ *on the* y- *and* x-*axes, the vector with initial phase* φ_0 *rotating with angular velocity* β.

Using the Euler formula [see (4), Sec. 7.5, Vol. I], the expression (8) may be rewritten as

$$z=Ae^{i(\beta t+\varphi_0)} \quad (9)$$

The imaginary and real parts of (9) are solutions of equation (4). Expression (9) is called the *complex solution* of (4). We can rewrite (9) as follows:

$$z = Ae^{i\varphi_0}e^{i\beta t} \tag{10}$$

The expression $Ae^{i\varphi_0}$ is called the *complex amplitude*. We denote it by A^*. Then the complex solution (10) can be rewritten as

$$z = A^*e^{i\beta t} \tag{11}$$

(4) Let $p \neq 0$ and $\frac{p^2}{4} < q$.

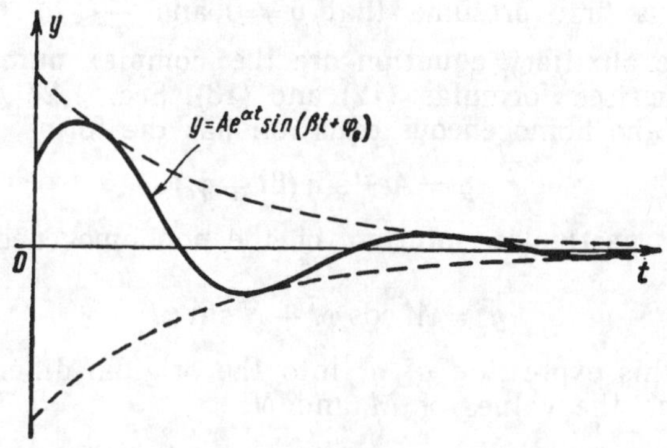

Fig. 29

In this case, the roots of the auxiliary equation are complex numbers:

$$k_1 = \alpha + i\beta, \quad k_2 = \alpha - i\beta$$

where

$$\alpha = -\frac{p}{2} < 0, \quad \beta = \sqrt{q - \frac{p^2}{4}}$$

The general solution has the form

$$y = e^{\alpha t}(C_1 \cos \beta t + C_2 \sin \beta t) \tag{12}$$

or

$$y = Ae^{\alpha t} \sin(\beta t + \varphi_0) \tag{13}$$

Here, for the amplitude we have to consider the quantity $Ae^{\alpha t}$ which depends on the time. Since $\alpha < 0$, it approaches zero as $t \to \infty$, which means that here we are dealing with *damped oscillations*. The graph of damped oscillations is shown in Fig. 29.

1.28 FORCED OSCILLATIONS

The equation of forced oscillations has the form

$$y'' + py' + qy = f(t) \quad (p > 0,\ q > 0;\ \text{see Sec. 1.26}) \tag{1}$$

Let us consider an important practical case when the disturbing external force is periodic and varies under the law

$$f(t) = a \sin \omega t$$

then the equation will have the form

$$y'' + py' + qy = a \sin \omega t \tag{1'}$$

(1) Let us first presume that $p \neq 0$ and $\frac{p^2}{4} < q$, that is, the roots of the auxiliary equation are the complex numbers $\alpha \pm i\beta$. In this case [see formulas (12) and (13), Sec. 1.27], the general solution of the homogeneous equation has the form

$$\bar{y} = Ae^{\alpha t} \sin(\beta t + \varphi_0) \tag{2}$$

We seek a particular solution of the nonhomogeneous equation in the form

$$y^* = M \cos \omega t + N \sin \omega t \tag{3}$$

Putting this expression of y^* into the original differential equation, we find the values of M and N:

$$M = \frac{-p\omega a}{(q-\omega^2)^2 + p^2\omega^2}, \qquad N = \frac{(q-\omega^2)\,a}{(q-\omega^2)^2 + p^2\omega^2}$$

Before putting these values of M and N into (3), let us introduce the new constants A^* and φ^*, setting

$$M = A^* \sin \varphi^*, \qquad N = A^* \cos \varphi^*$$

that is

$$A^* = \sqrt{M^2 + N^2} = \frac{a}{\sqrt{(q-\omega^2)^2 + p^2\omega^2}}, \qquad \tan \varphi^* = \frac{M}{N}$$

Then the particular solution of the nonhomogeneous equation may be written in the form

$$y^* = A^* \sin \varphi^* \cos \omega t + A^* \cos \varphi^* \sin \omega t = A^* \sin(\omega t + \varphi^*)$$

or, finally,

$$y^* = \frac{a}{\sqrt{(q-\omega^2)^2 + p^2\omega^2}} \sin(\omega t + \varphi^*)$$

The general solution of equation (1) is $y = \bar{y} + y^*$ or

$$y = Ae^{\alpha t} \sin(\beta t + \varphi_0) + \frac{a}{\sqrt{(q-\omega^2)^2 + p^2\omega^2}} \sin(\omega t + \varphi^*)$$

The first term of the sum on the right side (the solution of the homogeneous equation) represents damped oscillations; it diminishes with increasing t and, consequently, after some interval of time the second term (which determines the forced oscillations) will dominate. The frequency ω of these oscillations is equal to the frequency of the external force $f(t)$; the amplitude of the forced vibrations is the greater, the smaller p and the closer ω^2 is to q.

Let us investigate more closely the dependence of the amplitude of forced vibrations on the frequency ω for various values of p. For this, we denote the amplitude of forced vibrations by $D(\omega)$:

$$D(\omega)=\frac{a}{\sqrt{(q-\omega^2)^2+p^2\omega^2}}$$

Putting $q=\beta_1^2$ (for $p=0$, β_1 would be equal to its natural frequency), we have

$$D(\omega)=\frac{a}{\sqrt{(\beta_1^2-\omega^2)^2+p^2\omega^2}}=\frac{a}{\beta_1^2\sqrt{\left(1-\frac{\omega^2}{\beta_1^2}\right)^2+\frac{p^2}{\beta_1^2}\frac{\omega^2}{\beta_1^2}}}.$$

Introducing the notation

$$\frac{\omega}{\beta_1}=\lambda,\quad \frac{p}{\beta_1}=\gamma$$

where λ is the ratio of the frequency of the disturbing force to the frequency of free oscillations of the system, and the constant γ is independent of the disturbing force, we find that the magnitude of the amplitude will be expressed by the formula

$$\overline{D}(\lambda)=\frac{a}{\beta_1^2\sqrt{(1-\lambda^2)^2+\gamma^2\lambda^2}} \tag{4}$$

Let us find the maximum of this function. It will obviously be for that value of λ for which the denominator has a minimum. But the minimum of the function

$$\sqrt{(1-\lambda^2)^2+\gamma^2\lambda^2} \tag{5}$$

is reached when

$$\lambda=\sqrt{1-\frac{\gamma^2}{2}}$$

and is equal to

$$\gamma\sqrt{1-\frac{\gamma^2}{4}}$$

Hence, the maximum amplitude is equal to

$$\overline{D}_{\max}=\frac{a}{\beta_1^2\gamma\sqrt{1-\frac{\gamma^2}{4}}}$$

The graphs of the function $\overline{D}(\lambda)$ for various values of γ are shown in Fig. 30 (in constructing the graphs we put $a=1$, $\beta_1=1$ for the sake of definiteness). These curves are called *resonance curves*.

From formula (5) it follows that for small γ the maximum value of amplitude is attained for values of λ close to unity, that

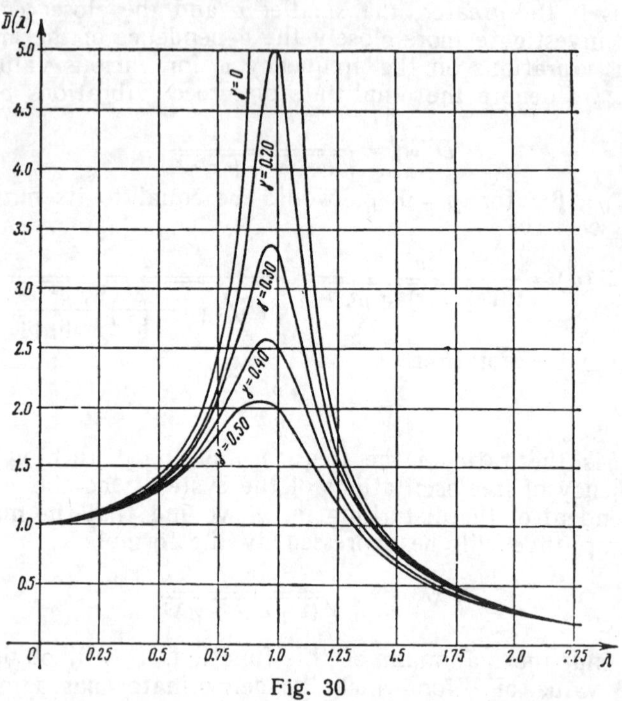

Fig. 30

is, when the frequency of the external force is close to the frequency of free oscillations. If $\gamma=0$ (hence, $p=0$), that is, if there is no resistance to motion, the amplitude of forced vibrations increases without bound as $\lambda \to 1$ or as $\omega \to \beta_1 = \sqrt{q}$:

$$\lim_{\substack{\lambda \to 1 \\ (\gamma=0)}} \overline{D}(\lambda) = \infty$$

At $\omega^2 = q$ we have resonance.

(2) Now let us suppose that $p=0$; that is, we consider the equation of elastic oscillations without resistance but with a periodic external force:

$$y'' + qy = a \sin \omega t \tag{6}$$

The general solution of the homogeneous equation is

$$\bar{y} = C_1 \cos \beta t + C_2 \sin \beta t \qquad (\beta^2 = q)$$

If $\beta \neq \omega$, that is, if the frequency of the external force is not equal to the natural frequency, then the particular solution of the nonhomogeneous equation will have the form

$$y^* = M \cos \omega t + N \sin \omega t$$

Putting this expression into the original equation, we find

$$M = 0, \qquad N = \frac{a}{q - \omega^2}$$

The general solution is

$$y = A \sin (\beta t + \varphi_0) + \frac{a}{q - \omega^2} \sin \omega t$$

Thus, motion results from the superposition of a natural oscillation with frequency β and a forced vibration with frequency ω.

If $\beta = \omega$, that is, the natural frequency coincides with the frequency of the external force, then function (3) is not a solution of equation (6). In this case, in accord with the results of Sec. 1.24, we have to seek the particular solution in the form

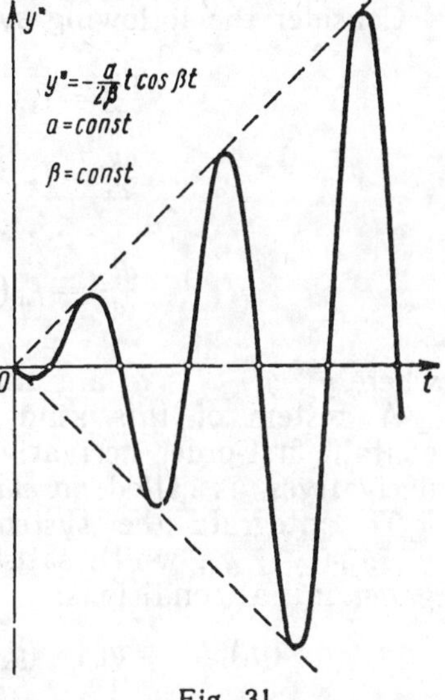

Fig. 31

$$y^* = t (M \cos \beta t + N \sin \beta t) \tag{7}$$

Substituting this expression into the equation, we find M and N:

$$M = -\frac{a}{2\beta}, \quad N = 0$$

Consequently,

$$y^* = -\frac{a}{2\beta} t \cos \beta t$$

The general solution will have the form

$$y = A \sin (\beta t + \varphi_0) - \frac{a}{2\beta} t \cos \beta t$$

The second term on the right side shows that in this case the amplitude increases without bound with the time t. This phenomenon, which occurs when the natural frequency of the system coincides with the frequency of the external force, is called *resonance*.

The graph of the function y^* is shown in Fig. 31.

1.29 SYSTEMS OF ORDINARY DIFFERENTIAL EQUATIONS

In the solution of many problems it is required to find the functions $y_1 = y_1(x)$, $y_2 = y_2(x)$, ..., $y_n = y_n(x)$, which satisfy a system of differential equations containing the argument x, the unknown functions y_1, y_2, ..., y_n and their derivatives.

Consider the following system of first-order equations:

$$\left.\begin{aligned} \frac{dy_1}{dx} &= f_1(x, y_1, y_2, \ldots, y_n) \\ \frac{dy_2}{dx} &= f_2(x, y_1, y_2, \ldots, y_n) \\ &\cdots\cdots\cdots\cdots\cdots \\ \frac{dy_n}{dx} &= f_n(x, y_1, y_2, \ldots, y_n) \end{aligned}\right\} \qquad (1)$$

where y_1, y_2, ..., y_n are unknown functions and x is the argument.

A system of this kind, where the left sides of the equations contain first-order derivatives, while the right sides do not contain derivatives, is called *normal*.

To integrate the system means to determine the functions y_1, y_2, ..., y_n, which satisfy the system of equations (1) and the given initial conditions:

$$(y_1)_{x=x_0} = y_{10}, \quad (y_2)_{x=x_0} = y_{20}, \quad \ldots, \quad (y_n)_{x=x_0} = y_{n0} \qquad (2)$$

Integration of a system like (1) is performed as follows. Differentiate the first equation of (1) with respect to x:

$$\frac{d^2y_1}{dx^2} = \frac{\partial f_1}{\partial x} + \frac{\partial f_1}{\partial y_1}\frac{dy_1}{dx} + \ldots + \frac{\partial f_1}{\partial y_n}\frac{dy_n}{dx}$$

Replacing the derivatives $\frac{dy_1}{dx}, \frac{dy_2}{dx}, \ldots, \frac{dy_n}{dx}$ with their expressions f_1, f_2, ..., f_n from equations (1), we get the equation

$$\frac{d^2y_1}{dx^2} = F_2(x, y_1, \ldots, y_n)$$

Differentiating this equation and then doing as before, we obtain

$$\frac{d^3y_1}{dx^3} = F_3(x, y_1, y_2, \ldots, y_n)$$

Continuing in the same fashion, we finally get the equation

$$\frac{d^ny_1}{dx^n} = F_n(x, y_1, \ldots, y_n)$$

We thus get the following system:

$$\left.\begin{aligned} \frac{dy_1}{dx} &= f_1(x, y_1, \ldots, y_n) \\ \frac{d^2y_1}{dx^2} &= F_2(x, y_1, \ldots, y_n) \\ &\ldots\ldots\ldots\ldots \\ \frac{d^ny_1}{dx^n} &= F_n(x, y_1, \ldots, y_n) \end{aligned}\right\} \qquad (3)$$

From the first $n-1$ equations we determine (if this is possible) $y_2, y_3, \ldots, y_n$ and express them in terms of x, y_1 and the derivatives $\frac{dy_1}{dx}, \frac{d^2y_1}{dx^2}, \ldots, \frac{d^{n-1}y_1}{dx^{n-1}}$:

$$\left.\begin{aligned} y_2 &= \varphi_2(x, y_1, y_1', \ldots, y_1^{(n-1)}) \\ y_3 &= \varphi_3(x, y_1, y_1', \ldots, y_1^{(n-1)}) \\ &\ldots\ldots\ldots\ldots \\ y_n &= \varphi_n(x, y_1, y_1', \ldots, y_1^{(n-1)}) \end{aligned}\right\} \qquad (4)$$

Putting these expressions into the last of the equations (3), we get an nth-order equation for determining y_1:

$$\frac{d^ny_1}{dx^n} = \Phi(x, y_1, y_1', \ldots, y_1^{(n-1)}) \qquad (5)$$

Solving this equation, we find y_1:

$$y_1 = \psi_1(x, C_1, C_2, \ldots, C_n) \qquad (6)$$

Differentiating the latter expression $n-1$ times, we find the derivatives $\frac{dy_1}{dx}, \frac{d^2y_1}{dx^2}, \ldots, \frac{d^{n-1}y_1}{dx^{n-1}}$ as functions of $x, C_1, C_2, \ldots, C_n$.

Substituting these functions into equations (4), we determine $y_2, y_3, \ldots, y_n$:

$$\left.\begin{aligned} y_2 &= \psi_2(x, C_1, C_2, \ldots, C_n) \\ &\ldots\ldots\ldots\ldots \\ y_n &= \psi_n(x, C_1, C_2, \ldots, C_n) \end{aligned}\right\} \qquad (7)$$

For this solution to satisfy the given initial conditions (2), it remains for us to find [from equations (6) and (7)] the appropriate values of the constants $C_1, C_2, \ldots, C_n$ (like we did in the case of a single differential equation).

Note 1. If the system (1) is linear in the unknown functions, then equation (5) is also linear.

Example 1. Integrate the system

$$\frac{dy}{dx} = y + z + x, \quad \frac{dz}{dx} = -4y - 3z + 2x \qquad \text{(a)}$$

with the initial conditions

$$(y)_{x=0} = 1, \quad (z)_{x=0} = 0 \qquad \text{(b)}$$

Solution. (1) Differentiating the first equation with respect to x, we have

$$\frac{d^2y}{dx^2}=\frac{dy}{dx}+\frac{dz}{dx}+1$$

Putting the expressions $\frac{dy}{dx}$ and $\frac{dz}{dx}$ from equations (a) into this equation, we get

$$\frac{d^2y}{dx^2}=(y+z+x)+(-4y-3z+2x)+1$$

or

$$\frac{d^2y}{dx^2}=-3y-2z+3x+1 \qquad \text{(c)}$$

(2) From the first equation of system (a) we find

$$z=\frac{dy}{dx}-y-x \qquad \text{(d)}$$

and put it into the equation just obtained; we get

$$\frac{d^2y}{dx^2}=-3y-2\left(\frac{dy}{dx}-y-x\right)+3x+1$$

or

$$\frac{d^2y}{dx^2}+2\frac{dy}{dx}+y=5x+1 \qquad \text{(e)}$$

The general solution of this equation is

$$y=(C_1+C_2x)\,e^{-x}+5x-9 \qquad \text{(f)}$$

and from (d) we have

$$z=(C_2-2C_1-2C_2x)\,e^{-x}-6x+14 \qquad \text{(g)}$$

Choosing the constants C_1 and C_2 so that the initial conditions (b) are satisfied,

$$(y)_{x=0}=1,\ (z)_{x=0}=0$$

we get, from equations (f) and (g),

$$1=C_1-9,\quad 0=C_2-2C_1+14$$

whence $C_1=10$ and $C_2=6$.

Thus, the solution that satisfies the given initial conditions (b) has the form

$$y=(10+6x)\,e^{-x}+5x-9,\quad z=(-14-12x)\,e^{-x}-6x+14$$

Note 2. In the foregoing we assumed that from the first $n-1$ equations of the system (3) it is possible to determine the functions $y_2, y_3, \ldots, y_n$. It may happen that the variables $y_2, \ldots, y_n$ are eliminated not from n, but from a smaller number of equations. Then to determine y_1 we will have an equation of order less than n.

Example 2. Integrate the system

$$\frac{dx}{dt}=y+z,\ \frac{dy}{dt}=x+z,\ \frac{dz}{dt}=x+y$$

Solution. Differentiating the first equation with respect to t, we find

$$\frac{d^2x}{dt^2}=\frac{dy}{dt}+\frac{dz}{dt}=(x+z)+(x+y)$$

$$\frac{d^2x}{dt^2}=2x+y+z$$

Eliminating the variables y and z from the equations

$$\frac{dx}{dt}=y+z, \quad \frac{d^2x}{dt^2}=2x+y+z$$

we get a second-order equation in x:

$$\frac{d^2x}{dt^2}-\frac{dx}{dt}-2x=0$$

Integrating this equation, we obtain its general solution:

$$x=C_1e^{-t}+C_2e^{2t} \tag{α}$$

Whence we find

$$\frac{dx}{dt}=-C_1e^{-t}+2C_2e^{2t} \text{ and } y=\frac{dx}{dt}-z=-C_1e^{-t}+2C_2e^{2t}-z \tag{β}$$

Putting into the third of the given equations the expressions that have been found for x and y, we get an equation for determining z:

$$\frac{dz}{dt}+z=3C_2e^{2t}$$

Integrating this equation, we find

$$z=C_3e^{-t}+C_2e^{2t} \tag{γ}$$

But then, from equation (β), we get

$$y=-(C_1+C_3)e^{-t}+C_2e^{2t} \tag{δ}$$

Equations (α), (γ), and (δ) give the general solution of the given system.

The differential equations of a system may contain higher-order derivatives. This then yields a system of differential equations of higher order.

For instance, the problem of the motion of a material point under the action of a force F reduces to a system of three second-order differential equations. Let F_x, F_y, F_z be the projections of the force F on the coordinate axes. The position of the point at any instant of time t is determined by its coordinates x, y, and z. Hence, x, y, z are functions of t. The projections of the velocity vector of the point on the axes will be $\frac{dx}{dt}$, $\frac{dy}{dt}$, $\frac{dz}{dt}$.

Suppose that the force F and, hence, its projections F_x, F_y, F_z depend on the time t, the coordinates x, y, z of the point, and on the velocity of motion of the point, that is, on $\frac{dx}{dt}$, $\frac{dy}{dt}$, $\frac{dz}{dt}$.

In this problem the following three functions are the sought-for functions:

$$x=x(t), \quad y=y(t), \quad z=z(t)$$

These functions are determined from equations of dynamics (Newton's law):

$$\left.\begin{aligned} m\frac{d^2x}{dt^2} &= F_x\left(t, x, y, z, \frac{dx}{dt}, \frac{dy}{dt}, \frac{dz}{dt}\right) \\ m\frac{d^2y}{dt^2} &= F_y\left(t, x, y, z, \frac{dx}{dt}, \frac{dy}{dt}, \frac{dz}{dt}\right) \\ m\frac{d^2z}{dt^2} &= F_z\left(t, x, y, z, \frac{dx}{dt}, \frac{dy}{dt}, \frac{dz}{dt}\right) \end{aligned}\right\} \tag{8}$$

We thus have a system of three second-order differential equations. In the case of plane motion, that is, motion in which the trajectory is a plane curve (lying, for example, in the xy-plane), we get a system of two equations for determining the functions $x(t)$ and $y(t)$:

$$m\frac{d^2x}{dt^2} = F_x\left(t, x, y, \frac{dx}{dt}, \frac{dy}{dt}\right) \tag{9}$$

$$m\frac{d^2y}{dt^2} = F_y\left(t, x, y, \frac{dx}{dt}, \frac{dy}{dt}\right) \tag{10}$$

It is possible to solve a system of differential equations of higher order by reducing it to a system of first-order equations. Using equations (9) and (10) as examples, we shall show how this is done. We introduce the notation

$$\frac{dx}{dt} = u, \qquad \frac{dy}{dt} = v$$

Then

$$\frac{d^2x}{dt^2} = \frac{du}{dt}, \qquad \frac{d^2y}{dt^2} = \frac{dv}{dt}$$

The system of two second-order equations (9) and (10) in two unknown functions $x(t)$ and $y(t)$ is replaced by a system of four first-order equations in four unknown functions x, y, u, v:

$$\frac{dx}{dt} = u$$

$$\frac{dy}{dt} = v$$

$$m\frac{du}{dt} = F_x(t, x, y, u, v)$$

$$m\frac{dv}{dt} = F_y(t, x, y, u, v)$$

We remark in conclusion that the general method that we have considered of solving the system may, in certain specific cases, be replaced by some artificial technique that gets the result faster.

Example 3. Find the general solution of the following system of differential equations:

$$\frac{d^2y}{dx^2} = z$$

$$\frac{d^2z}{dx^2} = y$$

Solution. Differentiate both sides of the first equation twice with respect to x:

$$\frac{d^4y}{dx^4}=\frac{d^2z}{dx^2}$$

But $\frac{d^2z}{dx^2}=y$, and so we get a fourth-order equation:

$$\frac{d^4y}{dx^4}=y$$

Integrating this equation, we obtain its general solution (see Sec. 1.22, Example 4):

$$y=C_1e^x+C_2e^{-x}+C_3\cos x+C_4\sin x$$

Finding $\frac{d^2y}{dx^2}$ from this equation and putting it into the first equation, we find z:

$$z=C_1e^x+C_2e^{-x}-C_3\cos x-C_4\sin x$$

1.30 SYSTEMS OF LINEAR DIFFERENTIAL EQUATIONS WITH CONSTANT COEFFICIENTS

Suppose we have the following system of differential equations:

$$\left.\begin{aligned}\frac{dx_1}{dt}&=a_{11}x_1+a_{12}x_2+\dots+a_{1n}x_n\\ \frac{dx_2}{dt}&=a_{21}x_1+a_{22}x_2+\dots+a_{2n}x_n\\ &\dots\dots\dots\dots\dots\\ \frac{dx_n}{dt}&=a_{n1}x_1+a_{n2}x_2+\dots+a_{nn}x_n\end{aligned}\right\}\tag{1}$$

where the coefficients a_{ij} are constants. Here, t is the argument, and $x_1(t), x_2(t), \dots, x_n(t)$ are the unknown functions. The system (1) is a *system of homogeneous linear differential equations with constant coefficients*.

As was pointed out in the preceding section, this system may be solved by reducing it to a single equation of order n, which in the given instance will be linear (this was indicated in Note 1 of the preceding section). But system (1) may be solved in another way, without reducing it to an equation of the nth order. This method makes it possible to analyze the character of the solutions more clearly.

We seek a particular solution of the system in the following form:

$$x_1=\alpha_1e^{kt},\quad x_2=\alpha_2e^{kt},\quad \dots,\quad x_n=\alpha_ne^{kt}\tag{2}$$

It is required to determine the constants α_1, α_2, $\dots$, α_n and k in such a way that the functions α_1e^{kt}, α_2e^{kt}, $\dots$, α_ne^{kt} should satisfy the system of equations (1). Putting them into system (1), we get:

$$\left.\begin{aligned} k\alpha_1 e^{kt} &= (a_{11}\alpha_1 + a_{12}\alpha_2 + \ldots + a_{1n}\alpha_n)\, e^{kt} \\ k\alpha_2 e^{kt} &= (a_{21}\alpha_1 + a_{22}\alpha_2 + \ldots + a_{2n}\alpha_n)\, e^{kt} \\ &\ldots\ldots\ldots\ldots\ldots\ldots\ldots \\ k\alpha_n e^{kt} &= (a_{n1}\alpha_1 + a_{n2}\alpha_2 + \ldots + a_{nn}\alpha_n)\, e^{kt} \end{aligned}\right\}$$

Cancel out e^{kt}. Transposing all terms to one side and collecting coefficients of $\alpha_1, \alpha_2, \ldots, \alpha_n$, we get a system of equations:

$$\left.\begin{aligned} (a_{11}-k)\alpha_1 + a_{12}\alpha_2 + \ldots + a_{1n}\alpha_n &= 0 \\ a_{21}\alpha_1 + (a_{22}-k)\,\alpha_2 + \ldots + a_{2n}\alpha_n &= 0 \\ \ldots\ldots\ldots\ldots\ldots\ldots\ldots\ldots \\ a_{n1}\alpha_1 + a_{n2}\alpha_2 + \ldots + (a_{nn}-k)\,\alpha_n &= 0 \end{aligned}\right\} \quad (3)$$

Choose $\alpha_1, \alpha_2, \ldots, \alpha_n$ and k such that will satisfy the system (3). This is a system of linear algebraic equations in $\alpha_1, \alpha_2, \ldots, \alpha_n$. Let us form the determinant of system (3):

$$\Delta(k) = \begin{vmatrix} a_{11}-k & a_{12} & \ldots & a_{1n} \\ a_{21} & a_{22}-k & \ldots & a_{2n} \\ \ldots & \ldots & \ldots & \ldots \\ a_{n1} & a_{n2} & \ldots & a_{nn}-k \end{vmatrix} \quad (4)$$

If k is such that the determinant Δ is different from zero, then system (3) has only trivial solutions $\alpha_1 = \alpha_2 = \ldots = \alpha_n = 0$ and, hence, formulas (2) yield only trivial solutions:

$$x_1(t) = x_2(t) = \ldots = x_n(t) \equiv 0$$

Thus, we obtain nontrivial solutions (2) only for k such that the determinant (4) vanishes. We arrive at an equation of order n for determining k:

$$\begin{vmatrix} a_{11}-k & a_{12} & \ldots & a_{1n} \\ a_{21} & a_{22}-k & \ldots & a_{2n} \\ \ldots & \ldots & \ldots & \ldots \\ a_{n1} & a_{n2} & \ldots & a_{nn}-k \end{vmatrix} = 0 \quad (5)$$

This equation is called the **auxiliary equation** of the system (1), and its roots are the **roots of the auxiliary equation**.

Let us consider a few cases.

I. The roots of the auxiliary equation are real and distinct. Denote by $k_1, k_2, \ldots, k_n$ the roots of the auxiliary equation. For each root k_i write the system (3) and determine the coefficients

$$\alpha_1^{(i)},\ \alpha_2^{(i)},\ \ldots,\ \alpha_n^{(i)}$$

It may be shown that one of them is arbitrary; it may be considered equal to unity. Thus we obtain:

for the root k_1 the following solution of the system (1)

$$x_1^{(1)}=\alpha_1^{(1)}e^{k_1t};\quad x_2^{(1)}=\alpha_2^{(1)}e^{k_1t},\ \ldots,\quad x_n^{(1)}=\alpha_n^{(1)}e^{k_1t}$$

for the root k_2 the solution of the system (1)

$$x_1^{(2)}=\alpha_1^{(2)}e^{k_2t},\quad x_2^{(2)}=\alpha_2^{(2)}e^{k_2t},\ \ldots,\quad x_n^{(2)}=\alpha_n^{(2)}e^{k_2t}$$

. . . .

for the root k_n the solution of the system (1)

$$x_1^{(n)}=\alpha_1^{(n)}e^{k_nt},\quad x_2^{(n)}=\alpha_2^{(n)}e^{k_nt},\ \ldots,\quad x_n^{(n)}=\alpha_n^{(n)}e^{k_nt}$$

By direct substitution into the equations we see that the system of functions

$$\left.\begin{aligned} x_1&=C_1\alpha_1^{(1)}e^{k_1t}+C_2\alpha_1^{(2)}e^{k_2t}+\ldots+C_n\alpha_1^{(n)}e^{k_nt}\\ x_2&=C_1\alpha_2^{(1)}e^{k_1t}+C_2\alpha_2^{(2)}e^{k_2t}+\ldots+C_n\alpha_2^{(n)}e^{k_nt}\\ &\ldots\ldots\ldots\ldots\ldots\ldots\ldots\ldots\\ x_n&=C_1\alpha_n^{(1)}e^{k_1t}+C_2\alpha_n^{(2)}e^{k_2t}+\ldots+C_n\alpha_n^{(n)}e^{k_nt}\end{aligned}\right\}\qquad(6)$$

where $C_1, C_2, \ldots, C_n$ are arbitrary constants, is likewise a solution of the system of differential equations (1). This is the **general solution of system** (1). It may readily be shown that one can find values of the constants such that the solution will satisfy the given initial conditions.

Example 1. Find the general solution of the system of equations

$$\frac{dx_1}{dt}=2x_1+2x_2,\quad \frac{dx_2}{dt}=x_1+3x_2$$

Solution. Form the auxiliary equation

$$\begin{vmatrix} 2-k & 2\\ 1 & 3-k\end{vmatrix}=0$$

or $k^2-5k+4=0$. Find its roots:

$$k_1=1,\quad k_2=4$$

Seek the solution of the system in the form

$$x_1^{(1)}=\alpha_1^{(1)}e^t,\quad x_2^{(1)}=\alpha_2^{(1)}e^t$$

and

$$x_1^{(2)}=\alpha_1^{(2)}e^{4t},\ x_2^{(2)}=\alpha_2^{(2)}e^{4t}$$

Form the system (3) for the root $k_1=1$ and determine $\alpha_1^{(1)}$ and $\alpha_2^{(1)}$:

$$\left.\begin{aligned}(2-1)\,\alpha_1^{(1)}+2\alpha_2^{(1)}&=0\\ 1\alpha_1^{(1)}+(3-1)\,\alpha_2^{(1)}&=0\end{aligned}\right\}$$

or

$$\begin{aligned}\alpha_1^{(1)}+2\alpha_2^{(1)}&=0\\ \alpha_1^{(1)}+2\alpha_2^{(1)}&=0\end{aligned}$$

whence $\alpha_2^{(1)}=-\frac{1}{2}\alpha_1^{(1)}$. Putting $\alpha_1^{(1)}=1$, we get $\alpha_2^{(1)}=-\frac{1}{2}$. Thus, we obtain the solution of the system:

$$x_1^{(1)}=e^t,\quad x_2^{(1)}=-\frac{1}{2}e^t$$

Now form system (3) for the root $k_2=4$ and determine $\alpha_1^{(2)}$ and $\alpha_2^{(2)}$:

$$-2\alpha_1^{(2)}+2\alpha_2^{(2)}=0$$
$$\alpha_1^{(2)}-\alpha_2^{(2)}=0$$

whence $\alpha_1^{(2)}=\alpha_2^{(2)}$ and $\alpha_1^{(2)}=1$, $\alpha_2^{(2)}=1$. We obtain the second solution of the system:

$$x_1^{(2)}=e^{4t},\quad x_2^{(2)}=e^{4t}$$

The general solution of the system will be [see (6)]

$$x_1=C_1e^t+C_2e^{4t}$$
$$x_2=-\frac{1}{2}C_1e^t+C_2e^{4t}$$

II. The roots of the auxiliary equation are distinct, but include complex roots. Among the roots of the auxiliary equation let there be two complex conjugate roots:

$$k_1=\alpha+i\beta,\quad k_2=\alpha-i\beta$$

To these roots will correspond the solutions

$$x_j^{(1)}=\alpha_j^{(1)}e^{(\alpha+i\beta)t}\qquad (j=1,\ 2,\ \ldots,\ n)\qquad (7)$$
$$x_j^{(2)}=\alpha_j^{(2)}e^{(\alpha-i\beta)t}\qquad (j=1,\ 2,\ \ldots,\ n)\qquad (8)$$

The coefficients $\alpha_j^{(1)}$ and $\alpha_j^{(2)}$ are determined from the system of equations (3).

Just as in Sec. 1.21, it may be shown that the real and imaginary parts of the complex solution are also solutions. We thus obtain two particular solutions:

$$\left.\begin{aligned}\bar{x}_j^{(1)}&=e^{\alpha t}(\lambda_j^{(1)}\cos\beta x+\lambda_j^{(2)}\sin\beta x)\\ \bar{x}_j^{(2)}&=e^{\alpha t}(\bar{\lambda}_j^{(1)}\sin\beta x+\bar{\lambda}_j^{(2)}\cos\beta x)\end{aligned}\right\}\qquad (9)$$

where $\lambda_j^{(1)}$, $\lambda_j^{(2)}$, $\bar{\lambda}_j^{(1)}$, $\bar{\lambda}_j^{(2)}$ are real numbers determined in terms of $\alpha_j^{(1)}$ and $\alpha_j^{(2)}$.

Appropriate combinations of functions (9) will enter into the general solution of the system.

Example 2. Find the general solution of the system

$$\frac{dx_1}{dt}=-7x_1+x_2$$

$$\frac{dx_2}{dt}=-2x_1-5x_2$$

Solution. Form the auxiliary equation

$$\begin{vmatrix} -7-k & 1 \\ -2 & -5-k \end{vmatrix}=0$$

or $k^2+12k+37=0$ and find its roots:

$$k_1=-6+i, \quad k_2=-6-i$$

Substituting $k_1=-6+i$ into the system (3), we find

$$\alpha_1^{(1)}=1, \quad \alpha_2^{(1)}=1+i$$

We write the solution (7):

$$x_1^{(1)}=1e^{(-6+i)t}, \quad x_2^{(1)}=(1+i)e^{(-6+i)t} \tag{7'}$$

Putting $k_2=-6-i$ into system (3), we find

$$\alpha_1^{(2)}=1, \quad \alpha_2^{(2)}=1-i$$

We get a second system of solutions (8):

$$x_1^{(2)}=e^{(-6-i)t}, \quad x_2^{(2)}=(1-i)e^{(-6-i)t} \tag{8'}$$

Rewrite the solution (7′):

$$x_1^{(1)}=e^{-6t}(\cos t+i\sin t)$$
$$x_2^{(1)}=(1+i)e^{-6t}(\cos t+i\sin t)$$

or

$$x_1^{(1)}=e^{-6t}\cos t+ie^{-6t}\sin t$$
$$x_2^{(1)}=e^{-6t}(\cos t-\sin t)+ie^{-6t}(\cos t+\sin t)$$

Rewrite the solution (8′):

$$x_1^{(2)}=e^{-6t}\cos t-ie^{-6t}\sin t$$
$$x_2^{(2)}=e^{-6t}(\cos t-\sin t)-ie^{-6t}(\cos t+\sin t)$$

For systems of particular solutions we can take the real parts and the imaginary parts separately:

$$\left.\begin{aligned} \bar{x}_1^{(1)}&=e^{-6t}\cos t, \quad &\bar{x}_2^{(1)}&=e^{-6t}(\cos t-\sin t) \\ \bar{x}_1^{(2)}&=e^{-6t}\sin t, \quad &\bar{x}_2^{(2)}&=e^{-6t}(\cos t+\sin t) \end{aligned}\right\} \tag{9'}$$

The general solution of the system is

$$x_1=C_1e^{-6t}\cos t+C_2e^{-6t}\sin t$$
$$x_2=C_1e^{-6t}(\cos t-\sin t)+C_2e^{-6t}(\cos t+\sin t)$$

By a similar method it is possible to find the solution of a system of linear differential equations of higher order with constant coefficients.

For instance, in mechanics and electric-circuit theory a study is made of the solution of a system of second-order differential equations:

$$\left.\begin{aligned} \frac{d^2x}{dt^2}&=a_{11}x+a_{12}y \\ \frac{d^2y}{dt^2}&=a_{21}x+a_{22}y \end{aligned}\right\} \tag{10}$$

Again we seek the solution in the form

$$x=\alpha e^{kt}, \quad y=\beta e^{kt}$$

Putting these expressions into system (10) and cancelling out e^{kt}, we get a system of equations for determining α, β and k:

$$\left.\begin{aligned}(a_{11}-k^2)\alpha+a_{12}\beta=0\\ a_{21}\alpha+(a_{22}-k^2)\beta=0\end{aligned}\right\} \qquad (11)$$

Nonzero α and β are determined only when the determinant of the system is equal to zero:

$$\begin{vmatrix} a_{11}-k^2 & a_{12} \\ a_{21} & a_{22}-k^2 \end{vmatrix}=0 \qquad (12)$$

This is the auxiliary equation of system (10); it is a fourth-order equation in k. Let k_1, k_2, k_3, and k_4 be its roots (we assume that the roots are distinct). For each root k_i of system (11) we find the values of α and β. The general solution, like (6), will have the form

$$x=C_1\alpha^{(1)}e^{k_1t}+C_2\alpha^{(2)}e^{k_2t}+C_3\alpha^{(3)}e^{k_3t}+C_4\alpha^{(4)}e^{k_4t}$$
$$y=C_1\beta^{(1)}e^{k_1t}+C_2\beta^{(2)}e^{k_2t}+C_3\beta^{(3)}e^{k_3t}+C_4\beta^{(4)}e^{k_4t}$$

If there are complex roots, then to each pair of complex roots in the general solution there will correspond expressions of the form (9).

Example 3. Find the general solution of the following system of differential equations

$$\frac{d^2x}{dt^2}=x-4y$$

$$\frac{d^2y}{dt^2}=-x+y$$

Solution. Write the auxiliary equation (12) and find its roots:

$$\begin{vmatrix} 1-k^2 & -4 \\ -1 & 1-k^2 \end{vmatrix}=0$$

$$k_1=i, \quad k_2=-i, \quad k_3=\sqrt{3}, \quad k_4=-\sqrt{3}.$$

We shall seek the solution in the form

$$x^{(1)}=\alpha^{(1)}e^{it}, \qquad y^{(1)}=\beta^{(1)}e^{it}$$
$$x^{(2)}=\alpha^{(2)}e^{-it}, \qquad y^{(2)}=\beta^{(2)}e^{-it},$$
$$x^{(3)}=\alpha^{(3)}e^{\sqrt{3}t}, \qquad y^{(3)}=\beta^{(3)}e^{\sqrt{3}t}$$
$$x^{(4)}=\alpha^{(4)}e^{-\sqrt{3}t}, \qquad y^{(4)}=\beta^{(4)}e^{-\sqrt{3}t}$$

From system (11) we find $\alpha^{(j)}$ and $\beta^{(j)}$:

$$\alpha^{(1)}=1,\ \beta^{(1)}=\frac{1}{2}$$

$$\alpha^{(2)}=1,\ \beta^{(2)}=\frac{1}{2}$$

$$\alpha^{(3)}=1,\ \beta^{(3)}=-\frac{1}{2}$$

$$\alpha^{(4)}=1,\ \beta^{(4)}=-\frac{1}{2}$$

We write out the complex solutions:

$$x^{(1)}=e^{it}=\cos t+i\sin t,\qquad y^{(1)}=\frac{1}{2}(\cos t+i\sin t)$$

$$x^{(2)}=e^{-it}=\cos t-i\sin t,\quad y^{(2)}=\frac{1}{2}(\cos t-i\sin t)$$

The real and imaginary parts separately form the solution:

$$\bar{x}^{(1)}=\cos t,\quad \bar{y}^{(1)}=\frac{1}{2}\cos t$$

$$\bar{x}^{(2)}=\sin t,\quad \bar{y}^{(2)}=\frac{1}{2}\sin t$$

We can now write the general solution:

$$x=C_1\cos t+C_2\sin t+C_3e^{\sqrt{3}t}+C_4e^{-\sqrt{3}t}$$

$$y=\frac{1}{2}C_1\cos t+\frac{1}{2}C_2\sin t-\frac{1}{2}C_3e^{\sqrt{3}t}-\frac{1}{2}C_4e^{-\sqrt{3}t}$$

Note. In this section we did not consider the case of multiple roots of the auxiliary equation. This question is dealt with in detail in "Lectures on the Theory of Ordinary Differential Equations" by I. G. Petrovsky.

1.31 ON LYAPUNOV'S THEORY OF STABILITY

Since the solutions of most differential equations and systems of equations are not expressible in terms of elementary functions or quadratures, use is made of approximate methods of integration in these cases when solving concrete differential equations. The elements of these methods were given in Sec. 1.3; some of these methods will also be considered in Secs. 1.32-1.34 and in Chapter 4.

The drawback of these methods lies in the fact that they yield only one particular solution; to obtain other particular solutions, one has to carry out all the calculations again. Knowing one particular solution does not permit us to draw conclusions about the character of the other solutions.

In many problems of mechanics and engineering it is sometimes important to know not the specific values of a solution for some concrete value of the argument, but the type of behaviour for changes in the argument and, in particular, for a boundless increase in the argument. For example, it is sometimes important to know whether the solutions that satisfy the given initial conditions are periodic, whether they approach some known function asymptotically, etc. These are the questions with which the qualitative theory of differential equations deals.

One of the basic problems of the qualitative theory is that of the stability of the solution or of the stability of motion; this problem was investigated in detail by the noted Russian mathematician A. M. Lyapunov (1857-1918).

Let there be given a system of differential equations:

$$\left.\begin{aligned} \frac{dx}{dt} &= f_1(t, x, y) \\ \frac{dy}{dt} &= f_2(t, x, y) \end{aligned}\right\} \tag{1}$$

Let $x = x(t)$ and $y = y(t)$ be the solutions of this system that satisfy the initial conditions

$$\left.\begin{aligned} x_{t=0} &= x_0 \\ y_{t=0} &= y_0 \end{aligned}\right\} \tag{1'}$$

Further, let $\bar{x} = \bar{x}(t)$ and $\bar{y} = \bar{y}(t)$ be the solutions of equation (1) that satisfy the initial conditions

$$\left.\begin{aligned} \bar{x}_{t=0} &= \bar{x}_0 \\ \bar{y}_{t=0} &= \bar{y}_0 \end{aligned}\right\} \tag{1''}$$

Definition. The solutions $x = x(t)$ and $y = y(t)$ that satisfy the equations (1) and the initial conditions (1′) are called Lyapunov *stable* as $t \to \infty$ if for every arbitrarily small $\varepsilon > 0$ there is a $\delta > 0$ such that for all values $t > 0$ the following inequalities are fulfilled:

$$\left.\begin{aligned} |\bar{x}(t) - x(t)| &< \varepsilon \\ |\bar{y}(t) - y(t)| &< \varepsilon \end{aligned}\right\} \tag{2}$$

if the initial data satisfy the inequalities

$$\left.\begin{aligned} |\bar{x}_0 - x_0| &< \delta \\ |\bar{y}_0 - y_0| &< \delta \end{aligned}\right\} \tag{3}$$

Let us figure out the meaning of this definition. From inequalities (2) and (3) it follows that for small variations in the initial conditions, the corresponding solutions differ but little for all positive values of t. If the system of differential equations is a system that describes some motion, then in the case of stability of solutions, the nature of the motions changes but slightly for small changes in the initial data.

Let us analyze an example of a first-order equation.

Suppose we have the differential equation:

$$\frac{dy}{dt} = -y + 1 \qquad \text{(a)}$$

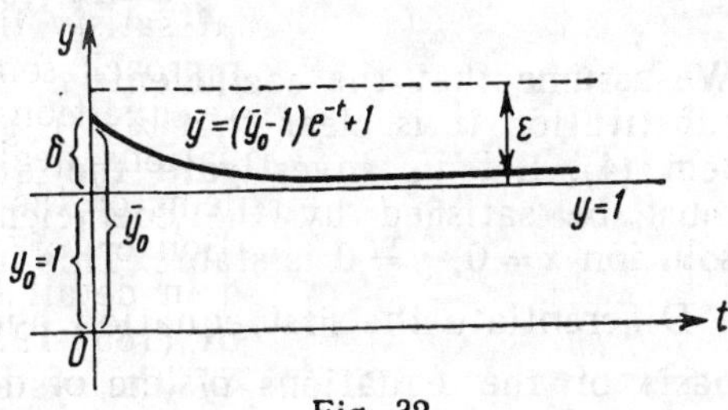

Fig. 32

The general solution of this equation is the function

$$y = Ce^{-t} + 1 \qquad \text{(b)}$$

Find a particular solution that satisfies the initial condition

$$y_{t=0} = 1 \qquad \text{(c)}$$

It is obvious that the solution $y=1$ results when $C=0$ (Fig. 32). Then find the particular solution that satisfies the initial condition

$$\bar{y}_{t=0} = \bar{y}_0$$

Find the value of C from equation (b):

$$\bar{y}_0 = C + 1$$

whence

$$C = \bar{y}_0 - 1$$

Putting this value of C into equation (b), we get

$$\bar{y} = (\bar{y}_0 - 1)\, e^{-t} + 1$$

The solution $y=1$ is obviously stable. Indeed,

$$\bar{y} - y = [(\bar{y}_0 - 1)\, e^{-t} + 1] - 1 = (\bar{y}_0 - 1)\, e^{-t} \to 0$$

when $t \to \infty$.

Hence, inequality (3) will be fulfilled for an arbitrary ε if the following inequality holds true:

$$(y_0 - 1) = \delta < \varepsilon$$

If the equations (1) describe motion and the argument t, the time, is being given implicitly, that is, if we have a system of the form

$$\frac{dx}{dt} = f_1(x, y)$$

$$\frac{dy}{dt} = f_2(x, y)$$

then this system is termed *autonomous*.

Let us also consider the following system of linear differential equations:

$$\left.\begin{aligned}\frac{dx}{dt}&=cx+gy\\ \frac{dy}{dt}&=ax+by\end{aligned}\right\}\qquad(4)$$

We assume that the coefficients a, b, c, g are constants; by direct substitution it is clear that $x=0$, $y=0$ is a solution of the system (4). Let us investigate the question of the conditions that must be satisfied by the coefficients of the system so that the solution $x=0$, $y=0$ is stable. This investigation is done as follows.

Differentiate the first equation and eliminate y and $\frac{dy}{dt}$ on the basis of the equations of the system

$$\frac{d^2x}{dt^2}=c\frac{dx}{dt}+g\frac{dy}{dt}=c\frac{dx}{dt}+g(ax+by)=c\frac{dx}{dt}+gax+b\left(\frac{dx}{dt}-cx\right)$$

or

$$\frac{d^2x}{dt^2}-(b+c)\frac{dx}{dt}-(ag-bc)x=0\qquad(5)$$

The auxiliary equation of the differential equation (5) is of the form

$$\lambda^2-(b+c)\lambda-(ag-bc)=0\qquad(6)$$

This equation may be written in determinantal form:

$$\begin{vmatrix}c-\lambda & g\\ a & b-\lambda\end{vmatrix}=0\qquad(7)$$

[see equation (4), Sec. 1.30].

We denote the roots of the auxiliary equation (7) by λ_1 and λ_2. As we shall see below, the stability or instability of the solutions of system (4) is determined by the nature of the roots λ_1 and λ_2.

Let us consider all possible cases.

I. The roots of the auxiliary equation are real, negative and distinct: $\lambda_1<0$, $\lambda_2<0$, $\lambda_1\neq\lambda_2$. From equation (5) we find

$$x=C_1e^{\lambda_1 t}+C_2e^{\lambda_2 t}$$

Knowing x, we find y from the first equation of (4). Thus, the solution of system (4) is of the form:

$$\left.\begin{aligned}x&=C_1e^{\lambda_1 t}+C_2e^{\lambda_2 t}\\ y&=[C_1(\lambda_1-c)e^{\lambda_1 t}+C_2(\lambda_2-c)e^{\lambda_2 t}]\frac{1}{g}\end{aligned}\right\}\qquad(8)$$

Note. If $g=0$ and $a\neq0$, then we form the equation (5) for the function y. Finding y, we then find x from the second equation of (4). The structure of the solutions (8) is preserved. But if $g=0$, $a=0$, then the solution of the system of equations becomes

$$x=C_1e^{ct}, \quad y=C_2e^{bt} \tag{8'}$$

In this case, the analysis of the character of the solutions is easier to carry out. Choose C_1 and C_2 so that the solutions (8) satisfy the initial conditions

$$x|_{t=0}=x_0, \quad y|_{t=0}=y_0$$

The solution satisfying the initial conditions will be

$$\left.\begin{aligned} x&=\frac{cx_0+gy_0-x_0\lambda_2}{\lambda_1-\lambda_2}e^{\lambda_1 t}+\frac{x_0\lambda_1-cx_0-gy_0}{\lambda_1-\lambda_2}e^{\lambda_2 t}\\ y&=\frac{1}{g}\left[\frac{cx_0+gy_0-x_0\lambda_2}{\lambda_1-\lambda_2}(\lambda_1-c)e^{\lambda_1 t}+\frac{x_0\lambda_1-cx_0-gy_0}{\lambda_1-\lambda_2}(\lambda_2-c)e^{\lambda_2 t}\right]\end{aligned}\right\} \tag{9}$$

From these equations it follows that for arbitrary $\varepsilon>0$, it is possible to select $|x_0|$ and $|y_0|$ so small that for all $t>0$ it will be true that $|x(t)|<\varepsilon$, $|y(t)|<\varepsilon$ since $e^{\lambda_1 t}<1$, $e^{\lambda_2 t}<1$.

We note that in this case

$$\left.\begin{aligned}\lim_{t\to+\infty} x(t)&=0\\ \lim_{t\to+\infty} y(t)&=0\end{aligned}\right\} \tag{10}$$

Consider the *xy*-plane. For the system of differential equations (4) and for the differential equation (5), this plane is termed the *phase plane*. We will consider the solutions (8) and (9) of system (4) as parametric equations of some curve in the phase plane *xOy*:

$$\left.\begin{aligned}x&=\bar{\varphi}(t, C_1, C_2)\\ y&=\bar{\psi}(t, C_1, C_2)\end{aligned}\right\} \tag{11}$$

$$\left.\begin{aligned}x&=\varphi(t, x_0, y_0)\\ y&=\psi(t, x_0, y_0)\end{aligned}\right\} \tag{12}$$

These curves are the *integral curves (solution curves)* of the differential equation

$$\frac{dy}{dx}=\frac{ax+by}{cx+gy} \tag{13}$$

which is obtained from the system (4) by dividing the right and left members by each other.

The origin, $O(0, 0)$, is a *singular point* of the differential equation (13), since this point does not belong to the domain of existence and uniqueness of the solution.

The nature of the solutions (9) and, generally, of the solutions of the system (4) is illustrated by the arrangement of the integral curves

$$\overline{F}(x, y, C)=0$$

which form the complete integral of the differential equation (13). The constant C is determined from the initial condition $y_{x=x_0}=y_0$. Substituting the value of C, we obtain the equation of the family in the form

$$F(x, y, x_0, y_0) \qquad (14)$$

In the case of solutions (9), the singular point is called a *stable nodal point*. We say that a point moving along the integral curve approaches a singular point without bound as $t \to +\infty$.

It is obvious that the relation (14) may be obtained by eliminating the parameter t from the system (12). We will not continue the analysis of the arrangement of integral curves near a singular point on the phase plane for all possible cases of roots of the auxiliary equation and will confine ourselves to an illustration of this fact in elementary instances that do not require unwieldy computations. We note that for arbitrary coefficients the behaviour of integral curves of the equation (13) near the origin is qualitatively the same as is now to be examined in the following examples.

Example 1. Investigate the stability of the solution $x=0$, $y=0$ of the system of equations

$$\frac{dx}{dt}=-x$$

$$\frac{dy}{dt}=-2y$$

Solution. The auxiliary equation is

$$\begin{vmatrix} -1-\lambda & 0 \\ 0 & -2-\lambda \end{vmatrix}=0$$

The roots of the auxiliary equation are

$$\lambda_1=-1, \quad \lambda_2=-2$$

In this case, the solutions (8′) will be

$$x=C_1e^{-t}, \quad y=C_2e^{-2t}$$

The solutions (9) will be

$$x=x_0e^{-t}, \quad y=y_0e^{-2t} \qquad (a)$$

Clearly, $x(t) \to 0$, and $y(t) \to 0$ as $t \to +\infty$. The solution $x=0$, $y=0$ is stable. Now let us examine the phase plane. Eliminating the parameter t from the equations (a), we get an equation of type (14).

$$\left(\frac{x}{x_0}\right)^2=\frac{y}{y_0} \qquad (b)$$

This is a family of parabolas (Fig. 33).

The equation of type (13) will, for the given example, be

$$\frac{dy}{dx}=\frac{2y}{x}$$

Integrating, we get

$$\ln|y|=2\ln|x|+\ln|C|. \tag{c}$$

$$y=Cx^2$$

Determine C from the condition

$$y_{x=x_0}=y_0, \qquad C=\frac{y_0}{x_0^2}$$

Substituting the value of C thus found into (c), we get the solution (b). The singular point O (0, 0) is a *stable nodal point*.

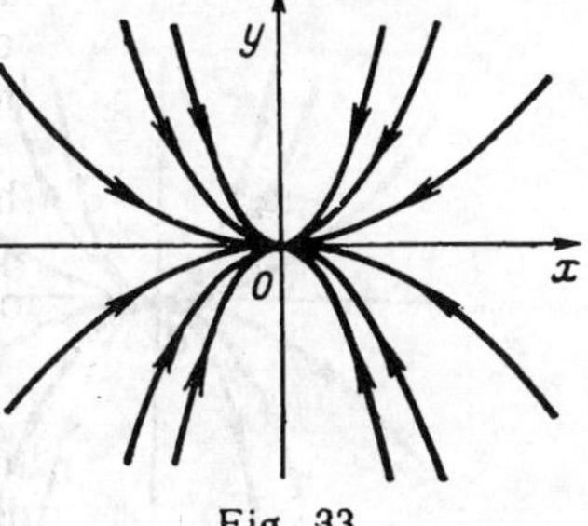

Fig. 33

II. The roots of the auxiliary equation are real, positive and distinct: $\lambda_1>0$, $\lambda_2>0$, $\lambda_1\neq\lambda_2$. In this case, the solutions are also expressed by the formulas (8) and (9). But in this case, for arbitrarily small $|x_0|$ and $|y_0|$ it will be true that $|x(t)|\to\infty$, $|y(t)|\to\infty$ as $t\to+\infty$, since $e^{\lambda_1 t}\to\infty$ and $e^{\lambda_2 t}\to\infty$ as $t\to+\infty$. On the phase plane, the singular point is an *unstable nodal point*: as $t\to+\infty$ the point on the integral curve recedes from the rest point $x=0$, $y=0$.

Example 2. Investigate the stability of the solutions of the system

$$\frac{dx}{dt}=x, \qquad \frac{dy}{dt}=2y$$

Solution. The auxiliary equation is

$$\begin{vmatrix} 1-\lambda & 0 \\ 0 & 2-\lambda \end{vmatrix}=0$$

Its solutions are

$$\lambda_1=1, \quad \lambda_2=2$$

The solution will be

$$x=x_0e^t, \quad y=y_0e^{2t}$$

The solution is unstable, since $|x(t)|\to\infty$, $|y(t)|\to\infty$ as $t\to+\infty$. Eliminating t, we obtain

$$\left(\frac{x}{x_0}\right)^2=\frac{y}{y_0}$$

(Fig. 34). The singular point O (0, 0) is an *unstable nodal point*

III. The roots of the auxiliary equation are real and of unlike sign, for example: $\lambda_1>0$, $\lambda_2<0$. From formulas (9) it follows that for arbitrarily small $|x_0|$ and $|y_0|$, if $cx_0+gy_0-x_0\lambda_2\neq0$, it will be true that $|x(t)| \to \infty$, $|y(t)| \to \infty$ as $t \to +\infty$. The solution is *unstable*. The singular point on the phase plane is termed a *saddle point*.

Example 3. Investigate the stability of the solution of the system

$$\frac{dx}{dt}=x, \quad \frac{dy}{dt}=-2y$$

Solution. The auxiliary equation is

$$\begin{vmatrix} 1-\lambda & 0 \\ 0 & -2-\lambda \end{vmatrix}=0$$

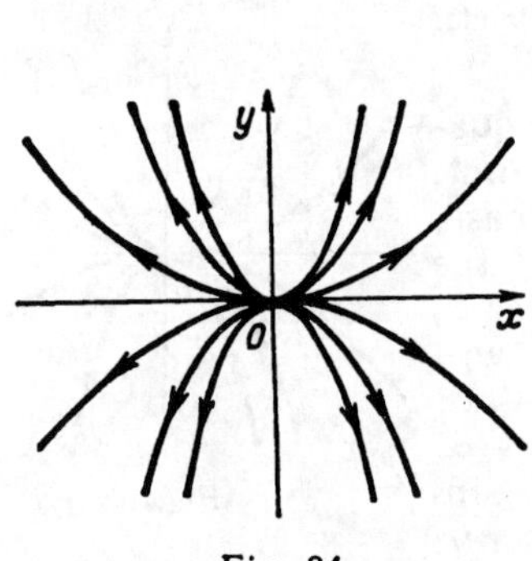

Fig. 34

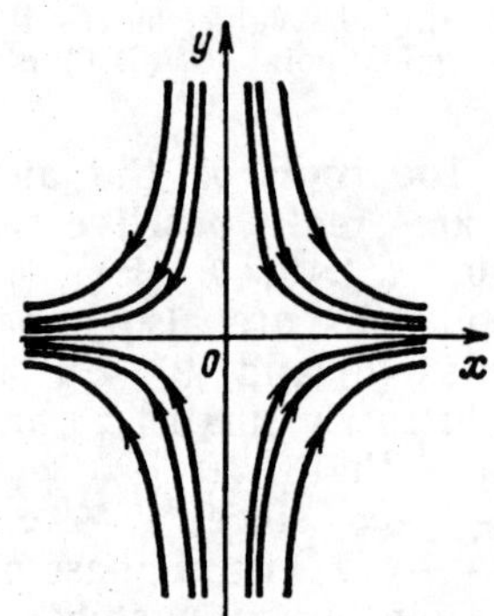

Fig. 35

hence, $\lambda_1=1$, $\lambda_2=-2$. The solution is

$$x=x_0e^{+t}, \quad y=y_0e^{-2t}$$

The solution is unstable. Eliminating the parameter t, we get a family of curves on the phase plane:

$$yx^2=y_0x_0^2$$

The singular point O (0, 0) is a *saddle point* (Fig. 35).

IV. The roots of the auxiliary equation are complex with negative real part: $\lambda_1=\alpha+i\beta$, $\lambda_2=\alpha-i\beta$ $(\alpha<0)$. The solution of system (4) is

$$\left.\begin{aligned} x&=e^{\alpha t}\,[C_1\cos\beta t+C_2\sin\beta t] \\ y&=\frac{1}{g}e^{\alpha t}\,[(\alpha C_1+\beta C_2-cC_1)\cos\beta t+(\alpha C_2-\beta C_1-cC_2)\sin\beta t] \end{aligned}\right\} \quad (15)$$

Introducing the notation

$$C=\sqrt{C_1^2+C_2^2}, \quad \sin\delta=\frac{C_1}{C}, \quad \cos\delta=\frac{C_2}{C}$$

we can rewrite equations (15) as

$$\left.\begin{aligned} x &= Ce^{\alpha t}\sin(\beta t+\delta) \\ y &= \frac{Ce^{\alpha t}}{g}\left[(\alpha-c)\sin(\beta t+\delta)+\beta\cos(\beta t+\delta)\right] \end{aligned}\right\} \qquad (16)$$

where C_1 and C_2 are arbitrary constants that are determined from the initial conditions: $x=x_0$, $y=y_0$ when $t=0$, and

$$x_0=C\sin\delta, \quad y_0=\frac{C}{g}\left[(\alpha-c)\sin\delta+\beta\cos\delta\right]$$

whence we find

$$C_1=x_0, \quad C_2=\frac{gy_0-x_0(\alpha-c)}{\beta} \qquad (17)$$

Note again that if $g=0$, then the form of the solution will be somewhat different, but the nature of the analysis does not change.

It is obvious that for arbitrary $\varepsilon>0$, given sufficiently small $|x_0|$ and $|y_0|$, the following relations will hold:

$$|x(t)|<\varepsilon, \quad |y(t)|<\varepsilon$$

The solution is *stable*. In this case, as $t\to+\infty$,

$$x(t)\to 0 \quad \text{and} \quad y(t)\to 0$$

changing sign an unlimited number of times. The singular point on the phase plane is called a *stable focal point*.

Example 4. Investigate the stability of the solution of the system of equations

$$\frac{dx}{dt}=-x+y$$

$$\frac{dy}{dt}=-x-y$$

Solution. Form the auxiliary equation and find its roots:

$$\begin{vmatrix} -1-\lambda & 1 \\ -1 & -1-\lambda \end{vmatrix}=0, \quad \lambda^2+2\lambda+2=0$$

$$\lambda_{1,2}=-1\pm 1i, \quad \alpha=-1, \ \beta=1$$

We find C_1 and C_2 from formulas (17): $C_1=x_0$, $C_2=y_0$. Substituting into (15), we get

$$\left.\begin{aligned} x &= e^{-t}(x_0\cos t+y_0\sin t) \\ y &= e^{-t}(y_0\cos t-x_0\sin t) \end{aligned}\right\} \qquad (A)$$

It is obvious that for arbitrary values of t,

$$|x|\leqslant|x_0|+|y_0|, \quad |y|\leqslant|x_0|+|y_0|$$

As $t\to+\infty$, $x(t)\to 0$ and $y(t)\to 0$, and the solution is stable.

Let us analyze the arrangement of the curves on the phase plane in this case. We transform expressions (A). Let

$$x_0 = M\cos\delta, \qquad y_0 = M\sin\delta$$

$$M = \sqrt{x_0^2 + y_0^2}, \qquad \tan\delta = \frac{y_0}{x_0}$$

Then equations (A) become

$$\left.\begin{aligned} x &= Me^{-t}\cos(\beta t - \delta) \\ y &= Me^{-t}\sin(\beta t - \delta) \end{aligned}\right\} \tag{B}$$

Pass to polar coordinates ρ and θ in the phase plane and establish the relationship $\rho = f(\theta)$. Equations (B) become

$$\left.\begin{aligned} \rho\cos\theta &= Me^{-t}\cos(\beta t - \delta) \\ \rho\sin\theta &= Me^{-t}\sin(\beta t - \delta) \end{aligned}\right\} \tag{C}$$

Squaring the right and left members and adding, we obtain

$$\rho^2 = M^2 e^{-2t}$$

or

$$\rho = Me^{-t} \tag{D}$$

How does t depend on θ? Dividing the terms of the lower equation of (C) by the corresponding terms of the upper equation, we get

$$\tan\theta = \tan(\beta t - \delta)$$

whence

$$t = \frac{\theta + \delta}{\beta}$$

Substituting into (D), we have

$$\rho = Me^{-\frac{\theta+\delta}{\beta}}$$

or

$$\rho = Me^{-\frac{\delta}{\beta} - \frac{\theta}{\beta}}$$

Putting $Me^{-\frac{\delta}{\beta}} = M_1$, we finally obtain

$$\rho = M_1 e^{-\frac{\theta}{\beta}} \tag{E}$$

This is a family of logarithmic spirals. In this case, a point moving along an integral curve approaches the origin as $t \to \infty$. The singular point $O(0, 0)$ is a *stable focal point*.

V. The roots of the auxiliary equation are complex with positive real part: $\lambda_1 = \alpha + i\beta$, $\lambda_2 = \alpha - i\beta$ $(\alpha > 0)$. Here the solution will also be expressed by the formulas (15), where $\alpha > 0$. For arbitrary initial conditions x_0 and y_0 $(\sqrt{x_0^2 + y_0^2} \neq 0)$, $|x(t)|$ and $|y(t)|$ can assume arbitrarily large values as $t \to +\infty$. The solution is *unstable*. The singular point in the phase plane is called an *unstable focal point*. A point on an integral curve recedes without bound from the origin of coordinates.

Example 5. Investigate the stability of the solution of the system of equations

$$\frac{dx}{dt}=x+y$$

$$\frac{dy}{dt}=-x+y$$

Solution. Form the auxiliary equation

$$\begin{vmatrix} 1-\lambda & 1 \\ -1 & 1-\lambda \end{vmatrix}=0, \quad \lambda^2-2\lambda+2=0$$

$$\lambda_1=1+i, \quad \lambda_2=1-i$$

Taking into account (17), the solution (15) in this case will be

$$x=e^t(x_0 \cos t+y_0 \sin t)$$

$$y=e^t(y_0 \cos t-x_0 \sin t)$$

In the phase plane, we obtain the curve in polar coordinates:

$$\rho=\overline{M}_1 e^{\theta/\beta}$$

The singular point is an *unstable focal point* (Fig. 36).

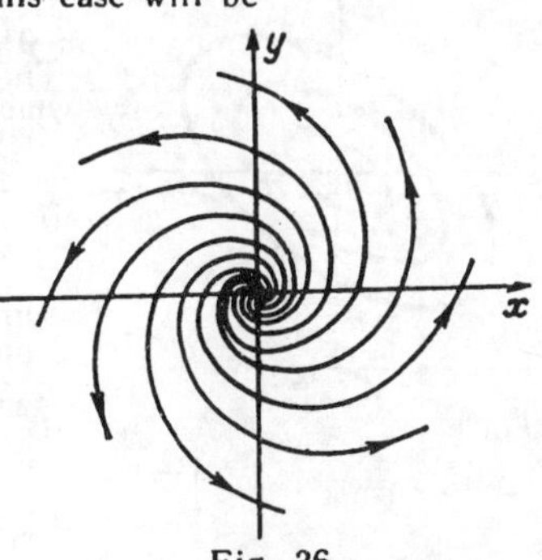

Fig. 36

VI. The roots of the auxiliary equation are pure imaginary: $\lambda_1=i\beta$, $\lambda_2=-i\beta$. The solutions (15) in this case assume the form

$$\left.\begin{aligned} x&=C_1 \cos \beta t+C_2 \sin \beta t \\ y&=\frac{1}{g}[(\beta C_2-cC_1)\cos \beta t+(-\beta C_1-cC_2)\sin \beta t] \end{aligned}\right\} \quad (18)$$

The constants C_1 and C_2 are found from formulas (17):

$$C_1=x_0, \quad C_2=\frac{gy_0+cx_0}{g} \quad (19)$$

Clearly, for arbitrary $\varepsilon>0$ and for all sufficiently small $|x_0|$ and $|y_0|$ it will be true that $|x(t)|<\varepsilon$, $|y(t)|<\varepsilon$ for arbitrary t. The solution is *stable*. Here, x and y are periodic functions of t.

In order to carry out the analysis of the integral curves on the phase plane, it is advisable to write the first equation of (18) as [see (16)]

$$\left.\begin{aligned} x&=C \sin(\beta t+\delta) \\ y&=\frac{C\beta}{g}\cos(\beta t+\delta)-\frac{Cc}{g}\sin(\beta t+\delta) \end{aligned}\right\} \quad (20)$$

where C and δ are arbitrary constants. From the expressions (20) it follows that x and y are periodic functions of t. Eliminate the parameter t from (20):

$$y=\frac{C\beta}{g}\sqrt{1-\frac{x^2}{C^2}}-\frac{c}{g}x$$

Eliminating the radical, we get

$$\left(y-\frac{c}{g}x\right)^2=\left(\frac{C\beta}{g}\right)^2\left(1-\frac{x^2}{C^2}\right) \tag{21}$$

This is a family of quadric curves (which are real) that depend on an arbitrary constant C. Neither of them *recedes to infinity*. Consequently, this is a family of ellipses around the origin (for $c=0$, the axes of the ellipses are parallel to the coordinate axes). The singular point is termed the *centre* (Fig. 37).

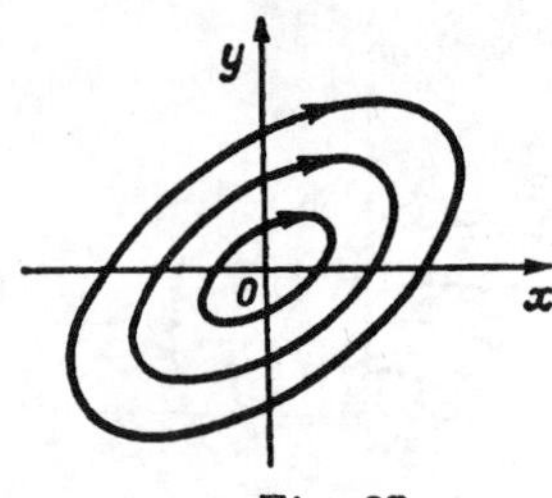

Fig. 37

Example 6. Investigate the stability of the solution of the system of equations

$$\frac{dx}{dt}=y, \quad \frac{dy}{dt}=-4x$$

Solution. Form the auxiliary equation and find its roots:

$$\begin{vmatrix} -\lambda & 1 \\ -4 & -\lambda \end{vmatrix}=0, \quad \lambda^2+4=0, \quad \lambda=\pm 2i$$

The solutions (20) will be

$$x=C\sin(2t+\delta),$$
$$y=2C\cos(2t+\delta)$$

Equation (21) will assume the form

$$y^2=4C^2\left(1-\frac{x^2}{C^2}\right), \quad \frac{y^2}{4C^2}+\frac{x^2}{C^2}=1$$

We have a system of ellipses on the phase plane, and the singular point is a *centre*.

VII. Let $\lambda_1=0$, $\lambda_2<0$. The solution (8) in this case becomes

$$\left.\begin{aligned} x&=C_1+C_2e^{\lambda_2 t} \\ y&=\frac{1}{g}\left[-C_1c+C_2(\lambda_2-c)\,e^{\lambda_2 t}\right] \end{aligned}\right\} \tag{22}$$

Clearly, for arbitrary $\varepsilon>0$ and for sufficiently small $|x_0|$ and $|y_0|$ it will be true that $|x(t)|<\varepsilon$, $|y(t)|<\varepsilon$ when $t>0$. Hence, the solution is *stable*.

Example 7. Investigate the stability of the solution of the system

$$\frac{dx}{dt}=0, \quad \frac{dy}{dt}=-y \tag{α}$$

Solution. We find the roots of the auxiliary equation

$$\begin{vmatrix} -\lambda & 0 \\ 0 & -1-\lambda \end{vmatrix}=0, \quad \lambda^2+\lambda=0, \quad \lambda_1=0,\ \lambda_2=-1$$

Here, $g=0$. The solutions are found directly by solving the system and without using the formulas (22):

$$x=C_1, \quad y=C_2e^{-t} \qquad (\beta)$$

The solution satisfying the initial conditions $x=x_0$, $y=y_0$ when $t=0$ is

$$x=x_0, \quad y=y_0e^{-t} \qquad (\gamma)$$

The solution is clearly *stable*. The differential equation on the phase plane will be $\frac{dx}{dy}=0$. The complete integral will be $x=C$. The integral curves are straight lines parallel to the y-axis.

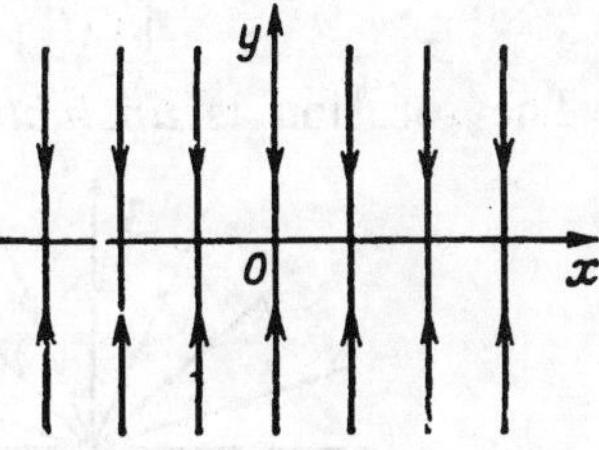

Fig. 38

From the equations (γ) it follows that the points along the integral curves approach the straight line $y=0$ (Fig. 38).

VIII. Let $\lambda_1=0$, $\lambda_2>0$. From formulas (22) or (8′) it follows that the solution is unstable since $|x(t)|+|y(t)|\to\infty$ as $t\to+\infty$.

IX. Let $\lambda_1=\lambda_2<0$. The solution is

$$\left.\begin{aligned} x &= (C_1+C_2t)\,e^{\lambda_1 t} \\ y &= \frac{1}{g}e^{\lambda_1 t}[C_1(\lambda_1-c)+C_2(1+\lambda_1 t-ct)] \end{aligned}\right\} \qquad (23)$$

Since $e^{\lambda_1 t}\to 0$ and $te^{\lambda_1 t}\to 0$ as $t\to+\infty$; then for an arbitrary $\varepsilon>0$ it is possible to choose C_1 and C_2 (by selecting x_0 and y_0) such that it will be true that $|x(t)|<\varepsilon$, $|y(t)|<\varepsilon$ for arbitrary $t>0$. The solution is thus *stable*, and $x(t)\to 0$ and $y(t)\to 0$ as $t\to+\infty$.

Example 8. Investigate the stability of the solution of the system

$$\frac{dx}{dt}=-x, \quad \frac{dy}{dt}=-y$$

Solution. We find the roots of the auxiliary equation:

$$\begin{vmatrix} -1-\lambda & 0 \\ 0 & -1-\lambda \end{vmatrix}=0, \quad (\lambda+1)^2=0, \quad \lambda_1=\lambda_2=-1$$

Here $g=0$. The solution of the system will be of the form (8′):

$$x=C_1e^{-t}, \quad y=C_2e^{-t}$$

and $x\to 0$, $y\to 0$ as $t\to+\infty$. The solution is stable. The family of curves on the phase plane will be

$$\frac{y}{x}=\frac{C_2}{C_1}=k, \quad \text{that is,} \quad y=kx$$

This is a family of straight lines passing through the origin. The points along the integral curves approach the origin. The singular point $O(0, 0)$ is a *nodal point* (Fig. 39).

Note that in the case of $\lambda_1 = \lambda_2 > 0$ the form of the solution (22) is preserved, but when $t \to +\infty$

$$|x(t)| \to \infty \quad \text{and} \quad |y(t)| \to \infty$$

The solution is *unstable*.

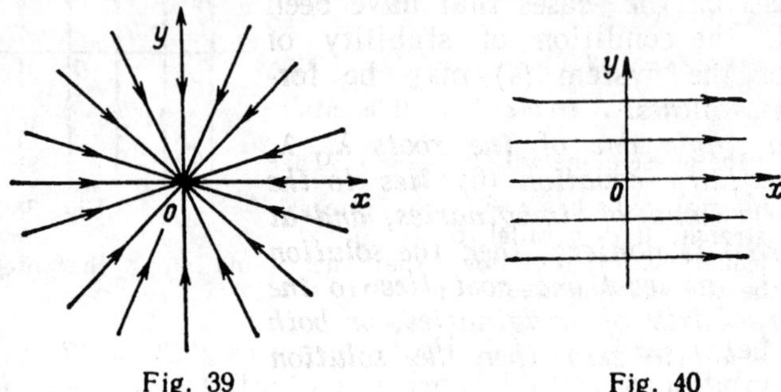

Fig. 39 Fig. 40

X. Let $\lambda_1 = \lambda_2 = 0$. Then

$$\left.\begin{aligned} x &= C_1 + C_2 t \\ y &= \frac{1}{g}[-cC_1 + C_2 - cC_2 t] \end{aligned}\right\} \tag{24}$$

Whence it is quite clear that $x \to \infty$ and $y \to \infty$ as $t \to +\infty$. The solution is *unstable*.

Example 9. Investigate the stability of the solution of the system of equations

$$\frac{dx}{dt} = y, \quad \frac{dy}{dt} = 0$$

Solution. Find the roots of the auxiliary equation

$$\begin{vmatrix} -\lambda & 1 \\ 0 & -\lambda \end{vmatrix} = 0, \quad \lambda_1^2 = 0, \quad \lambda_1 = \lambda_2 = 0$$

We find the solutions to be

$$y = C_2, \quad x = C_2 t + C_1$$

It is quite obvious that $x \to \infty$ as $t \to +\infty$. The solution is *unstable*. The equation on the phase plane is $\frac{dy}{dx} = 0$. The integral curves $y = C$ are straight lines parallel to the axis (Fig. 40). The singular point is called a *degenerate saddle point*.

To give a general criterion of the stability of solution of the system (4), we do as follows.

We write the roots of the auxiliary equation in the form of complex numbers:

$$\lambda_1 = \lambda_1^* + i\lambda_1^{**}$$
$$\lambda_2 = \lambda_2^* + i\lambda_2^{**}$$

(in the case of real roots, $\lambda_1^{**} = 0$ and $\lambda_2^{**} = 0$).

Let us take the plane of a complex variable $\lambda^*\lambda^{**}$ and display the roots of the auxiliary equation by points in this plane. Then, on the basis of the cases that have been considered, the condition of stability of solution of the system (4) may be formulated as follows.

If not a single one of the roots λ_1, λ_2 *of the auxiliary equation* (6) *lies to the right of the axis of imaginaries, and at least one root is nonzero, then the solution is stable; if at least one root lies to the right of the axis of imaginaries, or both roots are equal to zero, then the solution is unstable* (Fig. 41).

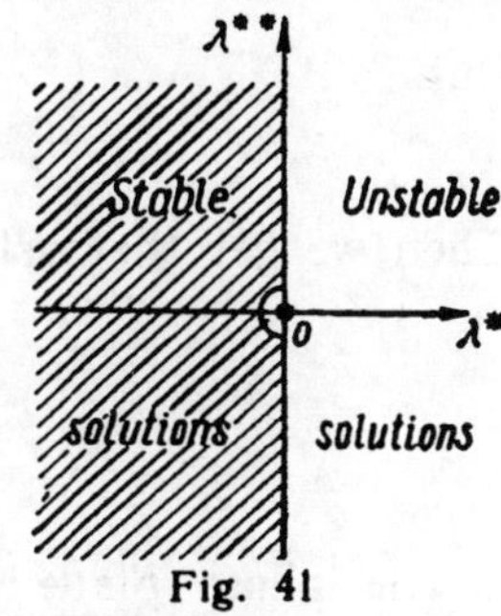

Fig. 41

Let us now consider a more general system of equations:

$$\left.\begin{aligned} \frac{dx}{dt} &= cx + gy + P(x, y) \\ \frac{dy}{dt} &= ax + by + Q(x, y) \end{aligned}\right\} \qquad (25)$$

But for exceptional cases, the solution of this system is not expressible in terms of elementary functions and quadratures.

To establish whether the solutions of this system are stable or unstable, they are compared with the solutions of a linear system. Suppose that for $x \to 0$ and $y \to 0$, the functions $P(x, y)$ and $Q(x, y)$ also approach zero and approach it faster than ρ, where $\rho = \sqrt{x^2 + y^2}$; in other words,

$$\lim_{\rho \to 0} \frac{P(x, y)}{\rho} = 0; \quad \lim_{\rho \to 0} \frac{Q(x, y)}{\rho} = 0$$

Then it may be proved that, save for the exceptional case, the solution of the system (25) will be stable when the solution of the system

$$\left.\begin{aligned} \frac{dx}{dt} &= cx + gy \\ \frac{dy}{dt} &= ax + by \end{aligned}\right\} \qquad (4)$$

is stable, and unstable when the solution of the system (4) is unstable. The exception is that case when both roots of the auxiliary equation lie on the axis of imaginaries; then the question of the stability or instability of solution of the system (25) is considerably more involved.

Lyapunov* investigated the question of the stability of solutions of systems of equations for rather general assumptions concerning the form of these equations.

In oscillation theory, one often has to deal with the equation

$$\frac{d^2x}{dt^2}=f\left(x, \frac{dx}{dt}\right) \tag{26}$$

Put

$$\frac{dx}{dt}=v \tag{27}$$

Then we get the system of equations

$$\left.\begin{aligned}\frac{dx}{dt}&=v\\ \frac{dv}{dt}&=f(x, v)\end{aligned}\right\} \tag{28}$$

The phase plane for this system is the xv-plane. The integral curves on the phase plane geometrically represent the velocity v as a function of the x-coordinate and give a pictorial and qualitative description of the variation of x and v. If the point $x=0$, $v=0$ is a singular point, then it determines a position of equilibrium.

Thus, for example, if the singular point of a system of equations is a centre, that is, the integral curves on the phase plane are closed curves circling the origin, then the motions described by equation (26) are undamped oscillations. If the singular point of the phase plane is a focal point (and then $|x|\to 0$, $|v|\to 0$ as $t\to\infty$), then the motions defined by equation (26) are damped oscillations. If the singular point is a nodal point or a saddle point (and this is the only singular point), then $x\to\pm\infty$ as $t\to\infty$. In this case a moving material point recedes to infinity.

If equation (26) is a linear equation of the form $\frac{d^2x}{dt^2}=ax+b\frac{dx}{dt}$, then the system (28) looks like

$$\frac{dx}{dt}=v, \quad \frac{dv}{dt}=ax+bv$$

This is a system of type (4). The point $x=0$, $v=0$ is a singular point, it defines a position of equilibrium. Note that variable x is not necessarily a mechanical displacement of a point. It may have a variety of physical meanings; for instance, it may represent a quantity describing electrical oscillations.

---

* A. M. Lyapunov, *The General Problem of Stability of Motion*, ONTI, 1935 (in Russian).

1.32 EULER'S METHOD OF APPROXIMATE SOLUTION OF FIRST-ORDER DIFFERENTIAL EQUATIONS

We shall consider two methods of numerical solution of first-order differential equation. In this section we consider **Euler's method.** Find (approximately) the solution of the equation

$$\frac{dy}{dx}=f(x, y) \tag{1}$$

on the interval $[x_0, b]$ that satisfies the initial condition at $x=x_0$ $y=y_0$. Divide the interval $[x_0, b]$ by the points $x_0, x_1, x_2, \ldots, x_n=b$ into n equal parts (here $x_0<x_1<x_2<\ldots<x_n$). Denote $x_1-x_0=$ $=x_2-x_1=\ldots=b-x_{n-1}=\Delta x=h$; hence,

$$h=\frac{b-x_0}{n}$$

Let $y=\varphi(x)$ be some approximate solution of equation (1) and

$$y_0=\varphi(x_0),\quad y_1=\varphi(x_1),\ \ldots,\ y_n=\varphi(x_n)$$

Put

$$\Delta y_0=y_1-y_0,\ \Delta y_1=y_2-y_1,\ \ldots,\ \Delta y_{n-1}=y_n-y_{n-1}$$

At each of the points $x_0, x_1, \ldots, x_n$ in equation (1) we replace the derivative with a ratio of finite differences:

$$\frac{\Delta y}{\Delta x}=f(x, y) \tag{2}$$

$$\Delta y=f(x, y)\,\Delta x \tag{2'}$$

When $x=x_0$ we have

$$\frac{\Delta y_0}{\Delta x}=f(x_0, y_0),\quad \Delta y_0=f(x_0, y_0)\,\Delta x$$

or

$$y_1-y_0=f(x_0, y_0)\,h$$

In this equation, x_0, y_0, h are known; thus we find

$$y_1=y_0+f(x_0, y_0)\,h$$

When $x=x_1$, equation (2′) takes the form

$$\Delta y_1=f(x_1, y_1)\,h$$

or

$$y_2-y_1=f(x_1, y_1)\,h$$
$$y_2=y_1+f(x_1, y_1)\,h$$

Here, x_1, y_1, h are known and y_2 is determined.
Similarly, we find

$$y_3 = y_2 + f(x_2, y_2) h$$
$$\cdots\cdots\cdots$$
$$y_{k+1} = y_k + f(x_k, y_k) h$$
$$\cdots\cdots\cdots$$
$$y_n = y_{n-1} + f(x_{n-1}, y_{n-1}) h$$

We have thus found the approximate values of the solution at the points $x_0, x_1, \ldots, x_n$. Connecting, in a coordinate plane, the points $(x_0, y_0), (x_1, y_1), \ldots, (x_n, y_n)$ by straight-line segments, we get a **polygonal line**—an approximate integral curve (Fig. 42). This line is called *Euler's polygon*.

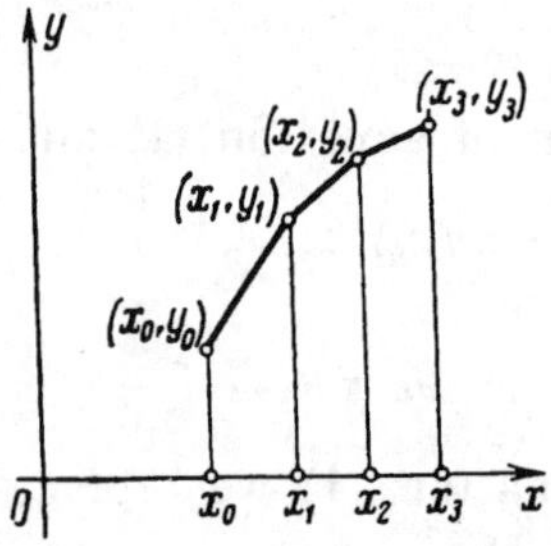

Fig. 42

Note. We denote by $y = \varphi_h(x)$ an approximate solution of equation (1), which corresponds to Euler's polygon when $\Delta x = h$. It may be proved * that if there exists a unique solution $y = \varphi^*(x)$ of equation (1) that satisfies the initial conditions and is defined on the interval $[x_0, b]$, then $\lim\limits_{h \to 0} |\varphi_h(x) - \varphi^*(x)| = 0$ for any x of the interval $[x_0, b]$.

Example. Find the approximate value (for $x = 1$) of the solution of the equation

$$y' = y + x$$

that satisfies the initial condition $y_0 = 1$ for $x_0 = 0$.

Solution. Divide the interval [0, 1] into 10 parts by the points $x_0 = 0$, 0.1, 0.2, ..., 1.0. Hence, $h = 0.1$. We seek the values $y_1, y_2, \ldots, y_n$ by formula (2'):

$$\Delta y_k = (y_k + x_k) h$$

or

$$y_{k+1} = y_k + (y_k + x_k) h$$

We thus get

$$y_1 = 1 + (1 + 0) \cdot 0.1 = 1 + 0.1 = 1.1$$
$$y_2 = 1.1 + (1.1 + 0.1) \cdot 0.1 = 1.22$$
$$\cdots\cdots\cdots\cdots$$

Tabulating the results as we solve, we get:

---

* For the proof see, for example, I. G. Petrovsky's *Lectures on the Theory of Ordinary Differential Equations*.

x_k	y_k	y_k+x_k	$\Delta y_k=(y_k+x_k)h$
$x_0=0$	1.000	1.000	0.100
$x_1=0.1$	1.100	1.200	0.120
$x_2=0.2$	1.220	1.420	0.142
$x_3=0.3$	1.362	1.620	0.162
$x_4=0.4$	1.524	1.924	0.1924
$x_5=0.5$	1.7164	2.2164	0.2216
$x_6=0.6$	1.9380	2.5380	0.2538
$x_7=0.7$	2.1918	2.8918	0.2892
$x_8=0.8$	2.4810	3.2810	0.3281
$x_9=0.9$	2.8091	3.7091	0.3709
$x_{10}=1.0$	3.1800		

We have found the approximate value $y\,|_{x=1}=3.1800$. The exact solution of this equation that satisfies the indicated initial conditions is

$$y=2e^x-x-1$$

Hence,

$$y\,|_{x=1}=2\,(e-1)=3.4366$$

The absolute error is 0.2566; the relative error is $\frac{0.2566}{3.4366}=0.075\approx 8\%$.

1.33 A DIFFERENCE METHOD FOR APPROXIMATE SOLUTION OF DIFFERENTIAL EQUATIONS BASED ON TAYLOR'S FORMULA. ADAMS METHOD

We once again seek the solution of the equation

$$y'=f\,(x,\ y) \tag{1}$$

on the interval $[x_0,\ b]$, which solution satisfies the initial condition $y=y_0$ when $x=x_0$. We introduce notation that will be needed later on. The approximate values of the solution at the points

$$x_0,\ x_1,\ x_2,\ \ldots,\ x_n$$

will be

$$y_0,\ y_1,\ y_2,\ \ldots,\ y_n$$

The first differences, or differences of the first order, are

$$\Delta y_0=y_1-y_0,\quad \Delta y_1=y_2-y_1,\ \ldots,\ \Delta y_{n-1}=y_n-y_{n-1}$$

The second differences, or differences of the second order, are

$$\Delta^2 y_0=\Delta y_1-\Delta y_0=y_2-2y_1+y_0$$
$$\Delta^2 y_1=\Delta y_2-\Delta y_1=y_3-2y_2+y_1$$
$$\ldots\ldots\ldots\ldots\ldots\ldots$$
$$\Delta^2 y_{n-2}=\Delta y_{n-1}-\Delta y_{n-2}=y_n-2y_{n-1}+y_{n-2}$$

Differences of the second differences are called differences of the third order, and so forth. We denote by $y'_0,\ y'_1,\ \ldots,\ y'_n$ the approximate values of the derivatives, and by $y''_0, y''_1, \ldots, y''_n$ the approximate values of the second derivatives, etc. Similarly we determine the first differences of the derivatives:

$$\Delta y'_0 = y'_1 - y'_0, \quad \Delta y'_1 = y'_2 - y'_1, \ldots, \Delta y'_{n-1} = y'_n - y'_{n-1}$$

the second differences of the derivatives:

$$\Delta^2 y'_0 = \Delta y'_1 - \Delta y'_0, \quad \Delta^2 y'_1 = \Delta y'_2 - \Delta y'_1, \ldots, \Delta^2 y'_{n-2} = \Delta y'_{n-1} - \Delta y'_{n-2}$$

and so on.

Write Taylor's formula for solving an equation in the neighbourhood of the point $x = x_0$ [see (6) Sec. 4.6, Vol. I]:

$$y = y_0 + \frac{x - x_0}{1} y'_0 + \frac{(x - x_0)^2}{1 \cdot 2} y''_0 + \ldots + \frac{(x - x_0)^m}{1 \cdot 2 \ldots m} y_0^{(m)} + R_m \quad (2)$$

In this formula y_0 is known, and the values $y'_0, y''_0, \ldots$ of the derivatives are found from equation (1) as follows. Putting the initial values x_0 and y_0 into the right side of equation (1), we find y'_0:

$$y'_0 = f(x_0, y_0)$$

Differentiating the terms of (1) with respect to x, we get

$$y'' = \frac{\partial f}{\partial x} + \frac{\partial f}{\partial y} y' \quad (3)$$

Substituting into the right side the values x_0, y_0, y'_0, we find

$$y''_0 = \left(\frac{\partial f}{\partial x} + \frac{\partial f}{\partial y} y' \right)_{x = x_0,\ y = y_0,\ y' = y'_0}$$

Once more differentiating (3) with respect to x and substituting the values x_0, y_0, y'_0, y''_0, we find y'''_0. Continuing in this fashion, we can find the values of the derivatives of any order for $x = x_0$.* All terms are known, except the remainder R_m in the right side of (2). Thus, neglecting the remainder, we can obtain an approximation of the solution for any value of x; the accuracy will depend upon the quantity $|x - x_0|$ and the number of terms in the expansion.

In the method given below, we determine by formula (2) only the first few values of y when $|x - x_0|$ is small. We determine the values y_1 and y_2 for $x_1 = x_0 + h$ and for $x_2 = x_0 + 2h$, taking four terms of the expansion (y_0 is known from the initial data):

$$y_1 = y_0 + \frac{h}{1} y'_0 + \frac{h^2}{1 \cdot 2} y''_0 + \frac{h^3}{3!} y'''_0 \quad (4)$$

$$y_2 = y_0 + \frac{2h}{1} y'_0 + \frac{(2h)^2}{1 \cdot 2} y''_0 + \frac{(2h)^3}{3!} y'''_0 \quad (4')$$

* From now on we shall assume that the function $f(x, y)$ is differentiable with respect to x and y as many times as is required by the reasoning.

We thus consider known three values* of the function: y_0, y_1, y_2. On the basis of these values and using equation (1), we find

$$y'_0=f(x_0, y_0), \quad y'_1=f(x_1, y_1), \quad y'_2=f(x_2, y_2)$$

Knowing, y'_0, y'_1, y'_2, it is possible to determine $\Delta y'_0$, $\Delta y'_1$, $\Delta^2 y'_0$. Tabulate the results of the computations:

x	y	y'	$\Delta y'$	$\Delta^2 y'$
x_0	y_0	y'_0		
			$\Delta y'_0$	
$x_1=x_0+h$	y_1	y'_1		$\Delta^2 y'_0$
			$\Delta y'_1$	
$x_2=x_0+2h$	y_2	y'_2		
...	...	...	...	...
$x_{k-2}=x_0+(k-2)h$	y_{k-2}	y'_{k-2}		
			$\Delta y'_{k-2}$	
$x_{k-1}=x_0+(k-1)h$	y_{k-1}	y'_{k-1}		$\Delta^2 y'_{k-2}$
			$\Delta y'_{k-1}$	
$x_k=x_0+kh$	y_k	y'_k		

Now suppose that we know the values of the solution

$$y_0, y_1, y_2, \ldots, y_k$$

From these values we can compute [using equation (1)] the values of the derivatives

$$y'_0, y'_1, y'_2, \ldots, y'_k$$

and, hence,

$$\Delta y'_0, \Delta y'_1, \ldots, \Delta y'_{k-1}$$

---

* If we were to seek the solution with greater accuracy, we would have to compute more than the first three values of y. This is dealt with in detail by Ya. S. Bezikovich in *Approximate Calculations*, Gostekhizdat, 1949 (in Russian).

and

$$\Delta^2 y'_0,\ \Delta^2 y'_1,\ \ldots,\ \Delta^2 y'_{k-2}$$

Let us determine the value of y_{k+1} from Taylor's formula setting $a=x_k$, $x=x_{k+1}=x_k+h$:

$$y_{k+1}=y_k+\frac{h}{1}y'_k+\frac{h^2}{1\cdot 2}y''_k+\frac{h^3}{1\cdot 2\cdot 3}y'''_k+\ldots+\frac{h^m}{m!}y_k^{(m)}+R_m$$

In our case we shall confine ourselves to four terms of the expansion:

$$y_{k+1}=y_k+\frac{h}{1}y'_k+\frac{h^2}{1\cdot 2}y''_k+\frac{h^3}{1\cdot 2\cdot 3}y'''_k \tag{5}$$

The unknowns in this formula are y''_k and y'''_k, which we shall try to determine by using the known first and second differences.

First, represent y'_{k-1} in Taylor's formula, putting $a=x_k$, $x-a=-h$:

$$y'_{k-1}=y'_k+\frac{(-h)}{1}y''_k+\frac{(-h)^2}{1\cdot 2}y'''_k \tag{6}$$

and y'_{k-2}, putting $a=x_k$, $x-a=-2h$:

$$y'_{k-2}=y'_k+\frac{(-2h)}{1}y''_k+\frac{(-2h)^2}{1\cdot 2}y'''_k \tag{7}$$

From (6) we find

$$y'_k-y'_{k-1}=\Delta y'_{k-1}=\frac{h}{1}y''_k-\frac{h^2}{1\cdot 2}y'''_k \tag{8}$$

Subtracting the terms of (7) from those of (6), we get

$$y'_{k-1}-y'_{k-2}=\Delta y'_{k-2}=\frac{h}{1}y''_k-\frac{3h^2}{2}y'''_k \tag{9}$$

From (8) and (9) we obtain

$$\Delta y'_{k-1}-\Delta y'_{k-2}=\Delta^2 y'_{k-2}=h^2 y'''_k$$

or

$$y'''_k=\frac{1}{h^2}\Delta^2 y'_{k-2} \tag{10}$$

Putting the expression y'''_k into (8), we get

$$y''_k=\frac{\Delta y'_{k-1}}{h}+\frac{\Delta^2 y'_{k-2}}{2h} \tag{11}$$

Thus, y''_k and y'''_k have been found. Putting expressions (10) and (11) into the expansion (5), we obtain

$$y_{k+1}=y_k+\frac{h}{1}y'_k+\frac{h}{2}\Delta y'_{k-1}+\frac{5h}{12}\Delta^2 y'_{k-2} \tag{12}$$

This is the so-called *Adams formula* with four terms. Formula (12) enables one to compute y_{k+1} when y_k, y_{k-1}, y_{k-2} are known. Thus, knowing y_0, y_1 and y_2 we can find y_3 and, further y_4, y_5, ...

Note 1. We state without proof that if there exists a unique solution of equation (1) on the interval $[x_0, b]$, which solution satisfies the initial conditions, then the error of the approximate values determined from formula (12) does not exceed, in absolute value, Mh^4, where M is a constant dependent on the length of the interval and the form of the function $f(x, y)$ and independent of the magnitude of h.

Note 2. If we want to obtain greater accuracy in our computations, we must take more terms than in expansion (5), and formula (12) will change accordingly. For instance, if in place of formula (5) we take a formula containing five terms to the right, that is, if we complete it with a term of order h^4, then in place of formula (12) we, in similar fashion, get the formula

$$y_{k+1} = y_k + \frac{h}{1} y'_k + \frac{h}{2} \Delta y'_{k-1} + \frac{5h}{12} \Delta^2 y'_{k-2} + \frac{3h}{8} \Delta^3 y'_{k-3}.$$

Here, y_{k+1} is determined by means of the values y_k, y_{k-1}, y_{k-2} and y_{k-3}. Thus, in order to begin computation using this formula we must know the first four values of the solution: y_0, y_1, y_2, y_3. When calculating these values from formulas of type (4) one should take five terms of the expansion.

Example 1. Find approximate values of the solution of the equation

$$y' = y + x$$

that satisfies the initial condition

$$y_0 = 1 \quad \text{when} \quad x_0 = 0$$

Determine the values of the solution for $x = 0.1$, 0.2, 0.3, 0.4.

Solution. First we find y_1 and y_2 using formulas (4) and (4'). From the equation and the initial data we get

$$y'_0 = (y + x)_{x=0} = y_0 + x_0 = 1 + 0 = 1$$

Differentiating the given equation, we have

$$y'' = y' + 1$$

Hence,

$$y''_0 = (y' + 1)_{x=0} = 1 + 1 = 2$$

Differentiating once again, we get

$$y''' = y''$$

Hence,

$$y'''_0 = y''_0 = 2$$

Substituting into (4) the values y_0, y'_0, y''_0 and $h = 0.1$, we get

$$y_1 = 1 + \frac{0.1}{1} \cdot 1 + \frac{(0.1)^2}{1 \cdot 2} \cdot 2 + \frac{(0.1)^3}{1 \cdot 2 \cdot 3} \cdot 2 = 1.1103$$

Similarly, for $h = 0.2$ we have

$$y_2 = 1 + \frac{0.2}{1} \cdot 1 + \frac{(0.2)^2}{1 \cdot 2} \cdot 2 + \frac{(0.2)^3}{1 \cdot 2 \cdot 3} \cdot 2 = 1.2427$$

Knowing y_0, y_1, y_2, we find (on the basis of the equation)

$$y_0' = y_0 + x_0 = 1$$
$$y_1' = y_1 + x_1 = 1.1103 + 0.1 = 1.2103$$
$$y_2' = y_2 + x_2 = 1.2427 + 0.2 = 1.4427$$
$$\Delta y_0' = 0.2103$$
$$\Delta y_1' = 0.2324$$
$$\Delta^2 y_0' = 0.0221$$

Tabulating the values obtained, we have

x	y	y'	$\Delta y'$	$\Delta^2 y'$
$x_0 = 0$	$y_0 = 1.0000$	$y_0' = 1$		
			$\Delta y_0' = 0.2103$	
$x_1 = 0.1$	$y_1 = 1.1103$	$y_1' = 1.2103$		$\Delta^2 y_0' = 0.0221$
			$\Delta y_1' = 0.2324$	
$x_2 = 0.2$	$y_2 = 1.2427$	$y_2' = 1.4427$		$\Delta^2 y_1' = 0.0244$
			$\Delta y_2' = 0.2568$	
$x_3 = 0.3$	$y_3 = 1.3995$	$y_3' = 1.6995$		
$x_4 = 0.4$	$y_4 = 1.5833$			

From formula (12) we find y_3:

$$y_3 = 1.2427 + \frac{0.1}{1} \cdot 1.4427 + \frac{0.1}{2} \cdot 0.2324 + \frac{5 \cdot (0.1)}{12} \cdot 0.0221 = 1.3995$$

We then find the values of y_3', $\Delta y_2'$, $\Delta^2 y_1'$. Again using formula (12) we find y_4.

$$y_4 = 1.3995 + \frac{0.1}{1} \cdot 1.6995 + \frac{0.1}{2} \cdot 0.2568 + \frac{5}{12} \cdot 0.1 \cdot 0.0244 = 1.5833$$

The exact expression of the solution of the given equation is

$$y = 2e^x - x - 1$$

Hence, $y_{x=0.4}=2e^{0.4}-0.4-1=1.58364$. The absolute error is 0.0003; the relative error, $\frac{0.0003}{1.5836}\approx 0.0002=0.02\%$ (In Euler's method, the absolute error of y_4 is 0.06, the relative error, $0.038=3.8\%$.)

Example 2. Find approximate values of the solution of the equation

$$y'=y^2+x^2$$

that satisfies the initial condition $y_0=0$ for $x_0=0$. Determine the values of the solution for $x=0.1$, 0.2, 0.3, 0.4.

Solution. We find

$$y_0'=0^2+0^2=0$$

$$y''_{x=0}=(2yy'+2x)_{x=0}=0$$

$$y'''_{x=0}=(2y'^2+2yy''+2)_{x=0}=2$$

By formulas (4) and (4′) we have

$$y_1=\frac{(0.1)^3}{3!}\cdot 2=0.0003,\qquad y_2=\frac{(0.2)^3}{3!}\cdot 2=0.0027$$

From the equation we find

$$y_0'=0,\qquad y_1'=0.0100,\qquad y_2'=0.0400$$

Using these data, we construct the first rows of the table, and then determine the values of y_3 and y_4 from formula (12).

x	y	y'	$\Delta y'$	$\Delta^2 y'$
$x_0=0$	$y_0=0$	$y_0'=0$		
			$\Delta y_0'=0.0100$	
$x_1=0.1$	$y_1=0.0003$	$y_1'=0.0100$		$\Delta^2 y_0'=0.0200$
			$\Delta y_1'=0.0300$	
$x_2=0.2$	$y_2=0.0027$	$y_2'=0.0400$		$\Delta^2 y_1'=0.0201$
			$\Delta y_2'=0.0501$	
$x_3=0.3$	$y_3=0.0090$	$y_3'=0.0901$		
$x_4=0.4$	$y_4=0.0204$			

Thus,

$$y_3=0.0027+\frac{0.1}{1}\cdot 0.0400+\frac{0.1}{2}\cdot 0.0300+\frac{5}{12}\cdot 0.1\cdot 0.0200=0.0090$$

$$y_4=0.0090+\frac{0.1}{1}\cdot 0.0901+\frac{0.1}{2}\cdot 0.0501+\frac{5}{12}\cdot 0.1\cdot 0.0201=0.0214$$

Note that, to four places, $y_4=0.0213$. (This may be obtained by other, more precise, methods with error estimation.)

1.34 AN APPROXIMATE METHOD FOR INTEGRATING SYSTEMS OF FIRST-ORDER DIFFERENTIAL EQUATIONS

The methods of approximate integration of differential equations considered in Secs. 1.32 and 1.33 are also applicable for solving systems of first-order differential equations. Here, we consider the difference method for solving systems of equations. We will be concerned with systems of two equations in two unknown functions.

It is required to find the solutions of a system of equations

$$\frac{dy}{dx}=f_1(x, y, z) \quad (1)$$

$$\frac{dz}{dx}=f_2(x, y, z) \quad (2)$$

that satisfy the initial conditions $y=y_0$, $z=z_0$ when $x=x_0$.

We determine the values of the functions y and z for values of the argument $x_0, x_1, x_2, \ldots, x_k, x_{k+1}, \ldots, x_n$. Again, let

$$x_{k+1}-x_k=\Delta x=h \quad (k=0, 1, 2, \ldots, n-1) \quad (3)$$

We denote the approximate values of the functions y and z, respectively, by

$$y_0, y_1, \ldots, y_k, y_{k+1}, \ldots, y_n$$

and

$$z_0, z_1, \ldots, z_k, z_{k+1}, \ldots, z_n$$

Write the recurrence formulas of type (12), Sec. 1.33:

$$y_{k+1}=y_k+\frac{h}{1}y'_k+\frac{h}{2}\Delta y'_{k-1}+\frac{5}{12}h\Delta^2 y'_{k-2} \quad (4)$$

$$z_{k+1}=z_k+\frac{h}{1}z'_k+\frac{h}{2}\Delta z'_{k-1}+\frac{5}{12}h\Delta^2 z'_{k-2} \quad (5)$$

To begin computations using these formulas we must know y_1, y_2; z_1, z_2 in addition to y_0 and z_0; we find these values from formulas of type (4) and (4′), Sec. 1.33:

$$y_1=y_0+\frac{h}{1}y'_0+\frac{h^2}{2}y''_0+\frac{h^3}{3!}y'''_0$$

$$y_2 = y_0 + \frac{2h}{1} y_0' + \frac{(2h)^2}{2} y_0'' + \frac{(2h)^3}{3!} y_0'''$$

$$z_1 = z_0 + \frac{h}{1} z_0' + \frac{h^2}{2} z_0'' + \frac{h^3}{3!} z_0'''$$

$$z_2 = z_0 + \frac{2h}{1} z_0' + \frac{(2h)^2}{2} z_0'' + \frac{(2h)^3}{3!} z_0'''$$

To apply these formulas, one has to know y_0', y_0'', y_0''', z_0', z_0'', z_0''', which we shall now determine. From (1) and (2) we find

$$y_0' = f_1(x_0, y_0, z_0)$$
$$z_0' = f_2(x_0, y_0, z_0)$$

Differentiating (1) and (2) and substituting the values of x_0, y_0, z_0, y_0' and z_0', we find

$$y_0'' = (y'')_{x=x_0} = \left(\frac{\partial f_1}{\partial x} + \frac{\partial f_1}{\partial y} y' + \frac{\partial f_1}{\partial z} z'\right)_{x=x_0}$$

$$z_0'' = (z'')_{x=x_0} = \left(\frac{\partial f_2}{\partial x} + \frac{\partial f_2}{\partial y} y' + \frac{\partial f_2}{\partial z} z'\right)_{x=x_0}$$

Differentiating once again, we find y_0''' and z_0'''. Knowing y_1, y_2, z_1, z_2, we find from the given equations (1) and (2),

$$y_1',\ y_2',\ z_1',\ z_2',\ \Delta y_0',\ \Delta y_1',\ \Delta^2 y_0',\ \Delta z_0',\ \Delta z_1',\ \Delta^2 z_0'$$

after which we can fill in the first five rows of the table:

x	y	y'	$\Delta y'$	$\Delta^2 y'$	z	z'	$\Delta z'$	$\Delta^2 z'$
x_0	y_0	y_0'			z_0	z_0'		
			$\Delta y_0'$				$\Delta z_0'$	
x_1	y_1	y_1'		$\Delta^2 y_0'$	z_1	z_1'		$\Delta^2 z_0'$
			$\Delta y_1'$				$\Delta z_1'$	
x_2	y_2	y_2'		$\Delta^2 y_1'$	z_2	z_2'		$\Delta^2 z_1'$
			$\Delta y_2'$				$\Delta z_2'$	
x_3	y_3	y_3'			z_3	z_3'		

From formulas (4) and (5) we find y_3 and z_3, and from equations (1) and (2) we find y'_3 and z'_3. Computing $\Delta y'_2$, $\Delta^2 y'_1$, $\Delta z'_2$, $\Delta^2 z'_1$, we find y_4, and z_4, etc., by applying formulas (4) and (5) once again.

Example. Approximate the solutions of the system

$$y' = z, \qquad z' = y$$

with initial conditions $y_0 = 0$ and $z_0 = 1$ for $x = 0$. Compute the values of the solutions for $x = 0, 0.1, 0.2, 0.3, 0.4$.

Solution. From the given equations, we find

$$y'_0 = z_{x=0} = 1$$
$$z'_0 = y_{x=0} = 0$$

Differentiating the given equations, we find

$$y''_0 = (y'')_{x=0} = (z')_{x=0} = 0$$
$$z''_0 = (z'')_{x=0} = (y')_{x=0} = 1$$
$$y'''_0 = (y''')_{x=0} = (z'')_{x=0} = 1$$
$$z'''_0 = (z''')_{x=0} = (y'')_{x=0} = 0$$

Using formulas of type (4) and (5), we find

$$y_1 = 0 + \frac{0.1}{1} \cdot 1 + \frac{(0.1)^2}{1 \cdot 2} \cdot 0 + \frac{(0.1)^3}{3!} \cdot 1 = 0.1002$$

$$y_2 = 0 + \frac{0.2}{1} \cdot 1 + \frac{(0.2)^2}{1 \cdot 2} \cdot 0 + \frac{(0.2)^3}{3!} \cdot 1 = 0.2013$$

$$z_1 = 1 + \frac{0.1}{1} \cdot 0 + \frac{(0.1)^2}{1 \cdot 2} \cdot 1 + \frac{(0.1)^3}{3!} \cdot 0 = 1.0050$$

$$z_2 = 1 + \frac{0.2}{1} \cdot 0 + \frac{(0.2)^2}{2!} \cdot 1 + \frac{(0.2)^3}{3!} \cdot 0 = 1.0200$$

Using the given equations, we find

$$y'_1 = 1.0050, \qquad z'_1 = 0.1002$$
$$y'_2 = 1.0200, \qquad z'_2 = 0.2013$$
$$\Delta y'_0 = 0.0050, \qquad \Delta z'_0 = 0.1002$$
$$\Delta y'_1 = 0.0150, \qquad \Delta z'_1 = 0.1011$$
$$\Delta^2 y'_0 = 0.0100, \qquad \Delta^2 z'_0 = 0.0009$$

and fill in the first five rows of the table (see p. 145).

From formulas (4) and (5) we find

$$y_3 = 0.2013 + \frac{0.1}{1} \cdot 1.0200 + \frac{0.1}{2} \cdot 0.0150 + \frac{5}{12} \cdot 0.1 \cdot 0.0100 = 0.3045$$

$$z_3 = 1.0200 + \frac{0.1}{1} \cdot 0.2013 + \frac{0.1}{2} \cdot 0.1011 + \frac{5}{12} \cdot 0.1 \cdot 0.0009 = 1.0452$$

x	y	y'	$\Delta y'$	$\Delta^2 y'$
$x_0=0$	$y_0=0$	$y_0'=1$		
			$\Delta y_0'=0.0050$	
$x_1=0.1$	$y_1=0.1002$	$y_1'=1.0050$		$\Delta^2 y_0'=0.0100$
			$\Delta y_1'=0.0150$	
$x_2=0.2$	$y_2=0.2013$	$y_2'=1.0200$		$\Delta^2 y_1'=0.0102$
			$\Delta y_2'=0.0252$	
$x_3=0.3$	$y_3=0.3045$	$y_3'=1.0452$		
$x_4=0.4$	$y_4=0.4107$			

x	z	z'	$\Delta z'$	$\Delta^2 z'$
$x_0=0$	$z_0=1$	$z_0'=0$		
			$\Delta z_0'=0.1002$	
$x_1=0.1$	$z_1=1.0050$	$z_1'=0.1002$		$\Delta^2 z_0'=0.0009$
			$\Delta z_1'=0.1014$	
$x_2=0.2$	$z_2=1.0200$	$z_2'=0.2013$		$\Delta^2 z_1'=0.0021$
			$\Delta z_2'=0.1032$	
$x_3=0.3$	$z_3=1.0452$	$z_3'=0.3045$		
$x_4=0.4$	$z_4=1.0809$			

and similarly

$$y_4=0.3045+\frac{0.1}{1}\cdot1.0452+\frac{0.1}{2}\cdot0.0252+\frac{5}{12}\cdot0.1\cdot0.0102=0.4107$$

$$z_4=1.0452+\frac{0.1}{1}\cdot0.3045+\frac{0.1}{2}\cdot0.1032+\frac{5}{12}\cdot0.1\cdot0.0021=1.0809$$

It is obvious that the exact solutions of the system of equations (the solutions satisfying the initial conditions) will be

$$y=\sinh x,\quad z=\cosh x$$

And so, solutions correct to the fifth decimal place are

$$y_4=\sinh 0.4=0.41075,\quad z_4=\cosh 0.4=1.08107$$

Note. Since higher-order equations and systems of equations in many cases reduce to a system of first-order equations, the method given above is applicable to the solution of such problems.

Exercises on Chapter 1

Show that the indicated functions, which depend on arbitrary constants, satisfy the corresponding differential equations:

Functions	Differential Equations
1. $y=\sin x-1+Ce^{-\sin x}$,	$\frac{dy}{dx}+y\cos x=\frac{1}{2}\sin 2x.$
2. $y=Cx+C-C^2$,	$\left(\frac{dy}{dx}\right)^2-\frac{dy}{dx}-x\frac{dy}{dx}+y=0.$
3. $y^2=2Cx+C^2$,	$y\left(\frac{dy}{dx}\right)^2+2x\frac{dy}{dx}-y=0.$
4. $y^2=Cx^2-\frac{a^2C}{1+C}$,	$xy\left[1-\left(\frac{dy}{dx}\right)^2\right]=(x^2-y^2-a^2)\frac{dy}{dx}.$
5. $y=C_1x+\frac{C_2}{x}+C_3$,	$\frac{d^3y}{dx^3}+\frac{3}{x}\frac{d^2y}{dx^2}=0.$
6. $y=(C_1+C_2x)e^{kx}+\frac{e^x}{(k-1)^2}$,	$\frac{d^2y}{dx^2}-2k\frac{dy}{dx}+k^2y=e^x.$
7. $y=C_1e^{a\arcsin x}+C_2e^{-a\arcsin x}$,	$(1-x^2)\frac{d^2y}{dx^2}-x\frac{dy}{dx}-a^2y=0.$
8. $y=\frac{C_1}{x}+C_2$,	$\frac{d^2y}{dx^2}+\frac{2}{x}\frac{dy}{dx}=0.$

Integrate the differential equations with variables separable:

9. $y\,dx-x\,dy=0$. *Ans.* $y=Cx$. **10.** $(1+u)v\,du+(1-v)u\,dv=0$. *Ans.* $\ln uv+u-v=C$. **11.** $(1+y)\,dx-(1-x)\,dy=0$. *Ans.* $(1+y)(1-x)=C$. **12.** $(t^2-xt^2)\frac{dx}{dt}+x^2+tx^2=0$. *Ans.* $\frac{t+x}{tx}+\ln\frac{x}{t}=C$. **13.** $(y-a)\,dx+x^2dy=0$. *Ans.* $(y-a)=Ce^{\frac{1}{x}}$. **14.** $z\,dt-(t^2-a^2)\,dz=0$. *Ans.* $z^{2a}=C\frac{t-a}{t+a}$. **15.** $\frac{dx}{dy}=\frac{1+x^2}{1+y^2}$. *Ans.* $x=\frac{y+C}{1-Cy}$. **16.** $(1+s^2)\,dt-\sqrt{t}\,ds=0$. *Ans.* $2\sqrt{t}-$

$-\arctan s=C$. **17.** $d\rho+\rho\tan\theta\,d\theta=0$. *Ans.* $\rho=C\cos\theta$. **18.** $\sin\theta\cos\varphi\,d\theta-\cos\theta\sin\varphi\,d\varphi=0$. *Ans.* $\cos\varphi=C\cos\theta$. **19.** $\sec^2\theta\tan\varphi\,d\theta+\sec^2\varphi\tan\theta\,d\varphi=0$. *Ans.* $\tan\theta\tan\varphi=C$. **20.** $\sec^2\theta\tan\varphi\,d\varphi+\sec^2\varphi\tan\theta\,d\theta=0$. *Ans.* $\sin^2\theta+\sin^2\varphi=C$. **21.** $(1+x^2)\,dy-\sqrt{1-y^2}\,dx=0$. *Ans.* $\arcsin y-\arctan x=C$. **22.** $\sqrt{1-x^2}\,dy-\sqrt{1-y^2}\,dx=0$ *Ans.* $y\sqrt{1-x^2}-x\sqrt{1-y^2}=C$. **23.** $3e^x\tan y\,dx+(1-e^x)\sec^2 y\,dy=0$. *Ans.* $\tan y=C(1-e^x)^3$. **24.** $(x-y^2x)\,dx+(y-x^2y)\,dy=0$. *Ans.* $x^2+y^2=x^2y^2+C$.

Problems in Setting up Differential Equations

25. Prove that a curve having the slope of the tangent to any point proportional to the abscissa of the point of tangency is a parabola. *Ans.* $y=ax^2+C$.

26. Find a curve passing through the point (0, —2) such that the slope of the tangent at any point of it is equal to the ordinate of this point increased by three units. *Ans.* $y=e^x-3$.

27. Find a curve passing through the point (1, 1) so that the slope of the tangent to the curve at any point is proportional to the square of the ordinate of this point. *Ans.* $k(x-1)y-y+1=0$.

28. Find a curve for which the slope of the tangent at any point is n times the slope of a straight line connecting this point with the origin. *Ans.* $y=Cx^n$.

29. Through the point (2, 1) draw a curve for which the tangent at any point coincides with the direction of the radius vector drawn from the origin to the same point. *Ans.* $y=\frac{1}{2}x$.

30. In polar coordinates, find the equation of a curve at each point of which the tangent of the angle between the radius vector and the tangent line is equal to the reciprocal of the radius vector with sign reversed. *Ans.* $r(\theta+C)=1$.

31. In polar coordinates, find the equation of a curve at each point of which the tangent of the angle formed by the radius vector and the tangent line is equal to the square of the radius vector. *Ans.* $r^2=2(\theta+C)$.

32. Prove that a curve with the property that all its normals pass through a constant point is a circle.

33. Find a curve such that at each point of it the length of the subtangent is equal to the doubled abscissa. *Ans.* $y=C\sqrt{x}$.

34. Find a curve for which the radius vector is equal to the length of the tangent between the point of tangency and the x-axis.

Solution. By hypothesis, $\left\|\frac{y}{y'}\right\|\sqrt{1+y'^2}=\sqrt{x^2+y^2}$, whence $\frac{dy}{y}=\pm\frac{dx}{x}$. Integrating, we get two families of curves: $y=Cx$ and $y=\frac{C}{x}$.

35. By Newton's law, the rate of cooling of some body in air is proportional to the difference between the temperature of the body and the temperature of the air. If the temperature of the air is 20°C and the body cools for 20 minutes from 100° to 60°C, how long will it take for its temperature to drop to 30°C?

Solution. The differential equation of the problem is $\frac{dT}{dt}=k(T-20)$. Integrating we find: $T-20=Ce^{kt}$; $T=100$ when $t=0$, $T=60$ when $t=20$; therefore, $C=80$, $40=Ce^{20k}$, $e^k=\left(\frac{1}{2}\right)^{\frac{1}{20}}$; consequently, $T=20+80\left(\frac{1}{2}\right)^{\frac{t}{20}}$. Assuming $T=30$, we find $t=60$ min.

36. During what time T will the water flow out of an opening 0.5 cm² at the bottom of a conic funnel 10 cm high with the vertex angle $d=60°$?

Solution. In two ways we calculate the volume of water that will flow out during the time between the instants t and $t+\Delta t$. Given a constant rate v, during 1 sec a cylinder of water with base 0.5 cm² and altitude h flows out, and during time Δt the outflow is the volume of water dv equal to

$$-dv=-0.5\,v\,dt=-0.3\sqrt{2gh}dt^{*}$$

On the other hand, due to the outflow, the height of the water receives a negative "increment" dh, and the differential of the volume of water outflow is

$$-dv=\pi r^2 dh=\frac{\pi}{3}(h+0{,}7)^2\,dh$$

Thus,

$$\frac{\pi}{3}(h+0.7)^2 dh=-0.3\sqrt{2gh}\,dt$$

whence

$$t=0.0315\,(10^{5/2}-h^{5/2})+0.0732\,(10^{3/2}-h^{3/2})+0.078\,(\sqrt{10}-\sqrt{h})$$

Setting $h=0$, we get the time of outflow $T=12.5$ sec.

37. The retarding action of friction on a disk rotating in a liquid is proportional to the angular velocity of rotation ω. Find the dependence of this angular velocity on the time if it is known that the disk begins rotating at 100 revolutions per minute and, after the elapse of one minute, rotates at 60 revolutions per minute. *Ans.* $\omega=100\left(\frac{3}{5}\right)^t$ rpm.

38. Suppose that in a vertical column of air the pressure at each level is due to the pressure of the above-lying layers. Find the dependence of the pressure on the height if it is known that at sea level this pressure is 1 kg per cm², while at 500 m above sea level it is 0.92 kg per cm².

Hint. Take advantage of the Boyle-Mariotte law, by virtue of which the density of a gas is proportional to the pressure. The differential equation of the problem is $dp=-kp\,dh$, whence $p=e^{-0.00017h}$. *Ans.* $p=e^{-0.00017h}$.

Integrate the following homogeneous differential equations:

39. $(y-x)\,dx+(y+x)\,dy=0$. *Ans.* $y^2+2xy-x^2=C$. **40.** $(x+y)\,dx+x\,dy=0$. *Ans.* $x^2+2xy=C$. **41.** $(x+y)\,dx+(y-x)\,dy=0$. *Ans.* $\ln(x^2+y^2)^{1/2}-\arctan\frac{y}{x}=C$. **42.** $x\,dy-y\,dx=\sqrt{x^2+y^2}\,dx$. *Ans.* $1+2Cy-C^2x^2=0$. **43.** $(8y+10x)\,dx+(5y+7x)\,dy=0$. *Ans.* $(x+y)^2(2x+y)^3=C$. **44.** $(2\sqrt{st}-s)\,dt+t\,ds=0$. *Ans.* $te^{\sqrt{\frac{s}{t}}}=C$ or $s=t\ln^2\frac{C}{t}$. **45.** $(t-s)\,dt+t\,ds=0$. *Ans.* $te^{\frac{s}{t}}=C$ or $s=t\ln\frac{C}{t}$. **46.** $xy^2\,dy=(x^3+y^3)\,dx$. *Ans.* $y=x\sqrt[3]{3\ln Cx}$. **47.** $x\cos\frac{y}{x}(y\,dx+x\,dy)=y\sin\frac{y}{x}(x\,dy-y\,dx)$. *Ans.* $xy\cos\frac{y}{x}=C$.

Integrate the differential equations that lead to homogeneous equations:

48. $(3y-7x+7)\,dx-(3x-7y-3)\,dy=0$. *Ans.* $(x+y-1)^5(x-y-1)^2=C$. **49.** $(x+2y+1)\,dx-(2x+4y+3)\,dy=0$. *Ans.* $\ln(4x+8y+5)+8y-4x=C$. **50.** $(x+2y+1)\,dx-(2x-3)\,dy=0$. *Ans.* $\ln(2x-3)-\frac{4y+5}{2x-3}=C$.

---

* The rate of outflow v of water from an opening a distance h from the free surface is given by the formula $v=0.6\sqrt{2gh}$, where g is the acceleration of gravity.

51. Determine the curve whose subnormal is the arithmetic mean between the abscissa and the ordinate. *Ans.* $(x-y)^2(x+2y)=C$.

52. Determine the curve in which the ratio of the y-intercept of the tangent to the radius vector is equal to a constant.

Solution. By hypothesis, $\dfrac{y-x\dfrac{dy}{dx}}{\sqrt{x^2+y^2}}=m$, whence $\left(\dfrac{x}{C}\right)^m-\left(\dfrac{C}{x}\right)^m=\dfrac{2y}{x}$.

53. Determine the curve in which the ratio of the x-intercept of the normal to the radius vector is equal to a constant.

Solution. It is given that $\dfrac{x+y\dfrac{dy}{dx}}{\sqrt{x^2+y^2}}=m$, whence $x^2+y^2=m^2(x-C)^2$.

54. Determine the curve in which the y-intercept of the tangent line is equal to $a\sec\theta$, where θ is the angle between the radius vector and the x-axis.

Solution. Since $\tan\theta=\dfrac{y}{x}$ and by hypothesis

$$y-x\frac{dy}{dx}=a\sec\theta$$

we obtain

$$y-x\frac{dy}{dx}=a\frac{\sqrt{x^2+y^2}}{x}$$

whence

$$y=\frac{x}{2}\left[e^{\frac{a}{x}+b}-e^{-\left(\frac{a}{x}+b\right)}\right]$$

55. Determine the curve for which the y-intercept of the normal drawn to some point of the curve is equal to the distance of this point from the origin.

Solution. The y-intercept of the normal is $y+\dfrac{x}{y'}$; therefore, by hypothesis, we have

$$y+\frac{x}{y'}=\sqrt{x^2+y^2}$$

whence

$$x^2=C(2y+C)$$

56. Find the shape of a mirror such that all rays emerging from a single point O are reflected parallel to the given direction.

Solution. For the x-axis we take the given direction, O is the origin. Let OM be the incident ray, MP the reflected ray, and MQ the normal to the desired curve:

$$\alpha=\beta, \qquad OM=OQ, \qquad NM=y$$

$$NQ=NO+OQ=-x+\sqrt{x^2+y^2}=y\cot\beta=y\frac{dy}{dx}$$

whence

$$y\,dy=(-x+\sqrt{x^2+y^2})\,dx$$

Integrating, we have

$$y^2=C^2+2Cx$$

Integrate the following linear differential equations:

57. $y'-\frac{2y}{x+1}=(x+1)^3$. *Ans.* $2y=(x+1)^4+C(x+1)^2$. 58. $y'-a\frac{y}{x}=\frac{x+1}{x}$. *Ans.* $y=Cx^a+\frac{x}{1-a}-\frac{1}{a}$. 59. $(x-x^3)y'+(2x^2-1)y-ax^3=0$. *Ans.* $y=ax+$ $+Cx\sqrt{1-x^2}$. 60. $\frac{ds}{dt}\cos t+s\sin t=1$. *Ans.* $s=\sin t+C\cos t$. 61. $\frac{ds}{dt}+s\cos t=$ $=\frac{1}{2}\sin 2t$. *Ans.* $s=\sin t-1+Ce^{-\sin t}$. 62. $y'-\frac{n}{x}y=e^x x^n$. *Ans.* $y=x^n(e^x+C)$.
63. $y'+\frac{n}{x}y=\frac{a}{x^n}$. *Ans.* $x^n y=ax+C$. 64. $y'+y=\frac{1}{e^x}$. *Ans.* $e^x y=x+C$.
65. $y'+\frac{1-2x}{x^2}y-1=0$. *Ans.* $y=x^2\left(1+Ce^{\frac{1}{x}}\right)$.

Integrate the Bernoulli equations:

66. $y'+xy=x^3y^3$. *Ans.* $y^2(x^2+1+Ce^{x^2})=1$. 67. $(1-x^2)y'-xy-axy^2=0$. *Ans.* $(C\sqrt{1-x^2}-a)y=1$. 68. $3y^2y'-ay^3-x-1=0$. *Ans.* $a^2y^3=Ce^{ax}-$ $-a(x+1)-1$. 69. $y'(x^2y^3+xy)=1$. *Ans.* $x\left[(2-y^2)e^{\frac{y^2}{2}}+C\right]=e^{\frac{y^2}{2}}$.
70. $(y\ln x-2)y\,dx=x\,dy$. *Ans.* $y(Cx^2+\ln x^2+1)=4$. 71. $y-y'\cos x=$ $=y^2\cos x(1-\sin x)$. *Ans.* $y=\frac{\tan x+\sec x}{\sin x+C}$.

Integrate the following exact differential equations:

72. $(x^2+y)\,dx+(x-2y)\,dy=0$. *Ans.* $\frac{x^3}{3}+yx-y^2=C$. 73. $(y-3x^2)\,dx-$ $-(4y-x)\,dy=0$. *Ans.* $2y^2-xy+x^3=C$. 74. $(y^3-x)y'=y$. *Ans.* $y^4=4xy+C$.
75. $\left[\frac{y^2}{(x-y)^2}-\frac{1}{x}\right]dx+\left[\frac{1}{y}-\frac{x^2}{(x-y)^2}\right]dy=0$. *Ans.* $\ln\frac{y}{x}-\frac{xy}{x-y}=C$.
76. $2(3xy^2+2x^3)\,dx+3(2x^2y+y^2)\,dy=0$. *Ans.* $x^4+3x^2y^2+y^3=C$.
77. $\frac{x\,dx+(2x+y)\,dy}{(x+y)^2}=0$. *Ans.* $\ln(x+y)-\frac{x}{x+y}=C$. 78. $\left(\frac{1}{x^2}+\frac{3y^2}{x^4}\right)dx=\frac{2y\,dy}{x^3}$. *Ans.* $x^2+y^2=Cx^3$. 79. $\frac{x^2\,dy-y^2\,dx}{(x-y)^2}=0$. *Ans.* $\frac{xy}{x-y}=C$. 80. $x\,dx+y\,dy=$ $=\frac{y\,dx-x\,dy}{x^2+y^2}$. *Ans.* $x^2+y^2-2\arctan\frac{x}{y}=C$.

81. Determine the curve that has the property that the product of the square of the distance of any point of it from the origin into the x-intercept of the normal at this point is equal to the cube of the abscissa of this point. *Ans.* $y^2(2x^2+y^2)=C$.

82. Find the envelope of the following families of lines: (a) $y=Cx+C^2$. *Ans.* $x^2+4y=0$. (b) $y=\frac{x}{C}+C^2$. *Ans.* $27x^2=4y^3$. (c) $\frac{x}{C}-\frac{y}{C^3}=2$. *Ans.* $27y=x^3$. (d) $C^2x+Cy-1=0$. *Ans.* $y^2+4x=0$. (e) $(x-C)^2+(y-C)^2=C^2$. *Ans.* $x=0$, $y=0$. (f) $(x-C)^2+y^2=4C$. *Ans.* $y^2=4x+4$. (g) $(x-C)^2+(y-C)^2=4$. *Ans.* $(x-y)^2=8$. (h) $Cx^2+C^2y=1$. *Ans.* $x^4+4y=0$.

83. A straight line is in motion so that the sum of the segments it cuts off on the axes is a constant a. Form the equation of the envelope of all positions of the straight line. *Ans.* $x^{1/2}+y^{1/2}=a^{1/2}$ (parabola).

84. Find the envelope of a family of straight lines on which the coordinate axes cut off a segment of constant length a. *Ans.* $x^{\frac{2}{3}}+y^{\frac{2}{3}}=a^{\frac{2}{3}}$.

85. Find the envelope of a family of circles whose diameters are the doubled ordinates of the parabola $y^2=2px$. *Ans.* $y^2=2p\left(x+\frac{p}{2}\right)$.

86. Find the envelope of a family of circles whose centres lie on the parabola $y^2=2px$; all the circles of the family pass through the vertex of this parabola. *Ans.* The cissoid $x^3+y^2(x+p)=0$.

87. Find the envelope of a family of circles whose diameters are chords of the ellipse $b^2x^2+a^2y^2=a^2b^2$ perpendicular to the x-axis. *Ans.* $\frac{x^2}{a^2+b^2}+\frac{y^2}{b^2}=1$.

88. Find the evolute of the ellipse $b^2x^2+a^2y^2=a^2b^2$ as the envelope of its normals. *Ans.* $(ax)^{2/3}+(by)^{2/3}=(a^2-b^2)^{\frac{2}{3}}$.

Integrate the following equations (Lagrange equations):

89. $y=2xy'+y'^2$. *Ans.* $x=\frac{C}{3p^2}-\frac{2}{3}p$, $y=\frac{2C-p^3}{3p}$.

90. $y=xy'^2+y'^2$. *Ans.* $y=(\sqrt{x+1}+C)^2$. Singular solution: $y=0$.
91. $y=x(1+y')+(y')^2$. *Ans.* $x=Ce^{-p}-2p+2$, $y=C(p+1)e^{-p}-p^2+2$
92. $y=yy'^2+2xy'$. *Ans.* $4Cx=4C^2-y^2$. 93. Find a curve with constant normal. *Ans.* $(x-C)^2+y^2=a^2$. Singular solution: $y=\pm a$.

Integrate the given Clairaut equations:

94. $y=xy'+y'-y'^2$. *Ans.* $y=Cx+C-C^2$. Singular solution: $4y=(x+1)^2$.
95. $y=xy'+\sqrt{1-y'^2}$. *Ans.* $y=Cx+\sqrt{1-C^2}$. Singular solution: $y^2-x^2=1$.
96. $y=xy'+y'$. *Ans.* $y=Cx+C$. 97. $y=xy'+\frac{1}{y'}$. *Ans.* $y=Cx+\frac{1}{C}$. Singular solution: $y^2=4x$. 98. $y=xy'-\frac{1}{y'^2}$. *Ans.* $y=Cx-\frac{1}{C^2}$. Singular solution: $y^3=-\frac{27}{4}x^2$.

99. The area of a triangle formed by the tangent to the sought-for curve and the coordinate axes is a constant. Find the curve. *Ans.* The equilateral hyperbola $4xy=\pm a^2$. Also, any straight line of the family $y=Cx\pm a\sqrt{C}$.

100. Find a curve such that the segment of its tangent between the coordinate axes is of constant length a. *Ans.* $y=Cx\pm\frac{aC}{\sqrt{1+C^2}}$. Singular solution: $x^{2/3}+y^{2/3}=a^{2/3}$.

101. Find a curve the tangents to which form, on the axes, segments whose sum is $2a$. *Ans.* $y=Cx-\frac{2aC}{1-C}$. Singular solution: $(y-x-2a)^2=8ax$.

102. Find curves for which the product of the distance of any tangent line to two given points is constant. *Ans.* Ellipses and hyperbolas. (Orthogonal and isogonal trajectories.)

103. Find the orthogonal trajectories of the family of curves $y=ax^n$. *Ans.* $x^2+ny^2=C$.

101. Find the orthogonal trajectories of the family of parabolas $y^2=2p(x-\alpha)$ (α is the parameter of the family). *Ans.* $y=Ce^{-\frac{x}{p}}$.

105. Find the orthogonal trajectories of the family of curves $x^2-y^2=\alpha$ (α is the parameter). *Ans.* $y=\frac{C}{x}$.

106. Find the orthogonal trajectories of the family of circles $x^2+y^2=2ax$. *Ans.* Circles: $y=C(x^2+y^2)$.

107. Find the orthogonal trajectories of equal parabolas tangent at the vertex of the given straight line. *Ans.* If $2p$ is the parameter of the parabolas and the

given straight line is the y-axis, then the equation of the trajectory will be $y+C=\frac{2}{3}\sqrt{\frac{2}{p}}x^{\frac{3}{2}}$

108. Find the orthogonal trajectories of the cissoids $y^2=\frac{x^3}{2a-x}$. *Ans.* $(x^2+y^2)^2=C(y^2+2x^2)$.

109. Find the orthogonal trajectories of the lemniscates $(x^2+y^2)^2=(x^2-y^2)a^2$. *Ans.* $(x^2+y^2)^2=Cxy$.

110. Find the isogonal trajectories of the family of curves: $x^2=2a(y-x\sqrt{3})$, where a is a variable parameter if the constant angle formed by the trajectories and the lines of the family is $\omega=60°$.

Solution. We find the differential equation of the family $y'=\frac{2y}{x}-\sqrt{3}$ and for y' substitute the expression $q=\frac{y'-\tan\omega}{1+y'\tan\omega}$. If $\omega=60°$, then $q=\frac{y'-\sqrt{3}}{1+\sqrt{3}y'}$ and we get the differential equation $\frac{y'-\sqrt{3}}{1+y'\sqrt{3}}=\frac{2y}{x}-\sqrt{3}$. The complete integral $y^2=C(x-y\sqrt{3})$ yields the desired family of trajectories.

111. Find the isogonal trajectories of the family of parabolas $y^2=4Cx$ when $\omega=45°$. *Ans.* $y^2-xy+2x^2=Ce^{\frac{6}{\sqrt{7}}\arctan\frac{2y-x}{x\sqrt{7}}}$.

112. Find the isogonal trajectories of the family of straight lines $y=Cx$ for the case $\omega=30°$, $45°$. *Ans.* The logarithmic spirals $\begin{cases} x^2+y^2=e^{2\sqrt{3}\arctan\frac{y}{x}} \\ x^2+y^2=e^{2\arctan\frac{y}{x}}. \end{cases}$

113. $y=C_1e^x+C_2e^{-x}$. Eliminate C_1 and C_2. *Ans.* $y''-y=0$.

114. Write the differential equation of all circles lying in one plane. *Ans.* $(1+y'^2)y'''-3y'y''^2=0$.

115. Write the differential equation of all central quadric curves whose principal axes coincide with the x- and y-axes. *Ans.* $x(yy''+y'^2)-y'y=0$.

116. Given the differential equation $y'''-2y''-y'+2y=0$ and its general solution $y=C_1e^x+C_2e^{-x}+C_3e^{2x}$.

It is required to: (1) verify that the given family of curves is indeed the general solution; (2) find a particular solution if for $x=0$ we have $y=1$, $y'=0$, $y''=-1$. *Ans.* $y=\frac{1}{6}(9e^x+e^{-x}-4e^{2x})$.

117. Given the differential equation $y''=\frac{1}{2y'}$ and its general solution $y=\pm\frac{2}{3}(x+C_1)^{\frac{3}{2}}+C_2$.

It is required to: (1) verify that the given family of curves is indeed the general solution; (2) find the integral curve passing through the point (1, 2) if the tangent at this point forms with the positive x-axis an angle of 45°. *Ans.* $y=\frac{2}{3}\sqrt{x^3}+\frac{4}{3}$.

Integrate some of the simpler types of differential equations of the second order that lead to first-order equations.

118. $xy'''=2$. *Ans.* $y=x^2\ln x+C_1x^2+C_2x+C_3$; pick out a particular solution that satisfies the following initial conditions: $x=1$, $y=1$, $y'=1$, $y''=3$.

Ans. $y=x^2\ln x+1$. **119.** $y^{(n)}=x^m$. *Ans.* $y=\frac{m!\,x^{m+n}}{(m+n)!}+C_1x^{n-1}+\ldots+C_{n-1}x+$ $+C_n$. **120.** $y''=a^2y$. *Ans.* $ax=\ln(ay+\sqrt{a^2y^2+C_1})+C_2$ or $y=C_1e^{ax}+C_2e^{-ax}$. **121.** $y''=\frac{a}{y^3}$. *Ans.* $(C_1x+C_2)^2=C_1y^2-a$.

In Nos. 122-125 pick out a particular solution that satisfies the following initial conditions: $x=0$, $y=-1$, $y'=0$.

122. $xy''-y'=x^2e^x$. *Ans.* $y=e^x(x-1)+C_1x^2+C_2$. Particular solution: $y=e^x(x-1)$. **123.** $yy''-(y')^2+(y')^3=0$. *Ans.* $y+C_1\ln y=x+C_2$. Particular solution: $y=-1$. **124.** $y''+y'\tan x=\sin 2x$. *Ans.* $y=C_2+C_1\sin x-x-\frac{1}{2}\sin 2x$. Particular solution: $y=2\sin x-\sin x\cos x-x-1$. **125.** $(y'')^2+(y')^2=a^2$. *Ans.* $y=C_2-a\cos(x+C_1)$. Particular solutions: $y=a-1-a\cos x$, $y=a\cos x-(a+1)$. (**Hint**. Parametric form: $y''=a\cos t$, $y'=a\sin t$.) **126.** $y''=\frac{1}{2y'}$. *Ans.* $y=$ $=\pm\frac{2}{3}(x+C_1)^{3/2}+C_2$. **127.** $y'''=y''^2$. *Ans.* $y=(C_1-x)[\ln(C_1-x)-1]+C_2x+C_3$. **128.** $y'y'''-3y''^2=0$. *Ans.* $x=C_1y^2+C_2y+C_3$.

Integrate the following linear differential equations with constant coefficients:

129. $y''=9y$. *Ans.* $y=C_1e^{3x}+C_2e^{-3x}$. **130.** $y''+y=0$. *Ans.* $y=A\cos x+$ $+B\sin x$. **131.** $y''-y'=0$. *Ans.* $y=C_1+C_2e^x$. **132.** $y''+12y=7y'$. *Ans.* $y=C_1e^{3x}+C_2e^{4x}$. **133.** $y''-4y'+4y=0$. *Ans.* $y=(C_1+C_2x)e^{2x}$. **134.** $y''+2y'+$ $+10y=0$. *Ans.* $y=e^{-x}(A\cos 3x+B\sin 3x)$. **135.** $y''+3y'-2y=0$. *Ans.* $y=C_1e^{\frac{-3+\sqrt{17}}{2}x}+C_2e^{\frac{-3-\sqrt{17}}{2}x}$. **136.** $4y''-12y'+9y=0$. *Ans.* $y=(C_1+C_2x)e^{\frac{3x}{2}}$. **137.** $y''+y'+y=0$. *Ans.* $y=e^{-\frac{1}{2}x}\left[A\cos\left(\frac{\sqrt{3}}{2}x\right)+B\sin\left(\frac{\sqrt{3}}{2}x\right)\right]$.

138. Two identical loads are suspended from the end of a spring. Find the motion imparted to one load if the other breaks loose. *Ans.* $x=a\cos\left(\sqrt{\frac{g}{a}}\,t\right)$, where a is the increase in length of the spring under the action of one load at rest.

139. A material point of mass m is attracted by each of two centres with a force proportional to the distance. The factor of proportionality is k. The distance between the centres is $2c$. At the initial instant the point lies on the line connecting the centres at a distance a from the middle. The initial velocity is zero. Find the law of motion of the point. *Ans.* $x=a\cos\left(\sqrt{\frac{2k}{m}}\,t\right)$.

140. $y^{IV}-5y''+4y=0$. *Ans.* $y=C_1e^x+C_2e^{-x}+C_3e^{2x}+C_4e^{-2x}$. **141.** $y'''-2y''-$ $-y'+2y=0$. *Ans.* $y=C_1e^{2x}+C_2e^x+C_3e^{-x}$. **142.** $y'''-3ay''+3a^2y'-a^3y=0$. *Ans.* $y=(C_1+C_2x+C_3x^2)e^{ax}$. **143.** $y^V-4y'''=0$. *Ans.* $y=C_1+C_2x+C_3x^2+$ $+C_4e^{2x}+C_5e^{-2x}$. **144.** $y^{IV}+2y''+9y=0$. *Ans.* $y=(C_1\cos\sqrt{2}x+C_2\sin\sqrt{2}x)e^{-x}+$ $+(C_3\cos\sqrt{2}x+C_4\sin\sqrt{2}x)e^x$. **145.** $y^{IV}-8y''+16y=0$. *Ans.* $y=C_1e^{2x}+$ $+C_2e^{-2x}+C_3xe^{2x}+C_4xe^{-2x}$. **146.** $y^{IV}+y=0$. *Ans.* $y=e^{\frac{x}{\sqrt{2}}}\left(C_1\cos\frac{x}{\sqrt{2}}+\right.$ $\left.+C_2\sin\frac{x}{\sqrt{2}}\right)+e^{-\frac{x}{\sqrt{2}}}\left(C_3\cos\frac{x}{\sqrt{2}}+C_4\sin\frac{x}{\sqrt{2}}\right)$.

147. $y^{IV}-a^4y=0$. Find the general solution and pick out a particular solution that satisfies the initial conditions for $x_0=0$, $y=1$, $y'=0$, $y''=-a^2$,

$y'''=0$. *Ans.* General solution: $y=C_1e^{ax}+C_2e^{-ax}+C_3\cos ax+C_4\sin ax$. Particular solution: $y_0=\cos ax$.

Integrate the following nonhomogeneous linear differential equations (find the general solution):

148. $y''-7y'+12y=x$. *Ans.* $y=C_1e^{3x}+C_2e^{4x}+\frac{12x+7}{144}$. **149.** $s''-a^2s=t+1$. *Ans.* $s=C_1e^{at}+C_2e^{-at}-\frac{t+1}{a^2}$. **150.** $y''+y'-2y=8\sin 2x$. *Ans.* $y=C_1e^x+C_2e^{-2x}-\frac{1}{5}(6\sin 2x+2\cos 2x)$. **151.** $y''-y=5x+2$. *Ans.* $y=C_1e^x+C_2e^{-x}-5x-2$. **152.** $s''-2as'+a^2s=e^t$ $(a\neq 1)$. *Ans.* $s=C_1e^{at}+C_2te^{at}+\frac{e^t}{(a-1)^2}$. **153.** $y''+6y'+5y=e^{2x}$. *Ans.* $y=C_1e^{-x}+C_2e^{-5x}+\frac{1}{21}e^{2x}$. **154.** $y''+9y=6e^{3x}$. *Ans.* $y=C_1\cos 3x+C_2\sin 3x+\frac{1}{3}e^{3x}$. **155.** $y''-3y'=2-6x$. *Ans.* $y=C_1+C_2e^{3x}+x^2$. **156.** $y''-2y'+3y=e^{-x}\cos x$. *Ans.* $y=e^x(A\cos\sqrt{2}x+B\sin\sqrt{2}x)+\frac{e^{-x}}{41}(5\cos x-4\sin x)$. **157.** $y''+4y=2\sin 2x$. *Ans.* $y=A\sin 2x+B\cos 2x-\frac{x}{2}\cos 2x$. **158.** $y'''-4y''+5y'-2y=2x+3$. *Ans.* $y=(C_1+C_2x)e^x+C_3e^{2x}-x-4$. **159.** $y^{IV}-a^4y=5a^4e^{ax}\sin ax$. *Ans.* $y=(C_1-\sin ax)e^{ax}+C_2e^{-ax}+C_3\cos ax+C_4\sin ax$. **160.** $y^{IV}+2a^2y''+a^4y=8\cos ax$. *Ans.* $y=(C_1+C_2x)\cos ax+(C_3+C_4x)\sin ax-\frac{x^2}{a^2}\cos ax$.

161. Find the integral curve of the equation $y''+k^2y=0$ that passes through the point $M(x_0, y_0)$, and is tangent at the point M to the straight line $y=ax$. *Ans.* $y=y_0\cos k(x-x_0)+\frac{a}{k}\sin k(x-x_0)$.

162. Find a solution of the equation $y''+2hy'+n^2y=0$ that satisfies the conditions $y=a$, $y'=C$ when $x=0$. *Ans.* For $h<n$ $y=e^{-hx}\left(a\cos\sqrt{n^2-h^2}x+\frac{C+ah}{\sqrt{n^2-h^2}}\sin\sqrt{n^2-h^2}x\right)$; for $h=n$, $y=e^{-hx}[(C+ah)x+a]$; for $h>n$,

$$y=\frac{C+a(h+\sqrt{h^2-n^2})}{2\sqrt{h^2-n^2}}e^{-(h-\sqrt{h^2-n^2})x}-\frac{C+a(h-\sqrt{h^2-n^2})}{2\sqrt{h^2-n^2}}e^{-(h-\sqrt{h^2-n^2})x}.$$

163. Find a solution of the equation $y''+n^2y=h\sin px$ $(p\neq n)$ that satisfies the conditions: $y=a$, $y'=C$ for $x=0$. *Ans.* $y=a\cos nx+\frac{C(n^2-p^2)-hp}{n(n^2-p^2)}\sin nx+\frac{h}{n^2-p^2}\sin px$.

164. A load weighing 4 kg is suspended from a spring and increases the length of the spring by 1 cm. Find the law of motion of the load if we assume that the upper end of the spring performs harmonic oscillations under the law $y=\sin\sqrt{100\,g}\,t$, where y is measured vertically.

Solution. Denoting by x the vertical coordinate of the load reckoned from the position of rest, we have

$$\frac{4}{g}\frac{d^2x}{dt^2}=-k(x-y-l)$$

where l is the length of the spring in the free state and $k=400$, as is evident from the initial conditions. Whence $\frac{d^2x}{dt^2}+100gx=100g\sin\sqrt{100g}t+100\,lg$. We must seek the particular integral of this equation in the form

$$t\,(C_1\cos\sqrt{100g}\,t+C_2\sin\sqrt{100g}\,t)+l$$

since the first term on the right enters into the solution of the homogeneous equation.

165. In Problem 139, the initial velocity is v_0 and the direction is perpendicular to the straight line connecting the centres. Find the trajectories.

Solution. If for the origin we take the mid-point between the centres, the differential equations of motion will be

$$m\frac{d^2x}{dt^2}=k\,(C-x)-k\,(C+x)=-2kx,\quad m\frac{d^2y}{dt^2}=-2ky$$

The initial data for $t=0$ are

$$x=a,\qquad \frac{dx}{dt}=0,\qquad y=0,\qquad \frac{dy}{dt}=v_0$$

Integrating, we find

$$x=a\cos\left(\sqrt{\frac{2k}{m}}\,t\right),\qquad y=v_0\sqrt{\frac{m}{2k}}\sin\left(\sqrt{\frac{2k}{m}}\,t\right)$$

Whence $\frac{x^2}{a^2}+\frac{y^2 2k}{mv_0^2}=1$ (ellipse).

166. A horizontal tube is in rotation about a vertical axis with constant angular velocity ω. A sphere inside the tube slides along it without friction. Find the law of motion of the sphere if at the initial instant it lies on the axis of rotation and has velocity v_0 (along the tube).

Hint. The differential equation of motion is $\frac{d^2r}{dt^2}=\omega^2 r$. The initial data are: $r=0$, $\frac{dr}{dt}=v_0$ for $t=0$. Integrating, we find

$$r=\frac{v_0}{2\omega}\,[e^{\omega t}+e^{-\omega t}]$$

Applying the method of variation of parameters, integrate the following differential equations:

167. $y''-7y'+6y=\sin x$. *Ans.* $y=C_1e^x+C_2e^{6x}+\frac{5\sin x+7\cos x}{74}$.

168. $y''+y=\sec x$. *Ans.* $y=C_1\cos x+C_2\sin x+x\sin x+\cos x\ln\cos x$.

169. $y''+y=\frac{1}{\cos 2x\sqrt{\cos 2x}}$. *Ans.* $y=C_1\cos x+C_2\sin x-\sqrt{\cos 2x}$.

Integrate the following systems of equations:

170. $\frac{dx}{dt}=y+1$, $\frac{dy}{dt}=x+1$. Pick out the particular solutions that satisfy the initial conditions $x=-2$, $y=0$ for $t=0$. *Ans.* $y=C_1e^t+C_2e^{-t}-1$, $x=C_1e^t-C_2e^{-t}-1$. Particular solution: $x^*=-e^{-t}-1$, $y^*=e^{-t}-1$.

171. $\frac{dx}{dt}=x-2y$, $\frac{dy}{dt}=x-y$. Pick out the particular solutions that satisfy the initial conditions: $x=1$, $y=1$ for $t=0$. *Ans.* $y=C_1\cos t+C_2\sin t$, $x=(C_1+C_2)\cos t+(C_2-C_1)\sin t$. Particular solution: $x^*=\cos t-\sin t$, $y^*=\cos t$.

172. $\begin{cases} 4\dfrac{dx}{dt}-\dfrac{dy}{dt}+3x=\sin t \\ \dfrac{dx}{dt}+y=\cos t. \end{cases}$ *Ans.* $x=C_1e^{-t}+C_2e^{-3t}$
$y=C_1e^{-t}+3C_2e^{-3t}+\cos t.$

173. $\begin{cases} \dfrac{d^2y}{dt^2}=x \\ \dfrac{d^2x}{dt^2}=y. \end{cases}$ *Ans.* $x=C_1e^t+C_2e^{-t}+C_3\cos t+C_4\sin t,$
$y=C_1e^t+C_2e^{-t}-C_3\cos t-C_4\sin t.$

174. $\begin{cases} \dfrac{d^2x}{dt^2}+\dfrac{dy}{dt}+x=e^t \\ \dfrac{dx}{dt}+\dfrac{d^2y}{dt^2}=1. \end{cases}$ *Ans.* $x=C_1+C_2t+C_3t^2-\dfrac{1}{6}t^3+e^t,$
$y=C_4-(C_1+2C_3)t-\dfrac{1}{2}(C_2-1)t^2-$
$-\dfrac{1}{3}C_3t^3+\dfrac{1}{24}t^4-e^t.$

175. $\begin{cases} \dfrac{dy}{dx}=z-y \\ \dfrac{dz}{dx}=-y-3z. \end{cases}$ *Ans.* $y=(C_1+C_2x)e^{-2x},$
$z=(C_2-C_1-C_2x)e^{-2x}.$

176. $\begin{cases} \dfrac{dy}{dx}+z=0 \\ \dfrac{dz}{dx}+4y=0. \end{cases}$ *Ans.* $y=C_1e^{2x}+C_2e^{-2x}$
$z=-2(C_1e^{2x}-C_2e^{-2x}).$

177. $\begin{cases} \dfrac{dy}{dx}+2y+z=\sin x \\ \dfrac{dz}{dx}-4y-2z=\cos x. \end{cases}$ *Ans.* $y=C_1+C_2x+2\sin x,$
$z=-2C_1-C_2(2x+1)-3\sin x-2\cos x.$

178. $\begin{cases} \dfrac{dx}{dt}=y+z \\ \dfrac{dy}{dt}=x+z \\ \dfrac{dz}{dt}=x+y. \end{cases}$ *Ans.* $x=C_1e^{-t}+C_2e^{2t},$
$y=C_3e^{-t}+C_2e^{2t},$
$z=-(C_1+C_3)e^{-t}+C_2e^{2t}.$

179. $\begin{cases} \dfrac{dy}{dx}=1-\dfrac{1}{z} \\ \dfrac{dz}{dx}=\dfrac{1}{y-x}. \end{cases}$ *Ans.* $z=C_2e^{C_1x},$
$y=x+\dfrac{1}{C_1C_2}e^{-C_1x}.$

180. $\begin{cases} \dfrac{dy}{dx}=\dfrac{x}{yz} \\ \dfrac{dz}{dx}=\dfrac{x}{y^2}. \end{cases}$ *Ans.* $\dfrac{z}{y}=C_1,\ zy^2-\dfrac{3}{2}x^2=C_2.$

Integrate the following different types of equations:

181. $yy''=y'^2+1$. *Ans.* $y=\frac{1}{2C_1}[e^{C_1(x-C_2)}+e^{-C_1(x-C_2)}]$. **182.** $\frac{x^2dy-y^2dx}{(x-y)^2}=0$. *Ans.* $\frac{xy}{x-y}=C$. **183.** $y=xy'^2+y'^2$. *Ans.* $y=(\sqrt{x+1}+C)^2$. Singular solutions: $y=0$, $x+1=0$. **184.** $y''+y=\sec x$. *Ans.* $y=C_1\cos x+C_2\sin x+x\sin x+\cos x\ln\cos x$. **185.** $(1+x^2)y'-xy-a=0$. *Ans.* $y=ax+C\sqrt{1+x^2}$. **186.** $x\cos\frac{y}{x}\frac{dy}{dx}=y\cos\frac{y}{x}-x$. *Ans.* $xe^{\sin\frac{y}{x}}=C$. **187.** $y''-4y=e^{2x}\sin 2x$. *Ans.* $y=C_1e^{-2x}+C_2e^{2x}-\frac{e^{2x}}{20}(\sin 2x+2\cos 2x)$. **188.** $xy'+y-y^2\ln x=0$. *Ans.* $(\ln x+1+Cx)\,y=1$. **189.** $(2x+2y-1)\,dx+(x+y-2)\,dy=0$. *Ans.* $2x+y-3\ln(x+y+1)=C$. **190.** $3e^x\tan y\,dx+(1-e^x)\sec^2 y\,dy=0$. *Ans.* $\tan y=C(1-e^x)^3$.

Investigate and determine whether the solution $x=0$, $y=0$ is stable for the following systems of differential equations:

191. $\begin{cases}\frac{dx}{dt}=2x-3y\\ \frac{dy}{dt}=5x+6y.\end{cases}$ *Ans.* Unstable.

192. $\begin{cases}\frac{dx}{dt}=-4x-10y\\ \frac{dy}{dt}=x-2y.\end{cases}$ *Ans.* Stable.

193. $\begin{cases}\frac{dx}{dt}=12x+18y\\ \frac{dy}{dt}=-8x-12y.\end{cases}$ *Ans.* Unstable.

194. Approximate the solution of the equation $y'=y^2+x$ that satisfies the initial condition $y=1$ when $x=0$. Find the values of the solution for x equal to 0.1, 0.2, 0.3, 0.4, 0.5. *Ans.* $y_{x=0.5}=2.235$.

195. Approximate the value of $y_{x=1.4}$ of a solution of the equation $y'+\frac{1}{x}y=e^x$ that satisfies the initial conditions $y=1$ when $x=1$. Compare the result obtained with the exact solution.

196. Find the approximate values of $x_{t=1.4}$ and $y_{t=1.4}$ of the solutions of a system of equations $\frac{dx}{dt}=y-x$, $\frac{dy}{dt}=-x-3y$ that satisfy the initial conditions $x=0$, $y=1$ when $t=1$. Compare the values obtained with the exact values.

CHAPTER 2

MULTIPLE INTEGRALS

2.1 DOUBLE INTEGRALS

In an xy-plane we consider a closed* domain D bounded by a line L.

In D let there be given a continuous function

$$z = f(x, y)$$

Using arbitrary lines we divide the domain D into n parts

$$\Delta s_1, \Delta s_2, \Delta s_3, \ldots, \Delta s_n$$

(Fig. 43) which we shall call subdomains. So as not to introduce new symbols we will denote by $\Delta s_1, \ldots, \Delta s_n$ both the subdomains and their areas. In each subdomain Δs_i (it is immaterial whether in the interior or on the boundary) take a point P_i; we will then have n points:

$$P_1, P_2, \ldots, P_n$$

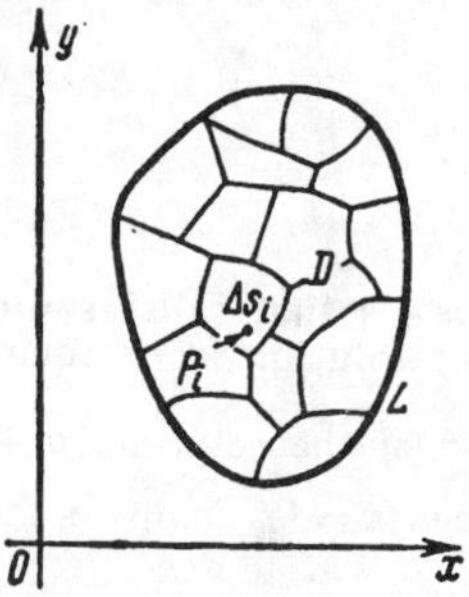

Fig. 43

We denote by $f(P_1), f(P_2), \ldots, f(P_n)$ the values of the functions at the chosen points and then form the sum of the products $f(P_i)\Delta s_i$:

$$V_n = f(P_1)\Delta s_1 + f(P_2)\Delta s_2 + \ldots + f(P_n)\Delta s_n = \sum_{i=1}^{n} f(P_i)\Delta s_i \qquad (1)$$

This is the *integral sum* of the function $f(x, y)$ in the domain D.

If $f \geqslant 0$ in D, then each term $f(P_i)\Delta s_i$ may be represented geometrically as the volume of a small cylinder with base Δs_i and altitude $f(P_i)$.

The sum V_n is the sum of the volumes of the indicated elementary cylinders, that is, the volume of a certain "step-like" solid (Fig. 44).

---

* A domain D is called *closed* if it is bounded by a closed line, and the points lying on the boundary are considered as belonging to D.

Consider an arbitrary sequence of integral sums formed by means of the function $f(x, y)$ for the given domain D,

$$V_{n_1}, V_{n_2}, \ldots, V_{n_k}, \ldots \qquad (2)$$

for different ways of partitioning D into subdomains Δs_i. We shall assume that the maximum diameter of the subdomains Δs_i approaches zero as $n_k \to \infty$. Then the following proposition, which we give without proof, holds true.

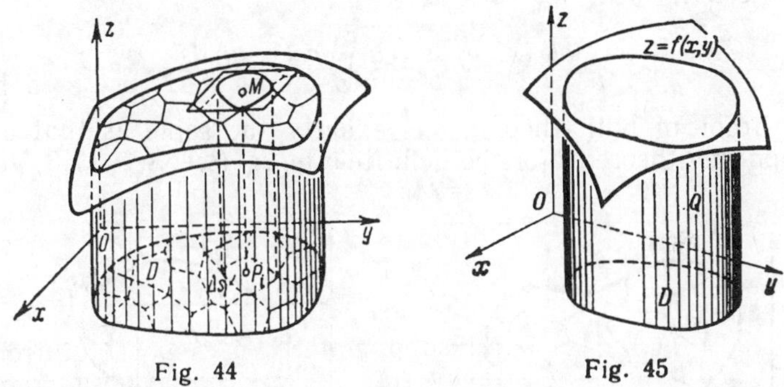

Fig. 44

Fig. 45

Theorem 1. *If a function $f(x, y)$ is continuous in a closed domain D, then the sequence (2) of integral sums (1) has a limit if the maximum diameter of the subdomains Δs_i approaches zero as $n_k \to \infty$. This limit is the same for any sequence of type (2), that is, it is independent either of the way D is partitioned into subdomains Δs_i or of the choice of the point P_i inside a subdomain Δs_i.*

This limit is called the *double integral* of the function $f(x, y)$ over D and is denoted by

$$\iint_D f(P)\, ds \quad \text{or} \quad \iint_D f(x, y)\, dx\, dy,$$

that is,

$$\lim_{\operatorname{diam} \Delta s_i \to 0} \sum_{i=1}^{n} f(P_i)\, \Delta s_i = \iint_D f(x, y)\, dx\, dy$$

This domain D is called the *domain (region) of integration*.

If $f(x, y) \geqslant 0$, then the double integral of $f(x, y)$ over D is equal to the volume of the solid Q bounded by surface $z = f(x, y)$, the plane $z = 0$, and a cylindrical surface whose generators are parallel to the z-axis, while the directrix is the boundary of the domain D (Fig. 45).

Now consider the following theorems about the double integral.

Theorem 2. *The double integral of a sum of two functions* $\varphi(x, y)+\psi(x, y)$ *over a domain* D *is equal to the sum of the double integrals over* D *of each of the functions taken separately:*

$$\iint_D [\varphi(x, y)+\psi(x, y)]\,ds = \iint_D \varphi(x, y)\,ds + \iint_D \psi(x, y)\,ds$$

Theorem 3. *A constant factor may be taken outside the double integral sign:*

if $a=\text{const}$, *then*

$$\iint_D a\varphi(x, y)\,ds = a\iint_D \varphi(x, y)\,ds$$

The proof of both theorems is exactly the same as that of the corresponding theorems for the definite integral (see. Sec. 11.3, Vol. I).

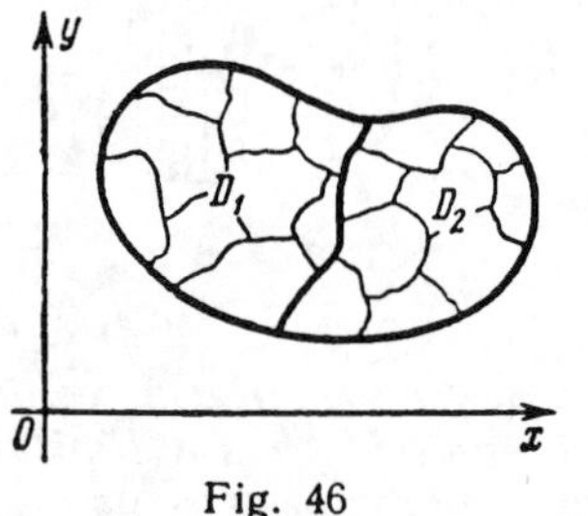

Fig. 46

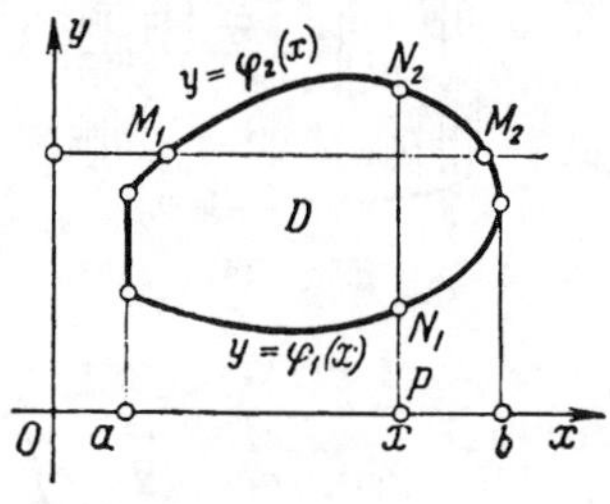

Fig. 47

Theorem 4. *If a region* D *is divided into two domains* D_1 *and* D_2 *without common interior points, and a function* $f(x, y)$ *is continuous at all points of* D, *then*

$$\iint_D f(x, y)\,dx\,dy = \iint_{D_1} f(x, y)\,dx\,dy + \iint_{D_2} f(x, y)\,dx\,dy \tag{3}$$

Proof. The integral sum over D may be given in the form (Fig. 46)

$$\sum_D f(P_i)\,\Delta s_i = \sum_{D_1} f(P_i)\,\Delta s_i + \sum_{D_2} f(P_i)\,\Delta s_i \tag{4}$$

where the first sum contains terms that correspond to the subdomains of D_1, the second, those corresponding to the subdomains of D_2. Indeed, since the double integral does not depend on the manner of partition, we divide D so that the common boundary of the domains D_1 and D_2 is a boundary of the subdomains Δs_i. Passing to the limit in (4) as $\Delta s_i \to 0$, we get (3). This theor[em] is obviously true for any number of terms.

2.2 CALCULATING DOUBLE INTEGRALS

Let a domain D lying in the xy-plane be such that any straight line parallel to one of the coordinate axes (for example, the y-axis) and passing through an **interior*** point of the domain, cuts the boundary of the domain at two points N_1 and N_2 (Fig. 47).

In this case we assume that the domain D is bounded by the lines: $y=\varphi_1(x)$, $y=\varphi_2(x)$, $x=a$, $x=b$ and that

$$\varphi_1(x) \leqslant \varphi_2(x), \quad a<b$$

while the functions $\varphi_1(x)$ and $\varphi_2(x)$ are continuous on the interval $[a, b]$. We shall call such a domain *regular in the y-direction*. The definition is similar for a domain *regular in the x-direction*.

A domain that is regular in both x- and y-directions we shall simply call a *regular domain*. In Fig. 47 we have a regular domain D.

Let the function $f(x, y)$ be continuous in D.

Consider the expression

$$I_D=\int_a^b\left(\int_{\varphi_1(x)}^{\varphi_2(x)} f(x, y)\,dy\right)dx$$

which we shall call a *twofold iterated* integral of $f(x, y)$ over D. In this expression we first calculate the integral in the parentheses (the integration is performed with respect to y while x is considered to be constant). The integration yields a continuous** function of x:

$$\Phi(x)=\int_{\varphi_1(x)}^{\varphi_2(x)} f(x, y)\,dy$$

Integrating this function with respect to x from a to b,

$$I_D=\int_a^b \Phi(x)\,dx$$

we get a certain constant.

Example. Calculate the twofold iterated integral

$$I_D=\int_0^1\left(\int_0^{x^2}(x^2+y^2)\,dy\right)dx$$

Solution. First calculate the inner integral (in brackets):

$$\Phi(x)=\int_0^{x^2}(x^2+y^2)\,dy=\left(x^2y+\frac{y^3}{3}\right)_0^{x^2}=x^2x^2+\frac{(x^2)^3}{3}=x^4+\frac{x^6}{3}$$

---

* An interior point of a domain is one that does not lie on its boundary.

** We do not prove here that the function $\Phi(x)$ is continuous.

Integrating the function obtained from 0 to 1, we find

$$\int_0^1 \left(x^4+\frac{x^6}{3}\right) dx=\left(\frac{x^5}{5}+\frac{x^7}{3\cdot 7}\right)\Bigg|_0^1=\frac{1}{5}+\frac{1}{21}=\frac{26}{105}$$

Determine the domain D. Here, D is the domain bounded by the lines (Fig. 48)

$$y=0, \quad x=0, \quad y=x^2, \quad x=1$$

It may happen that the domain D is such that one of the functions $y=\varphi_1(x)$, $y=\varphi_2(x)$ cannot be represented by one analytic

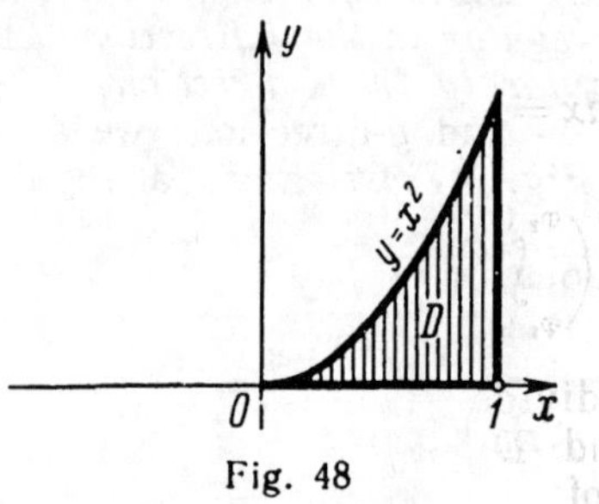

Fig. 48

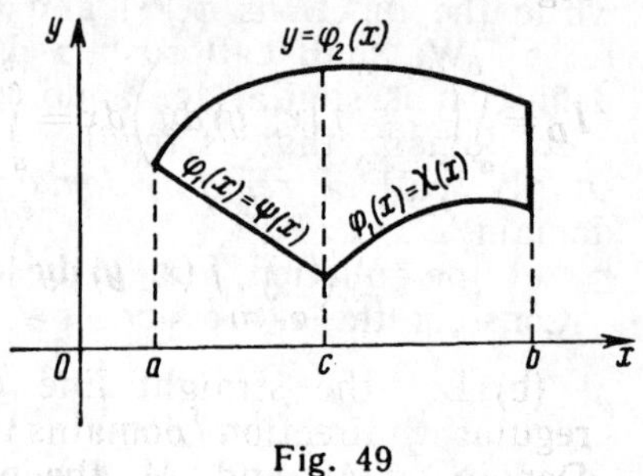

Fig. 49

expression over the entire range of x (from $x=a$ to $x=b$). For example, let $a<c<b$, and

$$\varphi_1(x)=\psi(x) \text{ on the interval } [a, c]$$

$$\varphi_1(x)=\chi(x) \text{ on the interval } [c, b]$$

where $\psi(x)$ and $\chi(x)$ are analytically specified functions (Fig. 49). Then the twofold iterated integral will be written as follows:

$$\int_a^b\left(\int_{\varphi_1(x)}^{\varphi_2(x)} f(x, y)\, dy\right) dx$$

$$=\int_a^c\left(\int_{\varphi_1(x)}^{\varphi_2(x)} f(x, y)\, dy\right) dx+\int_c^b\left(\int_{\varphi_1(x)}^{\varphi_2(x)} f(x, y)\, dy\right) dx$$

$$=\int_a^c\left(\int_{\psi(x)}^{\varphi_2(x)} f(x, y)\, dy\right) dx+\int_c^b\left(\int_{\chi(x)}^{\varphi_2(x)} f(x, y)\, dy\right) dx$$

The first of these equations is written on the basis of a familiar property of the definite integral, the second, by virtue of the fact that on the interval $[a, c]$ we have $\varphi_1(x)\equiv\psi(x)$, and on the interval $[c, b]$ we have $\varphi_1(x)\equiv\chi(x)$.

We would also have a similar notation for the twofold iterated integral if the function $\varphi_2(x)$ were defined by different analytic expressions on different subintervals of the interval $[a, b]$.

Let us establish some properties of a twofold iterated integral.

Property 1. *If a regular y-direction domain D is divided into two domains D_1 and D_2 by a straight line parallel to the y-axis or the x-axis, then the twofold iterated integral I_D over D will be equal to the sum of such integrals over D_1 and D_2; that is,*

$$I_D = I_{D_1} + I_{D_2} \tag{1}$$

Proof. (a) Let the straight line $x=c$ $(a<c<b)$ divide the region D into two regular y-direction domains* D_1 and D_2. Then

$$I_D = \int\limits_a^b \left(\int\limits_{\varphi_1(x)}^{\varphi_2(x)} f(x, y)\,dy \right) dx = \int\limits_a^b \Phi(x)\,dx = \int\limits_a^c \Phi(x)\,dx + \int\limits_c^b \Phi(x)\,dx$$

$$= \int\limits_a^c \left(\int\limits_{\varphi_1(x)}^{\varphi_2(x)} f(x, y)\,dy \right) dx + \int\limits_c^b \left(\int\limits_{\varphi_1(x)}^{\varphi_2(x)} f(x, y)\,dy \right) dx = I_{D_1} + I_{D_2}$$

(b) Let the straight line $y=h$ divide the domain D into two regular y-direction domains D_1 and D_2 as shown in Fig. 50. Denote by M_1 and M_2 the points of intersection of the straight line $y=h$ with the boundary L of D. Denote the abscissas of these points by a_1 and b_1.

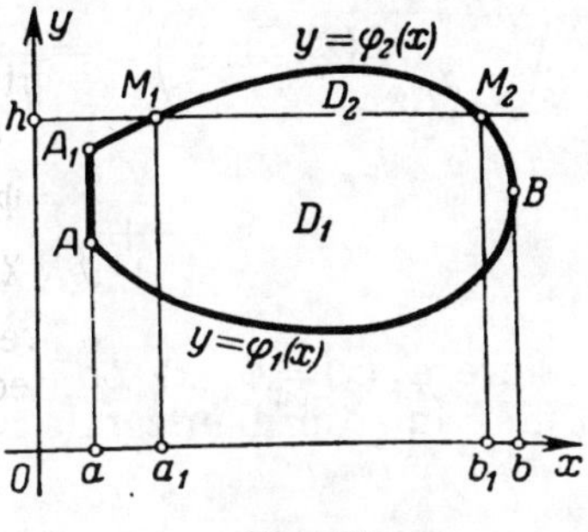

The domain D_1 is bounded by continuous lines:

(1) $y=\varphi_1(x)$;

(2) the curve $A_1M_1M_2B$, whose equation we shall conditionally write in the form

$$y = \varphi_1^*(x)$$

Fig. 50

having in view that $\varphi_1^*(x) = \varphi_2(x)$ when $a \leqslant x \leqslant a_1$ and when $b_1 \leqslant x \leqslant b$ and that

$$\varphi_1^*(x) = h \quad \text{when } a_1 \leqslant x \leqslant b_1;$$

(3) by the straight lines $x=a$, $x=b$.

The domain D_2 is bounded by the lines

$$y = \varphi_1^*(x), \quad y = \varphi_2(x), \quad \text{where } a_1 \leqslant x \leqslant b_1$$

---

* The fact that a part of the boundary of the domain D_1 (and of D_2) is a portion of the vertical straight line does not stop this domain from being regular in the y-direction: for a domain to be regular, it is only necessary that any vertical straight line passing through an interior point of the domain should have no more than two common points with the boundary.

We write the identity by applying to the inner integral the theorem on partitioning the interval of integration:

$$I_D = \int_a^b \left(\int_{\varphi_1(x)}^{\varphi_2(x)} f(x, y)\, dy \right) dx$$

$$= \int_a^b \left(\int_{\varphi_1(x)}^{\varphi_1^*(x)} f(x, y)\, dy + \int_{\varphi_1^*(x)}^{\varphi_2(x)} f(x, y)\, dy \right) dx$$

$$= \int_a^b \left(\int_{\varphi_1(x)}^{\varphi_1^*(x)} f(x, y)\, dy \right) dx + \int_a^b \left(\int_{\varphi_1^*(x)}^{\varphi_2(x)} f(x, y)\, dy \right) dx$$

We break up the latter integral into three integrals and apply to the outer integral the theorem on partitioning the interval of integration:

$$\int_a^b \left(\int_{\varphi_1^*(x)}^{\varphi_2(x)} f(x, y)\, dy \right) dx = \int_a^{a_1} \left(\int_{\varphi_1^*(x)}^{\varphi_2(x)} f(x, y)\, dy \right) dx$$

$$+ \int_{a_1}^{b_1} \left(\int_{\varphi_1^*(x)}^{\varphi_2(x)} f(x, y)\, dy \right) dx + \int_{b_1}^{b} \left(\int_{\varphi_1^*(x)}^{\varphi_2(x)} f(x, y)\, dy \right) dx$$

since $\varphi_1^*(x) = \varphi_2(x)$ on $[a, a_1]$ and on $[b_1, b]$, it follows that the first and third integrals are identically zero. Therefore,

$$I_D = \int_a^b \left(\int_{\varphi_1(x)}^{\varphi_1^*(x)} f(x, y)\, dy \right) dx + \int_{a_1}^{b_1} \left(\int_{\varphi_1^*(x)}^{\varphi_2(x)} f(x, y)\, dy \right) dx$$

Here, the first integral is a twofold iterated integral over D_1, the second, over D_2. Consequently,

$$I_D = I_{D_1} + I_{D_2}$$

The proof will be similar for any position of the cutting straight line M_1M_2. If M_1M_2 divides D into three or a larger number of domains, we get a relation similar to (1), in the first part of which we will have the appropriate number of terms.

Corollary. We can again divide each of the domains obtained (using a straight line parallel to the y-axis or x-axis) into regular y-direction domains, and we can apply to them equation (1).

Thus, D may be divided by straight lines parallel to the coordinate axes into any number of regular domains

$$D_1, D_2, D_3, \ldots, D_i$$

and the assertion that the *twofold iterated integral over D is equal to the sum of twofold iterated integrals over the subdomains* holds; that is (Fig. 51),

$$I_D = I_{D_1} + I_{D_2} + I_{D_3} + \ldots + I_{D_i} \quad (2)$$

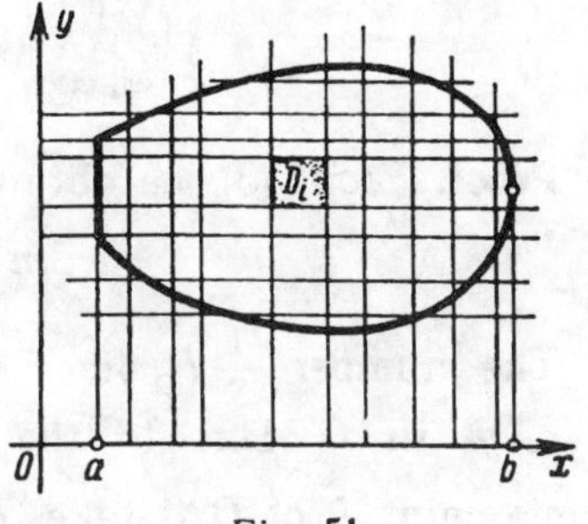

Fig. 51

Property 2 (Evaluation of an iterated integral). *Let m and M be the least and greatest values of the function f (x, y) in the domain D. Denote by S the area of D. Then we have the relation*

$$mS \leqslant \int_a^b \left(\int_{\varphi_1(x)}^{\varphi_2(x)} f(x, y)\, dy \right) dx \leqslant MS \quad (3)$$

Proof. Evaluate the inner integral denoting it by $\Phi(x)$:

$$\Phi(x) = \int_{\varphi_1(x)}^{\varphi_2(x)} f(x, y)\, dy \leqslant \int_{\varphi_1(x)}^{\varphi_2(x)} M\, dy = M[\varphi_2(x) - \varphi_1(x)]$$

We then have

$$I_D = \int_a^b \left(\int_{\varphi_1(x)}^{\varphi_2(x)} f(x, y)\, dy \right) dx \leqslant \int_a^b M[\varphi_2(x) - \varphi_1(x)]\, dx = MS$$

that is,

$$I_D \leqslant MS \quad (3')$$

Similarly

$$\Phi(x) = \int_{\varphi_1(x)}^{\varphi_2(x)} f(x, y)\, dy \geqslant \int_{\varphi_1(x)}^{\varphi_2(x)} m\, dy = m[\varphi_2(x) - \varphi_1(x)]$$

$$I_D = \int_a^b \Phi(x)\, dx \geqslant \int_a^b m[\varphi_2(x) - \varphi_1(x)]\, dx = mS$$

that is,

$$I_D \geqslant mS \quad (3'')$$

From the inequalities (3′) and (3″) follows the relation (3):

$$mS \leqslant I_D \leqslant MS$$

In the next section we will determine the geometric meaning of this theorem.

Property 3 (Mean-value theorem). *A twofold iterated integral I_D of a continuous function $f(x, y)$ over a domain D with area S is equal to the product of the area S by the value of the function at some point P in D; that is,*

$$\int_a^b \left(\int_{\varphi_1(x)}^{\varphi_2(x)} f(x, y)\, dy \right) dx = f(P)\, S \tag{4}$$

Proof. From (3) we obtain

$$m \leqslant \frac{1}{S} I_D \leqslant M$$

The number $\frac{1}{S} I_D$ lies between the greatest and least values of $f(x, y)$ in D. Due to the continuity of the function $f(x, y)$, at some point P of D it takes on a value equal to the number $\frac{1}{S} I_D$; that is,.

$$\frac{1}{S} I_D = f(P)$$

whence

$$I_D = f(P)\, S \tag{5}$$

2.3 CALCULATING DOUBLE INTEGRALS (CONTINUED)

Theorem. *The double integral of a continuous function $f(x, y)$ over a regular domain D is equal to the twofold iterated integral of this function over D; that is,* *

$$\iint_D f(x, y)\, dx\, dy = \int_a^b \left(\int_{\varphi_1(x)}^{\varphi_2(x)} f(x, y)\, dy \right) dx$$

Proof. Partition the domain D with straight lines parallel to the coordinate axes into n regular (rectangular) subdomains:

$$\Delta s_1, \ \Delta s_2, \ \ldots, \ \Delta s_n$$

By Property 1 [formula (2)] of the preceding section we have

$$I_D = I_{\Delta s_1} + I_{\Delta s_2} + \ldots + I_{\Delta s_n} = \sum_{i=1}^{n} I_{\Delta s_i} \tag{1}$$

We transform each of the terms on the right by the mean-value theorem for a twofold iterated integral:

$$I_{\Delta s_i} = f(P_i)\, \Delta s_i$$

---

* Here, we again assume that the domain D is regular in the y-direction and bounded by the lines $y = \varphi_1(x)$, $y = \varphi_2(x)$, $x = a$, $x = b$.

Then (1) takes the form

$$I_D = f(P_1)\Delta s_1 + f(P_2)\Delta s_2 + \ldots + f(P_n)\Delta s_n = \sum_{i=1}^{n} f(P_i)\Delta s_i \quad (2)$$

where P_i is some point of the subdomain Δs_i. On the right is the integral sum of the function $f(x, y)$ over D. From the existence theorem of a double integral it follows that the limit of this sum, as $n \to \infty$ and as the greatest diameter of the subdomains Δs_i approaches zero, exists and is equal to the double integral of $f(x, y)$ over D. The value of the double integral I_D on the left side of (2) does not depend on n. Thus, passing to the limit in (2), we obtain

$$I_D = \lim_{\operatorname{diam} \Delta s_i \to 0} \sum f(P_i)\Delta s_i = \iint_D f(x, y)\,dx\,dy$$

or

$$\iint_D f(x, y)\,dx\,dy = I_D \quad (3)$$

Writing out in full the expression of the twofold iterated integral I_D, we finally get

$$\iint_D f(x, y)\,dx\,dy = \int_a^b \left[\int_{\varphi_1(x)}^{\varphi_2(x)} f(x, y)\,dy\right] dx \quad (4)$$

Note 1. For the case where $f(x, y) \geqslant 0$, formula (4) has a pictorial geometric interpretation. Consider a solid bounded by a surface $z = f(x, y)$, a plane $z = 0$, and a cylindrical surface whose generators are parallel to the z-axis and the directrix of which is the boundary of the region D (Fig. 52). Calculate the volume of this solid V. It has already been shown that the volume of this solid is equal to the double integral of the function $f(x, y)$ over the domain D:

$$V = \iint_D f(x, y)\,dx\,dy \quad (5)$$

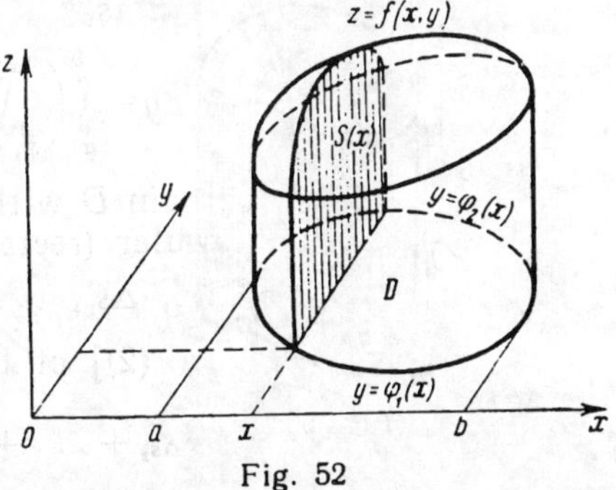

Fig. 52

Now let us calculate the volume of this solid using the results of Sec. 12.4, Vol. I, on the evaluation of the volume of a solid from the areas of parallel sections (slices). Draw the plane $x = \text{const}$ $(a < x < b)$ that cuts the solid. Calculate the area $S(x)$ of the figure obtained in the section $x = \text{const}$. This figure is a curvilinear trapezoid bounded by the lines $z = f(x, y)$ $(x = \text{const})$, $z = 0$, $y = \varphi_1(x)$, $y = \varphi_2(x)$. Hence, the area can be expressed by the integral

$$S(x)=\int_{\varphi_1(x)}^{\varphi_2(x)} f(x, y)\,dy \tag{6}$$

Knowing the areas of parallel sections, it is easy to find the volume of the solid:

$$V=\int_a^b S(x)\,dx$$

or, substituting expression (6) for the area $S(x)$, we get

$$V=\int_a^b\left(\int_{\varphi_1(x)}^{\varphi_2(x)} f(x, y)\,dy\right)dx \tag{7}$$

In formulas (5) and (7) the left sides are equal; and so the right sides are equal too:

$$\iint_D f(x, y)\,dx\,dy=\int_a^b\left(\int_{\varphi_1(x)}^{\varphi_2(x)} f(x, y)\,dy\right)dx$$

It is now easy to figure out the geometric meaning of theorem on evaluating a twofold iterated integral (Property 2, Sec. 2.2): the volume V of a solid bounded by the surface $z=f(x, y)$, the plane $z=0$, and a cylindrical surface whose directrix is the boun-

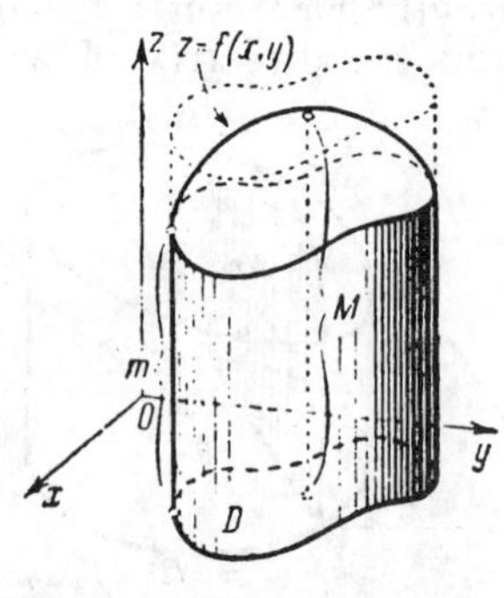

Fig. 53

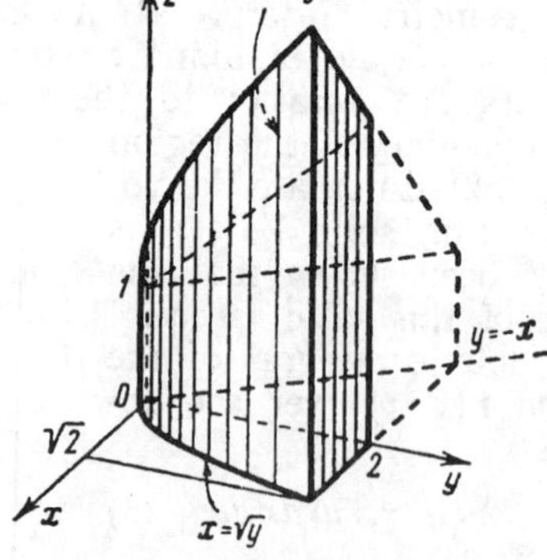

Fig. 54

dary of the region D, exceeds the volume of a cylinder with base area S and altitude m, but is less than the volume of a cylinder with base area S and altitude M [where m and M are the least and greatest values of the function $z=f(x, y)$ in the domain D (Fig. 53)]. This follows from the fact that the twofold iterated integral I_D is equal to the volume V of this solid.

Example 1. Evaluate the double integral $\iint\limits_D (4-x^2-y^2)\,dx\,dy$ if the domain D is bounded by the straight lines $x=0$, $x=1$, $y=0$, and $y=\frac{3}{2}$.

Solution. By the formula

$$V=\int_0^{3/2}\left[\int_0^1 (4-x^2-y^2)\,dx\right]dy=\int_0^{3/2}\left[4x-y^2x-\frac{x^3}{3}\right]_0^1 dy$$

$$=\int_0^{3/2}\left(4-y^2-\frac{1}{3}\right)dy=\left(4y-\frac{y^3}{3}-\frac{1}{3}y\right)\Big|_0^{3/2}=\frac{35}{8}$$

Example 2. Evaluate the double integral of the function $f(x, y)=1+x+y$ over a region bounded by the lines $y=-x$, $x=\sqrt{y}$, $y=2$, $z=0$ (Fig. 54).

Solution.

$$V=\int_0^2\left[\int_{-y}^{\sqrt{y}} (1+x+y)\,dx\right]dy=\int_0^2\left[x+xy+\frac{x^2}{2}\right]_{-y}^{\sqrt{y}} dy$$

$$=\int_0^2\left[\left(\sqrt{y}+y\sqrt{y}+\frac{y}{2}\right)-\left(-y-y^2+\frac{y^2}{2}\right)\right]dy$$

$$=\int_0^2\left[\sqrt{y}+\frac{3y}{2}+y\sqrt{y}+\frac{y^2}{2}\right]dy$$

$$=\left[\frac{2y^{\frac{3}{2}}}{3}+\frac{3y^2}{4}+\frac{2y^{\frac{5}{2}}}{5}+\frac{y^3}{6}\right]_0^2=\frac{44}{15}\sqrt{2}+\frac{13}{3}$$

Note 2. Let a regular x-direction domain D be bounded by the lines

$$x=\psi_1(y),\quad x=\psi_2(y),\quad y=c,\quad y=d$$

and let $\psi_1(y)\leqslant\psi_2(y)$ (Fig. 55).

In this case, obviously,

$$\iint\limits_D f(x, y)\,dx\,dy=\int_c^d\left(\int_{\psi_1(y)}^{\psi_2(y)} f(x, y)\,dx\right)dy \tag{8}$$

To evaluate the double integral we must represent it as a twofold iterated integral. As we have already seen, this may be done in two different ways: either by formula (4) or by formula (8). Depending upon the type of domain D or the integrand in each specific case, we choose one of the formulas to calculate the double integral.

Example 3. Change the order of integration in the integral

$$I=\int_0^1\left(\int_x^{\sqrt{x}} f(x,\,y)\,dy\right)dx$$

Solution. The domain of integration is bounded by the straight line $y=x$ and the parabola $y=\sqrt{x}$ (Fig. 56).

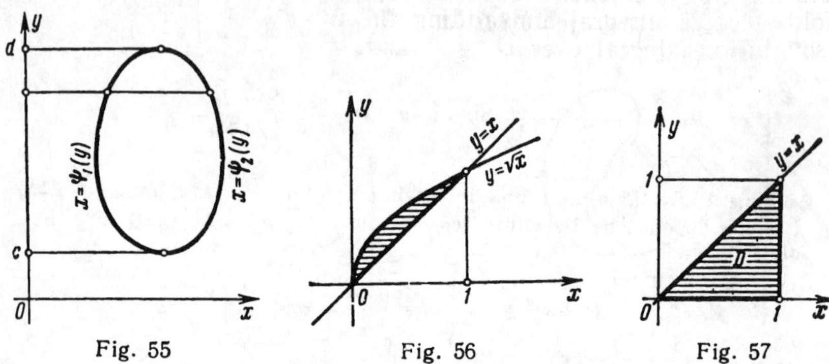

Fig. 55 Fig. 56 Fig. 57

Every straight line parallel to the x-axis cuts the boundary of the domain at no more than two points; hence, we can compute the integral by formula (8), setting

$$\psi_1(y)=y^2,\quad \psi_2(y)=y,\quad 0\leqslant y\leqslant 1$$

then

$$I=\int_0^1\left(\int_{y^2}^{y} f(x,\,y)\,dx\right)dy$$

Example 4. Evaluate $\iint\limits_D e^{\frac{y}{x}}\,ds$ if the domain D is a triangle bounded by the straight lines $y=x$, $y=0$, and $x=1$ (Fig. 57).

Solution. Replace this double integral by a twofold iterated integral using formula (4). [If we used formula (8), we would have to integrate the function $e^{\frac{y}{x}}$ with respect to x; but this integral is not expressible in terms of elementary functions]:

$$\iint\limits_D e^{\frac{y}{x}}\,ds=\int_0^1\left[\int_0^x e^{\frac{y}{x}}\,dy\right]dx=\int_0^1\left[xe^{\frac{y}{x}}\right]_0^x dx$$

$$=\int_0^1 x(e-1)\,dx=(e-1)\left.\frac{x^2}{2}\right|_0^1=\frac{e-1}{2}=0.859\ldots$$

Note 3. If the domain D is not regular either in the x-direction or the y-direction (that is, there exist vertical and horizontal straight lines which, while passing through interior points of the domain, cut the boundary of the domain at more than two points), then we cannot represent the double integral over this domain in the form of a twofold iterated integral. If we manage to partition the irregular domain D into a finite number of regular x-direction or y-direction domains $D_1, D_2, \ldots, D_n$, then, by evaluating the double integral over each of these subdomains by means of the twofold iterated integral and adding the results obtained, we get the sought-for integral over D.

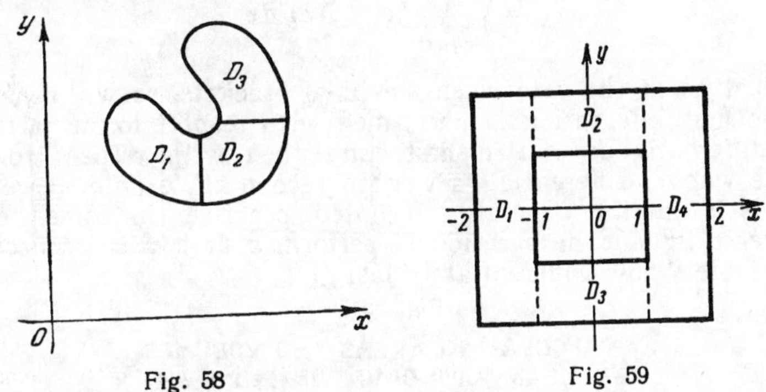

Fig. 58 Fig. 59

Fig. 58 is an example of how an irregular domain D may be divided into three regular subdomains D_1, D_2 and D_3.

Example 5. Evaluate the double integral

$$\iint_D e^{x+y}\,ds$$

over a domain D which lies between two squares with centre at the origin and with sides parallel to the axes of coordinates, if each side of the inner square is equal to 2 and that of the outer square is 4 (Fig. 59).

Solution. D is irregular. However, the straight lines $x=-1$ and $x=1$ divide it into four regular subdomains D_1, D_2, D_3, D_4. Therefore,

$$\iint_D e^{x+y}\,ds=\iint_{D_1} e^{x+y}\,ds+\iint_{D_2} e^{x+y}\,ds+\iint_{D_3} e^{x+y}\,ds+\iint_{D_4} e^{x+y}\,ds$$

Representing each of these integrals in the form of a twofold iterated integral, we find

$$\iint_D e^{x+y}\,ds=\int_{-2}^{-1}\left[\int_{-2}^{2} e^{x+y}\,dy\right]dx+\int_{-1}^{1}\left[\int_{1}^{2} e^{x+y}\,dy\right]dx$$

$$+\int_{-1}^{1}\left[\int_{-2}^{-1} e^{x+y}\,dy\right]dx+\int_{1}^{2}\left[\int_{-2}^{2} e^{x+y}\,dy\right]dx$$

$$=(e^2-e^{-2})(e^{-1}-e^{-2})+(e^2-e)(e-e^{-1})+(e^{-1}-e^{-2})(e-e^{-1})$$

$$+(e^2-e^{-2})(e^2-e)=(e^3-e^{-3})(e-e^{-1})=4\sinh 3\sinh 1$$

Note 4. From now on, when writing the twofold iterated integral

$$I_D=\int_a^b\left[\int_{\varphi_1(x)}^{\varphi_2(x)} f(x,\ y)\,dy\right]dx$$

we will drop the brackets containing the inner integral and will write

$$I_D=\int_a^b\int_{\varphi_1(x)}^{\varphi_2(x)} f(x,\ y)\,dy\,dx$$

Here, just as in the case when we have brackets, we will consider that the first integration is performed with respect to the variable whose differential is written first, and then with respect to the variable whose differential is written second. [We note, however, that this is not the generally accepted practice; in some books the reverse is done: integration is performed first with respect to the variable whose differential is last.*]

2.4 CALCULATING AREAS AND VOLUMES BY MEANS OF DOUBLE INTEGRALS

1. **Volume.** As we saw in Sec. 2.1, the volume V of a solid bounded by a surface $z=f(x,\ y)$, where $f(x,\ y)$ is a nonnegative function, by a plane $z=0$ and by a cylindrical surface whose directrix is the boundary of the domain D and the generators are parallel to the z-axis, is equal to the double integral of the function $f(x,\ y)$ over D:

$$V=\iint_D f(x,\ y)\,ds$$

Example 1. Calculate the volume of a solid bounded by the surfaces $x=0$, $y=0$, $x+y+z=1$, $z=0$ (Fig. 60).

Solution.

$$V=\iint_D (1-x-y)\,dy\,dx$$

where D is (in Fig. 60) the hatched triangular region in the xy-plane bounded by the straight lines $x=0$, $y=0$, and $x+y=1$. Putting the limits on the double

---

* The following notation is also sometimes used:

$$I_D=\int_a^b\left[\int_{\varphi_1}^{\varphi_2} f(x,\ y)\,dy\right]dx=\int_a^b dx\int_{\varphi_1}^{\varphi_2} f(x,\ y)\,dy$$

integral, we calculate the volume:

$$V=\int_0^1\int_0^{1-x}(1-x-y)\,dy\,dx=\int_0^1\left[(1-x)\,y-\frac{y^2}{2}\right]_0^{1-x}dx=\int_0^1\frac{1}{2}(1-x)^2\,dx=\frac{1}{6}$$

Thus, $V=\frac{1}{6}$ cubic units.

Note 1. If a solid, the volume of which is being sought, is bounded above by the surface $z=\Phi_2(x, y)\geqslant 0$, and below by the surface $z=\Phi_1(x, y)\geqslant 0$, and the domain D is the projection of

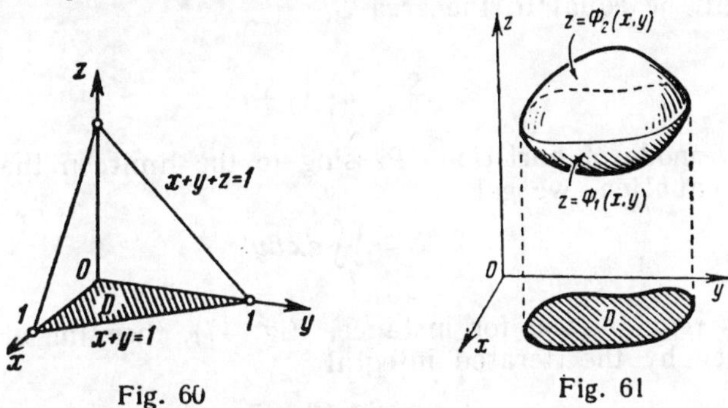

Fig. 60

Fig. 61

both surfaces on the xy-plane, then the volume V of this solid is equal to the difference between the volumes of the two "cylindrical" bodies; the first of these cylindrical bodies has the domain D for its lower base, and the surface $z=\Phi_2(x, y)$ for its upper base; the second body also has D for its lower base, and the surface $z=\Phi_1(x, y)$ for its upper base (Fig. 61).

Therefore, the volume V is equal to the difference between the two double integrals

$$V=\iint_D \Phi_2(x, y)\,ds-\iint_D \Phi_1(x, y)\,ds$$

or

$$V=\iint_D [\Phi_2(x, y)-\Phi_1(x, y)]\,ds \tag{1}$$

Further, it is easy to prove that formula (1) holds true not only for the case where $\Phi_1(x, y)$ and $\Phi_2(x, y)$ are nonnegative, but also where $\Phi_1(x, y)$ and $\Phi_2(x, y)$ are any continuous functions that satisfy the relationship

$$\Phi_2(x, y)\geqslant\Phi_1(x, y)$$

Note 2. If in the domain D the function $f(x, y)$ changes sign, then we divide the domain into two parts: (1) the subdomain D_1

where $f(x, y) \geqslant 0$; (2) the subdomain D_2 where $f(x, y) \leqslant 0$. Suppose the subdomains D_1 and D_2 are such that the double integrals over them exist. Then the integral over D_1 will, be positive and equal to the volume of the solid lying above the xy-plane. The integral over D_2 will be negative and equal, in absolute value, to the volume of the solid lying below the xy-plane. Thus, the integral over D will be expressed as the difference between the corresponding volumes.

2. **Calculating the area of a plane region.** If we form the integral sum of the function $f(x, y) \equiv 1$ over the domain D, then this sum will be equal to the area S,

$$S = \sum_{i=1}^{n} 1 \cdot \Delta s_i$$

for any mode of partition. Passing to the limit on the right side of the equation, we get

$$S = \iint_D dx\, dy$$

If D is regular (see, for instance, Fig. 47), then the area will be expressed by the iterated integral

$$S = \int_a^b \left[\int_{\varphi_1(x)}^{\varphi_2(x)} dy \right] dx$$

Performing the integration in the brackets, we obviously have

$$S = \int_a^b [\varphi_2(x) - \varphi_1(x)]\, dx$$

(cf. Sec. 12.1, Vol. I).

Example 2. Calculate the area of a region bounded by the curves

$$y = 2 - x^2, \quad y = x$$

Solution. Determine the points of intersection of the given curves (Fig. 62). At the point of intersection the ordinates are equal; that is,

$$x = 2 - x^2$$

whence

$$x^2 + x - 2 = 0$$

$$x_1 = -2$$

$$x_2 = 1$$

We get two points of intersection: $M_1(-2, -2)$, $M_2(1, 1)$. Hence, the required area is

$$S = \int_{-2}^{1} \left(\int_{x}^{2-x^2} dy \right) dx = \int_{-2}^{1} (2 - x^2 - x)\, dx = \left[2x - \frac{x^3}{3} - \frac{x^2}{2} \right]_{-2}^{1} = \frac{9}{2}$$

2.5 THE DOUBLE INTEGRAL IN POLAR COORDINATES

Suppose that in a polar coordinate system θ, ρ, a domain D is given such that each ray* passing through an interior point of the region cuts the boundary of D at no more than two points. Suppose that D is bounded by the curves $\rho=\Phi_1(\theta)$, $\rho=\Phi_2(\theta)$ and the rays $\theta=\alpha$ and $\theta=\beta$, where $\Phi_1(\theta)\leqslant\Phi_2(\theta)$ and $\alpha<\beta$ (Fig. 63). Again we shall call such a region a *regular* domain.

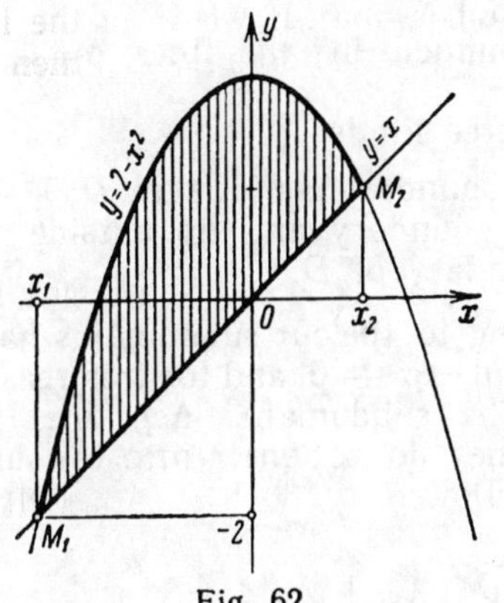

Fig. 62

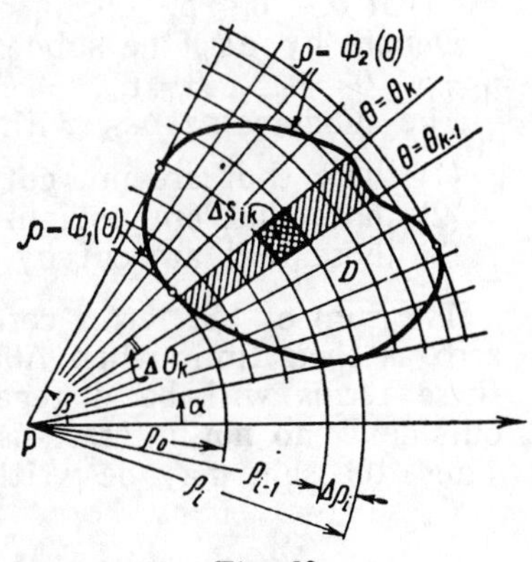

Fig. 63

Let there be given in D a continuous function of the coordinates θ and ρ:

$$z=F(\theta, \rho)$$

We divide D in some way into subdomains $\Delta s_1, \Delta s_2, \ldots, \Delta s_n$. Form the (integral) sum

$$V_n=\sum_{k=1}^{n} F(P_k)\,\Delta s_k \tag{1}$$

where P_k is some point in the subdomain Δs_k.

From the existence theorem of a double integral it follows that as the greatest diameter of the subdomain Δs_k approaches zero, there exists a limit V of the integral sum (1). By definition, this limit V is the double integral of the function $F(\theta, \rho)$ over the domain D:

$$V=\iint_D F(\theta, \rho)\,ds \tag{2}$$

Let us now evaluate this double integral.

---

* A *ray* is any half-line issuing from the coordinate origin, that is, from the pole P.

Since the limit of the sum is independent of the manner of partitioning D into subdomains Δs_k, we can divide the domain in a way that is most convenient. This most convenient (for purposes of calculation) manner will be to partition the domain by means of the rays $\theta=\theta_0$, $\theta=\theta_1$, $\theta=\theta_2$, ..., $\theta=\theta_n$ (where $\theta_0=\alpha$, $\theta_n=\beta$, $\theta_0<\theta_1<\theta_2<\ldots<\theta_n$) and the concentric circles $\rho=\rho_0$, $\rho=\rho_1$, ..., $\rho=\rho_m$ [where ρ_0 is equal to the least value of the function $\Phi_1(\theta)$, and ρ_m, to the greatest value of the function $\Phi_2(\theta)$ in the interval $\alpha\leqslant\theta\leqslant\beta$, $\rho_0<\rho_1<\ldots<\rho_m$].

Denote by Δs_{ik} the subdomain bounded by the lines $\rho=\rho_{i-1}$, $\rho=\rho_i$, $\theta=\theta_{k-1}$, $\theta=\theta_k$.

The subdomains Δs_{ik} will be of three kinds:

(1) those that are not cut by the boundary and lie in D;
(2) those that are not cut by the boundary and lie outside D;
(3) those that are cut by the boundary of D.

The sum of the terms corresponding to the cut subdomains have zero as their limit when $\Delta\theta_k\to 0$ and $\Delta\rho_i\to 0$ and for this reason these terms will be disregarded. The subdomains Δs_{ik} that lie outside D do not interest us since they do not enter into the sum. Thus, the sum may be written as follows:

$$V_n=\sum_{k=1}^{n}\left[\sum_{i}F(P_{ik})\,\Delta s_{ik}\right]$$

where P_{ik} is an arbitrary point of the subdomain Δs_{ik}.

The double summation sign here should be understood as meaning that we first perform the summation with respect to the index i, holding k fast (that is, we pick out all terms that correspond to the subdomains lying between two adjacent rays*). The outer summation sign signifies that we take together all the sums obtained in the first summation (that is, we sum with respect to the index k).

Let us find the expression of the area of the subdomain Δs_{ik} that is not cut by the boundary of the domain. It will be equal to the difference of the areas of the two sectors:

$$\Delta s_{ik}=\frac{1}{2}(\rho_i+\Delta\rho_i)^2\,\Delta\theta_k-\frac{1}{2}\rho_i^2\Delta\theta_k=\left(\rho_i+\frac{\Delta\rho_i}{2}\right)\Delta\rho_i\,\Delta\theta_k$$

or

$$\Delta s_{ik}=\rho_i^*\,\Delta\rho_i\Delta\theta_k,\quad \text{where } \rho_i<\rho_i^*<\rho_i+\Delta\rho_i$$

---

* Note that in summing with respect to the index i this index will not run through all values from 1 to m, because not all of the subdomains lying between the rays $\theta=\theta_k$ and $\theta=\theta_{k+1}$ belong to D.

Thus, the integral sum will have the form*

$$V_n = \sum_{k=1}^{n} \left[\sum_i F(\theta_k^*, \rho_i^*) \rho_i^* \Delta\rho_i \Delta\theta_k \right]$$

where $P(\theta_k^*, \rho_i^*)$ is a point of the subdomain Δs_{ik}. Now take the factor $\Delta\theta_k$ outside the sign of the inner sum (this is permissible since it is a common factor for all the terms of this sum):

$$V_n = \sum_{k=1}^{n} \left[\sum_i F(\theta_k^*, \rho_i^*) \rho_i^* \Delta\rho_i \right] \Delta\theta_k$$

Suppose that $\Delta\rho_i \to 0$ and $\Delta\theta_k$ remains constant. Then the expression in the brackets will tend to the integral

$$\int_{\Phi_1(\theta_k^*)}^{\Phi_2(\theta_k^*)} F(\theta_k^*, \rho) \rho \, d\rho$$

Now, assuming that $\Delta\theta_k \to 0$, we finally get **

$$V = \int_{\alpha}^{\beta} \left(\int_{\Phi_1(\theta)}^{\Phi_2(\theta)} F(\theta, \rho) \rho \, d\rho \right) d\theta \qquad (3)$$

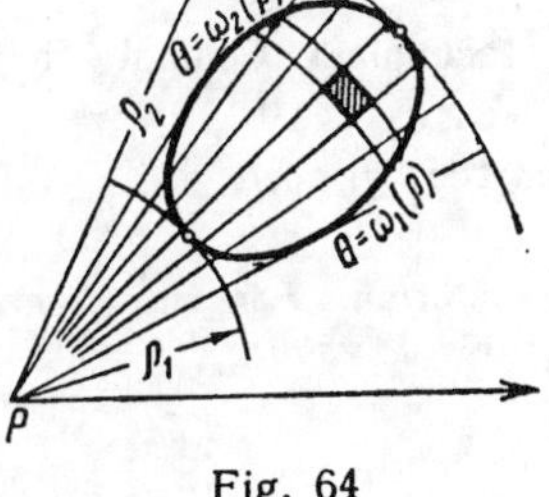

Fig. 64

Formula (3) is used to compute double integrals in polar coordinates.

If the first integration is performed with respect to θ and the second one to ρ, then we get the formula (Fig. 64)

$$V = \int_{\rho_1}^{\rho_2} \left(\int_{\omega_1(\rho)}^{\omega_2(\rho)} F(\theta, \rho) \, d\theta \right) \rho \, d\rho \qquad (3')$$

* We can consider the integral sum in this form because the limit of the sum does not depend on the position of the point inside the subdomain.

** Our derivation of formula (3) is not rigorous; in deriving this formula we first let $\Delta\rho_i$ approach zero, leaving $\Delta\theta_k$ constant, and only then made $\Delta\theta_k$ approach zero. This does not exactly correspond to the definition of a double integral, which we regard as the limit of an integral sum as the diameters of the subdomains approach zero (i.e., in the simultaneous approach to zero of $\Delta\theta_k$ and $\Delta\rho_i$). However, though the proof lacks rigour, the result is true [i.e., formula (3) is true]. This formula could be rigorously derived by the method used when considering the double integral in rectangular coordinates. We also note that this formula will be derived once again in Sec. 2.6 with different reasoning (as a particular case of the more general formula for transforming coordinates in a double integral).

Let it be required to compute the double integral of a function $f(x, y)$ over a domain D given in rectangular coordinates:

$$\iint_D f(x, y)\, dx\, dy$$

If D is regular in the polar coordinates θ, ρ, then the computation of the given integral can be reduced to computing the iterated integral in polar coordinates.

Indeed, since

$$x = \rho\cos\theta, \quad y = \rho\sin\theta$$
$$f(x, y) = f[\rho\cos\theta, \rho\sin\theta] = F(\theta, \rho),$$

it follows that

$$\iint_D f(x, y)\, dx\, dy = \int_\alpha^\beta \left(\int_{\Phi_1(\theta)}^{\Phi_2(\theta)} f[\rho\cos\theta, \rho\sin\theta]\rho\, d\rho \right) d\theta \tag{4}$$

Example 1. Compute the volume V of a solid bounded by the spherical surface

$$x^2 + y^2 + z^2 = 4a^2$$

and the cylinder

$$x^2 + y^2 - 2ay = 0$$

Solution. For the domain of integration here we can take the base of the cylinder $x^2 + y^2 - 2ay = 0$, that is, a circle with centre at $(0, a)$ and radius a. The equation of this circle may be written in the form $x^2 + (y-a)^2 = a^2$ (Fig. 65).

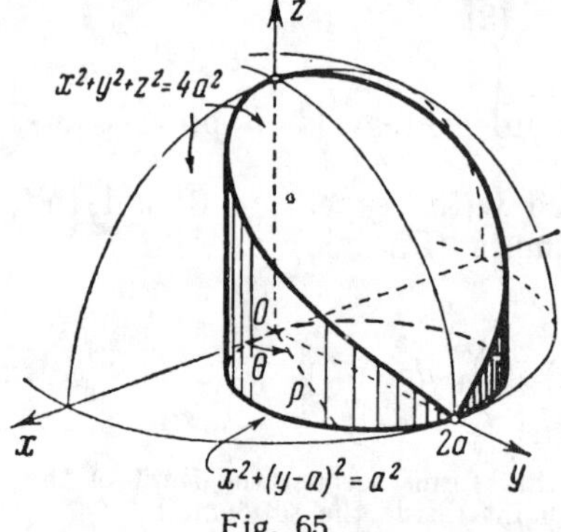

Fig. 65

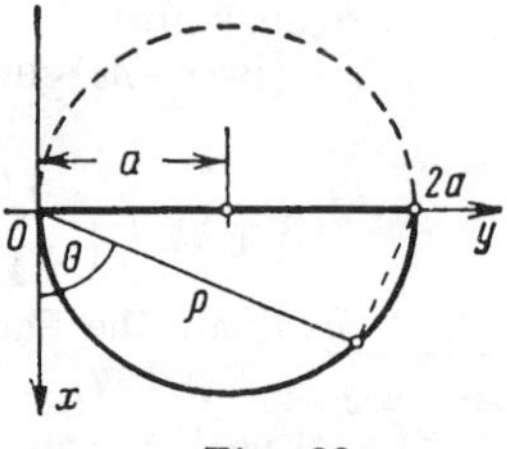

Fig. 66

We calculate $\frac{1}{4}$ of the required volume V, namely that part which is situated in the first octant. Then for the domain of integration we will have to take the semicircle whose boundaries are defined by the equations

$$x = \varphi_1(y) = 0, \quad x = \varphi_2(y) = \sqrt{2ay - y^2}$$
$$y = 0, \quad y = 2a$$

The integrand is

$$z = f(x, y) = \sqrt{ x^2 - y^2}$$

Consequently

$$\frac{1}{4}V=\int_0^{2a}\left(\int_0^{\sqrt{2ay-y^2}}\sqrt{4a^2-x^2-y^2}\,dx\right)dy$$

Transform the integral obtained to the polar coordinates θ, ρ:

$$x=\rho\cos\theta, \qquad y=\rho\sin\theta$$

Determine the limits of integration. To do so, write the equation of the given circle in polar coordinates; since

$$x^2+y^2=\rho^2$$
$$y=\rho\sin\theta$$

it follows that

$$\rho^2-2a\rho\sin\theta=0$$

or

$$\rho=2a\sin\theta$$

Hence, in polar coordinates (Fig. 66), the boundaries of the domain are defined by the equations

$$\rho=\Phi_1(\theta)=0, \quad \rho=\Phi_2(\theta)=2a\sin\theta, \quad \alpha=0, \quad \beta=\frac{\pi}{2}$$

and the integrand has the form

$$F(\theta,\rho)=\sqrt{4a^2-\rho^2}$$

Thus, we have

$$\frac{V}{4}=\int_0^{\frac{\pi}{2}}\left(\int_0^{2a\sin\theta}\sqrt{4a^2-\rho^2}\,\rho\,d\rho\right)d\theta=\int_0^{\frac{\pi}{2}}\left[-\frac{(4a^2-\rho^2)^{3/2}}{3}\right]_0^{2a\sin\theta}d\theta$$

$$=-\frac{1}{3}\int_0^{\frac{\pi}{2}}\left[(4a^2-4a^2\sin^2\theta)^{3/2}-(4a^2)^{3/2}\right]d\theta=\frac{8a^3}{3}\int_0^{\frac{\pi}{2}}(1-\cos^3\theta)\,d\theta$$

$$=\frac{4}{9}a^3(3\pi-4)$$

Example 2. Evaluate the Poisson integral

$$\int_{-\infty}^{+\infty}e^{-x^2}\,dx$$

Solution. First evaluate the integral $I_R=\iint\limits_D e^{-x^2-y^2}\,dx\,dy$, where the domain of integration D is the circle $x^2+y^2=R^2$ (Fig. 67).

Passing to the polar coordinates θ, ρ, we obtain

$$I_R=\int_0^{2\pi}\left(\int_0^R e^{-\rho^2}\rho\,d\rho\right)d\theta=-\frac{1}{2}\int_0^{2\pi}e^{-\rho^2}\Big|_0^R\,d\theta=\pi(1-e^{-R^2})$$

Now, if we increase the radius R without bound (that is, if we expand without limit the domain of integration), we get the so-called *improper iterated integral:*

$$\int_0^{2\pi}\left(\int_a^{\infty} e^{-\rho^2}\rho\,d\rho\right)d\theta=\lim_{R\to\infty}\int_0^{2\pi}\left(\int_0^{R} e^{-\rho^2}\rho\,d\rho\right)d\theta=\lim_{R\to\infty}\pi\left(1-e^{-R^2}\right)=\pi$$

We shall show that the integral $\iint\limits_{D} e^{-x^2-y^2}\,dx\,dy$ approaches the limit π if a domain D' of arbitrary form expands in such manner that finally any point of the plane is in D' and remains there (we shall conditionally indicate such an expansion of D' by the relationship $D'\longrightarrow\infty$).

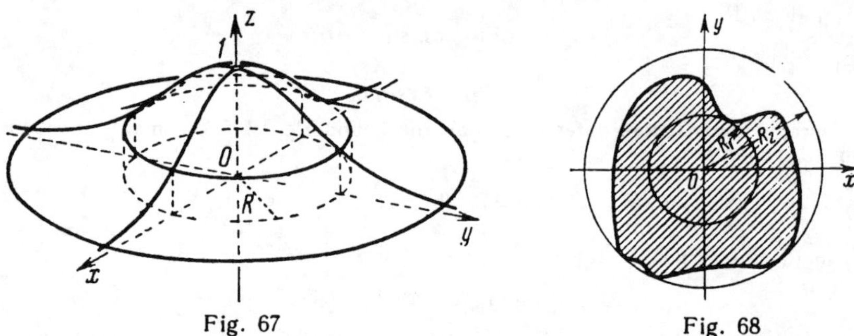

Fig. 67 Fig. 68

Let R_1 and R_2 be the least and greatest distances of the boundary of D' from the origin (Fig. 68).

Since the function $e^{-x^2-y^2}$ is everywhere greater than zero, the following inequalities hold:

$$I_{R_1}\leqslant\iint\limits_{D'} e^{-x^2-y^2}\,dx\,dy\leqslant I_{R_2}$$

or

$$\pi\left(1-e^{-R_1^2}\right)\leqslant\iint\limits_{D'} e^{-x^2-y^2}\,dx\,dy\leqslant\pi\left(1-e^{-R_2^2}\right)$$

Since as $D'\longrightarrow\infty$ it is obvious that $R_1\longrightarrow\infty$ and $R_2\longrightarrow\infty$, it follows that the extreme parts of the inequality tend to the same limit π. Hence, the middle term also approaches this limit; that is,

$$\lim_{D'\to\infty}\iint\limits_{D'} e^{-x^2-y^2}\,dx\,dy=\pi \tag{5}$$

As a particular instance, let D' be a square with side $2a$ and centre at the origin; then

$$\iint\limits_{D'} e^{-x^2-y^2}\,dx\,dy=\int_{-a}^{a}\int_{-a}^{a} e^{-x^2-y^2}\,dx\,dy$$

$$=\int_{-a}^{a}\int_{-a}^{a} e^{-x^2}e^{-y^2}\,dx\,dy=\int_{-a}^{a}\left(\int_{-a}^{a} e^{-x^2}e^{-y^2}\,dx\right)dy$$

Now take the factor e^{-y^2} outside the sign of the inner integral (this is permissible since e^{-y^2} does not depend on the variable of integration x). Then

$$\iint_{D'} e^{-x^2-y^2}\,dx\,dy = \int_{-a}^{a} e^{-y^2}\left(\int_{-a}^{a} e^{-x^2}\,dx\right)dy$$

Set $\int_{-a}^{a} e^{-x^2}\,dx = B_a$. This is a constant (dependent only on a); therefore,

$$\iint_{D'} e^{-x^2-y^2}\,dx\,dy = \int_{-a}^{a} e^{-y^2} B_a\,dy = B_a \int_{-a}^{a} e^{-y^2}\,dy$$

But the latter integral is likewise equal to $B_a\left(\text{because } \int_{-a}^{a} e^{-x^2}\,dx = \int_{-a}^{a} e^{-y^2}\,dy\right)$; thus,

$$\iint_{D'} e^{-x^2-y^2}\,dx\,dy = B_a B_a = B_a^2$$

We pass to the limit in this equation, by making a approach infinity (in the process, D' expands without limit):

$$\lim_{D'\to\infty}\iint_{D'} e^{-x^2-y^2}\,dx\,dy = \lim_{a\to\infty} B_a^2 = \lim_{a\to\infty}\left[\int_{-a}^{a} e^{-x^2}\,dx\right]^2 = \left[\int_{-\infty}^{+\infty} e^{-x^2}\,dx\right]^2$$

But, as has been proved [see (5)],

$$\lim_{D'\to\infty}\iint_{D'} e^{-x^2-y^2}\,dx\,dy = \pi$$

Hence,

$$\left[\int_{-\infty}^{\infty} e^{-x^2}\,dx\right]^2 = \pi$$

or

$$\int_{-\infty}^{\infty} e^{-x^2}\,dx = \sqrt{\pi}$$

This integral is frequently encountered in probability theory and in statistics. We remark that we would not be able to compute this integral directly (by means of an indefinite integral) because the antiderivative of e^{-x^2} is not expressible in terms of elementary functions.

2.6 CHANGE OF VARIABLES IN A DOUBLE INTEGRAL (GENERAL CASE)

In the *xy*-plane let there be a domain D bounded by a line L. Suppose that the coordinates x and y are functions of new variables u and v:

$$x=\varphi(u, v), \quad y=\psi(u, v) \tag{1}$$

Let the functions $\varphi(u, v)$ and $\psi(u, v)$ be single-valued and continuous, and let them have continuous derivatives in some domain D', which will be defined later on. Then by formulas (1) to each pair of values u and v there corresponds a unique pair of values x and y. Further, suppose that the functions φ and ψ are such that if we give x and y definite values in D, then by formulas (1) we will find definite values of u and v.

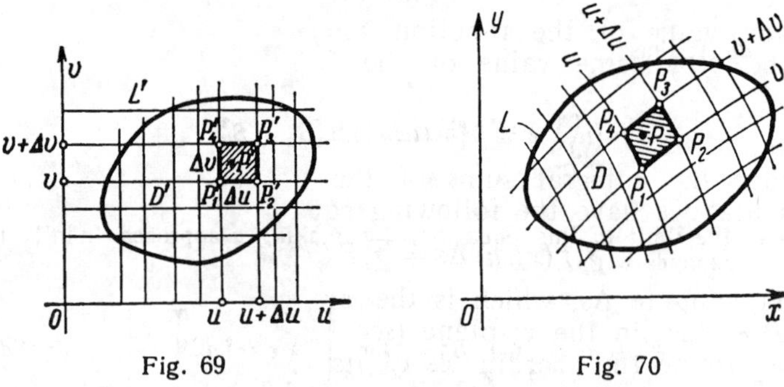

Fig. 69 Fig. 70

Consider a rectangular coordinate system Ouv (Fig. 69). From the foregoing it follows that with each point $P(x, y)$ in the *xy*-plane (Fig. 70) there is uniquely associated a point $P'(u, v)$ in the *uv*-plane with coordinates u, v, which are determined by formulas (1). The numbers u and v are called *curvilinear* coordinates of the point P.

If in the *xy*-plane a point describes a closed line L bounding the domain D, then in the *uv*-plane a corresponding point will trace out a closed line L' bounding a certain domain D'; and to each point of D' there will correspond a point of D.

Thus, the formulas (1) establish a *one-to-one correspondence between the points of the domains D and D'*, or, the *mapping*, by formulas (1), *of D onto D' is said to be one-to-one.*

In the domain D' let us consider a line $u=\text{const}$. By formulas (1) we find that in the *xy*-plane there will, generally speaking, be a certain curve corresponding to it. In exactly the same way, to each straight line $v=\text{const}$ of the *uv*-plane there will correspond some line in the *xy*-plane.

Let us divide D' (using the straight lines $u=\text{const}$ and $v=\text{const}$) into rectangular subdomains (we shall disregard subdomains that overlap the boundary of the region D'). Using suitable curves, divide D into certain curvilinear quadrangles (Fig. 70).

Consider, in the uv-plane, a rectangular subdomain $\Delta s'$ bounded by the straight lines $u=\text{const}$, $u+\Delta u=\text{const}$, $v=\text{const}$, $v+\Delta v=\text{const}$, and consider also the curvilinear subdomain Δs corresponding to it in the xy-plane. We denote the areas of these subdomains by $\Delta s'$ and Δs, respectively. Then, obviously,

$$\Delta s'=\Delta u\,\Delta v$$

Generally speaking, the areas Δs and $\Delta s'$ are different.

Suppose in D we have a continuous function

$$z=f(x, y)$$

To each value of the function $z=f(x, y)$ in D there corresponds the very same value of the function $z=F(u, v)$ in D', where

$$F(u, v)=f[\varphi(u, v), \quad \psi(u, v)]$$

Consider the integral sums of the function z over D. It is obvious that we have the following equation:

$$\sum f(x, y)\,\Delta s=\sum F(u, v)\,\Delta s \tag{2}$$

Let us compute Δs, which is the area of the curvilinear quadrangle $P_1P_2P_3P_4$ in the xy-plane (see Fig. 70).

We determine the coordinates of its vertices:

$$\left.\begin{array}{lll} P_1(x_1, y_1), & x_1=\varphi(u, v), & y_1=\psi(u, v) \\ P_2(x_2, y_2), & x_2=\varphi(u+\Delta u, v), & y_2=\psi(u+\Delta u, v) \\ P_3(x_3, y_3), & x_3=\varphi(u+\Delta u, v+\Delta v), & y_3=\psi(u+\Delta u, v+\Delta v) \\ P_4(x_4, y_4), & x_4=\varphi(u, v+\Delta v), & y_4=\psi(u, v+\Delta v) \end{array}\right\} \tag{3}$$

When computing the area of the curvilinear quadrangle $P_1P_2P_3P_4$ we shall consider the lines P_1P_2, P_2P_3, P_3P_4, P_4P_1 as parallel in pairs; we shall also replace the increments of the functions by corresponding differentials. We shall thus ignore infinitesimals of order higher than the infinitesimals Δu, Δv. Then formulas (3) will have the form

$$\left.\begin{array}{ll} x_1=\varphi(u, v); & y_1=\psi(u, v) \\ x_2=\varphi(u, v)+\dfrac{\partial\varphi}{\partial u}\Delta u, & y_2=\psi(u, v)+\dfrac{\partial\psi}{\partial u}\Delta u \\ x_3=\varphi(u, v)+\dfrac{\partial\varphi}{\partial u}\Delta u+\dfrac{\partial\varphi}{\partial v}\Delta v, & y_3=\psi(u, v)+\dfrac{\partial\psi}{\partial u}\Delta u+\dfrac{\partial\psi}{\partial v}\Delta v \\ x_4=\varphi(u, v)+\dfrac{\partial\varphi}{\partial v}\Delta v, & y_4=\psi(u, v)+\dfrac{\partial\psi}{\partial v}\Delta v \end{array}\right\} \tag{3'}$$

With these assumptions, the curvilinear quadrangle $P_1P_2P_3P_4$ may be regarded as a parallelogram. Its area Δs is approximately equal to the doubled area of the triangle $P_1P_2P_3$ and is found by the following formula of analytic geometry:

$$\Delta s \approx |(x_3-x_1)(y_3-y_2)-(x_3-x_2)(y_3-y_1)|$$
$$=\left|\left(\frac{\partial\varphi}{\partial u}\Delta u+\frac{\partial\varphi}{\partial v}\Delta v\right)\frac{\partial\psi}{\partial v}\Delta v-\frac{\partial\varphi}{\partial v}\Delta v\left(\frac{\partial\psi}{\partial u}\Delta u+\frac{\partial\psi}{\partial v}\Delta v\right)\right|$$
$$=\left|\frac{\partial\varphi}{\partial u}\frac{\partial\psi}{\partial v}\Delta u\,\Delta v-\frac{\partial\varphi}{\partial v}\frac{\partial\psi}{\partial u}\Delta u\Delta v\right|=\left|\frac{\partial\varphi}{\partial u}\frac{\partial\psi}{\partial v}-\frac{\partial\varphi}{\partial v}\frac{\partial\psi}{\partial u}\right|\Delta u\,\Delta v$$
$$=\left\|\begin{matrix}\frac{\partial\varphi}{\partial u} & \frac{\partial\varphi}{\partial v}\\ \frac{\partial\psi}{\partial u} & \frac{\partial\psi}{\partial v}\end{matrix}\right\|\Delta u\,\Delta v$$

Here, the outer vertical lines indicate that the absolute value of the determinant is taken. We introduce the notation

$$\left|\begin{matrix}\frac{\partial\varphi}{\partial u} & \frac{\partial\varphi}{\partial v}\\ \frac{\partial\psi}{\partial u} & \frac{\partial\psi}{\partial v}\end{matrix}\right|=I$$

Thus,

$$\Delta s\approx|I|\,\Delta s' \tag{4}$$

The determinant I is called the *functional determinant* of the functions $\varphi(u, v)$ and $\psi(u, v)$. It is also called the *Jacobian* after the German mathematician Jacobi.

Equation (4) is only approximate, because in the process of computing the area of Δs we neglected infinitesimals of higher order. However, the smaller the dimensions of the subdomains Δs and $\Delta s'$, the more exact will this equation be. And it becomes absolutely exact in the limit, when the diameters of the subdomains Δs and $\Delta s'$ approach zero:

$$|I|=\lim_{\text{diam }\Delta s'\to 0}\frac{\Delta s}{\Delta s'}$$

Let us now apply the equation obtained to an evaluation of the double integral. From (2) we can write

$$\sum f(x, y)\,\Delta s\approx\sum F(u, v)\,|I|\,\Delta s'$$

(the integral sum on the right is extended over the domain D'). Passing to the limit as diam $\Delta s'\to 0$, we get the exact equation

$$\iint\limits_D f(x, y)\,dx\,dy=\iint\limits_{D'} F(u, v)\,|I|\,du\,dv \tag{5}$$

This is the *formula for transformation of coordinates in a double integral*. It permits reducing the evaluation of a double integral over a domain D to the computation of a double integral over a domain D', which may simplify the problem. A rigorous proof of this formula was first given by the noted Russian mathematician M. V. Ostrogradsky.

Note. The transformation from rectangular coordinates to polar coordinates considered in the preceding section is a special case of change of variables in a double integral. Here, $u=\theta$, $v=\rho$:

$$x=\rho\cos\theta, \qquad y=\rho\sin\theta$$

The curve AB $(\rho=\rho_1)$ in the xy-plane (Fig. 71) is transformed into the straight line $A'B'$ in the $\theta\rho$-plane (Fig. 72). The curve

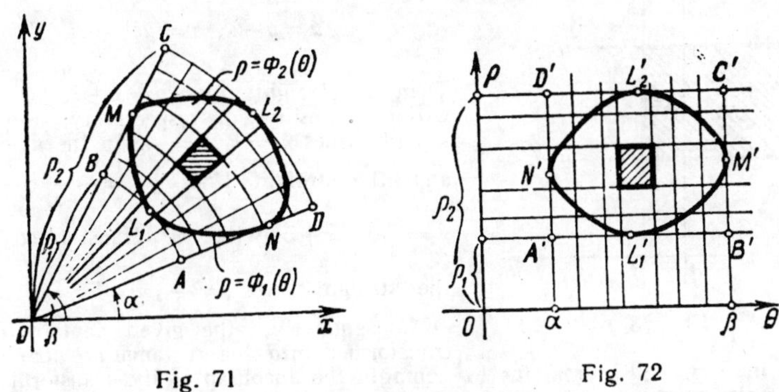

Fig. 71 Fig. 72

DC $(\rho=\rho_2)$ in the xy-plane is transformed into the straight line $D'C'$ in the $\theta\rho$-plane.

The straight lines AD and BC in the xy-plane are transformed into the straight lines $A'D'$ and $B'C'$ in the $\theta\rho$-plane. The curves L_1 and L_2 are transformed into the curves L_1' and L_2'.

Let us calculate the Jacobian of a transformation of the Cartesian coordinates x and y into the polar coordinates θ and ρ:

$$I=\begin{vmatrix} \frac{\partial x}{\partial \theta} & \frac{\partial x}{\partial \rho} \\ \frac{\partial y}{\partial \theta} & \frac{\partial y}{\partial \rho} \end{vmatrix} = \begin{vmatrix} -\rho\sin\theta & \cos\theta \\ \rho\cos\theta & \sin\theta \end{vmatrix} = -\rho\sin^2\theta-\rho\cos^2\theta=-\rho$$

Hence, $|I|=\rho$ and therefore

$$\iint\limits_D f(x, y)\,dx\,dy=\int\limits_\alpha^\beta\left(\int\limits_{\Phi_1(\theta)}^{\Phi_2(\theta)} F(\theta, \rho)\,\rho\,d\rho\right)d\theta$$

This was the formula that we derived in the preceding section.

Example. Let it be required to compute the double integral

$$\iint\limits_D (y-x)\,dx\,dy$$

over the region D in the xy-plane bounded by the straight lines

$$y=x+1,\quad y=x-3,\ y=-\frac{1}{3}x+\frac{7}{3},\quad y=-\frac{1}{3}x+5$$

It would be difficult to compute this double integral directly; however, a simple change of variables permits reducing this integral to one over a rectangle whose sides are parallel to the coordinate axes.

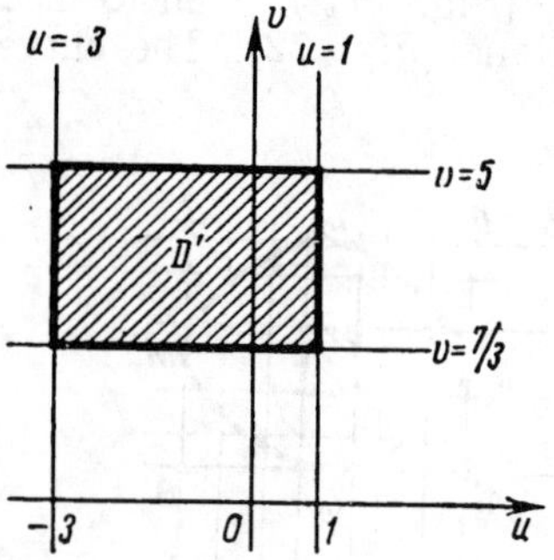

Set

$$u=y-x,\quad v=y+\frac{1}{3}x \qquad (6)$$

Then the straight lines $y=x+1$, $y=x-3$ will be transformed, respectively, into the straight lines $u=1$, $u=-3$ in the uv-plane; and the straight lines $y=-\frac{1}{3}x+\frac{7}{3}$, $y=-\frac{1}{3}x+5$ will be transformed into the straight lines $v=\frac{7}{3}$, $v=5$.

Fig. 73

Consequently, the given domain D is transformed into the rectangular domain D' shown in Fig. 73. It remains to compute the Jacobian of the transformation. To do this, express x and y in terms of u and v. Solving the system of equations (6), we obtain

$$x=-\frac{3}{4}u+\frac{3}{4}v,\quad y=\frac{1}{4}u+\frac{3}{4}v$$

Consequently,

$$I=\begin{vmatrix} \frac{\partial x}{\partial u} & \frac{\partial x}{\partial v} \\ \frac{\partial y}{\partial u} & \frac{\partial y}{\partial v} \end{vmatrix} = \begin{vmatrix} -\frac{3}{4} & \frac{3}{4} \\ \frac{1}{4} & \frac{3}{4} \end{vmatrix} = -\frac{9}{16}-\frac{3}{16}=-\frac{3}{4}$$

and the absolute value of the Jacobian is $|I|=\frac{3}{4}$. Therefore,

$$\iint\limits_D (y-x)\,dx\,dy=\iint\limits_{D'}\left[\left(+\frac{1}{4}u+\frac{3}{4}v\right)-\left(-\frac{3}{4}u+\frac{3}{4}v\right)\right]\frac{3}{4}\,du\,dv$$

$$=\iint\limits_{D'}\frac{3}{4}u\,du\,dv=\int\limits_{\frac{7}{3}}^{5}\int\limits_{-3}^{1}\frac{3}{4}u\,du\,dv=-8$$

2.7 COMPUTING THE AREA OF A SURFACE

Let it be required to compute the area of a surface bounded by a curve Γ (Fig. 74); the surface is defined by the equation $z=f(x, y)$, where the function $f(x, y)$ is continuous and has continuous partial derivatives. Denote the projection of Γ on the xy-plane by L. Denote by D the domain on the xy-plane bounded by the curve L.

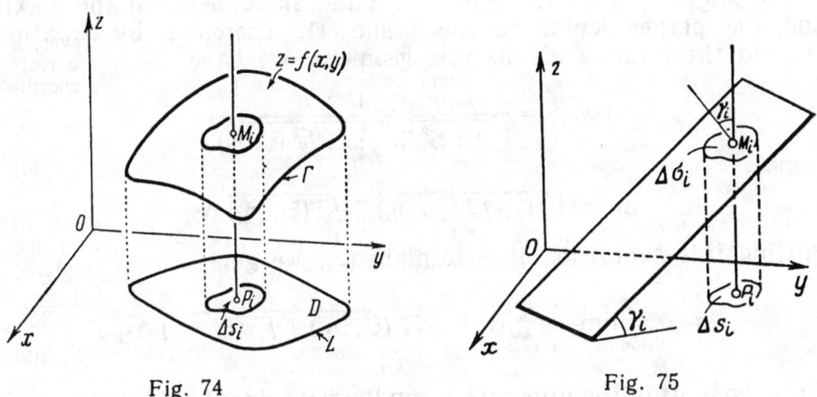

Fig. 74 Fig. 75

In arbitrary fashion, divide D into n elementary subdomains $\Delta s_1, \Delta s_2, \ldots, \Delta s_n$. In each subdomain Δs_i take a point $P_i(\xi_i, \eta_i)$. To the point P_i there will correspond, on the surface, a point

$$M_i[\xi_i, \eta_i, f(\xi_i, \eta_i)]$$

Through M_i draw a tangent plane to the surface. Its equation is of the form

$$z-z_i=f'_x(\xi_i, \eta_i)(x-\xi_i)+f'_y(\xi_i, \eta_i)(y-\eta_i) \tag{1}$$

(see Sec. 9.6, Vol. I). In this plane, pick out a subdomain $\Delta\sigma_i$ which is projected onto the xy-plane in the form of a subdomain Δs_i. Consider the sum of all the subdomains $\Delta\sigma_i$:

$$\sum_{i=1}^{n}\Delta\sigma_i$$

We shall call the limit σ of this sum, when the greatest of the diameters of the subdomains $\Delta\sigma_i$ approaches zero, the *area of the surface*; that is, by definition we set

$$\sigma=\lim_{\operatorname{diam}\Delta\sigma_i\to 0}\sum_{i=1}^{n}\Delta\sigma_i \tag{2}$$

Now let us calculate the area of the surface. Denote by γ_i the angle between the tangent plane and the xy-plane. Using a familiar formula of analytic geometry we can write (Fig. 75)

$$\Delta s_i = \Delta\sigma_i \cos\gamma_i$$

or

$$\Delta\sigma_i = \frac{\Delta s_i}{\cos\gamma_i} \tag{3}$$

The angle γ_i is at the same time the angle between the z-axis and the perpendicular to the plane (1). Therefore, by equation (1) and the formula of analytic geometry we have

$$\cos\gamma_i = \frac{1}{\sqrt{1+f_x'^2(\xi_i, \eta_i)+f_y'^2(\xi_i, \eta_i)}}$$

Hence,

$$\Delta\sigma_i = \sqrt{1+f_x'^2(\xi_i, \eta_i)+f_y'^2(\xi_i, \eta_i)}\,\Delta s_i$$

Putting this expression into formula (2), we get

$$\sigma = \lim_{\operatorname{diam}\Delta s_i \to 0} \sum_{i=1}^{n} \sqrt{1+f_x'^2(\xi_i, \eta_i)+f_y'^2(\xi_i, \eta_i)}\,\Delta s_i$$

Since the limit of the integral sum on the right side of the last equation is, by definition, the double integral $\iint\limits_D \sqrt{1+\left(\frac{\partial z}{\partial x}\right)^2+\left(\frac{\partial z}{\partial y}\right)^2}\,dx\,dy$, we finally get

$$\sigma = \iint\limits_D \sqrt{1+\left(\frac{\partial z}{\partial x}\right)^2+\left(\frac{\partial z}{\partial y}\right)^2}\,dx\,dy \tag{4}$$

This is the formula used to compute the area of the surface $z=f(x, y)$.

If the equation of the surface is given in the form

$$x=\mu(y, z) \text{ or in the form } y=\chi(x, z)$$

then the corresponding formulas for calculating the surface area are of the form

$$\sigma = \iint\limits_{D'} \sqrt{1+\left(\frac{\partial x}{\partial y}\right)^2+\left(\frac{\partial x}{\partial z}\right)^2}\,dy\,dz \tag{3'}$$

$$\sigma = \iint\limits_{D''} \sqrt{1+\left(\frac{\partial y}{\partial x}\right)^2+\left(\frac{\partial y}{\partial z}\right)^2}\,dx\,dz \tag{3''}$$

where D' and D'' are the domains in the xy-plane and the xz-plane in which the given surface is projected.

Example 1. Compute the surface area σ of the sphere

$$x^2+y^2+z^2=R^2$$

Solution. Compute the surface area of the upper half of the sphere:

$$z=\sqrt{R^2-x^2-y^2}$$

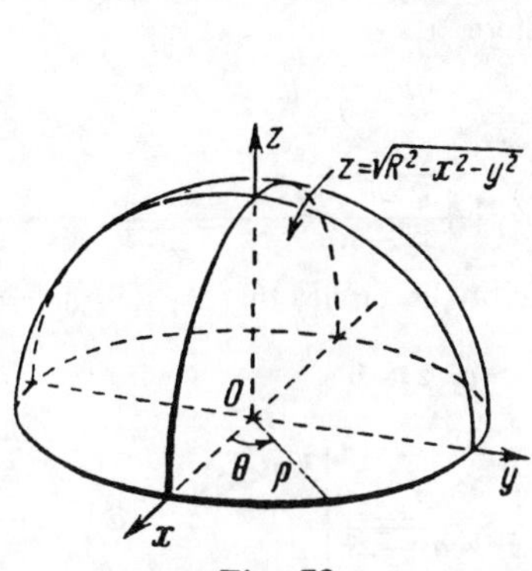

Fig. 76

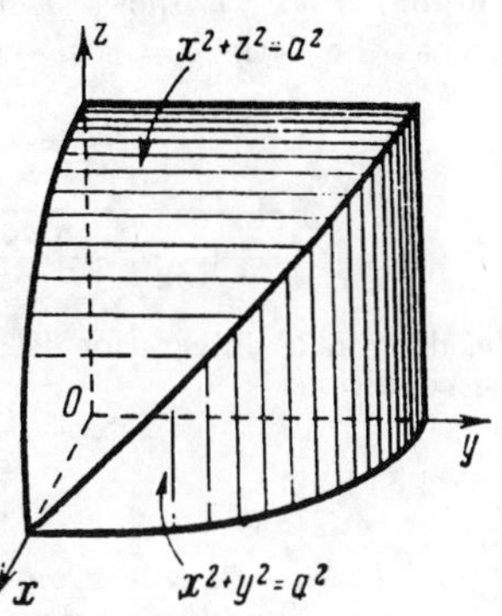

Fig. 77

Fig. 76). In this case

$$\frac{\partial z}{\partial x}=-\frac{x}{\sqrt{R^2-x^2-y^2}}$$

$$\frac{\partial z}{\partial y}=-\frac{y}{\sqrt{R^2-x^2-y^2}}$$

Hence,

$$\sqrt{1+\left(\frac{\partial z}{\partial x}\right)^2+\left(\frac{\partial z}{\partial y}\right)^2}=\sqrt{\frac{R^2}{R^2-x^2-y^2}}=\frac{R}{\sqrt{R^2-x^2-y^2}}$$

The domain of integration is defined by the condition

$$x^2+y^2\leqslant R^2$$

Thus, by formula (4) we will have

$$\frac{1}{2}\sigma=\int\limits_{-R}^{R}\left(\int\limits_{-\sqrt{R^2-x^2}}^{\sqrt{R^2-x^2}}\frac{R}{\sqrt{R^2-x^2-y^2}}dy\right)dx$$

To compute the double integral obtained let us make the transformation to polar coordinates. In polar coordinates the boundary of the domain of integration is determined by the equation $\rho=R$. Hence,

$$\sigma=2\int\limits_0^{2\pi}\left(\int\limits_0^R\frac{R}{\sqrt{R^2-\rho^2}}\rho\,d\rho\right)d\theta=2R\int\limits_0^{2\pi}\left[-\sqrt{R^2-\rho^2}\right]_0^R d\theta$$

$$=2R\int\limits_0^{2\pi}R\,d\theta=4\pi R^2$$

Example 2. Find the area of that part of the surface of the cylinder

$$x^2+y^2=a^2$$

which is cut out by the cylinder

$$x^2+z^2=a^2$$

Solution. Fig. 77 shows 1/8th of the desired surface. The equation of the surface has the form $y=\sqrt{a^2-x^2}$; therefore,

$$\frac{\partial y}{\partial x}=-\frac{x}{\sqrt{a^2-x^2}}, \quad \frac{\partial y}{\partial z}=0$$

$$\sqrt{1+\left(\frac{\partial y}{\partial x}\right)^2+\left(\frac{\partial y}{\partial z}\right)^2}=\sqrt{1+\frac{x^2}{a^2-x^2}}=\frac{a}{\sqrt{a^2-x^2}}$$

The domain of integration is a quarter of the circle, that is, it is determined by the conditions

$$x^2+z^2\leqslant a^2, \ x\geqslant 0, \ z\geqslant 0$$

Consequently,

$$\frac{1}{8}\sigma=\int_0^a\left(\int_0^{\sqrt{a^2-x^2}}\frac{a}{\sqrt{a^2-x^2}}dz\right)dx=a\int_0^a \frac{z}{\sqrt{a^2-x^2}}\Bigg|_0^{\sqrt{a^2-x^2}}dx=a\int_0^a dx=a^2$$

$$\sigma=8a^2$$

2.8 THE DENSITY DISTRIBUTION OF MATTER AND THE DOUBLE INTEGRAL

In a domain D, let a certain substance be distributed in such manner that there is a definite amount per unit area of D. We shall henceforward speak of the distribution of **mass**, although our reasoning will hold also for the case when speaking of the distribution of electric charge, quantity of heat, and so forth.

We consider an arbitrary subdomain Δs of the domain D. Let the mass of substance associated with this given subdomain be Δm. Then the ratio $\frac{\Delta m}{\Delta s}$ is called the mean surface density of the substance in the subdomain Δs.

Now let the subdomain Δs decrease and contract to the point $P(x, y)$. Consider the limit $\lim\limits_{\Delta s\to 0}\frac{\Delta m}{\Delta s}$. If this limit exists, then, generally speaking, it will depend on the position of the point P, that is, upon its coordinates x and y, and will be some function $f(P)$ of the point P. We shall call this limit the *surface density* of the substance at the point P:

$$\lim_{\Delta s\to 0}\frac{\Delta m}{\Delta s}=f(P)=f(x, y) \qquad (1)$$

Thus, the surface density is a function $f(x, y)$ of the coordinates of the point of the domain.

Conversely, in a domain D, let the surface density of some substance be given as a continuous function $f(P)=f(x, y)$ and let it be required to determine the total quantity of substance M contained in D. Divide D into subdomains Δs_i $(i=1, 2, \ldots, n)$ and in each subdomain take a point P_i; then $f(P_i)$ is the surface density at the point P_i.

To within higher-order infinitesimals, the product $f(P_i)\Delta s_i$ gives us the quantity of substance contained in the subdomain Δs_i, and the sum

$$\sum_{i=1}^{n} f(P_i)\Delta s_i$$

expresses approximately the total quantity of substance distributed in the domain D. But this is the integral sum of the function $f(P)$ in D. The exact value is obtained in the limit as $\Delta s_i \to 0$. Thus,*

$$M=\lim_{\Delta s_i \to 0} \sum_{i=1}^{n} f(P_i)\Delta s_i=\iint_D f(P)\,ds=\iint_D f(x, y)\,dx\,dy \qquad (2)$$

or the total quantity of substance in D is equal to the double integral (over D) of the density $f(P)=f(x, y)$ of this substance.

Example. Determine the mass of a circular lamina of radius R if the surface density $f(x, y)$ of the material at each point $P(x, y)$ is proportional to the distance of the point (x, y) from the centre of the circle, that is, if

$$f(x, y)=k\sqrt{x^2+y^2}$$

Solution. By formula (2) we have

$$M=\iint_D k\sqrt{x^2+y^2}\,dx\,dy$$

where the domain of integration D is the circle $x^2+y^2\leqslant R^2$.

Passing to polar coordinates, we obtain

$$M=k\int_0^{2\pi}\left(\int_0^R \rho\rho\,d\rho\right)d\theta=k2\pi\frac{\rho^3}{3}\Bigg|_0^R=\frac{2}{3}k\pi R^3$$

2.9 THE MOMENT OF INERTIA OF THE AREA OF A PLANE FIGURE

The moment of inertia I of a material point M of mass m relative to some point O is the product of the mass m by the square of its distance r from the point O:

$$I=mr^2$$

* The relationship $\Delta s_i \to 0$ is to be understood in the sense that the diameter of the subdomain Δs_i approaches zero.

The moment of inertia of a system of material points $m_1, m_2, \ldots, m_n$ relative to O is the sum of moments of inertia of the individual points of the system:

$$I=\sum_{i=1}^{n} m_i r_i^2$$

Let us determine the moment of inertia of a material plane figure D.

Let D be located in an xy-coordinate plane. Let us determine the moment of inertia of this figure relative to the origin, assuming that the surface density is everywhere equal to unity.

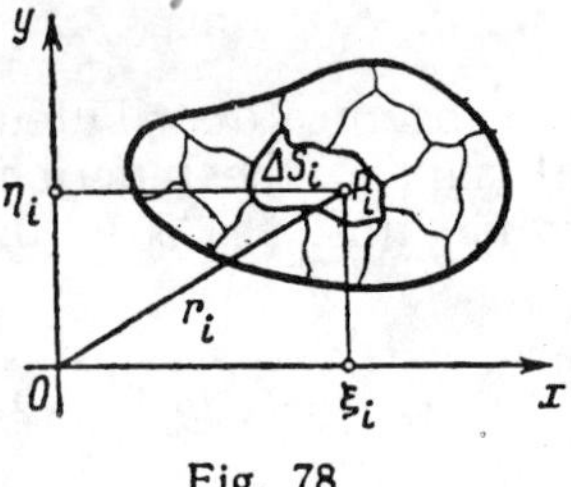

Fig. 78

Divide the domain D into elementary subdomains ΔS_i ($i=1, 2, \ldots, n$) (Fig. 78). In each subdomain take a point P_i with coordinates ξ_i, η_i. Let us call the product of the mass of the subdomain ΔS_i by the square of the distance $r_i^2=\xi_i^2+\eta_i^2$ an elementary moment of inertia ΔI_i of the subdomain ΔS_i:

$$\Delta I_i=(\xi_i^2+\eta_i^2)\,\Delta S_i$$

and let us form the sum of such moments:

$$\sum_{i=1}^{n}(\xi_i^2+\eta_i^2)\,\Delta S_i$$

This is the integral sum of the function $f(x, y)=x^2+y^2$ over the domain D.

We define the *moment of inertia of the figure* D as the limit of this sum when the diameter of each elementary subdomain ΔS_i approaches zero:

$$I_0=\lim_{\operatorname{diam}\Delta S_i\to 0}\sum_{i=1}^{n}(\xi_i^2+\eta_i^2)\,\Delta S_i$$

But the limit of this sum is the double integral $\iint\limits_D (x^2+y^2)\,dx\,dy$. Thus, the moment of inertia of the figure D relative to the origin is

$$I_0=\iint\limits_D (x^2+y^2)\,dx\,dy \tag{1}$$

where D is a domain which coincides with the given plane figure.

The integrals

$$I_{xx}=\iint\limits_D y^2\,dx\,dy \tag{2}$$

$$I_{yy}=\iint\limits_D x^2\,dx\,dy \tag{3}$$

are called, respectively, the *moments of inertia of the figure D relative to the x-axis and y-axis.*

Example 1. Compute the moment of inertia of the area of a circle D of radius R relative to the centre O.

Solution. By formula (1) we have

$$I_0=\iint_D (x^2+y^2)\,dx\,dy$$

To evaluate this integral we change to the polar coordinates θ, ρ. The equation of the circle in polar coordinates is $\rho=R$. Therefore

$$I_0=\int_0^{2\pi}\left(\int_0^R \rho^2\rho\,d\rho\right)d\theta=\frac{\pi R^4}{2}$$

Note. If the surface density γ is not equal to unity, but is some function of x and y, i.e., $\gamma=\gamma(x, y)$, then the mass of the subdomain ΔS_i will, to within infinitesimals of higher order, be equal to $\gamma(\xi_i, \eta_i)\Delta S_i$ and, for this reason, the moment of inertia of the plane figure relative to the origin will be

$$I_0=\iint_D \gamma(x, y)(x^2+y^2)\,dx\,dy \tag{1'}$$

Example 2. Compute the moment of inertia of a plane material figure D bounded by the lines $y^2=1-x$; $x=0$, $y=0$ relative to the y-axis if the surface density at each point is equal to y (Fig. 79).

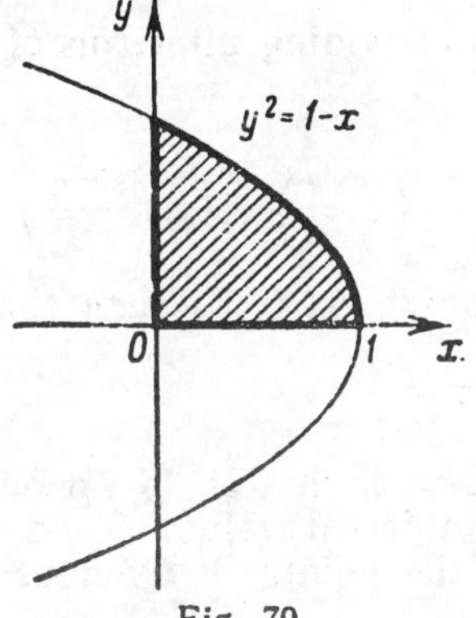

Fig. 79

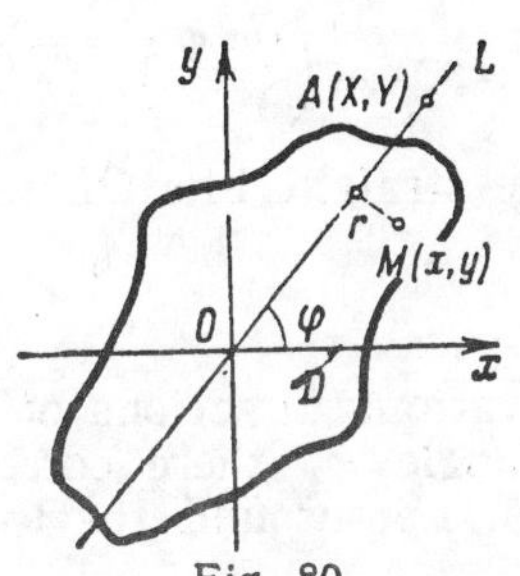

Fig. 80

Solution.

$$I_{yy}=\int_0^1\left(\int_0^{\sqrt{1-x}} yx^2\,dy\right)dx=\int_0^1 \frac{x^2y^2}{2}\Bigg|_0^{\sqrt{1-x}}dx=\frac{1}{2}\int_0^1 x^2(1-x)\,dx=\frac{1}{24}$$

Ellipse of inertia. Let us determine the moment of inertia of the area of a plane figure D relative to some axis OL that passes through the point O, which we shall take as the coordinate origin. Denote by φ the angle formed by the straight line OL with the positive x-axis (Fig. 80).

The normal equation of OL is

$$x \sin\varphi - y\cos\varphi = 0$$

The distance r of some point $M(x, y)$ from this line is

$$r = |x\sin\varphi - y\cos\varphi|$$

The moment of inertia I of the area of D relative to OL is expressed, by definition, by the integral

$$I = \iint_D r^2\,dx\,dy = \iint_D (x\sin\varphi - y\cos\varphi)^2\,dx\,dy$$

$$= \sin^2\varphi \iint_D x^2\,dx\,dy - 2\sin\varphi\cos\varphi \iint_D xy\,dx\,dy + \cos^2\varphi \iint_D y^2\,dx\,dy$$

Therefore

$$I = I_{yy}\sin^2\varphi - 2I_{xy}\sin\varphi\cos\varphi + I_{xx}\cos^2\varphi \tag{4}$$

here, $I_{yy} = \iint_D x^2\,dx\,dy$ is the moment of inertia of the figure relative to the y-axis, $I_{xx} = \iint_D y^2\,dx\,dy$ is the moment of inertia relative to the x-axis, and $I_{xy} = \iint_D xy\,dx\,dy$. Dividing all terms of the last equation by I, we get

$$1 = I_{xx}\left(\frac{\cos\varphi}{\sqrt{I}}\right)^2 - 2I_{xy}\left(\frac{\sin\varphi}{\sqrt{I}}\right)\left(\frac{\cos\varphi}{\sqrt{I}}\right) + I_{yy}\left(\frac{\sin\varphi}{\sqrt{I}}\right)^2 \tag{5}$$

On the straight line OL take a point $A(X, Y)$ such that

$$OA = \frac{1}{\sqrt{I}}$$

To the various directions of the OL-axis, that is, to various values of the angle φ, there correspond different values I and different points A. Let us find the locus of the points A. Obviously,

$$X = \frac{1}{\sqrt{I}}\cos\varphi, \quad Y = \frac{1}{\sqrt{I}}\sin\varphi$$

By virtue of (5), the quantities X and Y are connected by the relation

$$1 = I_{xx}X^2 - 2I_{xy}XY + I_{yy}Y^2 \tag{6}$$

Thus, the locus of points $A(X, Y)$ is a second-degree curve (6). We shall prove that this curve is an ellipse.

The following inequality established by the Russian mathematician Bunyakovsky * holds true:

$$\left(\iint_D xy\,dx\,dy\right)^2 <$$
$$< \left(\iint_D x^2\,dx\,dy\right)\left(\iint_D y^2\,dx\,dy\right)$$

or

$$I_{xx}I_{yy} - I_{xy}^2 > 0$$

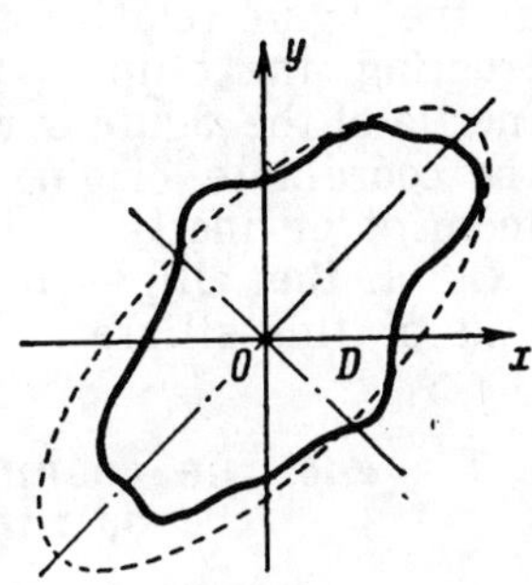

Fig. 81

Thus, the discriminant of the curve (6) is positive and, consequently, the curve is an ellipse (Fig. 81). This ellipse is called the **ellipse of inertia.** The notion of an ellipse of inertia is very important in mechanics.

We note that the lengths of the axes of the ellipse of inertia and its position in the plane depend on the shape of the given plane figure. Since the distance from the origin to some point A

---

* To prove Bunyakovsky's (also spelt Buniakowski) inequality, we consider the following obvious inequality:

$$\iint_D [f(x, y) - \lambda\varphi(x, y)]^2\,dx\,dy \geqslant 0$$

where λ is a constant. The equality sign is possible only when $f(x, y) - \lambda\varphi(x, y) \equiv 0$; that is, if $f(x, y) = \lambda\varphi(x, y)$. If we assume that $\frac{f(x, y)}{\varphi(x, y)} \neq \text{const} = \lambda$, then the inequality sign will always hold. Thus, removing brackets under the integral sign, we obtain

$$\iint_D f^2(x, y)\,dx\,dy - 2\lambda \iint_D f(x, y)\,\varphi(x, y)\,dx\,dy + \lambda^2 \iint_D \varphi^2(x, y)\,dx\,dy > 0$$

Consider the expression on the left as a function of λ. This is a second-degree polynomial that never vanishes; hence, its roots are complex, and this will occur when the discriminant formed of the coefficients of the quadratic polynomial is negative, that is,

$$\left(\iint_D f\varphi\,dx\,dy\right)^2 - \iint_D f^2\,dx\,dy \iint_D \varphi^2\,dx\,dy < 0$$

or

$$\left(\iint_D f\varphi\,dx\,dy\right)^2 < \iint_D f^2\,dx\,dy \iint_D \varphi^2\,dx\,dy$$

This is *Bunyakovsky's inequality*.

In our case, $f(x, y) = x$, $\varphi(x, y) = y$, $\frac{x}{y} \neq \text{const}$.

Bunyakovsky's inequality is widely used in various fields of mathematics. In many textbooks it is called Schwarz' inequality. Bunyakovsky published it (among other important inequalities) in 1859. Schwarz published his work 16 years later, in 1875.

of the ellipse is equal to $\frac{1}{\sqrt{I}}$, where I is the moment of inertia of the figure relative to the OA-axis, it follows that, after constructing the ellipse, we can readily calculate the moment of inertia of the figure D relative to some straight line passing through the coordinate origin. In particular, it is easy to see that the moment of inertia of the figure will be least relative to the major axis of the ellipse of inertia and greatest relative to the minor axis of this ellipse.

2.10 THE COORDINATES OF THE CENTRE OF GRAVITY OF THE AREA OF A PLANE FIGURE

In Sec. 12.8, Vol. I, it was stated that the coordinates of the centre of gravity of a system of material points $P_1, P_2, \ldots, P_n$ with masses $m_1, m_2, \ldots, m_n$ are defined by the formulas

$$x_C = \frac{\sum x_i m_i}{\sum m_i}, \quad y_C = \frac{\sum y_i m_i}{\sum m_i} \tag{1}$$

Let us now determine the coordinates of the centre of gravity of a plane figure D. Divide this figure into very small subdomains ΔS_i. If the surface density is taken equal to unity, then the mass of a subdomain will be equal to its area. If it is approximately taken that the entire mass of subdomain ΔS_i is concentrated in some point of it, $P_i(\xi_i, \eta_i)$, the figure D may be regarded as a **system of material points**. Then, by formulas (1), the coordinates of the centre of gravity of this figure will be **approximately** determined by the equations

$$x_C \approx \frac{\sum_{i=1}^{n} \xi_i \Delta S_i}{\sum_{i=1}^{n} \Delta S_i}; \quad y_C \approx \frac{\sum_{i=1}^{n} \eta_i \Delta S_i}{\sum_{i=1}^{n} \Delta S_i}.$$

In the limit, as $\Delta S_i \to 0$, the integral sums in the numerators and denominators of the fractions will pass into double integrals, and we will obtain exact formulas for computing the coordinates of the centre of gravity of a plane figure:

$$x_C = \frac{\iint\limits_D x\,dx\,dy}{\iint\limits_D dx\,dy}, \quad y_C = \frac{\iint\limits_D y\,dx\,dy}{\iint\limits_D dx\,dy} \tag{2}$$

These formulas, which have been derived for a plane figure with surface density 1, obviously hold true also for a figure with any other density γ constant at all points.

If, however, the surface density is variable,

$$\gamma = \gamma(x, y)$$

then the corresponding formulas will have the form

$$x_C = \frac{\iint\limits_D \gamma(x, y)\, x\, dx\, dy}{\iint\limits_D \gamma(x, y)\, dx\, dy}, \qquad y_C = \frac{\iint\limits_D \gamma(x, y)\, y\, dx\, dy}{\iint\limits_D \gamma(x, y)\, dx\, dy}$$

The expressions $M_y = \iint\limits_D \gamma(x, y)\, x\, dx\, dy$ and $M_x = \iint\limits_D \gamma(x, y)\, y\, dx\, dy$ are called *static moments* of the plane figure D relative to the y-axis and x-axis.

The integral $\iint \gamma(x, y)\, dx\, dy$ expresses the quantity of **mass** of the figure in question.

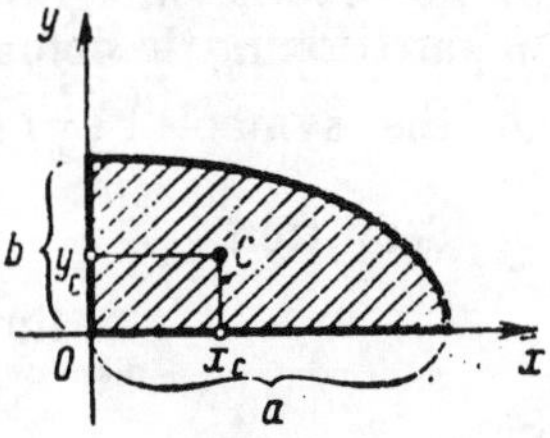

Example. Determine the coordinates of the centre of gravity of a quarter of the ellipse (Fig. 82)

$$\frac{x^2}{a^2} + \frac{y^2}{b^2} = 1$$

Fig. 82

assuming that the surface density at all points is equal to 1.

Solution. By formulas (2) we have

$$x_C = \frac{\int\limits_0^a \left(\int\limits_0^{\frac{b}{a}\sqrt{a^2-x^2}} x\, dy \right) dx}{\int\limits_0^a \left(\int\limits_0^{\frac{b}{a}\sqrt{a^2-x^2}} dy \right) dx} = \frac{\frac{b}{a} \int\limits_0^a \sqrt{a^2-x^2}\, x\, dx}{\frac{1}{4}\pi ab} = \frac{-\frac{b}{a} \cdot \frac{1}{3} (a^2-x^2)^{3/2} \Big|_0^a}{\frac{1}{4}\pi ab} = \frac{4a}{3\pi}$$

$$y_C = \frac{\int\limits_0^a \left(\int\limits_0^{\frac{b}{a}\sqrt{a^2-x^2}} y\, dy \right) dx}{\frac{1}{4}\pi ab} = \frac{4b}{3\pi}$$

2.11 TRIPLE INTEGRALS

Let there be given, in space, a certain domain V bounded by a closed surface S. Let some continuous function $f(x, y, z)$, where x, y, z are the rectangular coordinates of a point of the domain, be given in V and on its boundary. For clarity, if $f(x, y, z) \geqslant 0$, we can regard this function as the density distribution of some substance in the domain V.

Divide V, in arbitrary fashion, into subdomains Δv_i; the symbol Δv_i will denote not only the domain itself, but its volume as well. Within the limits of each subdomain Δv_i, choose an arbitrary point P_i and denote by $f(P_i)$ the value of the function f at this point. Form a sum of the type

$$\sum f(P_i)\,\Delta v_i \tag{1}$$

and increase without bound the number of subdomains Δv_i so that the largest diameter of Δv_i should approach zero.* If the function $f(x, y, z)$ is continuous, sums of type (1) will have a limit. This limit is to be understood in the same sense as for the definition of the double integral.** It is not dependent either on the manner of partitioning the domain V or on the choice of points P_i; it is designated by the symbol $\iiint\limits_V f(P)\,dv$ and is called a *triple integral*. Thus, by definition,

$$\lim_{\text{diam}\,\Delta v_i \to 0} \sum f(P_i)\,\Delta v_i = \iiint\limits_V f(P)\,dv$$

or

$$\iiint\limits_V f(P)\,dv = \iiint\limits_V f(x, y, z)\,dx\,dy\,dz \tag{2}$$

If $f(x, y, z)$ is considered the volume density of distribution of a substance over the domain V, then the integral (2) yields the mass of the entire substance contained in V.

2.12 EVALUATING A TRIPLE INTEGRAL

Suppose that a spatial (three-dimensional) domain V bounded by a closed surface S possesses the following properties:

(1) every straight line parallel to the z-axis and drawn through an interior (that is, not lying on the boundary S) point of the domain V cuts the surface S at two points;

(2) the entire domain V is projected on the xy-plane into a regular (two-dimensional) domain D;

(3) any part of the domain V cut off by a plane parallel to any one of the coordinate planes (Oxy, Oxz, Oyz) likewise possesses Properties 1 and 2.

* The diameter of a subdomain Δv_i is the maximum distance between points lying on the boundary of the subdomain.

** This theorem of the existence of a limit of integral sums (that is, of the existence of a triple integral) for any function continuous in a closed domain V (including the boundary) is accepted without proof.

We shall call the domain V that possesses the indicated properties a *regular* three-dimensional domain.

To illustrate, an ellipsoid, a rectangular parallelepiped, a tetrahedron, and so on are examples of regular three-dimensional domains. An instance of an irregular three-dimensional domain is given in Fig. 83. In this section we will consider only regular domains.

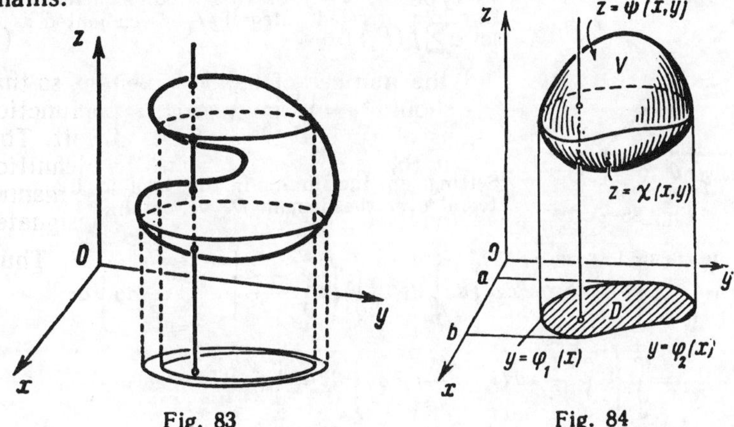

Fig. 83 Fig. 84

Let the surface bounding V below have the equation $z=\chi(x, y)$, and the surface bounding this domain above, the equation $z=\psi(x, y)$ (Fig. 84).

We introduce the concept of a **threefold iterated** integral I_V, over the domain V, of a function of three variables $f(x, y, z)$ defined and continuous in V. Suppose that the domain D is the projection of the domain V onto the xy-plane bounded by the curves

$$y=\varphi_1(x), \quad y=\varphi_2(x), \quad x=a, \quad x=b$$

Then *a threefold iterated integral* of the function $f(x, y, z)$ over V is defined as follows:

$$I_V=\int_a^b\left[\int_{\varphi_1(x)}^{\varphi_2(x)}\left\{\int_{\chi(x, y)}^{\psi(x, y)} f(x, y, z)\,dz\right\} dy\right] dx \qquad (1)$$

We note that as a result of integration with respect to z and substitution of limits in the braces (inner brackets) we get a function of x and y. We then compute the double integral of this function over the domain D as has already been done.

The following is an example of the evaluation of a threefold iterated integral.

Example 1. Compute the threefold iterated integral of the function $f(x, y, z) = xyz$ over the domain V bounded by the planes

$$x=0, \quad y=0, \quad z=0, \quad x+y+z=1$$

Solution. This domain is regular, it is bounded above and below by the planes $z=0$ and $z=1-x-y$ and is projected on the xy-plane into a regular plane domain D, which is a triangle bounded by the straight lines $x=0$, $y=0$, $y=1-x$ (Fig. 85). Therefore, the threefold iterated integral I_V is computed as follows:

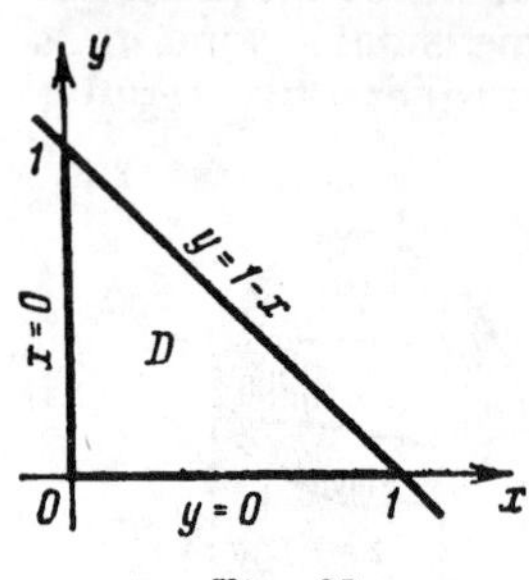

Fig. 85

$$I_V = \iint_D \left[\int_0^{1-x-y} xyz\, dz \right] d\sigma$$

Setting up the limits in the twofold iterated integral over the domain D, we obtain

$$I_V = \int_0^1 \left\{ \int_0^{1-x} \left[\int_0^{1-x-y} xyz\, dz \right] dy \right\} dx = \int_0^1 \left\{ \int_0^{1-x} \frac{xyz^2}{2} \Bigg|_{z=0}^{z=1-x-y} dy \right\} dx$$

$$= \int_0^1 \left\{ \int_0^{1-x} \frac{1}{2} xy (1-x-y)^2\, dy \right\} dx = \int_0^1 \frac{x}{24} (1-x)^4\, dx = \frac{1}{720}$$

Let us now consider some of the properties of a threefold iterated integral.

Property 1. *If a domain V is divided into two domains V_1 and V_2 by a plane parallel to one of the coordinate planes, then the threefold iterated integral over V is equal to the sum of the threefold iterated integrals over the domains V_1 and V_2.*

The proof of this property is exactly the same as that for twofold iterated integrals. We shall not repeat it.

Corollary. *For any kind of partition of the domain V into a finite number of subdomains $V_1, \ldots, V_n$ by planes parallel to the coordinate planes, we have the equality*

$$I_V = I_{V_1} + I_{V_2} + \ldots + I_{V_n}$$

Property 2 (Theorem on the evaluation of a threefold iterated integral). *If m and M are, respectively, the smallest and largest values of the function $f(x, y, z)$ in the domain V, we have the inequality*

$$mV \leqslant I_V \leqslant MV$$

where V is the volume of the given domain and I_V is a threefold iterated integral of the function $f(x, y, z)$ over V.

Proof. Let us first evaluate the inside integral in the iterated integral $I_V = \iint\limits_D \left[\int\limits_{\chi(x,\,y)}^{\psi(x,\,y)} f(x, y, z)\, dz \right] d\sigma$:

$$\int\limits_{\chi(x,\,y)}^{\psi(x,\,y)} f(x, y, z)\, dz \leqslant \int\limits_{\chi(x,\,y)}^{\psi(x,\,y)} M\, dz = M \int\limits_{\chi(x,\,y)}^{\psi(x,\,y)} dz = Mz \Big|_{\chi(x,\,y)}^{\psi(x,\,y)}$$
$$= M[\psi(x, y) - \chi(x, y)]$$

Thus, the inside integral does not exceed the expression $M[\psi(x, y) - \chi(x, y)]$. Therefore, by virtue of the theorem of Sec. 2.1 on double integrals, we get (denoting by D the projection of the domain V on the xy-plane)

$$I_V = \iint\limits_D \left[\int\limits_{\chi(x,\,y)}^{\psi(x,\,y)} f(x, y, z)\, dz \right] d\sigma \leqslant \iint\limits_D M[\psi(x, y) - \chi(x, y)]\, d\sigma$$
$$= M \iint\limits_D [\psi(x, y) - \chi(x, y)]\, d\sigma$$

But the latter iterated integral is equal to the double integral of the function $\psi(x, y) - \chi(x, y)$ and, consequently, is equal to the volume of the domain which lies between the surfaces $z = \chi(x, y)$ and $z = \psi(x, y)$, that is, to the volume of the domain V. Therefore,

$$I_V \leqslant MV$$

It is similarly proved that $I_V \geqslant mV$. Property 2 is thus proved.

Property 3 (Mean-value theorem). *The threefold iterated integral I_V of a continuous function $f(x, y, z)$ over a domain V is equal to the product of its volume V by the value of the function at some point P of V; that is,*

$$I_V = \int\limits_a^b \left\{ \int\limits_{\varphi_1(x)}^{\varphi_2(x)} \left[\int\limits_{\chi(x,\,y)}^{\psi(x,\,y)} f(x, y, z)\, dz \right] dy \right\} dx = f(P)\, V \qquad (2)$$

The proof of this property is carried out in the same way as that for a twofold iterated integral [see Sec. 2.2, Property 3, formula (4)].

We can now prove the theorem for evaluating a triple integral.

Theorem. *The triple integral of a function $f(x, y, z)$ over a regular domain V is equal to a threefold iterated integral over the same domain; that is,*

$$\iiint\limits_V f(x, y, z)\, dv = \int\limits_a^b \left\{ \int\limits_{\varphi_1(x)}^{\varphi_2(x)} \left[\int\limits_{\chi(x,\,y)}^{\psi(x,\,y)} f(x, y, z)\, dz \right] dy \right\} dx$$

Proof. Divide the domain V by planes parallel to the coordinate planes into n regular subdomains:

$$\Delta v_1, \Delta v_2, \ldots, \Delta v_n$$

As done above, denote by I_V the threefold iterated integral of the function $f(x, y, z)$ over the domain V, and by $I_{\Delta v_i}$ the threefold iterated integral of this function over the subdomain Δv_i. Then by the corollary of Property 1 we can write the equation

$$I_V = I_{\Delta v_1} + I_{\Delta v_2} + \ldots + I_{\Delta v_n} \tag{3}$$

We transform each of the terms on the right by formula (2):

$$I_V = f(P_1)\Delta v_1 + f(P_2)\Delta v_2 + \ldots + f(P_n)\Delta v_n \tag{4}$$

where P_i is some point of the subdomain Δv_i.

On the right side of this equation is an integral sum. It is assumed that the function $f(x, y, z)$ is continuous in V; and for this reason the limit of this sum, as the largest diameter of Δv_i approaches zero, exists and is equal to the triple integral of the function $f(x, y, z)$ over V. Thus, passing to the limit in (4), as diam $\Delta v_i \to 0$, we get

$$I_V = \iiint\limits_V f(x, y, z)\, dv$$

or, finally, interchanging the expressions on the right and left,

$$\iiint\limits_V f(x, y, z)\, dv = \int\limits_a^b \left\{ \int\limits_{\varphi_1(x)}^{\varphi_2(x)} \left[\int\limits_{\chi(x, y)}^{\psi(x, y)} f(x, y, z)\, dz \right] dy \right\} dx$$

Thus, the theorem is proved.

Here, $z = \chi(x, y)$ and $z = \psi(x, y)$ are the equations of the surfaces bounding the regular domain V below and above. The lines $y = \varphi_1(x)$, $y = \varphi_2(x)$, $x = a$, $x = b$ bound the domain D, which is the projection of V onto the xy-plane.

Note. As in the case of the double integral, we can form a threefold iterated integral with a different order of integration with respect to the variables and with other limits, if, of course, the shape of the domain V permits this.

Computing the volume of a solid by means of a threefold iterated integral. If the integrand $f(x, y, z) = 1$, then the triple integral over the domain V expresses the **volume of** V:

$$V = \iiint\limits_V dx\, dy\, dz \tag{5}$$

Example 2. Compute the volume of the ellipsoid

$$\frac{x^2}{a^2}+\frac{y^2}{b^2}+\frac{z^2}{c^2}=1$$

Solution. The ellipsoid (Fig. 86) is bounded below by the surface $z=-c\sqrt{1-\frac{x^2}{a^2}-\frac{y^2}{b^2}}$, and above by the surface $z=c\sqrt{1-\frac{x^2}{a^2}-\frac{y^2}{b^2}}$. The projection of this ellipsoid on the xy-plane (domain D) is an ellipse, $\frac{x^2}{a^2}+\frac{y^2}{b^2}=1$. Hence, reducing the computation of volume to that of a threefold iterated integral, we obtain

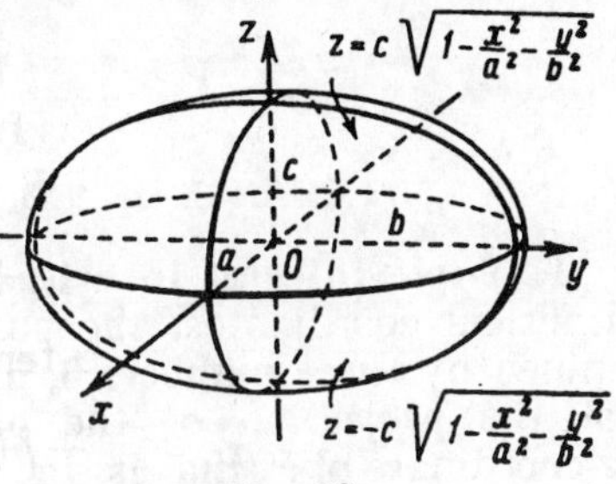

Fig. 86

$$V=\int_{-a}^{a}\left[\int_{-b\sqrt{1-\frac{x^2}{a^2}}}^{b\sqrt{1-\frac{x^2}{a^2}}}\left(\int_{-c\sqrt{1-\frac{x^2}{a^2}-\frac{y^2}{b^2}}}^{c\sqrt{1-\frac{x^2}{a^2}-\frac{y^2}{b^2}}}dz\right)dy\right]dx$$

$$=2c\int_{-a}^{a}\left[\int_{-b\sqrt{1-\frac{x^2}{a^2}}}^{b\sqrt{1-\frac{x^2}{a^2}}}\sqrt{1-\frac{x^2}{a^2}-\frac{y^2}{b^2}}\,dy\right]dx$$

When computing the inside integral, x is held constant. Make the substitution:

$$y=b\sqrt{1-\frac{x^2}{a^2}}\sin t,\qquad dy=b\sqrt{1-\frac{x^2}{a^2}}\cos t\,dt$$

The variable y varies from $-b\sqrt{1-\frac{x^2}{a^2}}$ to $b\sqrt{1-\frac{x^2}{a^2}}$; therefore t varies from $-\frac{\pi}{2}$ to $\frac{\pi}{2}$. Putting new limits in the integral, we get

$$V=2c\int_{-a}^{a}\left[\int_{-\frac{\pi}{2}}^{\frac{\pi}{2}}\sqrt{\left(1-\frac{x^2}{a^2}\right)-\left(1-\frac{x^2}{a^2}\right)\sin^2 t}\;\; b\sqrt{1-\frac{x^2}{a^2}}\cos t\,dt\right]dx$$

$$=2cb\int_{-a}^{a}\left[\left(1-\frac{x^2}{a^2}\right)\int_{-\frac{\pi}{2}}^{\frac{\pi}{2}}\cos^2 t\,dt\right]dx=\frac{cb\pi}{a^2}\int_{-a}^{a}(a^2-x^2)\,dx=\frac{4\pi abc}{3}$$

Hence,

$$V = \frac{4}{3}\pi abc$$

If $a=b=c$, we get the volume of the sphere:

$$V = \frac{4}{3}\pi a^3$$

2.13 CHANGE OF VARIABLES IN A TRIPLE INTEGRAL

1. Triple integral in cylindrical coordinates. In the case of cylindrical coordinates, the position of a point P in space is determined by three numbers θ, ρ, z, where θ and ρ are polar coordinates of the projection of the point P on the xy-plane and z is the z-coordinate of P, that is, the distance of the point to the xy-plane—with the plus sign if the point lies above the xy-plane, and with the minus sign if below the xy-plane (Fig. 87).

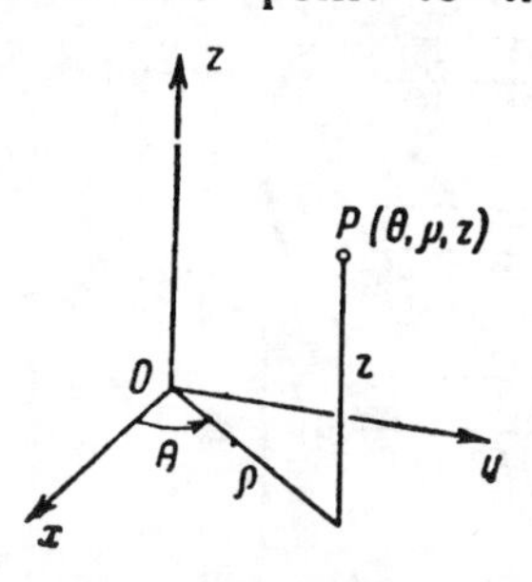

Fig. 87

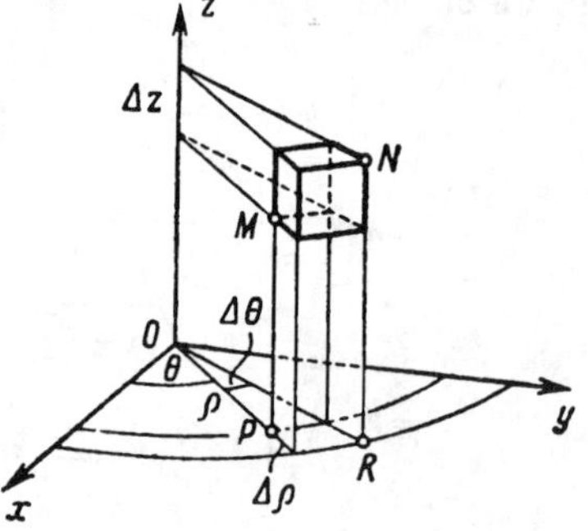

Fig. 88

In this case, we divide the given three-dimensional domain V into elementary volumes by the coordinate surfaces $\theta=\theta_i$, $\rho=\rho_j$, $z=z_k$ (half-planes adjoining the z-axis, circular cylinders whose axis coincides with the z-axis, planes perpendicular to the z-axis). The curvilinear "prism" shown in Fig. 88 is a volume element. The base area of this prism is equal, to within infinitesimals of higher order, to $\rho\Delta\theta\,\Delta\rho$, the altitude is Δz (to simplify notation we drop the indices i, j, k). Thus, $\Delta v = \rho\Delta\theta\,\Delta\rho\,\Delta z$. Hence, the triple integral of the function $F(\theta, \rho, z)$ over the domain V has the form

$$I = \iiint\limits_V F(\theta, \rho, z)\,\rho\,d\theta\,d\rho\,dz \qquad (1)$$

The limits of integration are determined by the shape of the domain V.

If a triple integral of the function $f(x, y, z)$ is given in rectangular coordinates, it can readily be changed to a triple integral in cylindrical coordinates. Indeed, noting that

$$x=\rho\cos\theta, \quad y=\rho\sin\theta, \quad z=z$$

we have

$$\iiint\limits_V f(x, y, z)\,dx\,dy\,dz=\iiint\limits_V F(\theta, \rho, z)\,\rho\,d\theta\,d\rho\,dz$$

where

$$f(\rho\cos\theta, \rho\sin\theta, z)=F(\theta, \rho, z)$$

Example. Determine the mass M of a hemisphere of radius R with centre at the origin, if the density F of its substance at each point (x, y, z) is proportional to the distance of this point from the base, that is, $F=kz$.

Solution. The equation of the upper part of the hemisphere

$$z=\sqrt{R^2-x^2-y^2}$$

in cylindrical coordinates has the form

$$z=\sqrt{R^2-\rho^2}$$

Hence,

$$M=\iiint\limits_V kz\rho\,d\theta\,d\rho\,dz=\int\limits_0^{2\pi}\left[\int\limits_0^R\left(\int\limits_0^{\sqrt{R^2-\rho^2}} kz\,dz\right)\rho\,d\rho\right]d\theta$$

$$=\int\limits_0^{2\pi}\left[\int\limits_0^R \frac{kz^2}{2}\Bigg|_0^{\sqrt{R^2-\rho^2}}\rho\,d\rho\right]d\theta=\int\limits_0^{2\pi}\left[\int\limits_0^R \frac{k}{2}(R^2-\rho^2)\,\rho\,d\rho\right]d\theta$$

$$=\frac{k}{2}\int\limits_0^{2\pi}\left[\frac{R^4}{2}-\frac{R^4}{4}\right]d\theta=\frac{k}{2}\,\frac{R^4}{4}\,2\pi=\frac{k\pi R^4}{4}$$

2. Triple integral in spherical coordinates. In spherical coordinates, the position of a point P in space is determined by three numbers, θ, r, φ, where r is the distance of the point from the origin, the so-called radius vector of the point, φ is the angle between the radius vector and the z-axis, θ is the angle between the projection of the radius vector on the xy-plane and the x-axis reckoned from this axis in a positive sense (counterclockwise) (Fig. 89). For any point of space we have

$$0\leqslant r<\infty, \quad 0\leqslant\varphi\leqslant\pi; \quad 0\leqslant\theta\leqslant 2\pi$$

Divide the domain V into volume elements Δv by the coordinate surfaces $r=\text{const}$ (spheres), $\varphi=\text{const}$ (conic surfaces with vertices at origin), $\theta=\text{const}$ (half-planes passing through the z-axis). To within infinitesimals of higher order, the volume element Δv may be considered a parallelepiped with edges of length Δr, $r\Delta\varphi$,

$r \sin\varphi\, \Delta\theta$. Then the volume element is equal (see Fig. 90) to

$$\Delta v = r^2 \sin\varphi\, \Delta r\, \Delta\theta\, \Delta\varphi$$

The triple integral of a function $F(\theta, r, \varphi)$ over a domain V has the form

$$I = \iiint_V F(\theta, r, \varphi)\, r^2 \sin\varphi\, dr\, d\theta\, d\varphi \qquad (1')$$

The limits of integration are determined by the shape of the domain V. From Fig. 89 it is easy to establish the expressions of

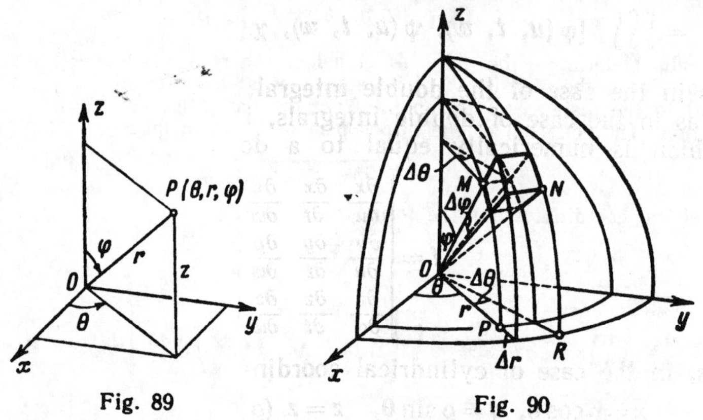

Fig. 89 Fig. 90

Cartesian coordinates in terms of spherical coordinates:

$$x = r \sin\varphi \cos\theta$$
$$y = r \sin\varphi \sin\theta$$
$$z = r \cos\varphi$$

For this reason, the formula for transforming a triple integral from Cartesian coordinates to spherical coordinates has the form

$$\iiint_V f(x, y, z)\, dx\, dy\, dz$$

$$= \iiint_V f[r \sin\varphi \cos\theta,\ r \sin\varphi \sin\theta,\ r \cos\varphi]\, r^2 \sin\varphi\, dr\, d\theta\, d\varphi$$

3. General change of variables in a triple integral. Transformations from Cartesian coordinates to cylindrical and spherical coordinates in a triple integral represent special cases of the general transformation of coordinates in space.

Let the functions

$$x = \varphi(u, t, w)$$
$$y = \psi(u, t, w)$$
$$z = \chi(u, t, w)$$

map, in one-to-one manner, a domain V in Cartesian coordinates x, y, z onto a domain V' in curvilinear coordinates u, t, w. Let the volume element Δv of V be carried over to the volume element $\Delta v'$ of V' and let

$$\lim_{\Delta v' \to 0} \frac{\Delta v}{\Delta v'} = |I|$$

Then

$$\iiint_V f(x, y, z)\, dx\, dy\, dz$$

$$= \iiint_{V'} f[\varphi(u, t, w), \psi(u, t, w), \chi(u, t, w)]\, |I|\, du\, dt\, dw$$

As in the case of the double integral, I is called the *Jacobian*; and as in the case of double integrals, it may be proved that the Jacobian is numerically equal to a determinant of order three:

$$I = \begin{vmatrix} \frac{\partial x}{\partial u} & \frac{\partial x}{\partial t} & \frac{\partial x}{\partial w} \\ \frac{\partial y}{\partial u} & \frac{\partial y}{\partial t} & \frac{\partial y}{\partial w} \\ \frac{\partial z}{\partial u} & \frac{\partial z}{\partial t} & \frac{\partial z}{\partial w} \end{vmatrix}$$

Thus, in the case of cylindrical coordinates we have

$$x = \rho\cos\theta, \quad y = \rho\sin\theta, \quad z = z \quad (\rho = u, \ \theta = t, \ z = w)$$

$$I = \begin{vmatrix} \cos\theta & -\rho\sin\theta & 0 \\ \sin\theta & \rho\cos\theta & 0 \\ 0 & 0 & 1 \end{vmatrix} = \rho$$

In the case of spherical coordinates we have

$$x = r\sin\varphi\cos\theta, \quad y = r\sin\varphi\sin\theta, \quad z = r\cos\varphi \quad (r = u, \ \varphi = t, \ \theta = w)$$

$$I = \begin{vmatrix} \sin\varphi\cos\theta & r\cos\varphi\cos\theta & -r\sin\varphi\sin\theta \\ \sin\varphi\sin\theta & r\cos\varphi\sin\theta & r\sin\varphi\cos\theta \\ \cos\varphi & -r\sin\varphi & 0 \end{vmatrix} = r^2\sin\varphi$$

2.14 THE MOMENT OF INERTIA AND THE COORDINATES OF THE CENTRE OF GRAVITY OF A SOLID

1. The moment of inertia of a solid. The moments of inertia of a point $M(x, y, z)$ of mass m relative to the coordinate axes Ox, Oy and Oz (Fig. 91) are expressed, respectively, by the formulas

$$I_{xx} = (y^2 + z^2)\, m$$

$$I_{yy} = (x^2 + z^2)\, m, \quad I_{zz} = (x^2 + y^2)\, m$$

The moments of inertia of a **solid** are expressed by the corresponding integrals. For instance, the moment of inertia of a solid relative to the z-axis is expressed by the integral $I_{zz} = \iiint\limits_V (x^2+y^2)\gamma(x, y, z)\,dx\,dy\,dz$, where $\gamma(x, y, z)$ is the density of the substance.

Example 1. Compute the moment of inertia of a right circular cylinder of altitude $2h$ and radius R relative to the diameter of its median section, considering the density constant and equal to γ_0.

Solution. Choose a coordinate system as follows: direct the z-axis along the axis of the cylinder, and put the origin of coordinates at its centre of symmetry (Fig. 92).

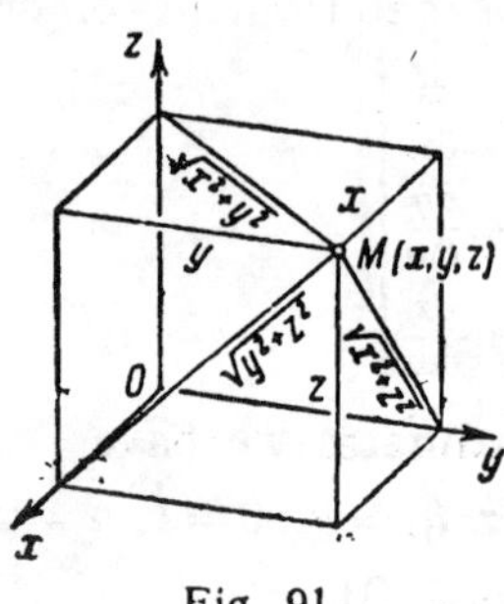

Fig. 91

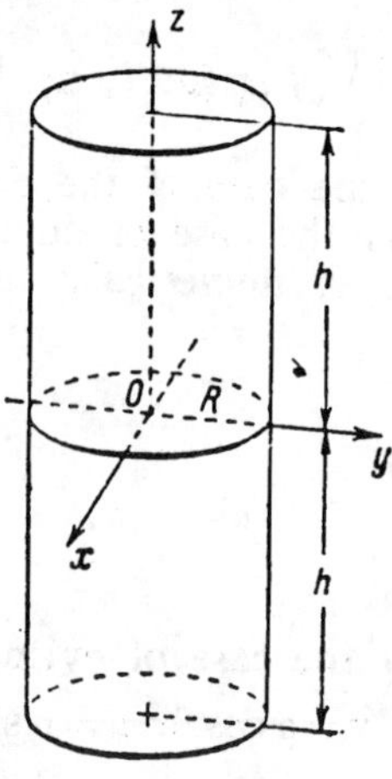

Fig. 92

Then the problem reduces to computing the moment of inertia of the cylinder relative to the x-axis:

$$I_{xx} = \iiint\limits_V (y^2+z^2)\gamma_0\,dx\,dy\,dz$$

Changing to cylindrical coordinates, we obtain

$$I_{xx} = \gamma_0 \int\limits_0^{2\pi} \left\{ \int\limits_0^R \left[\int\limits_{-h}^{h} (z^2+\rho^2\sin^2\theta)\,dz \right] \rho\,d\rho \right\} d\theta$$

$$= \gamma_0 \int\limits_0^{2\pi} \left\{ \int\limits_0^R \left[\frac{2h^3}{3} + 2h\rho^2\sin^2\theta \right] \rho\,d\rho \right\} d\theta = \gamma_0 \int\limits_0^{2\pi} \left\{ \frac{2h^3}{3}\frac{R^2}{2} + \frac{2hR^4}{4}\sin^2\theta \right\} d\theta$$

$$= \gamma_0 \left[\frac{2h^3R^2}{6} 2\pi + \frac{2hR^4}{4}\pi \right] = \gamma_0\pi hR^2 \left[\frac{2}{3}h^2 + \frac{R^2}{2} \right]$$

2. The coordinates of the centre of gravity of a solid. Like what we had in Sec. 12.8, Vol. I, for plane figures, the coordinates of the centre of gravity of a solid are expressed by the formulas

$$x_C = \frac{\iiint\limits_V x\gamma(x, y, z)\,dx\,dy\,dz}{\iiint\limits_V \gamma(x, y, z)\,dx\,dy\,dz}, \quad y_C = \frac{\iiint\limits_V y\gamma(x, y, z)\,dx\,dy\,dz}{\iiint\limits_V \gamma(x, y, z)\,dx\,dy\,dz}$$

$$z_C = \frac{\iiint\limits_V z\gamma(x, y, z)\,dx\,dy\,dz}{\iiint\limits_V \gamma(x, y, z)\,dx\,dy\,dz}$$

where $\gamma(x, y, z)$ is the density.

Example 2. Determine the coordinates of the centre of gravity of the upper half of a sphere of radius R with centre at the origin, assuming the density γ_0 constant.

Solution. The hemisphere is bounded by the surfaces

$$z = \sqrt{R^2 - x^2 - y^2}, \quad z = 0$$

The z-coordinate of its centre of gravity is given by the formula

$$z_C = \frac{\iiint\limits_V z\gamma_0\,dx\,dy\,dz}{\iiint\limits_V \gamma_0\,dx\,dy\,dz}$$

Changing to spherical coordinates, we get

$$z_C = \frac{\gamma_0 \int\limits_0^{2\pi} \left[\int\limits_0^{\frac{\pi}{2}} \left(\int\limits_0^R r\cos\varphi r^2 \sin\varphi\,dr \right) d\varphi \right] d\theta}{\gamma_0 \int\limits_0^{2\pi} \left[\int\limits_0^{\frac{\pi}{2}} \left(\int\limits_0^R r^2 \sin\varphi\,dr \right) d\varphi \right] d\theta} = \frac{2\pi \frac{R^4}{4} \frac{1}{2}}{\frac{4}{6}\pi R^3} = \frac{3}{8} R$$

Obviously, by virtue of the symmetry of the hemisphere, $x_c = y_c = 0$.

2.15 COMPUTING INTEGRALS DEPENDENT ON A PARAMETER

Consider an integral dependent on the parameter α:

$$I(\alpha) = \int\limits_a^b f(x, \alpha)\,dx$$

(We examined such integrals in Sec. 11.10, Vol. I.) We state without proof that if a function $f(x, \alpha)$ is continuous with respect to

x over an interval $[a, b]$ and with respect to α over an interval $[\alpha_1, \alpha_2]$, then the function

$$I(\alpha) = \int_a^b f(x, \alpha)\, dx$$

is a continuous function on $[\alpha_1, \alpha_2]$. Consequently, the function $I(\alpha)$ may be integrated with respect to α over the interval $[\alpha_1, \alpha_2]$:

$$\int_{\alpha_1}^{\alpha_2} I(\alpha)\, d\alpha = \int_{\alpha_1}^{\alpha_2} \left[\int_a^b f(x, \alpha)\, dx \right] d\alpha$$

The expression on the right is an iterated integral of the function $f(x, \alpha)$ over a rectangle situated in the plane $xO\alpha$. We can change the order of integration in this integral:

$$\int_{\alpha_1}^{\alpha_2} \left[\int_a^b f(x, \alpha)\, dx \right] d\alpha = \int_a^b \left[\int_{\alpha_1}^{\alpha_2} f(x, \alpha)\, d\alpha \right] dx$$

This formula shows that for integration of an integral dependent on a parameter α, it is sufficient to integrate the element of integration with respect to the parameter α. This formula is also useful when computing definite integrals.

Example. Compute the integral

$$\int_0^\infty \frac{e^{-ax} - e^{-bx}}{x}\, dx \quad (a > 0,\ b > 0)$$

This integral is not expressible in terms of elementary functions. To evaluate it, we consider another integral that may be readily computed:

$$\int_0^\infty e^{-\alpha x}\, dx = \frac{1}{\alpha} \quad (\alpha > 0)$$

Integrating this equation between the limits $\alpha = a$ and $\alpha = b$, we get

$$\int_a^b \left[\int_0^\infty e^{-\alpha x}\, dx \right] d\alpha = \int_a^b \frac{d\alpha}{\alpha} = \ln \frac{b}{a}$$

Changing the order of integration in the first integral, we rewrite this equation in the following form:

$$\int_0^\infty \left[\int_a^b e^{-\alpha x}\, d\alpha \right] dx = \ln \frac{b}{a}$$

whence, computing the inner integral, we get

$$\int_0^\infty \frac{e^{-ax} - e^{-bx}}{x}\, dx = \ln \frac{b}{a}$$

Exercises on Chapter 2

Evaluate the integrals: *

1. $\int_0^1\int_1^2 (x^2+y^2)\,dx\,dy$. *Ans.* $\frac{8}{3}$. 2. $\int_3^4\int_1^2 \frac{dy\,dx}{(x+y)^2}$. *Ans.* $\ln\frac{25}{24}$. 3. $\int_1^2\int_x^{x\sqrt{3}} xy\,dx\,dy$. *Ans.* $\frac{15}{4}$. 4. $\int_0^{2\pi}\int_{a\sin\theta}^{a} r\,dr\,d\theta$. *Ans.* $\frac{1}{2}\pi a^2$. 5. $\int_0^a\int_{\frac{x}{a}}^{x} \frac{x\,dy\,dx}{x^2+y^2}$. *Ans.* $\frac{\pi a}{4}-a\arctan\frac{1}{a}$.

6. $\int_0^a\int_{y-a}^{2y} xy\,dx\,dy$. *Ans.* $\frac{11a^4}{24}$. 7. $\int_{\frac{b}{2}}^{b}\int_0^{\pi/2} \rho\,d\theta\,d\rho$. *Ans.* $\frac{3}{16}\pi b^2$.

Determine the limits of integration for the integral $\iint\limits_D f(x, y)\,dx\,dy$ where the domain of integration is bounded by the lines:

8. $x=2$, $x=3$, $y=-1$, $y=5$. *Ans.* $\int_2^3\int_{-1}^5 f(x, y)\,dy\,dx$. 9. $y=0$, $y=1-x^2$. *Ans.* $\int_{-1}^1\int_0^{1-x^2} f(x, y)\,dy\,dx$. 10. $x^2+y^2=a^2$. *Ans.* $\int_{-a}^{a}\int_{-\sqrt{a^2-x^2}}^{\sqrt{a^2-x^2}} f(x, y)\,dy\,dx$.

11. $y=\frac{2}{1+x^2}$, $y=x^2$. *Ans.* $\int_{-1}^{1}\int_{x^2}^{\frac{2}{1+x^2}} f(x, y)\,dy\,dx$. 12. $y=0$, $y=a$, $y=x$, $y=x-2a$. *Ans.* $\int_0^a\int_y^{y+2a} f(x, y)\,dx\,dy$.

Change the order of integration in the integrals:

13. $\int_1^2\int_3^4 f(x, y)\,dy\,dx$. *Ans.* $\int_3^4\int_1^2 f(x, y)\,dx\,dy$. 14. $\int_0^1\int_{x^3}^{\sqrt{x}} f(x, y)\,dy\,dx$. *Ans.* $\int_0^1\int_{y^2}^{\sqrt[3]{y}} f(x, y)\,dx\,dy$. 15. $\int_0^a\int_0^{\sqrt{2ay-y^2}} f(x, y)\,dx\,dy$. *Ans.* $\int_0^a\int_{a-\sqrt{a^2-x^2}}^{a} f(x, y)\,dy\,dx$.

---

* If the integral is written as $\int_M^N\int_K^L f(x, y)\,dx\,dy$ then, as has already been stated, we can consider that the first integration is performed with respect to the variable whose differential occupies the first place; that is,

$$\int_M^N\int_K^L f(x, y)\,dx\,dy=\int_M^N\left(\int_K^L f(x, y)\,dx\right)dy$$

16. $\int_{-1}^{1}\int_{0}^{\sqrt{1-x^2}} f(x, y)\,dy\,dx$. *Ans.* $\int_{0}^{1}\int_{-\sqrt{1-y^2}}^{\sqrt{1-y^2}} f(x, y)\,dx\,dy$.

17. $\int_{0}^{1}\int_{-\sqrt{1-y^2}}^{1-y} f(x, y)\,dx\,dy$. *Ans.* $\int_{-1}^{0}\int_{0}^{\sqrt{1-x^2}} f(x, y)\,dy\,dx+\int_{0}^{1}\int_{0}^{1-x} f(x, y)\,dy\,dx$.

Compute the following integrals by changing to polar coordinates:

18. $\int_{0}^{a}\int_{0}^{\sqrt{a^2-x^2}} \sqrt{a^2-x^2-y^2}\,dy\,dx$. *Ans.* $\int_{0}^{\frac{\pi}{2}}\int_{0}^{a} \sqrt{a^2-\rho^2}\,\rho\,d\rho\,d\theta=\frac{\pi}{6}a^3$.

19. $\int_{0}^{a}\int_{0}^{\sqrt{a^2-y^2}} (x^2+y^2)\,dx\,dy$. *Ans.* $\int_{0}^{\frac{\pi}{2}}\int_{0}^{a} \rho^3\,d\rho\,d\theta=\frac{\pi a^4}{8}$. 20. $\int_{0}^{\infty}\int_{0}^{\infty} e^{-(x^2+y^2)}dy\,dx$.

Ans. $\int_{0}^{\frac{\pi}{2}}\int_{0}^{\infty} e^{-\rho^2}\rho\,d\rho\,d\theta=\frac{\pi}{4}$. 21. $\int_{0}^{2a}\int_{0}^{\sqrt{2ax-x^2}} dy\,dx$. *Ans.* $\int_{0}^{\frac{\pi}{2}}\int_{0}^{2a\cos\theta} \rho\,d\rho\,d\theta=\frac{\pi a^2}{2}$.

Transform the double integrals by introducing new variables u and v connected with x and y by the formulas $x=u-uv$, $y=uv$:

22. $\int_{0}^{e}\int_{\alpha x}^{\beta x} f(x, y)\,dy\,dx$. *Ans.* $\int_{\frac{\alpha}{1+\alpha}}^{\frac{\beta}{1+\beta}}\int_{0}^{\frac{e}{1+v}} f(u-uv, uv)\,u\,du\,dv$. 23. $\int_{0}^{c}\int_{0}^{b} f(x, y)\,dy\,dx$.

Ans. $\int_{0}^{\frac{b}{b+c}}\int_{0}^{\frac{c}{1-v}} f(u-uv, uv)\,u\,du\,dv+\int_{\frac{b}{b+c}}^{b}\int_{0}^{\frac{b}{v}} f(u-uv, uv)\,u\,du\,dv$.

Calculating Areas by Means of Double Integrals

24. Compute the area of a figure bounded by the parabola $y^2=2x$ and the straight line $y=x$. *Ans.* $\frac{2}{3}$.

25. Compute the area of a figure bounded by the curves $y^2=4ax$, $x+y=3a$, $y=0$. *Ans.* $\frac{10}{3}a^2$.

26. Compute the area of a figure bounded by the curves $x^{1/2}+y^{1/2}=a^{1/2}$, $x+y=a$. *Ans.* $\frac{a^2}{3}$.

27. Compute the area of a figure bounded by the curves $y=\sin x$, $y=\cos x$, $x=0$. *Ans.* $\sqrt{2}-1$.

28. Compute the area of a loop of the curve $\rho=a\sin 2\theta$. *Ans.* $\frac{\pi a^2}{8}$.

29. Compute the entire area bounded by the lemniscate $\rho^2=a^2\cos 2\varphi$. *Ans.* a^2.

30. Compute the area of a loop of the curve $\left(\frac{x^2}{a^2}+\frac{y^2}{b^2}\right)=\frac{2xy}{c^2}$.

Hint. Change to new variables $x=\rho a\cos\theta$ and $y=\rho b\sin\theta$. *Ans.* $\frac{a^2b^2}{c^2}$.

Calculating Volumes

31. Compute the volumes of solids bounded by the following surfaces: $\frac{x}{a}+\frac{y}{b}+\frac{z}{c}=1$, $x=0$, $y=0$, $z=0$. *Ans.* $\frac{abc}{6}$. **32.** $z=0$, $x^2+y^2=1$, $x+y+$ $+z=3$. *Ans.* 3π. **33.** $(x-1)^2+(y-1)^2=1$, $xy=z$, $z=0$. *Ans.* π. **34.** x^2+y^2- $-2ax=0$, $z=0$, $x^2+y^2=z^2$. *Ans.* $\frac{32}{9}a^3$. **35.** $y=x^2$, $x=y^2$, $z=0$, $z=12+y-x^2$. *Ans.* $\frac{549}{144}$.

36. Compute the volumes of solids bounded by the coordinate planes, the plane $2x+3y-12=0$ and the cylinder $z=\frac{1}{2}y^2$. *Ans.* 16.

37. Compute the volumes of solids bounded by a circular cylinder of radius a, whose axis coincides with the z-axis, the coordinate planes and the plane $\frac{x}{a}+\frac{z}{a}=1$. *Ans.* $a^3\left(\frac{\pi}{4}-\frac{1}{3}\right)$.

38. Compute the volumes of solids bounded by the cylinders $x^2+y^2=a^2$, $x^2+z^2=a^2$. *Ans.* $\frac{16}{3}a^3$. **39.** $y^2+z^2=x$, $x=y$, $z=0$. *Ans.* $\frac{\pi}{64}$. **40.** $x^2+y^2+z^2=$ $=a^2$, $x^2+y^2=R^2$, $a>R$. *Ans.* $\frac{4}{3}\pi\left[a^3-(\sqrt{a^2-R^2})^3\right]$. **41.** $az=x^2+y^2$, $z=0$, $x^2+y^2=2ax$. *Ans.* $\frac{3}{2}\pi a^3$. **42.** $\rho^2=a^2\cos 2\theta$, $x^2+y^2+z^2=a^2$, $z=0$. (Compute the volume that is interior with respect to the cylinder.) *Ans.* $\frac{1}{9}a^3(3\pi+20-16\sqrt{2})$.

Calculating Surface Areas

43. Compute the area of that part of the surface of the cone $x^2+y^2=z^2$ which is cut out by the cylinder $x^2+y^2=2ax$. *Ans.* $2\pi a^2\sqrt{2}$.

44. Compute the area of that part of the plane $x+y+z=2a$ which lies in the first octant and is bounded by the cylinder $x^2+y^2=a^2$. *Ans.* $\frac{\pi a^2}{4}\sqrt{3}$.

45. Compute the surface area of a spherical segment (minor) if the radius of the sphere is a and the radius of the base of the segment is b. *Ans.* $2\pi(a^2-a\sqrt{a^2-b^2})$.

46. Find the area of that part of the surface of the sphere $x^2+y^2+z^2=a^2$ which is cut out by the surface of the cylinder $\frac{x^2}{a^2}+\frac{y^2}{b^2}=1$ $(a>b)$. *Ans.* $4\pi a^2-8a^2\arcsin\frac{\sqrt{a^2-b^2}}{a}$.

47. Find the surface area of a solid that is the common part of two cylinders $x^2+y^2=a^2$, $y^2+z^2=a^2$. *Ans.* $16a^2$.

48. Compute the area of that part of the surface of the cylinder $x^2+y^2=2ax$ which lies between the plane $z=0$ and the cone $x^2+y^2=z^2$. *Ans.* $8a^2$.

49. Compute the area of that part of the surface of the cylinder $x^2+y^2=a^2$ which lies between the plane $z=mx$ and the plane $z=0$. *Ans.* $2ma^2$.

50. Compute the area of that part of the surface of the paraboloid $y^2+z^2=2ax$ which lies between the parabolic cylinder $y^2=ax$ and the plane $x=a$. *Ans.* $\frac{1}{3}\pi a^2(3\sqrt{3}-1)$.

Computing the Mass, the Coordinates of the Centre of Gravity, and the Moment of Inertia of Plane Solids

(In Problems 51-62 and 64 we consider the surface density constant and equal to unity.)

51. Determine the mass of a lamina in the shape of a circle of radius a if the density at any point P is inversely proportional to the distance of P from the axis of the cylinder (the proportionality factor is K). *Ans.* $2\pi aK$.

52. Compute the coordinates of the centre of gravity of an equilateral triangle if we take its altitude for the x-axis and the vertex of the triangle for the coordinate origin. *Ans.* $x=\frac{a\sqrt{3}}{3}$, $y=0$.

53. Find the coordinates of the centre of gravity of a circular sector of radius a, taking the bisector of its angle as the x-axis. The angle of spread of the sector is 2α. *Ans.* $x_C=\frac{2a\sin\alpha}{3\alpha}$, $y_C=0$.

54. Find the coordinates of the centre of gravity of the upper half of the circle $x^2+y^2=a^2$. *Ans.* $x_C=0$, $y_C=\frac{4a}{3\pi}$.

55. Find the coordinates of the centre of gravity of the area of one arch of the cycloid $x=a(t-\sin t)$, $y=a(1-\cos t)$. *Ans.* $x_C=a\pi$, $y_C=\frac{5a}{6}$.

56. Find the coordinates of the centre of gravity of the area bounded by a loop of the curve $\rho^2=a^2\cos 2\theta$. *Ans.* $x_C=\frac{\pi a\sqrt{2}}{8}$, $y_C=0$.

57. Find the coordinates of the centre of gravity of the area of the cardioid $\rho=a(1+\cos\theta)$. *Ans.* $x_C=\frac{5a}{6}$, $y_C=0$.

58. Compute the moment of inertia of the area of a rectangle bounded by the straight lines $x=0$, $x=a$, $y=0$, $y=b$ relative to the origin. *Ans.* $\frac{ab(a^2+b^2)}{3}$.

59. Compute the moment of inertia of the ellipse $\frac{x^2}{a^2}+\frac{y^2}{b^2}=1$: (a) relative to the y-axis; (b) relative to the origin. *Ans.* (a) $\frac{\pi a^3 b}{4}$; (b) $\frac{\pi ab}{4}(a^2+b^2)$.

60. Compute the moment of inertia of the area of the circle $\rho=2a\cos\theta$ relative to the pole. *Ans.* $\frac{3}{2}\pi a^4$.

61. Compute the moment of inertia of the area of the cardioid $\rho=a(1-\cos\theta)$ relative to the pole. *Ans.* $\frac{35\pi a^4}{16}$.

62. Compute the moment of inertia of the area of the circle $(x-a)^2+(y-b)^2=2a^2$ relative to the y-axis. *Ans.* $3\pi a^4$.

63. The density at any point of a square lamina with side a is proportional to the distance of this point from one of the vertices of the square. Compute the moment of inertia of the lamina relative to the side passing through this vertex. *Ans.* $\frac{1}{40}ka^5[7\sqrt{2}+3\ln(\sqrt{2}+1)]$, where k is the proportionality factor.

64. Compute the moment of inertia of the area of a figure, bounded by the parabola $y^2=ax$ and the straight line $x=a$, relative to the straight line $y=-a$. *Ans.* $\frac{8}{5}a^4$.

Triple Integrals

65. Compute $\iiint \frac{dx\,dy\,dz}{(x+y+z+1)^3}$ if the domain of integration is bounded by the coordinate planes and the plane $x+y+z=1$. *Ans.* $\frac{\ln 2}{2}-\frac{5}{16}$.

66. Evaluate $\int\limits_0^a\left[\int\limits_0^x\left(\int\limits_0^y xyz\,dz\right)dy\right]dx$. *Ans.* $\frac{a^6}{48}$.

67. Compute the volume of a solid bounded by the sphere $x^2+y^2+z^2=4$ and the surface of the paraboloid $x^2+y^2=3z$. *Ans.* $\frac{19}{6}\pi$.

68.* Compute the coordinates of the centre of gravity and the moments of inertia of a pyramid bounded by the planes $x=0$, $y=0$, $z=0$, $\frac{x}{a}+\frac{y}{b}+\frac{z}{c}=1$. *Ans.* $x_C=\frac{a}{4}$, $y_C=\frac{b}{4}$, $z_C=\frac{c}{4}$; $I_x=\frac{a^3bc}{60}$, $I_y=\frac{b^3ac}{60}$, $I_z=\frac{c^3ab}{60}$, $I_0=\frac{abc}{60}(a^2+$ $+b^2+c^2)$.

69. Compute the moment of inertia of a circular right cone relative to its axis. *Ans.* $\frac{1}{10}\pi hr^4$, where h is the altitude and r is the radius of the base of the cone.

70. Compute the volume of a solid bounded by a surface with equation $(x^2+y^2+z^2)^2=a^3x$. *Ans.* $\frac{1}{3}\pi a^3$.

71. Compute the moment of inertia of a circular cone relative to the diameter of the base. *Ans.* $\frac{\pi hr^2}{60}(2h^2+3r^2)$.

72. Compute the coordinates of the centre of gravity of a solid lying between a sphere of radius a and a conic surface with angle at the vertex 2α, if the vertex of the cone coincides with the centre of the sphere. *Ans.* $x_C=0$, $y_C=0$, $z_C=\frac{3}{8}a(1+\cos\alpha)$ (the z-axis is the axis of the cone, and the vertex lies at the origin).

73. Compute the coordinates of the centre of gravity of a solid bounded by a sphere of radius a and by two planes passing through the centre of the sphere and forming an angle of 60°. *Ans.* $\rho=\frac{9}{16}a$, $\theta=0$, $\varphi=\frac{\pi}{2}$ (the line of intersection of the planes is taken for the z-axis, the centre of the sphere for the origin; ρ, θ, φ are spherical coordinates).

74. Using the equation $\frac{1}{\sqrt{x}}=\frac{2}{\sqrt{\pi}}\int\limits_0^\infty e^{-\alpha^2x}\,d\alpha$ $(\alpha>0)$, compute the integrals $\int\limits_0^\infty \frac{\cos x\,dx}{\sqrt{x}}$ and $\int\limits_0^\infty \frac{\sin x\,dx}{\sqrt{x}}$. *Ans.* $\sqrt{\frac{\pi}{2}}$, $\sqrt{\frac{\pi}{2}}$.

---

* In Problems 68, 69 and 71 to 73 we consider the density constant and equal to unity.

CHAPTER 3

LINE INTEGRALS AND SURFACE INTEGRALS

3.1 LINE INTEGRALS

Let a point $P(x, y)$ be in motion along some plane curve L from the point M to the point N. To P is applied a force $\boldsymbol{F}$ which varies in magnitude and direction with the motion of P; it is thus some function of the coordinates of P:

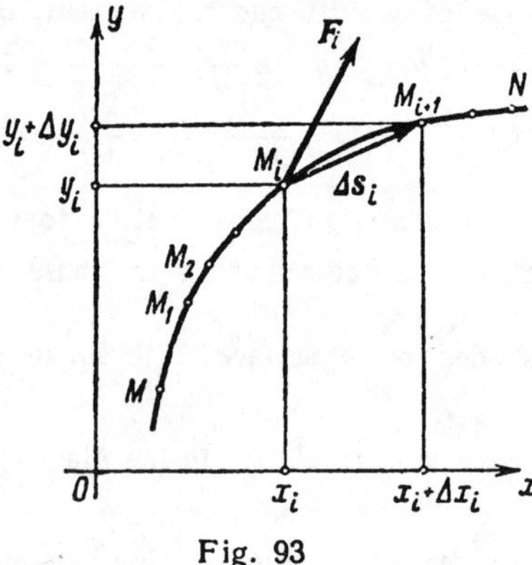

Fig. 93

$$\boldsymbol{F}=\boldsymbol{F}(P)$$

Let us compute the work A of the force $\boldsymbol{F}$ as the point P is translated from M to N (Fig. 93). To do this, we divide the curve MN into n arbitrary parts by the points $M=M_0, M_1, M_2, \ldots, M_n=N$ in the direction from M to N and we denote by $\Delta\boldsymbol{s}_i$ the vector $\overline{M_iM_{i+1}}$. We denote by $\boldsymbol{F}_i$ the magnitude of the force $\boldsymbol{F}$ at the point M_i. Then the scalar product $\boldsymbol{F}_i\Delta\boldsymbol{s}_i$ may be regarded as an approximate expression of the work of the force $\boldsymbol{F}$ along the arc $\widehat{M_iM_{i+1}}$:

$$A_i \approx \boldsymbol{F}_i\Delta\boldsymbol{s}_i$$

Let

$$\boldsymbol{F}=X(x, y)\boldsymbol{i}+Y(x, y)\boldsymbol{j}$$

where $X(x, y)$ and $Y(x, y)$ are the projections of the vector $\boldsymbol{F}$ on the x- and y-axes. Denoting by Δx_i and Δy_i the increments of the coordinates x_i and y_i when changing from the point M_i to the point M_{i+1}, we get

$$\Delta\boldsymbol{s}_i=\Delta x_i\boldsymbol{i}+\Delta y_i\boldsymbol{j}$$

Hence,

$$\boldsymbol{F}_i\Delta\boldsymbol{s}_i=X(x_i, y_i)\,\Delta x_i+Y(x_i, y_i)\,\Delta y_i$$

The approximate value of the work A of the force $\boldsymbol{F}$ over the entire curve MN will be

$$A \approx \sum_{i=1}^{n}\boldsymbol{F}_i\Delta\boldsymbol{s}_i=\sum_{i=1}^{n}[X(x_i, y_i)\,\Delta x_i+Y(x_i, y_i)\,\Delta y_i] \qquad (1)$$

Without making any precise statements, we shall say that if the expression on the right has a limit as $\Delta s_i \to 0$ (here, obviously, $\Delta x_i \to 0$ and $\Delta y_i \to 0$), then this limit expresses the work of the force $\boldsymbol{F}$ over the curve L from the point M to the point N:

$$A = \lim_{\substack{\Delta x_i \to 0 \\ \Delta y_i \to 0}} \sum_{i=1}^{n} [X(x_i, y_i)\Delta x_i + Y(x_i, y_i)\Delta y_i] \tag{2}$$

The limit* on the right is called the *line integral* of $X(x, y)$ and $Y(x, y)$ along the curve L and is denoted by

$$A = \int_L X(x, y)\,dx + Y(x, y)\,dy \tag{3}$$

or

$$A = \int_{(M)}^{(N)} X(x, y)\,dx + Y(x, y)\,dy \tag{3'}$$

Limits of sums of type (2) frequently occur in mathematics and mechanics; here, $X(x, y)$ and $Y(x, y)$ are regarded as functions of two variables in some domain D.

The letters M and N, which take the place of the limits of integration, are in brackets to signify that they are not numbers but symbols of the end points of the curve over which the line integral is taken. The direction along the curve L from M to N is called the *sense of integration*.

If the curve L is a space curve, then the line integral of three functions $X(x, y, z)$, $Y(x, y, z)$, $Z(x, y, z)$ is defined similarly:

$$\int_L X(x, y, z)\,dx + Y(x, y, z)\,dy + Z(x, y, z)\,dz$$

$$= \lim_{\substack{\Delta x_k \to 0 \\ \Delta y_k \to 0 \\ \Delta z_k \to 0}} \sum_{k=1}^{n} X(x_k, y_k, z_k)\Delta x_k + Y(x_k, y_k, z_k)\Delta y_k + Z(x_k, y_k, z_k)\Delta z_k$$

The letter L under the integral sign indicates that the integration is performed along the curve L.

We note two properties of a line integral.

Property 1. A line integral is determined by the element of integration, the form of the curve of integration, and the sense of integration.

---

* Here the limit of the integral sum is to be understood in the same sense as in the case of the definite integral, see Sec. 11.2 (Vol. I).

A line integral changes sign when the sense of integration is reversed, since in that case the vector $\Delta \boldsymbol{s}$ and, hence, its projections Δx and Δy, change sign.

Property 2. Divide the curve L by the point K into pieces L_1 and L_2 so that $\overset{\frown}{MN} = \overset{\frown}{MK} + \overset{\frown}{KN}$ (Fig. 94). Then, from formula (1) it follows directly that

$$\int_{(M)}^{(N)} X\,dx + Y\,dy = \int_{(M)}^{(K)} X\,dx + Y\,dy + \int_{(K)}^{(N)} X\,dx + Y\,dy$$

This relation holds for any number of terms.

It will further be noted that the definition of a line integral holds true also for the case when the curve L is closed.

In this case, the initial and terminal points of the curve coincide. Therefore, in the case of a closed curve we cannot write $\int_{(M)}^{(N)} X\,dx + Y\,dy$, but only $\int_{L} X\,dx + Y\,dy$; and we have to indicate the **direction of circulation** (sense of description) over the closed curve L. The line integral over a **closed contour** L is frequently denoted also by the symbol $\oint_{L} X\,dx + Y\,dy$.

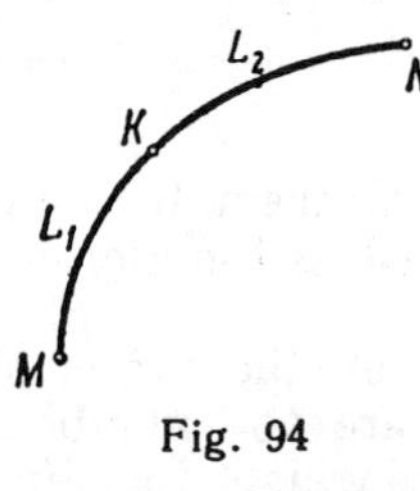

Fig. 94

Note. We arrived at the concept of a line integral while considering the problem of the work of a force $\boldsymbol{F}$ on a curved path L.

Here, at all points of the curve L the force $\boldsymbol{F}$ was given as a vector function $\boldsymbol{F}$ of the coordinates of the point of application (x, y); the projections of the variable vector $\boldsymbol{F}$ on the coordinate axes are equal to the scalar (numerical, that is) functions $X(x, y)$ and $Y(x, y)$. For this reason, a line integral of the form $\int_{L} X\,dx + Y\,dy$ may be regarded as an integral of the **vector** function $\boldsymbol{F}$ given by the projections X and Y.

The integral of a vector function $\boldsymbol{F}$ along a curve L is denoted by the symbol

$$\int_{L} \boldsymbol{F}\,d\boldsymbol{s}$$

If the vector $\boldsymbol{F}$ is defined by its projections X, Y, Z then this integral is equal to the line integral

$$\int_{L} X\,dx + Y\,dy + Z\,dz$$

As a particular instance, if the vector $\boldsymbol{F}$ lies in the xy-plane, then the integral of this vector is equal to

$$\int_L X\,dx + Y\,dy$$

When the line integral of a vector function $\boldsymbol{F}$ is taken along a closed curve L, this line integral is also called a *circulation* of the vector $\boldsymbol{F}$ over the closed contour L.

3.2 EVALUATING A LINE INTEGRAL

In this section we shall make more precise the concept of the limit of the sum (1) of Sec. 3.1 and in this connection we shall make more precise the concept of the line integral and indicate a method for calculating it.

Let a curve L be represented by equations in parametric form:

$$x = \varphi(t), \quad y = \psi(t)$$

Consider the arc of the curve MN (Fig. 95). Let the points M and N correspond to the values of the parameter α and β. Divide the arc MN into subarcs Δs_i by the points $M_1(x_1, y_1)$, $M_2(x_2, y_2)$, ..., $M_n(x_n, y_n)$, and put $x_i = \varphi(t_i)$, $y_i = \psi(t_i)$.

Consider the line integral

$$\int_L X(x, y)\,dx + Y(x, y)\,dy \qquad (1)$$

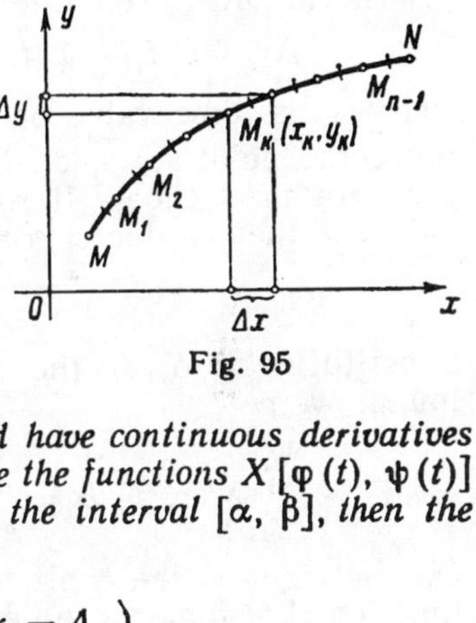

Fig. 95

defined in the preceding section. We give without proof the **existence theorem of a line integral.** *If the functions $\varphi(t)$ and $\psi(t)$ are continuous and have continuous derivatives $\varphi'(t)$ and $\psi'(t)$, and also continuous are the functions $X[\varphi(t), \psi(t)]$ and $Y[\varphi(t), \psi(t)]$ as functions of t on the interval $[\alpha, \beta]$, then the following limits exist:*

$$\left.\begin{aligned} \lim_{\Delta x_i \to 0} \sum_{i=1}^{n} X(\bar{x}_i, \bar{y}_i)\,\Delta x_i = A \\ \lim_{\Delta y_i \to 0} \sum_{i=1}^{n} Y(\bar{x}_i, \bar{y}_i)\,\Delta y_i = B \end{aligned}\right\} \qquad (2)$$

where $\bar{x}_i$ and *$\bar{y}_i$ are the coordinates of some point lying on the arc Δs_i. These limits do not depend on the way the arc L is divided into subarcs Δs_i, provided that $\Delta s_i \to 0$, and do not depend on the choice of the point $\bar{M}_i(\bar{x}_i, \bar{y}_i)$ on the subarc Δs_i;* they are called *line* *integrals* and are denoted as

$$A=\int_L X(x, y)\,dx$$
$$B=\int_L Y(x, y)\,dy \tag{2'}$$

Note. From this theorem it follows that the sums defined in the preceding section, where the points $\overline{M}_i(\overline{x}_i, \overline{y}_i)$ are the extremities of the subarc Δs_i and the manner of partition of the arc L into subarcs Δs_i is arbitrary, approach the same limit—the line integral.

This theorem makes it possible to develop a method for computing a line integral.

Thus, by definition, we have

$$\int_{(M)}^{(N)} X(x, y)\,dx=\lim_{\Delta x_i\to 0}\sum_{i=1}^{n} X(\overline{x}_i, \overline{y}_i)\,\Delta x_i \tag{3}$$

where

$$\Delta x_i=x_i-x_{i-1}=\varphi(t_i)-\varphi(t_{i-1})$$

Transform this latter difference by the Lagrange formula

$$\Delta x_i=\varphi(t_i)-\varphi(t_{i-1})=\varphi'(\tau_i)(t_i-t_{i-1})=\varphi'(\tau_i)\,\Delta t_i$$

where τ_i is some value of t that lies between the values t_{i-1} and t_i. Since the point $\overline{x}_i$, $\overline{y}_i$ on the subarc Δs_i may be chosen at pleasure, we shall choose it so that its coordinates correspond to the value of the parameter τ_i:

$$\overline{x}_i=\varphi(\tau_i),\quad \overline{y}_i=\psi(\tau_i)$$

Substituting into (3) the values of $\overline{x}_i$, $\overline{y}_i$ and Δx_i that we have found, we get

$$\int_{(M)}^{(N)} X(x, y)\,dx=\lim_{\Delta t_i\to 0}\sum_{i=1}^{n} X[\varphi(\tau_i)\,\psi(\tau_i)]\,\varphi'(\tau_i)\,\Delta t_i$$

On the right is the limit of the integral sum for the continuous function of a single variable $X[\varphi(t), \psi(t)]\,\varphi'(t)$ on the interval $[\alpha, \beta]$.

Hence, this limit is equal to the definite integral of the function:

$$\int_{(M)}^{(N)} X(x, y)\,dx=\int_{\alpha}^{\beta} X[\varphi(t), \psi(t)]\,\varphi'(t)\,dt$$

In analogous fashion we get the formula

$$\int_{(M)}^{(N)} Y(x, y)\,dy=\int_{\alpha}^{\beta} Y[\varphi(t), \psi(t)]\,\psi'(t)\,dt$$

Adding these equations term by term, we obtain

$$\int_{(M)}^{(N)} X(x, y)\,dx + Y(x, y)\,dy = \int_{\alpha}^{\beta} \{X[\varphi(t), \psi(t)]\varphi'(t) + Y[\varphi(t), \psi(t)]\psi'(t)\}\,dt \tag{4}$$

This is the desired formula for computing a line integral.

In similar manner we compute the line integral

$$\int X\,dx + Y\,dy + Z\,dz$$

over the space curve defined by the equations $x = \varphi(t)$, $y = \psi(t)$, $z = \chi(t)$.

Example 1. Compute the line integral of three functions: x^3, $3zy^2$, $-x^2y$ (or, what is the same thing, of the vector function $x^3\mathbf{i} + 3zy^2\mathbf{j} - x^2y\mathbf{k}$) along the straight-line segment from the point $M(3, 2, 1)$ to the point $N(0, 0, 0)$ (Fig. 96).

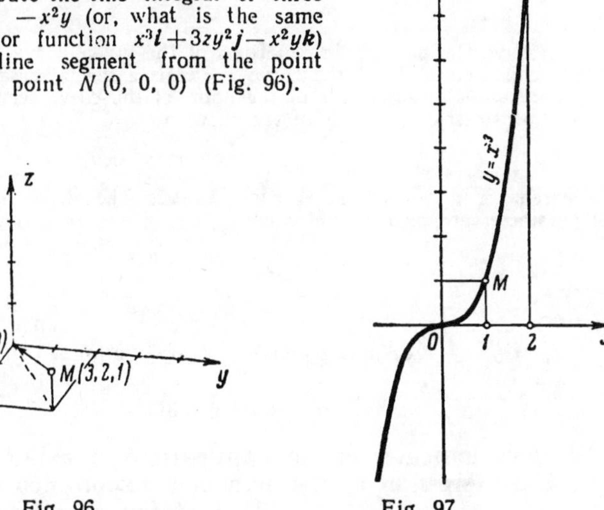

Fig. 96 Fig. 97

Solution. To find the parametric equations of the line MN, along which the integration is to be performed, we write the equation of the straight line that passes through the given two points:

$$\frac{x}{3} = \frac{y}{2} = \frac{z}{1}$$

and denote all these ratios by a single letter t; then we get the equations of the straight line in parametric form:

$$x = 3t, \quad y = 2t, \quad z = t$$

Here, obviously, to the origin of the segment MN corresponds the value of the parameter $t = 1$, and to the terminus of the segment, the value $t = 0$. The derivatives of x, y, z with respect to the parameter t (which will be needed for evaluating the line integral) are easily found:

$$x'_t = 3, \quad y'_t = 2, \quad z'_t = 1$$

Now the desired line integral may be computed by formula (4):

$$\int_{(M)}^{(N)} x^3\,dx+3zy^2\,dy-x^2y\,dz=\int_1^0 [(3t)^3\cdot 3+3t\,(2t)^2\cdot 2-(3t)^2\cdot 2t\cdot 1]\,dt$$

$$=\int_1^0 87t^3\,dt=-\frac{87}{4}$$

Example 2. Evaluate the line integral of a pair of functions, $6x^2y$, and $10xy^2$, along a plane curve $y=x^3$ from the point $M(1, 1)$ to the point $N(2, 8)$ (Fig. 97).

Solution. To compute the required integral

$$\int_{(M)}^{(N)} 6x^2y\,dx+10xy^2\,dy$$

we must have the parametric equations of the given curve. However, the explicitly defined equation of the curve $y=x^3$ is a special case of the parametric equation: here, the abscissa x of the point of the curve serves as the parameter, and the parametric equations of the curve are

$$x=x,\quad y=x^3$$

The parameter x varies from $x_1=1$ to $x_2=2$. The derivatives with respect to the parameter are readily evaluated:

$$x'_x=1,\quad y'_x=3x^2$$

Hence,

$$\int_{(M)}^{(N)} 6x^2y\,dx+10xy^2\,dy=\int_1^2 [6x^2x^3\cdot 1+10xx^6\cdot 3x^2]\,dx=\int_1^2 (6x^5+30x^9)\,dx$$

$$=[x^6+3x^{10}]_1^2=3132$$

We now indicate certain applications of a line integral.

1. **The expression of the area of a region bounded by a curve in terms of a line integral.** In an xy-plane let there be given a domain D (bounded by a contour L) such that any straight line parallel to one of the coordinate axes and passing through an interior point of the domain cuts the boundary L of the domain in no more than two points (which means that D is regular) (Fig. 98).

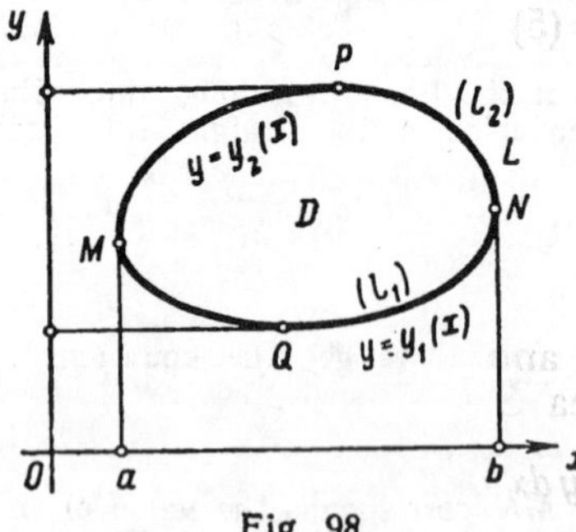

Fig. 98

Suppose that the domain D is projected on the x-axis in the interval $[a, b]$, and it is bounded below by the curve (l_1):

$$y=y_1(x)$$

and above by the curve (l_2):

$$y=y_2(x)$$
$$[y_1(x)\leqslant y_2(x)]$$

Then the area of the domain D is

$$S=\int_a^b y_2(x)dx-\int_a^b y_1(x)\,dx$$

But the first integral is a line integral over the curve l_2 (MPN), since $y=y_2(x)$ is the equation of this curve; hence,

$$\int_a^b y_2(x)\,dx=\int_{MPN} y\,dx$$

The second integral is a line integral over the curve l_1 ($\widehat{MQN}$), that is,

$$\int_a^b y_1(x)\,dx=\int_{MQN} y\,dx.$$

By Property 1 of the line integral we have

$$\int_{MPN} y\,dx=-\int_{NPM} y\,dx$$

Hence,

$$S=-\int_{NPM} y\,dx-\int_{MQN} y\,dx=-\int_L y\,dx \tag{5}$$

Here, the curve L is traced in a **counterclockwise** direction.

If part of the boundary L is the segment M_1M, parallel to the y-axis, then $\int_{(M_1)}^{(M)} y\,dx=0$, and equation (5) holds true in this case as well (Fig. 99).

Similarly, it may be shown that

$$S=\int_L x\,dy \tag{6}$$

Adding (5) and (6) term by term and dividing by 2, we get another formula for computing the area S:

$$S=\frac{1}{2}\int_L x\,dy-y\,dx \tag{7}$$

Example 3. Compute the area of the ellipse

$$x=a\cos t,\quad y=b\sin t$$

Solution. By formula (7) we find

$$S=\frac{1}{2}\int_{0}^{2\pi}[a\cos t\, b\cos t-b\sin t\,(-a\sin t)]\,dt=\pi ab$$

We note that formula (7) and formulas (5) and (6) as well hold true also for areas whose boundaries are cut in more than two points by lines parallel to the coordinate axes (Fig. 100).

To prove this, we divide the given domain (Fig. 100) into two regular domains by the line l^*.

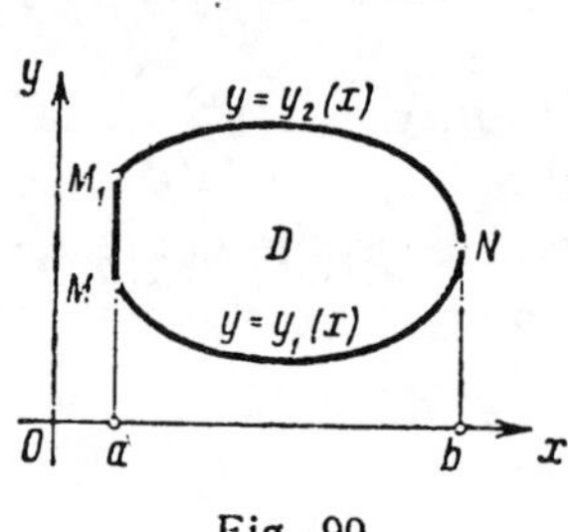

Fig. 99

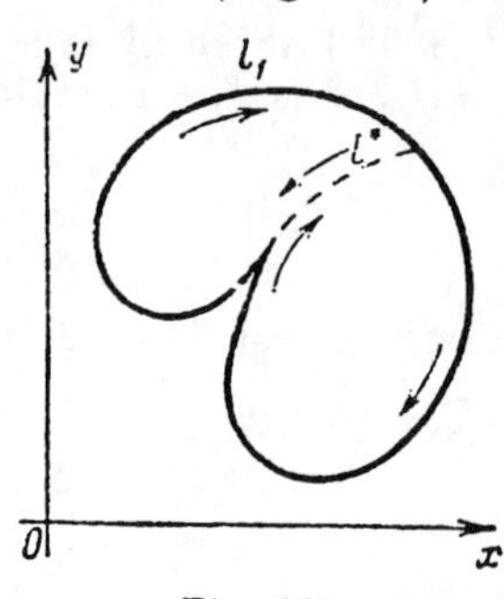

Fig. 100

Formula (7) holds for each of these domains. Adding the left and right sides, we get (on the left) the area of the given domain, on the right, a line integral taken over the entire boundary with coefficient $1/2$ since the line integral over the division line l^* is taken twice: in the direct and reverse senses; hence, it is equal to zero.

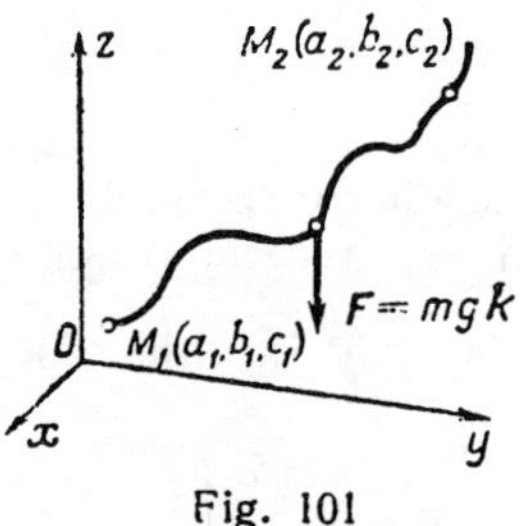

Fig. 101

2. **Computing the work of a variable force F on some curved path L.** As was shown at the beginning of Sec. 3.1, the work done by a force $\boldsymbol{F}=X(x, y, z)\boldsymbol{i}+Y(x, y, z)\boldsymbol{j}+Z(x, y, z)\boldsymbol{k}$ along a line $L=MN$ is equal to the line integral

$$A=\int_{(M)}^{(N)} X(x, y, z)\,dx+Y(x, y, z)\,dy+Z(x, y, z)\,dz$$

Let us consider an instance that shows how to calculate the work of a force in concrete cases.

Example 4. Determine the work A of the force of gravity $\boldsymbol{F}$ when a mass m is translated from the point $M_1(a_1, b_1, c_1)$ to the point $M_2(a_2, b_2, c_2)$ along an arbitrary path L (Fig. 101).

Solution. The projections of the force of gravity $\boldsymbol{F}$ on the coordinate axes are

$$X=0,\quad Y=0,\quad Z=-mg$$

Hence, the desired work is

$$A=\int_{(M_1)}^{(M_2)} X\,dx+Y\,dy+Z\,dz=\int_{c_1}^{c_2}(-mg)\,dz=mg\,(c_1-c_2)$$

Consequently, in this case the line integral is independent of the path of integration and dependent only on the initial and terminal points. More precisely, the work of the force of gravity is dependent only on the difference between the heights of the terminal and initial points of the path.

3.3 GREEN'S FORMULA

Let us establish a connection between a double integral over some plane domain D and the line integral around the boundary L of that domain.

In an xy-plane, let there be given a domain D, which is regular both in the direction of the x-axis and the y-axis, bounded by a closed contour L. Let the domain be bounded below by the curve $y=y_1(x)$, and above by the curve $y=y_2(x)$, $y_1(x)\leqslant y_2(x)$ $(a\leqslant x\leqslant b)$ (Fig. 98).

Together, both these curves represent the closed contour L. Let there be given, in D, continuous functions $X(x, y)$ and $Y(x, y)$ that have continuous partial derivatives. We consider the integral

$$\iint_D \frac{\partial X(x, y)}{\partial y}\,dx\,dy$$

Representing it in the form of a twofold iterated integral, we find

$$\iint_D \frac{\partial X}{\partial y}\,dx\,dy=\int_a^b\left[\int_{y_1(x)}^{y_2(x)}\frac{\partial X}{\partial y}\,dy\right]dx=\int_a^b X(x, y)\Big|_{y_1(x)}^{y_2(x)}\,dx$$

$$=\int_a^b [X(x, y_2(x))-X(x, y_1(x))]\,dx \qquad (1)$$

We note that the integral

$$\int_a^b X(x, y_2(x))\,dx$$

is numerically equal to the line integral

$$\int_{MPN} X(x, y)\,dx$$

taken along the curve MPN, whose equations, in parametric form, are

$$x=x,\quad y=y_2(x)$$

where x is the parameter.

Thus

$$\int_a^b X(x, y_2(x))\,dx = \int_{MPN} X(x, y)\,dx \tag{2}$$

Similarly, the integral

$$\int_a^b X(x, y_1(x))\,dx$$

is numerically equal to the line integral along the arc MQN:

$$\int_a^b X(x, y_1(x))\,dx = \int_{MQN} X(x, y)\,dx \tag{3}$$

Substituting expressions (2) and (3) into formula (1), we obtain

$$\iint_D \frac{\partial X}{\partial y}\,dx\,dy = \int_{MPN} X(x, y)\,dx - \int_{MQN} X(x, y)\,dx \tag{4}$$

But

$$\int_{MQN} X(x, y)\,dx = -\int_{NQM} X(x, y)\,dx$$

(see Sec. 3.1, Property 1). And so formula (4) may be written thus:

$$\iint_D \frac{\partial X}{\partial y}\,dx\,dy = \int_{MPN} X(x, y)\,dx + \int_{NQM} X(x, y)\,dx$$

But the sum of the line integrals on the right is equal to the line integral taken along the entire closed curve L in the clockwise direction. Hence, the last equation can be reduced to the form

$$\iint_D \frac{\partial X}{\partial y}\,dx\,dy = \int_{L \text{ (in the clockwise sense)}} X(x, y)\,dx \tag{5}$$

If part of the boundary is the segment l_3 parallel to the y-axis, then $\int_{l_3} X(x, y)\,dx = 0$, and equation (5) holds true in this case as well.

Analogously, we find

$$\iint_D \frac{\partial Y}{\partial x}\,dx\,dy = -\int_{L \text{ (in the clockwise sense)}} Y(x, y)\,dy \tag{6}$$

Subtracting (6) from (5), we obtain

$$\iint_D \left(\frac{\partial X}{\partial y} - \frac{\partial Y}{\partial x}\right)dx\,dy = \int_{L \text{ (in the clockwise sense)}} X\,dx + Y\,dy$$

If the contour is traversed in the counterclockwise sense, then *

$$\iint_D \left(\frac{\partial Y}{\partial x}-\frac{\partial X}{\partial y}\right) dx\,dy=\int_L X\,dx+Y\,dy$$

This is *Green's formula*, named after the English physicist and mathematician G. Green (1793-1841). **

We assumed that the domain D is regular. But, as in the area problem (see Sec. 3.2), it may be shown that this formula holds true for any domain that may be divided into regular domains.

3.4 CONDITIONS FOR A LINE INTEGRAL TO BE INDEPENDENT OF THE PATH OF INTEGRATION

Consider the line integral

$$\int_{(M)}^{(N)} X\,(dx)+Y\,dy$$

taken around some plane curve L connecting the points M and N. We assume that the functions $X(x, y)$ and $Y(x, y)$ have continuous partial derivatives in the domain D under consideration. Let us find out under what conditions the line integral above is independent of the shape of the curve L and is dependent only on the position of the initial and terminal points M and N.

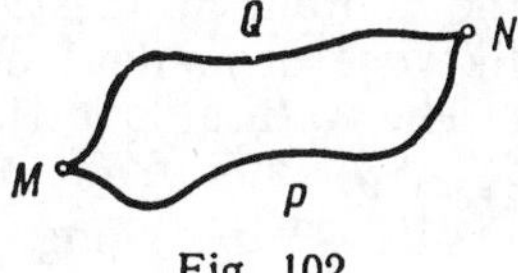

Fig. 102

Consider two arbitrary curves MPN and MQN lying in a given domain D and connecting the points M and N (Fig. 102). Let

$$\int_{MPN} X\,dx+Y\,dy=\int_{MQN} X\,dx+Y\,dy \tag{1}$$

that is,

$$\int_{MPN} X\,dx+Y\,dy-\int_{MQN} X\,dx+Y\,dy=0$$

Then, on the basis of Properties 1 and 2 of line integrals (Sec. 3.1) we have

$$\int_{MPN} X\,dx+Y\,dy+\int_{NQM} X\,dx+Y\,dy=0$$

* If in a line integral along a closed contour the direction of circulation is not indicated, it is assumed to be counterclockwise. If the direction of circulation is clockwise, this must be specified.

** This formula is a special case of a more general formula discovered by the Russian mathematician M. V. Ostrogradsky.

which is a line integral around the closed contour L:

$$\int_L X\,dx+Y\,dy=0 \tag{2}$$

In this formula, the line integral is taken around the closed contour L, which is made up of the curves MPN and NQM. This contour L may obviously be considered arbitrary.

Thus, from the condition that for any two points M and N the line integral is independent of the shape of the curve connecting them and is dependent only on the position of these points, it follows that the *line integral along any closed contour is equal to zero*.

The converse conclusion is also true: if a line integral around any closed contour is equal to zero, then this line integral is independent of the shape of the curve connecting the two points, and *depends only upon the position of the points*. Indeed, equation (1) follows from equation (2).

In Example 4 of Sec. 3.2, the line integral is independent of the path of integration; in Example 3 the line integral depends on the path of integration because there the integral around the closed contour is not equal to zero, but yields an area bounded by the contour in question; in Examples 1 and 2 the line integrals are likewise dependent on the path of integration.

The natural question arises: what conditions must the functions $X(x, y)$ and $Y(x, y)$ satisfy in order that the line integral $\int X\,dx+Y\,dy$ along any closed contour be equal to zero. The answer is given by the following theorem.

Theorem. *At all points of some domain D, let the functions $X(x, y)$, $Y(x, y)$, together with their partial derivatives $\frac{\partial X(x, y)}{\partial y}$ and $\frac{\partial Y(x, y)}{\partial x}$ be continuous. Then, for the line integral along any closed contour L lying in this domain to be zero, that is, for*

$$\int_L X(x, y)\,dx+Y(x, y)\,dy=0 \tag{2'}$$

it is necessary and sufficient that the equation

$$\frac{\partial X}{\partial y}=\frac{\partial Y}{\partial x} \tag{3}$$

holds at all points of D.

Proof. Consider an arbitrary closed contour L in D and write Green's formula for it:

$$\iint_D \left(\frac{\partial Y}{\partial x}-\frac{\partial X}{\partial y}\right)dx\,dy=\int_L X\,dx+Y\,dy$$

If condition (3) is fulfilled, then the double integral on the left is identically zero and, hence,

$$\int_L X\,dx + Y\,dy = 0$$

This proves the **sufficiency** of condition (3).

Now we prove the **necessity** of this condition; that is, we prove that if (2) is fulfilled for any closed curve L in the domain D, then condition (3) is also fulfilled at each point of this domain.

Let us assume, on the contrary, that equation (2) holds true, that is,

$$\int_L X\,dx + Y\,dy = 0$$

and that condition (3) does not hold,

$$\frac{\partial Y}{\partial x} - \frac{\partial X}{\partial y} \neq 0$$

at least in one point. For example, suppose at some point $P(x_0, y_0)$ we have the inequality

$$\frac{\partial Y}{\partial x} - \frac{\partial X}{\partial y} > 0$$

Since there is a continuous function on the left, it will be positive and greater than some number $\delta > 0$ at all points of some sufficiently small domain D' containing the point $P(x_0, y_0)$. Take the double integral of the difference $\frac{\partial Y}{\partial x} - \frac{\partial X}{\partial y}$ over this domain. It will have a positive value. Indeed,

$$\iint_{D'} \left(\frac{\partial Y}{\partial x} - \frac{\partial X}{\partial y} \right) dx\,dy > \iint_{D'} \delta\,dx\,dy = \delta \iint_{D'} dx\,dy = \delta D' > 0$$

But by Green's formula the left side of the last inequality is equal to a line integral along the boundary L' of D', which, by assumption, is zero. Hence, the last inequality contradicts condition (2) and therefore the assumption that $\frac{\partial Y}{\partial x} - \frac{\partial X}{\partial y}$ is different from zero in at least one point is not correct. Whence it follows that

$$\frac{\partial Y}{\partial x} - \frac{\partial X}{\partial y} = 0$$

at all points of the given domain D.

The proof of the theorem is thus complete.

In Sec. 1.9, it was proved that fulfilment of the condition

$$\frac{\partial Y(x, y)}{\partial x} = \frac{\partial X(x, y)}{\partial y}$$

is tantamount to the fact that the expression $X\,dx+Y\,dy$ is an **exact differential of some function** $u(x, y)$, or

$$X\,dx+Y\,dy=du(x, y)$$

and

$$X(x, y)=\frac{\partial u}{\partial x}, \quad Y(x, y)=\frac{\partial u}{\partial y}$$

But in this case the vector

$$\boldsymbol{F}=X\boldsymbol{i}+Y\boldsymbol{j}=\frac{\partial u}{\partial x}\boldsymbol{i}+\frac{\partial u}{\partial y}\boldsymbol{j}$$

is the gradient of the function $u(x, y)$; the function $u(x, y)$, the gradient of which is equal to the vector $X\boldsymbol{i}+Y\boldsymbol{j}$, is called the *potential* of this vector.

We shall prove that *in this case the line integral* $I=\int\limits_{(M)}^{(N)} X\,dx+Y\,dy$ *along any curve L connecting the points M and N is equal to the difference between the values of the function u at these points:*

$$\int\limits_{(M)}^{(N)} X\,dx+Y\,dy=\int\limits_{(M)}^{(N)} du(x, y)=u(N)-u(M)$$

Proof. If $X\,dx+Y\,dy$ is the exact differential of the function $u(x, y)$, then $X=\frac{\partial u}{\partial x}$, $Y=\frac{\partial u}{\partial y}$ and the line integral takes on the form

$$I=\int\limits_{(M)}^{(N)} \frac{\partial u}{\partial x}dx+\frac{\partial u}{\partial y}dy$$

To evaluate this integral we write the parametric equations of the curve L connecting the points M and N:

$$x=\varphi(t), \qquad y=\psi(t)$$

We shall assume that to the value of the parameter $t=t_0$ corresponds the point M, and to $t=T$, the point N. Then the line integral reduces to the following definite integral:

$$I=\int\limits_{t_0}^{T} \left[\frac{\partial u}{\partial x}\frac{dx}{dt}+\frac{\partial u}{\partial y}\frac{dy}{dt}\right] dt$$

The expression in the brackets is a function of t, and this function is the total derivative of the function $u[\varphi(t), \psi(t)]$ with respect to t. Therefore

$$I=\int\limits_{t_0}^{T} \frac{du}{dt}dt=u[\varphi(t), \psi(t)]\Big|_{t_0}^{T}=u[\varphi(t), \psi(t)]$$
$$-u[\varphi(t_0), \psi(t_0)]=u(N)-u(M)$$

As we see, *the line integral of an exact differential is independent of the shape of the curve along which the integration is performed.*

We have a similar assertion for a line integral over a space curve (see below, Sec. 3.7).

Note. It is sometimes necessary to consider line integrals of some function $X(x, y)$ along the arc length L:

$$\int_L X(x, y)\, ds = \lim_{\Delta s_i \to 0} \sum_{i=1}^{n} X(x_i, y_i)\, \Delta s_i \tag{4}$$

where ds is the differential arc length. Such integrals are evaluated in similar fashion to the line integrals considered above. Let the curve L be represented by the parametric equations

$$x = \varphi(t), \quad y = \psi(t)$$

where $\varphi(t)$, $\psi(t)$, $\varphi'(t)$, $\psi'(t)$ are continuous functions of t.

Let α and β be values of the parameter t corresponding to the initial and terminal points of the arc L.

Since

$$ds = \sqrt{\varphi'(t)^2 + \psi'(t)^2}\, dt,$$

we get a formula for evaluating integral (4):

$$\int_L X(x, y)\, ds = \int_\alpha^\beta X[\varphi(t), \psi(t)] \sqrt{\varphi'(t)^2 + \psi'(t)^2}\, dt$$

We can consider the line integral along the arc of the space curve $x = \varphi(t)$, $y = \psi(t)$, $z = \chi(t)$:

$$\int_L X(x, y, z)\, ds = \int_\alpha^\beta X[\varphi(t), \psi(t), \chi(t)] \sqrt{\varphi'(t)^2 + \psi'(t)^2 + \chi'(t)^2}\, dt$$

By the use of line integrals along an arc we can determine, for example, the coordinates of the centre of gravity of lines.

Reasoning as in Sec. 12.8, Vol. I, we obtain a formula for evaluating the coordinates of the centre of gravity of a space curve:

$$x_C = \frac{\int_L x\, ds}{\int_L ds}, \quad y_C = \frac{\int_L y\, ds}{\int_L ds}, \quad z_C = \frac{\int_L z\, ds}{\int_L ds} \tag{5}$$

Example. Find the coordinates of the centre of gravity of one turn of the helix

$$x = a\cos t, \quad y = a\sin t, \quad z = bt \quad (0 \leqslant t < 2\pi)$$

if its linear density is constant.

Solution. Applying formula (5), we find

$$x_C=\frac{\int\limits_0^{2\pi} a\cos t\sqrt{a^2\sin^2 t+a^2\cos^2 t+b^2}\,dt}{\int\limits_0^{2\pi}\sqrt{a^2\sin^2 t+a^2\cos^2 t+b^2}\,dt}$$

$$=\frac{\int\limits_0^{2\pi} a\cos t\sqrt{a^2+b^2}\,dt}{\int\limits_0^{2\pi}\sqrt{a^2+b^2}\,dt}=\frac{a\cdot 0}{2\pi}=0$$

Similarly, $y_C=0$,

$$z_C=\frac{\int\limits_0^{2\pi} bt\sqrt{a^2\sin^2 t+a^2\cos^2 t+b^2}\,dt}{2\pi\sqrt{a^2+b^2}}=\frac{b\cdot 4\pi^2}{2\cdot 2\pi}=\pi b$$

Thus, the coordinates of the centre of gravity of one turn of the helix are

$$x_C=0,\quad y_C=0,\quad z_C=\pi b$$

3.5 SURFACE INTEGRALS

Let a domain V be given in a rectangular xyz-coordinate system. Let a surface σ bounded by a certain space curve λ be given in V.

With respect to the surface σ we shall assume that at each point P of it the positive direction of the normal is determined by the unit vector $\boldsymbol{n}(P)$, the direction cosines of which are continuous functions of the coordinates of the surface points.

At each point of the surface let there be defined a vector

$$\boldsymbol{F}=X(x,\ y,\ z)\,\boldsymbol{i}+Y(x,\ y,\ z)\,\boldsymbol{j}+Z(x,\ y,\ z)\,\boldsymbol{k}$$

where X, Y, Z are continuous functions of the coordinates.

Divide the surface in some way into subdomains $\Delta\sigma_i$. In each one take an arbitrary point P_i and consider the sum

$$\sum_i (\boldsymbol{F}(P_i)\,\boldsymbol{n}(P_i))\,\Delta\sigma_i \tag{1}$$

where $\boldsymbol{F}(P_i)$ is the value of the vector $\boldsymbol{F}$ at the point P_i of the subdomain $\Delta\sigma_i$; $\boldsymbol{n}(P_i)$ is the unit normal vector at this point and $\boldsymbol{Fn}$ is the scalar product of these vectors.

The limit of the sum (1) extended over all subdomains $\Delta\sigma_i$ as the diameters of all such subdomains approach zero is called the *surface integral* and is denoted by the symbol

$$\iint_{\sigma} \boldsymbol{Fn}\, d\sigma$$

Thus, by definition,*

$$\lim_{\text{diam}\, \Delta\sigma_i \to 0} \sum \boldsymbol{F}_i \boldsymbol{n}_i\, \Delta\sigma_i = \iint_{\sigma} \boldsymbol{Fn}\, d\sigma \tag{2}$$

Each term of the sum (1)

$$\boldsymbol{F}_i \boldsymbol{n}_i \Delta\sigma_i = F_i \Delta\sigma_i \cos(\boldsymbol{n}_i, \boldsymbol{F}_i) \tag{3}$$

may be interpreted mechanically as follows: this product is equal to the volume of a cylinder with base $\Delta\sigma_i$ and altitude $F_i \cos(\boldsymbol{n}_i, \boldsymbol{F}_i)$. If the vector $\boldsymbol{F}$ is the rate of flow of a liquid through the surface σ, then the product (3) is equal to the quantity of liquid flowing through the subdomain $\Delta\sigma_i$ in unit time in the direction of the vector $\boldsymbol{n}_i$ (Fig. 103).

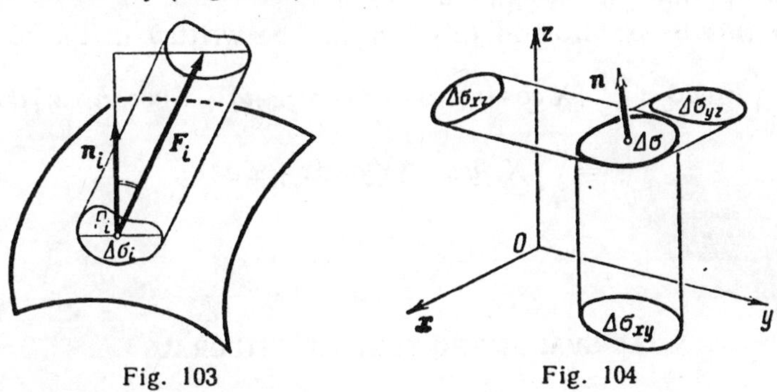

Fig. 103 Fig. 104

The expression $\iint_{\sigma} \boldsymbol{Fn}\, d\sigma$ yields the total quantity of liquid flowing in unit time through the surface σ in the positive direction if by the vector $\boldsymbol{F}$ we assume the flow-rate vector of the liquid at the given point. Therefore, the surface integral (2) is called the *flux of the vector field* $\boldsymbol{F}$ *through the surface* σ.

From the definition of a surface integral it follows that if the surface σ is divided into the parts $\sigma_1, \sigma_2, \ldots, \sigma_k$, then

$$\iint_{\sigma} \boldsymbol{Fn}\, d\sigma = \iint_{\sigma_1} \boldsymbol{Fn}\, d\sigma + \iint_{\sigma_2} \boldsymbol{Fn}\, d\sigma + \ldots + \iint_{\sigma_k} \boldsymbol{Fn}\, d\sigma$$

---

* If the surface σ is such that at each point of it there exists a tangent plane that constantly varies as the point P is translated over the surface, and if the vector function $\boldsymbol{F}$ is continuous on this surface, then this limit exists (we accept this existence theorem of a surface integral without proof).

Let us express the unit vector $\boldsymbol{n}$ in terms of its projections on the coordinate axes:

$$\boldsymbol{n} = \cos(n, x)\,\boldsymbol{i} + \cos(n, y)\,\boldsymbol{j} + \cos(n, z)\,\boldsymbol{k}$$

Substituting into the integral (2) the expressions of the vectors $\boldsymbol{F}$ and $\boldsymbol{n}$ in terms of their projections, we get

$$\iint_{\sigma} \boldsymbol{Fn}\, d\sigma = \iint_{\sigma} [X\cos(n, x) + Y\cos(n, y) + Z\cos(n, z)]\, d\sigma \quad (2')$$

The product $\Delta\sigma\cos(n, z)$ is the projection of the subdomain $\Delta\sigma$ on the xy-plane (Fig. 104); an analogous assertion holds true for the following products as well:

$$\Delta\sigma\cos(n, x) = \Delta\sigma_{yz}, \quad \Delta\sigma\cos(n, y) = \Delta\sigma_{xz}, \quad \Delta\sigma\cos(n, z) = \Delta\sigma_{xy} \quad (4)$$

where $\Delta\sigma_{yz}$, $\Delta\sigma_{xz}$, $\Delta\sigma_{xy}$ are the projections of the subdomain $\Delta\sigma$ on the appropriate coordinate planes.

On this basis, integral (2′) can also be written in the form

$$\iint_{\sigma} \boldsymbol{Fn}\, d\sigma = \iint_{\sigma} [X\cos(n, x) + Y\cos(n, y) + Z\cos(n, z)]\, d\sigma$$

$$= \iint_{\sigma} X\,dy\,dz + Y\,dz\,dx + Z\,dx\,dy \quad (2'')$$

3.6 EVALUATING SURFACE INTEGRALS

Computing the integral over a curved surface reduces to evaluating a double integral over a plane region.

To illustrate, the following is a method of computing the integral

$$\iint_{\sigma} Z\cos(n, z)\, d\sigma$$

Let the surface σ be such that any straight line parallel to the z-axis cuts it in one point. Then the equation of the surface may be written in the form

$$z = f(x, y)$$

Denoting by D the projection of the surface σ on the xy-plane, we get (by the definition of a surface integral)

$$\iint_{\sigma} Z(x, y, z)\cos(n, z)\, d\sigma = \lim_{\operatorname{diam}\Delta\sigma_i \to 0} \sum_{i=1}^{n} Z(x_i, y_i, z_i)\cos(n_i, z)\,\Delta\sigma_i$$

Noting, further, the last of formulas (4), Sec. 3.5, we obtain

$$\iint_{\sigma} Z\cos(n,\ z)\,d\sigma = \lim_{\operatorname{diam}\Delta\sigma_{xy}\to 0}\sum_{i=1}^{n} Z(x_i,\ y_i,\ f(x_i,\ y_i))\,(\Delta\sigma_{xy})_i$$

$$=\pm \lim_{\operatorname{diam}\Delta\sigma\to 0}\sum_{i=1}^{n} Z(x_i,\ y_i,\ f(x_i,\ y_i))\,|\,\Delta\sigma_{xy}\,|_i$$

the last expression is the integral sum for a double integral of the function $Z(x,\ y,\ f(x,\ y))$ over the domain D. Therefore,

$$\iint_{\sigma} Z\cos(n,\ z)\,d\sigma = \pm\iint_{D} Z(x,\ y,\ f(x,\ y))\,dx\,dy$$

The plus sign in front of the double integral is taken if $\cos(n,\ z)\geqslant 0$, the minus sign, if $\cos(n,\ z)\leqslant 0$.

If the surface σ does not satisfy the condition indicated at the beginning of this section, then it is divided into parts that satisfy this condition, and the integral is computed over each part separately.

The following integrals are computed in similar fashion:

$$\iint_{\sigma} X\cos(n,\ x)\,d\sigma,\qquad \iint_{\sigma} Y\cos(n,\ y)\,d\sigma$$

The foregoing proof justifies the notation of a surface integral in the form of (2″), Sec. 3.5.

Here, the right side of (2″) may be regarded as the sum of double integrals over the appropriate projections of the region σ and the signs of these double integrals (or, otherwise stated, the signs of the products $dy\,dz$, $dx\,dz$, $dx\,dy$) are taken in accord with the foregoing rule.

Example 1. Let a closed surface σ be such that any straight line parallel to the z-axis cuts it in no more than two points.

Consider the integral

$$\iint_{\sigma} z\cos(n,\ z)\,d\sigma$$

We shall call the outer normal the positive direction of the normal.

In this case, the surface may be divided into two parts: lower and upper; their equations are, respectively,

$$z=f_1(x,\ y)\quad\text{and}\quad z=f_2(x,\ y)$$

Denote by D the projection of σ on the xy-plane (Fig. 105); then

$$\iint_{\sigma} z\cos(n,\ z)\,d\sigma = \iint_{D} f_2(x\ \ y)\,dx\,dy - \iint_{D} f_1(x,\ y)\,dx\,dy$$

The minus sign in the second integral is taken because in a surface integral the sign of $dx\,dy$ on a surface $z=f_1(x,\ y)$ must be taken negative, since for it $\cos(n,\ z)$ is negative.

But the difference between the integrals on the right in the last formula yields a volume bounded by the surface σ. This means that the volume of the solid bounded by the closed surface σ is equal to the following integral over the surface

$$V=\iint_{\sigma} z\cos(n, z)\,d\sigma$$

Example 2. A positive electric charge e placed at the coordinate origin creates a vector field such that at each point of space the vector $\boldsymbol{F}$ is defined by the Coulomb law as

$$\boldsymbol{F}=k\frac{e}{r^2}\boldsymbol{r}$$

where r is the distance of the given point from the origin, $\boldsymbol{r}$ is the unit vector directed along the radius vector of the given point (Fig. 106), and k is a constant coefficient. Determine the vector-field flux through a sphere of radius R with centre at the origin of coordinates.

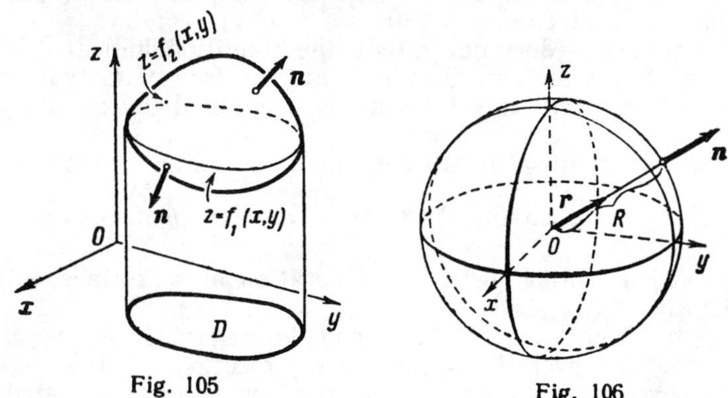

Fig. 105

Fig. 106

Solution. Taking into account that $r=R=\text{const}$, we will have

$$\iint_{\sigma} k\frac{e}{r^2}\boldsymbol{rn}\,d\sigma=\frac{ke}{R^2}\iint_{\sigma}\boldsymbol{rn}\,d\sigma$$

But the last integral is equal to the area of the surface σ. Indeed, by the definition of an integral (noting that $\boldsymbol{rn}=1$), we obtain

$$\iint_{\sigma}\boldsymbol{rn}\,d\sigma=\lim_{\Delta\sigma_k\to 0}\sum \boldsymbol{r}_k\boldsymbol{n}_k\Delta\sigma_k=\lim_{\Delta\sigma_k\to 0}\sum\Delta\sigma_k=\sigma$$

Hence, the flux is $\frac{ke}{R^2}\sigma=\frac{ke}{R^2}\cdot 4\pi R^2=4\pi ke$.

3.7 STOKES' FORMULA

Let there be a surface σ such that any straight line parallel to the z-axis cuts it in one point. Denote by λ the boundary of the surface σ. Take the positive direction of the normal $\boldsymbol{n}$ so that it forms an acute angle with the positive z-axis (Fig. 107).

Let the equation of the surface be $z=f(x, y)$. The direction cosines of the normal are expressed by the formulas (see Sec. 9.6, Vol. I):

$$\left.\begin{aligned}\cos(n, x) &= \frac{-\frac{\partial f}{\partial x}}{\sqrt{1+\left(\frac{\partial f}{\partial x}\right)^2+\left(\frac{\partial f}{\partial y}\right)^2}}\\ \cos(n, y) &= \frac{-\frac{\partial f}{\partial y}}{\sqrt{1+\left(\frac{\partial f}{\partial x}\right)^2+\left(\frac{\partial f}{\partial y}\right)^2}}\\ \cos(n, z) &= \frac{1}{\sqrt{1+\left(\frac{\partial f}{\partial x}\right)^2+\left(\frac{\partial f}{\partial y}\right)^2}}\end{aligned}\right\} \quad (1)$$

We shall assume that the surface σ lies entirely in some domain V. Let there be a function $X(x, y, z)$ given in V that is continuous together with its first-order partial derivatives. Consider the line integral along the curve λ:

$$\int_{\lambda} X(x, y, z)\,dx$$

On the curve λ, $z=f(x, y)$, where x, y are the coordinates of the points of the curve L, which is the projection of λ on the xy-plane (Fig. 107). Thus, we can write the equation

$$\int_{\lambda} X(x, y, z)\,dx = \int_{L} X(x, y, f(x, y))\,dx \quad (2)$$

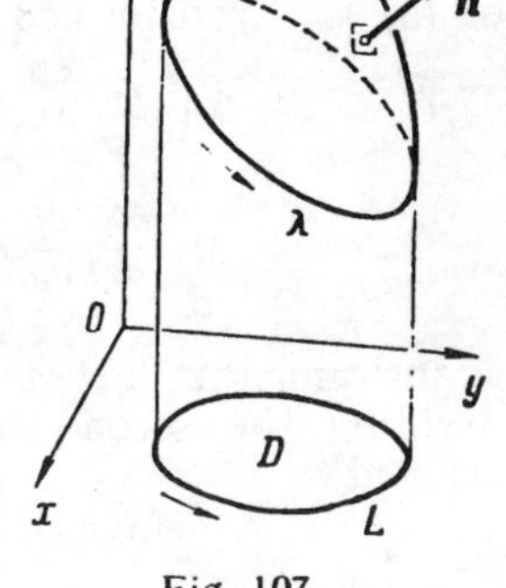

Fig. 107

The last integral is a line integral along L. Transform this integral by Green's formula, putting

$$X(x, y, f(x, y)) = \overline{X}(x, y), \qquad 0=\overline{Y}(x, y)$$

Substituting into Green's formula the expressions of $\overline{X}$ and $\overline{Y}$, we obtain

$$-\iint_{D} \frac{\partial X(x, y, f(x, y))}{\partial y}\,dx\,dy = \int_{L} X(x, y, f(x, y))\,dx \quad (3)$$

where the domain D is bounded by the curve L. On the basis of the derivative of the composite function $X(x, y, f(x, y))$, where y enters both directly and in terms of the function $z=f(x, y)$, we find

$$\frac{\partial X(x, y, f(x, y))}{\partial y} = \frac{\partial X(x, y, z)}{\partial y} + \frac{\partial X(x, y, z)}{\partial z}\frac{\partial f(x, y)}{\partial y} \quad (4)$$

Substituting expression (4) into the left side of (3), we obtain

$$-\iint_D \left[\frac{\partial X(x, y, z)}{\partial y} + \frac{\partial X(x, y, z)}{\partial z} \cdot \frac{\partial f(x, y)}{\partial y}\right] dx\,dy$$
$$= \int_L X(x, y, f(x, y))\,dx$$

Taking into account (2), the last equation may be rewritten as

$$\int_\lambda X(x, y, z)\,dx = -\iint_D \frac{\partial X}{\partial y}\,dx\,dy - \iint_D \frac{\partial X}{\partial z}\frac{\partial f}{\partial y}\,dx\,dy \qquad (5)$$

The last two integrals can be transformed into surface integrals. Indeed, from formula (2″), Sec. 3.5, it follows that if we have some function $A(x, y, z)$, the following equation is true:

$$\iint_\sigma A(x, y, z)\cos(n, z)\,d\sigma = \iint_D A\,dx\,dy$$

On the basis of this equation, the integrals on the right side of (5) are transformed as follows:

$$\left.\begin{aligned} \iint_D \frac{\partial X}{\partial y}\,dx\,dy &= \iint_\sigma \frac{\partial X}{\partial y}\cos(n, z)\,d\sigma \\ \iint_D \frac{\partial X}{\partial z}\frac{\partial f}{\partial y}\,dx\,dy &= \iint_\sigma \frac{\partial X}{\partial z}\frac{\partial f}{\partial y}\cos(n, z)\,d\sigma \end{aligned}\right\} \qquad (6)$$

Transform the last integral using formulas (1) of this section: dividing the second of these equations by the third termwise, we find

$$\frac{\cos(n, y)}{\cos(n, z)} = -\frac{\partial f}{\partial y}$$

or

$$\frac{\partial f}{\partial y}\cos(n, z) = -\cos(n, y)$$

Hence,

$$\iint_D \frac{\partial X}{\partial z}\frac{\partial f}{\partial y}\,dx\,dy = -\iint_\sigma \frac{\partial X}{\partial z}\cos(n, y)\,d\sigma \qquad (7)$$

Substituting expressions (6) and (7) into equation (5), we get

$$\int_\lambda X(x, y, z)\,dx = -\iint_\sigma \frac{\partial X}{\partial y}\cos(n, z)\,d\sigma + \iint_\sigma \frac{\partial X}{\partial z}\cos(n, y)\,d\sigma \quad (8)$$

The direction of circulation of the contour λ must agree with the chosen direction of the positive normal n. Namely, if an observer looks from the end of the normal, he sees the circulation along the curve λ as being counterclockwise.

Formula (8) holds true for any surface if this surface can be divided into parts whose equations have the form $z=f(x, y)$.

Similarly, we can write the formulas

$$\int_\lambda Y(x, y, z)\,dy=\iint_\sigma \left[-\frac{\partial Y}{\partial z}\cos(n, x)+\frac{\partial Y}{\partial x}\cos(n, z)\right]d\sigma \quad (8')$$

$$\int_\lambda Z(x, y, z)\,dz=\iint_\sigma \left[-\frac{\partial Z}{\partial x}\cos(n, y)+\frac{\partial Z}{\partial y}\cos(n, x)\right]d\sigma \quad (8'')$$

Adding the left and right sides of (8), (8'), and (8''), we get the formula

$$\int_\lambda X\,dx+Y\,dy+Z\,dz=\iint_\sigma \left[\left(\frac{\partial Z}{\partial y}-\frac{\partial Y}{\partial z}\right)\cos(n, x)\right.$$
$$\left.+\left(\frac{\partial X}{\partial z}-\frac{\partial Z}{\partial x}\right)\cos(n, y)+\left(\frac{\partial Y}{\partial x}-\frac{\partial X}{\partial y}\right)\cos(n, z)\right]d\sigma \quad (9)$$

This formula is called *Stokes' formula* after the English physicist and mathematician G. H. Stokes (1819-1903). It establishes a relationship between the integral over the surface σ and the line integral along the boundary λ of this surface, the curve λ being traversed according to the same rule as that given earlier.

The vector $\boldsymbol{B}$, defined by the projections

$$B_x=\frac{\partial Z}{\partial y}-\frac{\partial Y}{\partial z}, \quad B_y=\frac{\partial X}{\partial z}-\frac{\partial Z}{\partial x}, \quad B_z=\frac{\partial Y}{\partial x}-\frac{\partial X}{\partial y},$$

is called the *curl* of the vector function $\boldsymbol{F}=X\boldsymbol{i}+Y\boldsymbol{j}+Z\boldsymbol{k}$ and is denoted by the symbol curl $\boldsymbol{F}$.

Thus, in **vector** notation, formula (9) will have the form

$$\int_\lambda \boldsymbol{F}\,d\boldsymbol{s}=\iint_\sigma \boldsymbol{n}\,\mathrm{curl}\,\boldsymbol{F}\,d\sigma \quad (9')$$

and Stokes' theorem is formulated thus:

The circulation of a vector around the contour of some surface is equal to the flux of the curl through this surface.

Note. If the surface σ is a piece of plane parallel to the *xy*-plane, then $\Delta z=0$, and we get Green's formula as a special case of Stokes' formula.

From formula (9) it follows that if

$$\frac{\partial Y}{\partial x}-\frac{\partial X}{\partial y}=0, \quad \frac{\partial Z}{\partial y}-\frac{\partial Y}{\partial z}=0, \quad \frac{\partial X}{\partial z}-\frac{\partial Z}{\partial x}=0 \quad (10)$$

then the line integral along any closed space curve λ is zero:

$$\int_\lambda X\,dx+Y\,dy+Z\,dz=0 \quad (11)$$

Whence it follows that the line integral is independent of the shape of the curve of integration.

As in the case of a plane curve, it may be shown that the indicated conditions are not only sufficient but also necessary.

In the fulfilment of these conditions, the expression under the integral sign is an exact differential of some function $u(x, y, z)$:

$$X\,dx + Y\,dy + Z\,dz = du(x, y, z)$$

and, consequently,

$$\int_{(M)}^{(N)} X\,dx + Y\,dy + Z\,dz = \int_{(M)}^{(N)} du = u(N) - u(M)$$

This is proved exactly like the corresponding formula for a function of two variables (see Sec. 3.4).

Example 1. Write the basic equations of the dynamics of a material point:

$$m\frac{dv_x}{dt} = X, \quad m\frac{dv_y}{dt} = Y, \quad m\frac{dv_z}{dt} = Z$$

Here, m is the mass of the point, X, Y, Z are the projections of a force, acting on the point, onto the coordinate axes; $v_x = \frac{dx}{dt}$, $v_y = \frac{dy}{dt}$, $v_z = \frac{dz}{dt}$ are the projections of velocity v onto the axes.

Multiply the left and right sides of these equations by the expressions

$$v_x dt = dx, \quad v_y dt = dy, \quad v_z dt = dz$$

Adding the given equations term by term, we obtain

$$m(v_x dv_x + v_y dv_y + v_z dv_z) = X\,dx + Y\,dy + Z\,dz$$

$$m\frac{1}{2}d(v_x^2 + v_y^2 + v_z^2) = X\,dx + Y\,dy + Z\,dz$$

Since $v_x^2 + v_y^2 + v_z^2 = v^2$, we can write

$$d\left(\frac{1}{2}mv^2\right) = X\,dx + Y\,dy + Z\,dz$$

Take the integral along the trajectory connecting the points M_1 and M_2:

$$\frac{1}{2}mv_2^2 - \frac{1}{2}mv_1^2 = \int_{(M_1)}^{(M_2)} X\,dx + Y\,dy + Z\,dz$$

where v_1 and v_2 are the velocities at the points M_1 and M_2.

This last equation expresses the work-kinetic energy theorem: the increase in kinetic energy when passing from one point to another is equal to the work of the force acting on the mass m.

Example 2. Determine the work of the force of Newtonian attraction to a fixed centre of mass m in the translation of unit mass from $M_1(a_1, b_1, c_1)$ to $M_2(a_2, b_2, c_2)$.

Solution. Let the origin be in the fixed centre of attraction. Denote by $\boldsymbol{r}$ the radius vector of the point M (Fig. 108) corresponding to an arbitrary position of unit mass, and by $\boldsymbol{r}^0$ the unit vector directed along the vector $\boldsymbol{r}$. Then $\boldsymbol{F}=-\frac{km}{r^2}\boldsymbol{r}^0$, where k is the gravitation constant. The projections of the force $\boldsymbol{F}$ on the coordinate axes will be

$$X=-km\frac{1}{r^2}\frac{x}{r}, \quad Y=-km\frac{1}{r^2}\frac{y}{r}, \quad Z=-km\frac{1}{r^2}\frac{z}{r}$$

Then the work of the force $\boldsymbol{F}$ over the path M_1M_2 is

$$A=-km\int_{(M_1)}^{(M_2)}\frac{x\,dx+y\,dy+z\,dz}{r^3}$$

$$=-km\int_{(M_1)}^{(M_2)}\frac{r\,dr}{r^3}=km\int_{(M_1)}^{(M_2)}d\left(\frac{1}{r}\right)$$

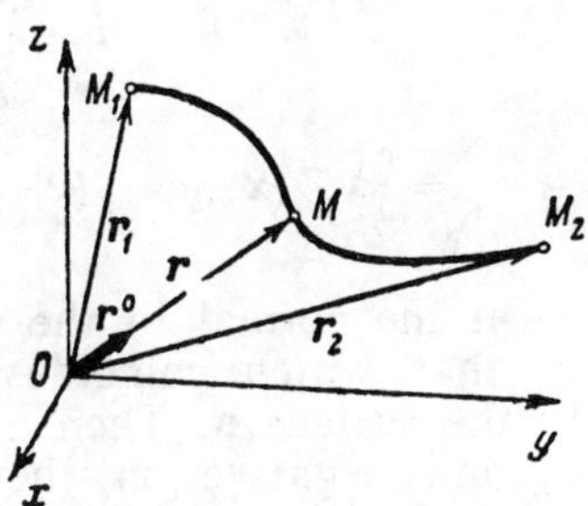

Fig. 108

(since $r^2=x^2+y^2+z^2$, it follows that $r\,dr=x\,dx+y\,dy+z\,dz$). If we denote by r_1 and r_2 the lengths of the radius vectors of the points M_1 and M_2, then

$$A=km\left(\frac{1}{r_2}-\frac{1}{r_1}\right)$$

Thus, here again the line integral does not depend on the shape of the curve of integration, but only on the position of the initial and terminal points. The function $u=\frac{km}{r}$ is called the *potential* of the gravitational field generated by the mass m. In the given case,

$$X=\frac{\partial u}{\partial x}, \quad Y=\frac{\partial u}{\partial y}, \quad Z=\frac{\partial u}{\partial z}, \quad A=u(M_2)-u(M_1)$$

that is, the work done in moving unit mass is equal to the difference between the values of the potential at the terminal and initial points.

3.8 OSTROGRADSKY'S FORMULA

Let there be given, in space, a regular three-dimensional domain V bounded by a closed surface σ and projected on an xy-plane into a regular two-dimensional domain D. We shall assume that the surface σ may be divided into three parts σ_1, σ_2, and σ_3 such that the equations of the first two have the form

$$z=f_1(x, y) \quad \text{and} \quad z=f_2(x, y)$$

where $f_1(x, y)$ and $f_2(x, y)$ are functions continuous in D and the third part σ_3 is a cylindrical surface with generator parallel to the z-axis.

Consider the integral

$$I=\iiint_V \frac{\partial Z(x, y, z)}{\partial z}\,dx\,dy\,dz$$

First perform the integration with respect to z:

$$I=\iint_D\left(\int_{f_1(x, y)}^{f_2(x, y)} \frac{\partial Z}{\partial z}\,dz\right)dx\,dy$$

$$=\iint_D Z(x, y, f_2(x, y))\,dx\,dy-\iint_D Z(x, y, f_1(x, y))\,dx\,dy \qquad (1)$$

On the normal to the surface, choose a definite direction, namely that which coincides with the direction of the outer normal to the surface σ. Then $\cos(n, z)$ will be positive on the surface σ_2 and negative on the surface σ_1; on the surface σ_3 it will be zero.

The double integrals on the right of (1) are equal to the corresponding surface integrals:

$$\iint_D Z(x, y, f_2(x, y))\,dx\,dy=\iint_{\sigma_2} Z(x, y, z)\cos(n, z)\,d\sigma \qquad (2')$$

$$\iint_D Z(x, y, f_1(x, y))\,dx\,dy=\iint_{\sigma_1} Z(x, y, z)\,(-\cos(n, z))\,d\sigma$$

In the last integral we wrote $[-\cos(n, z)]$ because the elements of surface σ_1 and σ_2 and the element of area Δs of the domain D are connected by the relation $\Delta s=\Delta\sigma[-\cos(n, z)]$, since the angle (n, z) is obtuse.

Thus,

$$\iint_D Z(x, y, f_1(x, y))\,dx\,dy=-\iint_{\sigma_1} Z(x, y, f_1(x, y))\cos(n, z)\,d\sigma \qquad (2'')$$

Substituting (2′) and (2″) into (1), we obtain

$$\iiint_V \frac{\partial Z(x, y, z)}{\partial z}\,dx\,dy\,dz$$

$$=\iint_{\sigma_2} Z(x, y, z)\cos(n, z)\,d\sigma+\iint_{\sigma_1} Z(x, y, z)\cos(n, z)\,d\sigma$$

For the sake of convenience in subsequent formulas, we shall rewrite the last equation as follows $\left(\text{adding } \iint_{\sigma_3} Z(x, y, z)\cos(n, z)\,d\sigma=0,\right.$

since the equation $\cos(n, z)=0$ is fulfilled on the surface σ_3):

$$\iiint_V \frac{\partial Z(x, y, z)}{\partial z}\, dx\, dy\, dz$$

$$= \iint_{\sigma_2} Z\cos(n, z)\, d\sigma + \iint_{\sigma_1} Z\cos(n, z)\, d\sigma + \iint_{\sigma_3} Z\cos(n, z)\, d\sigma$$

But the sum of integrals on the right of this equation is an integral over the entire closed surface σ; therefore,

$$\iiint_V \frac{\partial Z}{\partial z}\, dx\, dy\, dz = \iint_\sigma Z(x, y, z)\cos(n, z)\, d\sigma$$

Analogously, we can obtain the relations

$$\iiint_V \frac{\partial Y}{\partial y}\, dx\, dy\, dz = \iint_\sigma Y(x, y, z)\cos(n, y)\, d\sigma$$

$$\iiint_V \frac{\partial X}{\partial x}\, dx\, dy\, dz = \iint_\sigma X(x, y, z)\cos(n, x)\, d\sigma,$$

Adding together the last three equations term by term, we get *Ostrogradsky's formula*:*

$$\iiint_V \left(\frac{\partial X}{\partial x} + \frac{\partial Y}{\partial y} + \frac{\partial Z}{\partial z}\right) dx\, dy\, dz$$

$$= \iint_\sigma (X\cos(n, x) + Y\cos(n, y) + Z\cos(n, z))\, d\sigma \tag{2}$$

The expression $\frac{\partial X}{\partial x} + \frac{\partial Y}{\partial y} + \frac{\partial Z}{\partial z}$ is called the *divergence* of the vector (or the divergence of the vector function):

$$\boldsymbol{F} = X\boldsymbol{i} + Y\boldsymbol{j} + Z\boldsymbol{k}$$

and is denoted by the symbol div $\boldsymbol{F}$:

$$\text{div}\ \boldsymbol{F} = \frac{\partial X}{\partial x} + \frac{\partial Y}{\partial y} + \frac{\partial Z}{\partial z}$$

We note that this formula holds true for any domain which may be divided into subdomains that satisfy the conditions indicated at the beginning of this section.

Let us examine a hydromechanical interpretation of this formula.

Let the vector $\boldsymbol{F} = X\boldsymbol{i} + Y\boldsymbol{j} + Z\boldsymbol{k}$ be the velocity vector of a liquid flowing through the domain V. Then the surface integral in

* This formula (sometimes called the Ostrogradsky-Gauss formula) was discovered by the noted Russian mathematician M. V. Ostrogradsky (1801-1861) and published in 1828 in an article entitled "A Note on the Theory of Heat".

formula (2) is an integral of the projection of the vector $\boldsymbol{F}$ on the outer normal $\boldsymbol{n}$; it yields the quantity of liquid flowing out of domain V through the surface σ in unit time (or flowing into V if the integral is negative). This quantity is expressed in terms of the triple integral of div $\boldsymbol{F}$.

If div $\boldsymbol{F} \equiv 0$, then the double integral over any closed surface is equal to zero, that is, the quantity of liquid flowing out of (or into) something through any closed surface σ will be zero (no sources). More precisely, the quantity of liquid flowing into a domain is equal to the quantity of liquid flowing out of the domain.

In vector notation, Ostrogradsky's formula has the form

$$\iiint_V \operatorname{div} \boldsymbol{F}\, dv = \iint_\sigma \boldsymbol{F}\boldsymbol{n}\, ds \tag{1'}$$

and reads: *the integral of the divergence of a vector field* $\boldsymbol{F}$ *extended over some volume is equal to the vector flux through the surface bounding the given volume.*

3.9 THE HAMILTONIAN OPERATOR AND SOME APPLICATIONS

Suppose we have a function $u = u(x, y, z)$. At each point of the domain in which the function $u(x, y, z)$ is defined and differentiable, the following gradient is determined:

$$\operatorname{grad} u = \boldsymbol{i}\frac{\partial u}{\partial x} + \boldsymbol{j}\frac{\partial u}{\partial y} + \boldsymbol{k}\frac{\partial u}{\partial z} \tag{1}$$

The gradient of the function $u(x, y, z)$ is sometimes denoted as follows:

$$\nabla u = \boldsymbol{i}\frac{\partial u}{\partial x} + \boldsymbol{j}\frac{\partial u}{\partial y} + \boldsymbol{k}\frac{\partial u}{\partial z} \tag{2}$$

The symbol ∇ is read "del".

(1) It is convenient to write equation (2) symbolically as

$$\nabla u = \left(\boldsymbol{i}\frac{\partial}{\partial x} + \boldsymbol{j}\frac{\partial}{\partial y} + \boldsymbol{k}\frac{\partial}{\partial z}\right) u \tag{2'}$$

and to consider the symbol

$$\nabla = \boldsymbol{i}\frac{\partial}{\partial x} + \boldsymbol{j}\frac{\partial}{\partial y} + \boldsymbol{k}\frac{\partial}{\partial z} \tag{3}$$

as a "symbolic vector". This symbolic vector is called the *Hamiltonian operator* or *del operator* (∇-*operator*). From formulas (2) and (2′) it follows that "multiplication" of the symbolic vector ∇ by the scalar function u gives the gradient of this function:

$$\nabla u = \operatorname{grad} u \tag{4}$$

(2) We can form the scalar product of the symbolic vector ∇ by the vector $\boldsymbol{F}=\boldsymbol{i}X+\boldsymbol{j}Y+\boldsymbol{k}Z$:

$$\nabla\boldsymbol{F}=\left(\boldsymbol{i}\frac{\partial}{\partial x}+\boldsymbol{j}\frac{\partial}{\partial y}+\boldsymbol{k}\frac{\partial}{\partial z}\right)(\boldsymbol{i}X+\boldsymbol{j}Y+\boldsymbol{k}Z)$$
$$=\frac{\partial}{\partial x}X+\frac{\partial}{\partial y}Y+\frac{\partial}{\partial z}Z=\frac{\partial X}{\partial x}+\frac{\partial Y}{\partial y}+\frac{\partial Z}{\partial z}=\operatorname{div}\boldsymbol{F}$$

(see Sec. 3.8). Thus,

$$\nabla\boldsymbol{F}=\operatorname{div}\boldsymbol{F} \tag{5}$$

(3) Form the vector product of the symbolic vector ∇ by the vector $\boldsymbol{F}=\boldsymbol{i}X+\boldsymbol{j}Y+\boldsymbol{k}Z$:

$$\nabla\times\boldsymbol{F}=\left(\boldsymbol{i}\frac{\partial}{\partial x}+\boldsymbol{j}\frac{\partial}{\partial y}+\boldsymbol{k}\frac{\partial}{\partial z}\right)\times(\boldsymbol{i}X+\boldsymbol{j}Y+\boldsymbol{k}Z)$$
$$=\begin{vmatrix}\boldsymbol{i} & \boldsymbol{j} & \boldsymbol{k}\\ \frac{\partial}{\partial x} & \frac{\partial}{\partial y} & \frac{\partial}{\partial z}\\ X & Y & Z\end{vmatrix}=\boldsymbol{i}\begin{vmatrix}\frac{\partial}{\partial y} & \frac{\partial}{\partial z}\\ Y & Z\end{vmatrix}-\boldsymbol{j}\begin{vmatrix}\frac{\partial}{\partial x} & \frac{\partial}{\partial z}\\ X & Z\end{vmatrix}+\boldsymbol{k}\begin{vmatrix}\frac{\partial}{\partial x} & \frac{\partial}{\partial y}\\ X & Y\end{vmatrix}$$
$$=\boldsymbol{i}\left(\frac{\partial Z}{\partial y}-\frac{\partial Y}{\partial z}\right)-\boldsymbol{j}\left(\frac{\partial Z}{\partial x}-\frac{\partial X}{\partial z}\right)+\boldsymbol{k}\left(\frac{\partial Y}{\partial x}-\frac{\partial X}{\partial y}\right)$$
$$=\boldsymbol{i}\left(\frac{\partial Z}{\partial y}-\frac{\partial Y}{\partial z}\right)+\boldsymbol{j}\left(\frac{\partial X}{\partial z}-\frac{\partial Z}{\partial x}\right)+\boldsymbol{k}\left(\frac{\partial Y}{\partial x}-\frac{\partial X}{\partial y}\right)=\operatorname{curl}\boldsymbol{F}$$

(see Sec. 3.7). Thus

$$\nabla\times\boldsymbol{F}=\operatorname{curl}\boldsymbol{F} \tag{6}$$

From the foregoing it follows that vector operations may be greatly condensed by the use of the symbolic vector ∇. Let us consider several more formulas.

(4) The vector field $\boldsymbol{F}(x, y, z)=\boldsymbol{i}X+\boldsymbol{j}Y+\boldsymbol{k}Z$ is called a *potential vector field* if the vector $\boldsymbol{F}$ is the gradient of some scalar function $u(x, y, z)$:

$$\boldsymbol{F}=\operatorname{grad} u$$

or

$$\boldsymbol{F}=\boldsymbol{i}\frac{\partial u}{\partial x}+\boldsymbol{j}\frac{\partial u}{\partial y}+\boldsymbol{k}\frac{\partial u}{\partial z}$$

In this case the projections of the vector $\boldsymbol{F}$ will be

$$X=\frac{\partial u}{\partial x},\quad Y=\frac{\partial u}{\partial y},\quad Z=\frac{\partial u}{\partial z}$$

From these equations it follows (see Sec. 8.12, Vol. I) that

$$\frac{\partial X}{\partial y}=\frac{\partial Y}{\partial x},\quad \frac{\partial Y}{\partial z}=\frac{\partial Z}{\partial y},\quad \frac{\partial X}{\partial z}=\frac{\partial Z}{\partial x}$$

or

$$\frac{\partial X}{\partial y}-\frac{\partial Y}{\partial x}=0,\quad \frac{\partial Y}{\partial z}-\frac{\partial Z}{dy}=0,\quad \frac{\partial X}{\partial z}-\frac{\partial Z}{\partial x}=0$$

Hence, for the vector $\boldsymbol{F}$ under consideration,

$$\operatorname{curl} \boldsymbol{F}=0$$

Thus, we get

$$\operatorname{curl}(\operatorname{grad} u)=0 \tag{7}$$

Applying the del operator ∇, we can write (7) as follows [on the basis of (4) and (6)]:

$$(\nabla \times \nabla u)=0 \tag{7'}$$

Taking advantage of the property that for multiplication of a vector product by a scalar it is sufficient to multiply the scalar by one of the factors, we write

$$(\nabla \times \nabla) u=0 \tag{7''}$$

Here, the del operator again has the properties of an ordinary vector; the vector product of a vector into itself is zero.

The vector field $\boldsymbol{F}(x, y, z)$, for which $\operatorname{curl} \boldsymbol{F}=0$, is called *irrotational*. From (7) it follows that every potential field is irrotational.

The converse also holds: if some vector field $\boldsymbol{F}$ is irrotational, then it is potential. The truth of this statement follows from reasoning given at the end of Sec. 3.7.

(5) A vector field $\boldsymbol{F}(x, y, z)$ for which

$$\operatorname{div} \boldsymbol{F}=0$$

that is, a vector field in which there are no sources (see Sec. 3.8) is called *solenoidal*. We shall prove that

$$\operatorname{div}(\operatorname{curl} \boldsymbol{F})=0 \tag{8}$$

or that the rotational field is source-free.

Indeed, if $\boldsymbol{F}=\boldsymbol{i}X+\boldsymbol{j}Y+\boldsymbol{k}Z$, then

$$\operatorname{curl} \boldsymbol{F}=\boldsymbol{i}\left(\frac{\partial Z}{\partial y}-\frac{\partial Y}{\partial z}\right)+\boldsymbol{j}\left(\frac{\partial X}{\partial z}-\frac{\partial Z}{\partial x}\right)+\boldsymbol{k}\left(\frac{\partial Y}{\partial x}-\frac{\partial X}{\partial y}\right)$$

and therefore

$$\operatorname{div}(\operatorname{curl} \boldsymbol{F})=\frac{\partial}{\partial x}\left(\frac{\partial Z}{\partial y}-\frac{\partial Y}{\partial z}\right)+\frac{\partial}{\partial y}\left(\frac{\partial X}{\partial z}-\frac{\partial Z}{\partial x}\right)+\frac{\partial}{\partial z}\left(\frac{\partial Y}{\partial x}-\frac{\partial X}{\partial y}\right)=0$$

Using the del operator, we can write equation (8) as

$$\nabla(\nabla \times \boldsymbol{F})=0 \tag{8'}$$

The left side of this equation may be regarded as a vector-scalar (mixed) product of three vectors: ∇, ∇, $\boldsymbol{F}$, of which two are the same. This product is obviously equal to zero.

(6) Let there be a scalar field $u=u(x, y, z)$. Determine the gradient field:

$$\operatorname{grad} u=\boldsymbol{i}\frac{\partial u}{\partial x}+\boldsymbol{j}\frac{\partial u}{\partial y}+\boldsymbol{k}\frac{\partial u}{\partial z}$$

Then find

$$\operatorname{div}(\operatorname{grad} u)=\frac{\partial}{\partial x}\left(\frac{\partial u}{\partial x}\right)+\frac{\partial}{\partial y}\left(\frac{\partial u}{\partial y}\right)+\frac{\partial}{\partial z}\left(\frac{\partial u}{\partial z}\right)$$

or

$$\operatorname{div}(\operatorname{grad} u)=\frac{\partial^2 u}{\partial x^2}+\frac{\partial^2 u}{\partial y^2}+\frac{\partial^2 u}{\partial z^2} \tag{9}$$

The right member of this expression is denoted by

$$\Delta u=\frac{\partial^2 u}{\partial x^2}+\frac{\partial^2 u}{\partial y^2}+\frac{\partial^2 u}{\partial z^2} \tag{10}$$

or, symbolically,

$$\Delta u=\left(\frac{\partial^2}{\partial x^2}+\frac{\partial^2}{\partial y^2}+\frac{\partial^2}{\partial z^2}\right)u$$

The symbol

$$\Delta=\frac{\partial^2}{\partial x^2}+\frac{\partial^2}{\partial y^2}+\frac{\partial^2}{\partial z^2}$$

is termed the *Laplacian operator*.

Hence, (9) may be written as

$$\operatorname{div}(\operatorname{grad} u)=\Delta u \tag{11}$$

Using the del operator ∇ we can write (11) as

$$(\nabla\nabla u)=\Delta u, \text{ i.e., } \Delta=\nabla^2 \tag{11'}$$

We note that the equation

$$\frac{\partial^2 u}{\partial x^2}+\frac{\partial^2 u}{\partial y^2}+\frac{\partial^2 u}{\partial z^2}=0 \tag{12}$$

or

$$\Delta u=0 \tag{12'}$$

is called the *Laplace equation*. The function that satisfies the Laplace equation is called a *harmonic function*.

Exercises on Chapter 3

Compute the following line integrals:

1. $\int y^2\,dx+2xy\,dy$ over the circle $x=a\cos t$, $y=a\sin t$. *Ans.* 0.

2. $\int y\,dx-x\,dy$ over an arc of the ellipse $x=a\cos t$, $y=b\sin t$. *Ans.* $-2\pi ab$.

3. $\int\left(\frac{x}{x^2+y^2}dx-\frac{y}{x^2+y^2}dy\right)$ over a circle with centre at the origin. *Ans.* 0.

4. $\int\left(\frac{y\,dx+x\,dy}{x^2+y^2}\right)$ over a segment of the straight line $y=x$ from $x=1$ to $x=2$. *Ans.* $\ln 2$.

5. $\int yz\,dx+xz\,dy+xy\,dz$ over an arc of the helix $x=a\cos t$, $y=a\sin t$, $z=kt$ as t varies from 0 to 2π. *Ans.* 0.

6. $\int x\,dy - y\,dx$ over an arch of the astroid $x = a\cos^3 t$, $y = a\sin^3 t$. *Ans* $\frac{3}{4}\pi a^2$ (the double area of the astroid).

7. $\int x\,dy - y\,dx$ over the loop of the folium of Descartes $x = \frac{3at}{1+t^3}$, $y = \frac{3at^2}{1+t^3}$. *Ans.* $3a^2$ (the double area of the region bounded by the indicated loop).

8. $\int x\,dy - y\,dx$ over the curve $x = a(t - \sin t)$, $y = a(1 - \cos t)$ $(0 \leqslant t \leqslant 2\pi)$. *Ans.* $-6\pi a^2$ (the double area of the region bounded by one arch of a cycloid and the x-axis).

Prove that:

9. $\operatorname{grad}(c\varphi) = c \operatorname{grad} \varphi$, where c is a constant.
10. $\operatorname{grad}(c_1\varphi + c_2\psi) = c_1 \operatorname{grad} \varphi + c_2 \operatorname{grad} \psi$, where c_1 and c_2 are constants.
11. $\operatorname{grad}(\varphi\psi) = \varphi \operatorname{grad} \psi + \psi \operatorname{grad} \varphi$.

12. Find $\operatorname{grad} r$, $\operatorname{grad} r^2$, $\operatorname{grad} \frac{1}{r}$, $\operatorname{grad} f(r)$, where $r = \sqrt{x^2+y^2+z^2}$. *Ans.* $\frac{\mathbf{r}}{r}$, $2\mathbf{r}$, $-\frac{\mathbf{r}}{r^3}$, $f'(r)\frac{\mathbf{r}}{r}$.

13. Prove that $\operatorname{div}(\mathbf{A} + \mathbf{B}) = \operatorname{div} \mathbf{A} + \operatorname{div} \mathbf{B}$.
14. Compute $\operatorname{div} \mathbf{r}$, where $\mathbf{r} = x\mathbf{i} + y\mathbf{j} + z\mathbf{k}$. *Ans.* 3.
15. Compute $\operatorname{div}(\mathbf{A}\varphi)$, where $\mathbf{A}$ is a vector function and φ is a scalar function. *Ans.* $\varphi \operatorname{div} \mathbf{A} + (\operatorname{grad} \varphi\, \mathbf{A})$.
16. Compute $\operatorname{div}(\mathbf{r} \cdot \mathbf{c})$ where $\mathbf{c}$ is a constant vector. *Ans.* $\frac{(\mathbf{c} \cdot \mathbf{r})}{r}$.
17. Compute $\operatorname{div} \mathbf{B}(\mathbf{r}\mathbf{A})$. *Ans.* $\mathbf{AB}$.

Prove that:

18. $\operatorname{rot}(c_1\mathbf{A}_1 + c_2\mathbf{A}_2) = c_1 \operatorname{rot} \mathbf{A}_1 + c_2 \operatorname{rot} \mathbf{A}_2$, where c_1 and c_2 are constants.
19. $\operatorname{rot}(\mathbf{A}c) = \operatorname{grad} \mathbf{A} \times \mathbf{c}$, where $\mathbf{c}$ is a constant vector.
20. $\operatorname{rot}\operatorname{rot} \mathbf{A} = \operatorname{grad}\operatorname{div} \mathbf{A} - \Delta \mathbf{A}$.
21. $\mathbf{A} \times \operatorname{rot} \varphi = \operatorname{rot}(\varphi \mathbf{A})$.

Surface Integrals

22. Prove that $\iint \cos(\mathbf{n}, z)\, d\sigma = 0$ if σ is a closed surface and $\mathbf{n}$ is a normal to it.

23. Find the moment of inertia of the surface of a segment of a sphere with equation $x^2 + y^2 + z^2 = R^2$ cut off by the plane $z = H$ relative to the z-axis. *Ans.* $\frac{2\pi R}{3}(2R^3 - 3R^2H + H^3)$.

24. Find the moment of inertia of the surface of the paraboloid of revolution $x^2 + y^2 = 2cz$ cut off by the plane $z = c$ relative to the z-axis. *Ans.* $4\pi c^4 \frac{6\sqrt{3}+1}{15}$.

25. Compute the coordinates of the centre of gravity of a part of the surface of the cone $x^2 + y^2 = \frac{R^2}{H^2} z^2$ cut off by the plane $z = H$. *Ans.* 0, 0, $\frac{2}{3}H$.

26. Compute the coordinates of the centre of gravity of a segment of the surface of the sphere $x^2 + y^2 + z^2 = R^2$ cut off by the plane $z = H$. *Ans.* $\left(0, 0, \frac{R+H}{2}\right)$.

27. Find $\iint\limits_{\sigma} [x\cos(nx)+y\cos(ny)+z\cos(nz)]\,d\sigma$, where σ is a closed surface. *Ans.* $3V$, where V is the volume of the solid bounded by the surface σ.

28. Find $\iint\limits_{S} z\,dx\,dy$, where S is the external side of the sphere $x^2+y^2+z^2=R^2$. *Ans.* $\frac{4}{3}\pi R^3$.

29. Find $\iint\limits_{S} x^2\,dy\,dz+y^2\,dz\,dx+z^2\,dx\,dy$, where S is the external side of the surface of the sphere $x^2+y^2+z^2=R^2$. *Ans.* $2\pi R^4$.

30. Find $\iint\limits_{S} \sqrt{x^2+y^2}\,ds$, where S is the lateral surface of the cone $\frac{x^2}{a^2}+\frac{y^2}{a^2}-\frac{z^2}{b^2}=0$, $0\leqslant z\leqslant b$. *Ans.* $\frac{2\pi a^2\sqrt{a^2+b^2}}{3}$.

31. Using the Stokes formula, transform the integral $\int\limits_{L} y\,dx+z\,dy+x\,dz$. *Ans.* $-\iint\limits_{S} (\cos\alpha+\cos\beta+\cos\gamma)\,ds$.

Find the line integrals, applying the Stokes formula and directly:

32. $\int\limits_{L} (y+z)\,dx+(z+x)\,dy+(x+y)\,dz$, where L is the circle $x^2+y^2+z^2=a^2$, $x+y+z=0$. *Ans.* 0. **33.** $\int\limits_{L} x^2y^3\,dx+dy+z\,dz$, where L is the circle $x^2+y^2=R^2$, $z=0$. *Ans.* $-\frac{\pi R^6}{8}$.

Applying the Ostrogradsky formula, transform the surface integrals into volume integrals:

34. $\iint\limits_{S} (x\cos\alpha+y\cos\beta+z\cos\gamma)\,ds$. *Ans.* $\iiint\limits_{V} 3\,dx\,dy\,dz=3V$.

35. $\iint\limits_{S} (x^2+y^2+z^2)(dy\,dz+dx\,dz+dx\,dy)$. *Ans.* $2\iiint\limits_{V} (x+y+z)\,dx\,dy\,dz$.

36. $\iint\limits_{S} xy\,dx\,dy+yz\,dy\,dz+zx\,dz\,dx$. *Ans.* 0. **37.** $\iint\limits_{S} \frac{\partial u}{\partial x}\,dy\,dz+\frac{\partial u}{\partial y}\,dx\,dz+\frac{\partial u}{\partial z}\,dx\,dy$. *Ans.* $\iiint\limits_{V} \left(\frac{\partial^2 u}{\partial x^2}+\frac{\partial^2 u}{\partial y^2}+\frac{\partial^2 u}{\partial z^2}\right)dx\,dy\,dz$.

Using the Ostrogradsky formula compute the following integrals:

38. $\iint\limits_{S} (x\cos\alpha+y\cos\beta+z\cos\gamma)\,ds$, where S is the surface of the ellipsoid $\frac{x^2}{a^2}+\frac{y^2}{b^2}+\frac{z^2}{c^2}=1$. *Ans.* $4\pi abc$. **39.** $\iint\limits_{S} (x^3\cos\alpha+y^3\cos\beta+z^3\cos\gamma)\,ds$, where S is the surface of the sphere $x^2+y^2+z^2=R^2$. *Ans.* $\frac{12}{5}\pi R^5$. **40.** $\iint\limits_{S} x^2\,dy\,dz+$

$+y^2\,dz\,dx+z^2\,dx\,dy$, where S is the surface of the cone $\frac{x^2}{a^2}+\frac{y^2}{a^2}-\frac{z^2}{b^2}=0$ $(0\leqslant z\leqslant b)$. *Ans.* $\frac{\pi a^2b^2}{2}$. **41.** $\iint\limits_S x\,dy\,dz+y\,dx\,dz+z\,dx\,dy$, where S is the surface of the cylinder $x^2+y^2=a^2$, $-H\leqslant z\leqslant H$. *Ans.* $3\pi a^2H$. **42.** Prove the identity $\iint\limits_D\left(\frac{\partial^2u}{\partial x^2}+\frac{\partial^2u}{\partial y^2}\right)dx\,dy=\int\limits_C\frac{\partial u}{\partial n}\,ds$, where C is a contour bounding the domain D, and $\frac{\partial u}{\partial n}$ is the directional derivative of the outer normal.

Solution.

$$\iint\limits_D\left(\frac{\partial X}{\partial x}+\frac{\partial Y}{\partial y}\right)dx\,dy=\int\limits_C -Y\,dx+X\,dy=\int\limits_C[-Y\cos(s,\ x)+X\sin(s,\ x)]\,ds$$

where $(s,\ x)$ is the angle between the tangent line to the contour C and the x-axis. If we denote by $(n,\ x)$ the angle between the normal and the x-axis, then $\sin(s,\ x)=\cos(n,\ x)$, $\cos(s,\ x)=-\sin(n,\ x)$. Hence,

$$\iint\limits_D\left(\frac{\partial X}{\partial x}+\frac{\partial Y}{\partial y}\right)dx\,dy=\int\limits_C[X\cos(n,\ x)+Y\sin(n,\ x)]\,ds$$

Setting $X=\frac{\partial u}{\partial x}$; $Y=\frac{\partial u}{\partial y}$, we get

$$\iint\limits_D\left(\frac{\partial^2u}{\partial x^2}+\frac{\partial^2u}{\partial y^2}\right)dx\,dy=\int\limits_C\left[\frac{\partial u}{\partial x}\cos(n,\ x)+\frac{\partial u}{\partial y}\sin(n,\ x)\right]ds$$

or

$$\iint\limits_D\left[\frac{\partial^2u}{\partial x^2}+\frac{\partial^2u}{\partial y^2}\right]dx\,dy=\int\limits_C\frac{\partial u}{\partial n}\,ds$$

The expression $\frac{\partial^2u}{\partial x^2}+\frac{\partial^2u}{\partial y^2}$ is called the Laplacian operator.

43. Prove the identity (called *Green's formula*)

$$\iiint\limits_V(v\Delta u-u\Delta v)\,dx\,dy\,dz=\iint\limits_\sigma\left(v\frac{\partial u}{\partial n}-u\frac{\partial v}{\partial n}\right)d\sigma$$

where u and v are continuous functions with continuous derivatives up to the second order in the domain D.

The symbols Δu and Δv denote

$$\Delta u=\frac{\partial^2u}{\partial x^2}+\frac{\partial^2u}{\partial y^2}+\frac{\partial^2u}{\partial z^2},\qquad \Delta v=\frac{\partial^2v}{\partial x^2}+\frac{\partial^2v}{\partial y^2}+\frac{\partial^2v}{\partial z^2}$$

Solution. In the formula

$$\iiint\limits_V\left(\frac{\partial X}{\partial x}+\frac{\partial Y}{\partial y}+\frac{\partial Z}{\partial z}\right)dx\,dy\,dz=\iint\limits_\sigma[X\cos(n,\ x)+Y\cos(n,\ y)+Z\cos(n,\ z)]\,d\sigma$$

we put

$$X=vu'_x-uv'_x$$
$$Y=vu'_y-uv'_y$$
$$Z=vu'_z-uv'_z$$

Then

$$\frac{\partial X}{\partial x}+\frac{\partial Y}{\partial y}+\frac{\partial Z}{\partial z}=v\,(u''_{xx}+u''_{yy}+u''_{zz})-u\,(v''_{xx}+v''_{yy}+v''_{zz})=v\Delta u-u\Delta v$$

$$X\cos(n,\ x)+Y\cos(n,\ y)+Z\cos(n,\ z)$$
$$=v[u'_x\cos(n,\ x)+u'_y\cos(n,\ y)+u'_z\cos(n,\ z)]$$
$$-u[v'_x\cos(n,\ x)+v'_y\cos(n,\ y)+v'_z\cos(n,\ z)]=v\frac{\partial u}{\partial n}-u\frac{\partial v}{\partial n}$$

Hence,

$$\iiint\limits_V (v\Delta u-u\Delta v)\,dx\,dy\,dz=\iint\limits_\sigma \left(v\frac{\partial u}{\partial n}-u\frac{\partial v}{\partial n}\right)d\sigma$$

44. Prove the identity

$$\iiint\limits_V \Delta u\,dx\,dy\,dz=\iint\limits_\sigma \frac{\partial u}{\partial n}\,d\sigma$$

where $\Delta u=\frac{\partial^2 u}{\partial x^2}+\frac{\partial^2 u}{\partial y^2}+\frac{\partial^2 u}{\partial z^2}$.

Solution. In Green's formula, which was derived in the preceding problem, put $v=1$. Then $\Delta v=0$, and we get the desired identity.

45. If $u(x,\ y,\ z)$ is a harmonic function in some domain, that is, a function which at every point of the domain satisfies the Laplace equation

$$\frac{\partial^2 u}{\partial x^2}+\frac{\partial^2 u}{\partial y^2}+\frac{\partial^2 u}{\partial z^2}=0$$

then

$$\iint\limits_\sigma \frac{\partial u}{\partial n}\,d\sigma=0$$

where σ is a closed surface.

Solution. This follows directly from the formula of Problem 44.

46. Let $u(x,\ y,\ z)$ be a harmonic function in some domain V and let there be, in V, a sphere $\bar{\sigma}$ with centre at the point $M(x_1,\ y_1,\ z_1)$ and with radius R. Prove that

$$u(x_1,\ y_1,\ z_1)=\frac{1}{4\pi R^2}\iint\limits_{\bar{\sigma}} u\,d\sigma$$

Solution. Consider the domain Ω bounded by two spheres σ, $\bar{\sigma}$ of radii R and ρ $(\rho < R)$ with centres at the point $M(x_1,\ y_1,\ z_1)$. Apply Green's formula (found in Problem 43) to this domain taking for u the above-indicated function, and for the function v,

$$v=\frac{1}{r}=\frac{1}{\sqrt{(x-x_1)^2+(y-y_1)^2+(z-z_1)^2}}$$

By direct differentiation and substitution we are convinced that $\frac{\partial^2 v}{\partial x^2}+\frac{\partial^2 v}{\partial y^2}+\frac{\partial^2 v}{\partial z^2}=0$. Consequently,

$$\iint\limits_{\sigma+\bar{\sigma}} \left(\frac{1}{r}\frac{\partial u}{\partial n}-u\frac{\partial \frac{1}{r}}{\partial n}\right)d\sigma=0$$

or

$$\iint\limits_{\underline{\sigma}}\left(\frac{1}{r}\frac{\partial u}{\partial n}-u\frac{\partial\left(\frac{1}{r}\right)}{\partial n}\right)d\sigma-\iint\limits_{\bar{\sigma}}\left(\frac{1}{r}\frac{\partial u}{\partial n}-u\frac{\partial\left(\frac{1}{r}\right)}{\partial n}\right)d\sigma=0$$

On the surfaces $\bar{\sigma}$ and $\underline{\sigma}$ the quantity $\frac{1}{r}$ is constant $\left(\frac{1}{R}\text{ and }\frac{1}{\rho}\right)$ and so can be taken outside the integral sign. By virtue of the result obtained in Problem 45, we have

$$\frac{1}{R}\iint\limits_{\underline{\sigma}}\frac{\partial u}{\partial n}d\sigma=0,\quad \frac{1}{\rho}\iint\limits_{\bar{\sigma}}\frac{\partial u}{\partial n}d\sigma=0$$

Hence,

$$-\iint\limits_{\underline{\sigma}}u\frac{\partial\left(\frac{1}{r}\right)}{\partial n}d\sigma+\iint\limits_{\bar{\sigma}}u\frac{\partial\left(\frac{1}{r}\right)}{\partial n}d\sigma=0$$

but

$$\frac{\partial\left(\frac{1}{r}\right)}{\partial n}=\frac{d\left(\frac{1}{r}\right)}{dr}=-\frac{1}{r^2}$$

Therefore,

$$+\iint\limits_{\underline{\sigma}}u\frac{1}{r^2}d\sigma-\iint\limits_{\bar{\sigma}}u\frac{1}{r^2}d\sigma=0$$

or

$$\frac{1}{\rho^2}\iint\limits_{\underline{\sigma}}u\,d\sigma=\frac{1}{R^2}\iint\limits_{\bar{\sigma}}u\,d\sigma \tag{1}$$

Apply the mean-value theorem to the integral on the left:

$$\frac{1}{\rho^2}\iint\limits_{\underline{\sigma}}u\,d\sigma=\frac{u(\xi,\eta,\zeta)}{\rho^2}\iint\limits_{\bar{\sigma}}d\sigma \tag{2}$$

where $u(\xi, \eta, \zeta)$ is a point on the surface of a sphere of radius ρ with centre at the point $M(x_1, y_1, z_1)$.

We make ρ approach zero; then $u(\xi, \eta, \zeta)\to u(x_1, y_1, z_1)$:

$$\frac{1}{\rho^2}\iint\limits_{\bar{\sigma}}d\sigma=\frac{4\pi\rho^2}{\rho^2}=4\pi$$

Hence, as $\rho\to 0$, we get

$$\frac{1}{\rho^2}\iint\limits_{\bar{\sigma}}u\,d\sigma\to u(x_1, y_1, z_1)\,4\pi$$

Further, since the right side of (1) is independent of ρ, it follows that as $\rho\to 0$ we finally get

$$\frac{1}{R^2}\iint\limits_{\bar{\sigma}}u\,d\sigma=4\pi u(x_1, y_1, z_1)$$

or

$$u(x_1, y_1, z_1)=\frac{1}{4\pi R^2}\iint\limits_{\bar{\sigma}}u\,d\sigma$$

CHAPTER 4

SERIES

4.1 SERIES. SUM OF A SERIES

Definition 1. Suppose we have an infinite sequence of numbers*

$$u_1, u_2, u_3, \ldots, u_n, \ldots$$

The expression

$$u_1+u_2+u_3+\ldots+u_n+\ldots \tag{1}$$

is called a *numerical series*. Here, the numbers $u_1, u_2, \ldots, u_n, \ldots$ are called the *terms of the series*.

Definition 2. The sum of a finite number of terms (the first n terms) of a series is called the *nth partial sum of the series*:

$$s_n=u_1+u_2+\ldots+u_n$$

Consider the partial sums

$$\begin{aligned} s_1 &= u_1 \\ s_2 &= u_1+u_2 \\ s_3 &= u_1+u_2+u_3 \\ &\ldots\ldots\ldots\ldots \\ s_n &= u_1+u_2+u_3+\ldots+u_n \end{aligned}$$

If there exists a finite limit

$$s=\lim_{n\to\infty} s_n$$

it is called the *sum of the series* (1) and we say that the *series converges*.

If $\lim\limits_{n\to\infty} s_n$ does not exist (for example, $s_n\to\infty$ as $n\to\infty$), then we say that the *series* (1) *diverges and has no sum.*

Example. Consider the series

$$a+aq+aq^2+\ldots+aq^{n-1}+\ldots \tag{2}$$

This is a *geometric progression* with first term a and ratio q $(a\neq 0)$.

---

* A sequence is considered specified if we know the law by which it is possible to determine any term u_n for a given n.

The sum of the first n terms of the geometric progression is (when $q \neq 1$)

$$s_n = \frac{a - aq^n}{1-q}$$

or

$$s_n = \frac{a}{1-q} - \frac{aq^n}{1-q}$$

(1) If $|q| < 1$, then $q^n \to 0$ as $n \to \infty$ and, consequently,

$$\lim_{n\to\infty} s_n = \lim_{n\to\infty}\left(\frac{a}{1-q} - \frac{aq^n}{1-q}\right) = \frac{a}{1-q}$$

Hence, in the case of $|q| < 1$, the series (2) converges and its sum is

$$s = \frac{a}{1-q}$$

(2) If $|q| > 1$, then $|q^n| \to \infty$ as $n \to \infty$ and then $\frac{a-aq^n}{1-q} \to \pm\infty$ as $n \to \infty$, that is, $\lim\limits_{n\to\infty} s_n$ does not exist. Thus, when $|q| > 1$, the series (2) diverges.

(3) If $q = 1$, then the series (2) has the form

$$a + a + a + \ldots$$

In this case

$$s_n = na, \quad \lim_{n\to\infty} s_n = \infty$$

and the series diverges.

(4) If $q = -1$, then the series (2) has the form

$$a - a + a - a + \ldots$$

In this case

$$s_n = \begin{cases} 0 \text{ when } n \text{ is even} \\ a \text{ when } n \text{ is odd} \end{cases}$$

Thus, s_n has no limit and the series diverges.

Thus, a geometric progression (with first term different from zero) converges only when the ratio of the progression is less than unity in absolute value.

Theorem 1. *If a series obtained from a given series (1) by suppression of some of its terms converges, then the given series itself converges. Conversely, if a given series converges, then a series obtained from the given series by suppression of several terms also converges.*

In other words, *the convergence of a series is not affected by the suppression of a finite number of its terms.*

Proof. Let s_n be the sum of the first n terms of the series (1), c_k, the sum of k suppressed terms (we note that for a sufficiently large n, all suppressed terms are contained in the sum s_n), and σ_{n-k}, the sum of the terms of the series that enter into the sum s_n but do not enter into c_k. Then we have

$$s_n = c_k + \sigma_{n-k}$$

where c_k is a constant that is independent of n.

From this relation it follows that if $\lim\limits_{n\to\infty} \sigma_{n-k}$ exists, then $\lim\limits_{n\to\infty} s_n$ exists as well; if $\lim\limits_{n\to\infty} s_n$ exists, then $\lim\limits_{n\to\infty} \sigma_{n-k}$ also exists; which proves the theorem.

We conclude this section with two simple properties of series.

Theorem 2. *If a series*

$$a_1 + a_2 + \ldots \tag{3}$$

converges and its sum is s, then the series

$$ca_1 + ca_2 + \ldots \tag{4}$$

where c is some fixed number, also converges, and its sum is cs.

Proof. Denote the nth partial sum of the series (3) by s_n, and that of the series (4), by σ_n. Then

$$\sigma_n = ca_1 + \ldots + ca_n = c(a_1 + \ldots + a_n) = cs_n$$

It is now clear that the limit of the nth partial sum of the series (4) exists, since

$$\lim_{n\to\infty} \sigma_n = \lim_{n\to\infty} (cs_n) = c \lim_{n\to\infty} s_n = cs$$

Thus, the series (4) converges and its sum is equal to cs.

Theorem 3. *If the series*

$$a_1 + a_2 + \ldots \tag{5}$$

and

$$b_1 + b_2 + \ldots \tag{6}$$

converge and their sums, respectively, are $\bar{s}$ and $\bar{\bar{s}}$, then the series

$$(a_1 + b_1) + (a_2 + b_2) + \ldots \tag{7}$$

and

$$(a_1 - b_1) + (a_2 - b_2) + \ldots \tag{8}$$

also converge and their sums are $\bar{s} + \bar{\bar{s}}$ and $\bar{s} - \bar{\bar{s}}$, respectively.

Proof. We prove the convergence of the series (7). Denoting its nth partial sum by σ_n and the nth partial sums of the series (5) and (6) by $\bar{s}_n$ and $\bar{\bar{s}}_n$, respectively, we get

$$\begin{aligned}\sigma_n &= (a_1 + b_1) + \ldots + (a_n + b_n) \\ &= (a_1 + \ldots + a_n) + (b_1 + \ldots + b_n) = \bar{s}_n + \bar{\bar{s}}_n\end{aligned}$$

Passing to the limit in this equation as $n \to \infty$, we get

$$\lim_{n\to\infty} \sigma_n = \lim_{n\to\infty} (\bar{s}_n + \bar{\bar{s}}_n) = \lim_{n\to\infty} \bar{s}_n + \lim_{n\to\infty} \bar{\bar{s}}_n = \bar{s} + \bar{\bar{s}}$$

Thus, the series (7) converges and its sum is $\bar{s} + \bar{\bar{s}}$.

Similarly we can prove that the series (8) also converges and its sum is equal to $\bar{s}-\bar{\bar{s}}$.

We say that the series (7) and (8) were obtained by means of **termwise addition** or, respectively, **termwise subtraction** of the series (5) and (6).

4.2 NECESSARY CONDITION FOR CONVERGENCE OF A SERIES

One of the basic questions, when investigating series, is that of whether the given series converges or diverges. We shall establish sufficient conditions for one to decide this question. We shall also examine the necessary condition for convergence of a series; in other words, we shall establish a condition for which the series will diverge if it is not fulfilled.

Theorem. *If a series converges, its nth term approaches zero as n becomes infinite.*

Proof. Let the series

$$u_1+u_2+u_3+\ldots+u_n+\ldots$$

converge; that is, let us have

$$\lim_{n\to\infty} s_n = s$$

where s is the sum of the series (a finite fixed number). But then we also have the equation

$$\lim_{n\to\infty} s_{n-1} = s$$

since $(n-1)$ also tends to infinity as $n\to\infty$. Subtracting the second equation from the first termwise, we obtain

$$\lim_{n\to\infty} s_n - \lim_{n\to\infty} s_{n-1} = 0$$

or

$$\lim_{n\to\infty} (s_n - s_{n-1}) = 0$$

But

$$s_n - s_{n-1} = u_n$$

Hence,

$$\lim_{n\to\infty} u_n = 0$$

which is what was to be proved.

Corollary. *If the nth term of a series does not tend to zero as $n\to\infty$, then the series diverges.*

Example. The series

$$\frac{1}{3}+\frac{2}{5}+\frac{3}{7}+\ldots+\frac{n}{2n+1}+\ldots$$

diverges, since

$$\lim_{n\to\infty} u_n=\lim_{n\to\infty}\left(\frac{n}{2n+1}\right)=\frac{1}{2}\neq 0$$

We stress the fact that this condition is only a necessary condition, but not a sufficient condition; in other words, *from the fact that the nth term approaches zero, it does not follow that the series converges*, for the series may diverge.

For example, the so-called *harmonic series*

$$1+\frac{1}{2}+\frac{1}{3}+\frac{1}{4}+\ldots+\frac{1}{n}+\ldots$$

diverges, although

$$\lim_{n\to\infty} u_n=\lim_{n\to\infty}\frac{1}{n}=0$$

To prove this, write the harmonic series in more detail:

$$1+\frac{1}{2}+\underbrace{\frac{1}{3}+\frac{1}{4}}+\underbrace{\frac{1}{5}+\frac{1}{6}+\frac{1}{7}+\frac{1}{8}}+\underbrace{\frac{1}{9}+\frac{1}{10}+\frac{1}{11}+\frac{1}{12}+\frac{1}{13}+\frac{1}{14}+\frac{1}{15}+\frac{1}{16}}+\frac{1}{17}+\ldots \tag{1}$$

We also write the auxiliary series

$$1+\frac{1}{2}+\overbrace{\frac{1}{4}+\frac{1}{4}}+\overbrace{\frac{1}{8}+\frac{1}{8}+\frac{1}{8}+\frac{1}{8}}+\overbrace{\frac{1}{16}+\frac{1}{16}+\frac{1}{16}+\frac{1}{16}+\frac{1}{16}+\frac{1}{16}+\frac{1}{16}+\frac{1}{16}}+\overbrace{\frac{1}{32}+\ldots+\frac{1}{32}}^{16\text{ terms}}+\ldots \tag{2}$$

The series (2) is constructed as follows: its first term is equal to unity, its second is $^1/_2$, its third and fourth are $^1/_4$, the fifth to the eighth terms are equal to $^1/_8$, the terms 9 to 16 are equal to $^1/_{16}$, the terms 17 to 32 are equal to $^1/_{32}$, etc.

Denote by $s_n^{(1)}$ the sum of the first n terms of the harmonic series (1) and by $s_n^{(2)}$ the sum of the first n terms of the series (2).

Since each term of the series (1) is greater than the corresponding term of the series (2) or equal to it, then for $n>2$

$$s_n^{(1)}>s_n^{(2)} \tag{3}$$

We compute the partial sums of the series (2) for values of n equal to 2^1, 2^2, 2^3, 2^4, 2^5:

$$s_2=1+\frac{1}{2}=1+1\cdot\frac{1}{2}$$

$$s_4 = 1 + \frac{1}{2} + \left(\frac{1}{4} + \frac{1}{4}\right) = 1 + \frac{1}{2} + \frac{1}{2} = 1 + 2\cdot\frac{1}{2}$$

$$s_8 = 1 + \frac{1}{2} + \left(\frac{1}{4} + \frac{1}{4}\right) + \left(\frac{1}{8} + \frac{1}{8} + \frac{1}{8} + \frac{1}{8}\right) = 1 + 3\cdot\frac{1}{2}$$

$$s_{16} = 1 + \frac{1}{2} + \left(\frac{1}{4} + \frac{1}{4}\right) + \underbrace{\left(\frac{1}{8} + \ldots + \frac{1}{8}\right)}_{\text{4 terms}} + \underbrace{\left(\frac{1}{16} + \ldots + \frac{1}{16}\right)}_{\text{8 terms}} = 1 + 4\cdot\frac{1}{2}$$

$$s_{32} = 1 + \frac{1}{2} + \left(\frac{1}{4} + \frac{1}{4}\right) + \underbrace{\left(\frac{1}{8} + \ldots + \frac{1}{8}\right)}_{\text{4 terms}} + \underbrace{\left(\frac{1}{16} + \ldots + \frac{1}{16}\right)}_{\text{8 terms}} + \underbrace{\left(\frac{1}{32} + \ldots + \frac{1}{32}\right)}_{\text{16 terms}} = 1 + 5\cdot\frac{1}{2}$$

in the same way we find that $s_{2^6} = 1 + 6\cdot\frac{1}{2}$, $s_{2^7} = 1 + 7\cdot\frac{1}{2}$ and, generally, $s_{2^k} = 1 + k\cdot\frac{1}{2}$.

Thus, for sufficiently large k, the partial sums of the series (2) can be made greater than any positive number; that is,

$$\lim_{n\to\infty} s_n^{(2)} = \infty$$

but then from the relation (3) it also follows that

$$\lim_{n\to\infty} s_n^{(1)} = \infty$$

which means that the harmonic series (1) diverges.

4.3 COMPARING SERIES WITH POSITIVE TERMS

Suppose we have two series with **positive** terms:

$$u_1 + u_2 + u_3 + \ldots + u_n + \ldots \tag{1}$$

$$v_1 + v_2 + v_3 + \ldots + v_n + \ldots \tag{2}$$

For them the following assertions hold true.

Theorem 1. *If the terms of the series* (1) *do not exceed the corresponding terms of the series* (2); *that is,*

$$u_n \leqslant v_n \quad (n = 1, 2, \ldots) \tag{3}$$

and the series (2) *converges, then the series* (1) *also converges.*

Proof. Denote by s_n and σ_n, respectively, the partial sums of the first and second series:

$$s_n = \sum_{i=1}^{n} u_i, \quad \sigma_n = \sum_{i=1}^{n} v_i$$

From the condition (3) it follows that

$$s_n \leqslant \sigma_n \tag{4}$$

Since the series (2) converges, its partial sum has a limit σ:

$$\lim_{n\to\infty} \sigma_n = \sigma$$

From the fact that the terms of the series (1) and (2) are positive, it follows that $\sigma_n < \sigma$, and then, by virtue of (4),

$$s_n < \sigma$$

We have thus proved that the partial sums s_n are bounded. We note that as n increases, the partial sum s_n increases, and from the fact that the sequence of partial sums is bounded and increases, it follows that it has a limit *

$$\lim_{n\to\infty} s_n = s$$

and it is obvious that

$$s \leqslant \sigma$$

Using Theorem 1, we can judge of the convergence of certain series.

Example 1. The series

$$1+\frac{1}{2^2}+\frac{1}{3^3}+\frac{1}{4^4}+\ldots+\frac{1}{n^n}+\ldots$$

converges because its terms are smaller than the corresponding terms of the series

$$1+\frac{1}{2^2}+\frac{1}{2^3}+\frac{1}{2^4}+\ldots+\frac{1}{2^n}+\ldots$$

But the latter series converges because its terms, beginning with the second, form a geometric progression with common ratio $\frac{1}{2}$. The sum of this series is equal to $1\frac{1}{2}$. Hence, by virtue of Theorem 1, the given series also converges, and its sum does not exceed $1\frac{1}{2}$.

Theorem 2. *If the terms of the series* (1) *are not smaller than the corresponding terms of the series* (2); *that is,*

$$u_n \geqslant v_n \tag{5}$$

and the series (2) *diverges, then the series* (1) *also diverges.*

* To convince ourselves that the variable s_n has a limit, let us recall a condition for the existence of a limit of a sequence (see Theorem 7, Sec. 2.5, Vol. I): "if a variable is bounded and increases, it has a limit." Here, the sequence of sums s_n is bounded and increases. Hence it has a limit, i.e., the series converges.

Proof. From condition (5) it follows that

$$s_n \geqslant \sigma_n \tag{6}$$

Since the terms of the series (2) are positive, its partial sum σ_n increases with increasing n, and since it diverges, it follows that

$$\lim_{n\to\infty} \sigma_n = \infty$$

But then, by virtue of (6),

$$\lim_{n\to\infty} s_n = \infty$$

i.e., the series (1) diverges.

Example 2. The series

$$1+\frac{1}{\sqrt{2}}+\frac{1}{\sqrt{3}}+\ldots+\frac{1}{\sqrt{n}}+\ldots$$

diverges because its terms (from the second on) are greater than the corresponding terms of the harmonic series

$$1+\frac{1}{2}+\frac{1}{3}+\ldots+\frac{1}{n}+\ldots$$

which, as we know, diverges.

Note. Both the conditions that we have proved (Theorems 1 and 2) hold only for series with positive terms. They also hold true when some of the terms of the first or second series are zero. But these conditions do not hold if some of the terms of the series are negative numbers.

4.4 D'ALEMBERT'S TEST

Theorem (d'Alembert's test). *If in a series with positive terms*

$$u_1+u_2+u_3+\ldots+u_n+\ldots \tag{1}$$

the ratio of the $(n+1)$*th term to the nth term, as* $n\to\infty$, *has a (finite) limit* l, *that is,*

$$\lim_{n\to\infty} \frac{u_{n+1}}{u_n} = l \tag{2}$$

then:

(1) *the series converges for* $l<1$

(2) *the series diverges for* $l>1$.

(For $l=1$, the theorem does not yield the convergence or divergence of the series.)

Proof. (1) Let $l<1$. Consider a number q ($l<q<1$) (Fig. 109).

From the definition of a limit and relation (2) it follows that for all values of n after a certain integer N, that is, for $n\geqslant N$, we will have the inequality

$$\frac{u_{n+1}}{u_n} < q \tag{2'}$$

Indeed, since the quantity $\frac{u_{n+1}}{u_n}$ tends to the limit l, the difference between the quantity $\frac{u_{n+1}}{u_n}$ and the number l may (after a certain N) be made less (in absolute value) than any positive number, in particular less than $q-l$; that is,

$$\left|\frac{u_{n+1}}{u_n} - l\right| < q - l$$

Inequality (2') follows from this last inequality. Writing inequality (2') for various values of n, from N onwards, we get

$$\left.\begin{array}{l} u_{N+1} < qu_N \\ u_{N+2} < qu_{N+1} < q^2 u_N \\ u_{N+3} < qu_{N+2} < q^3 u_N \\ \dots\dots\dots\dots \end{array}\right\} \tag{3}$$

Now consider the two series

$$u_1 + u_2 + u_3 + \dots + u_N + u_{N+1} + u_{N+2} + \dots \tag{1}$$

$$u_N + qu_N + q^2 u_N + \dots \tag{1'}$$

The series (1') is a geometric progression with positive common ratio $q < 1$. Hence, this series converges. The terms of the

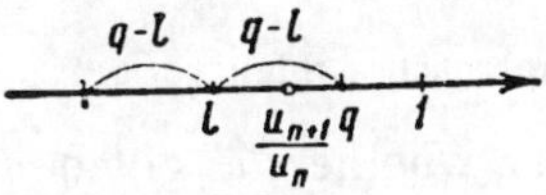

Fig. 109

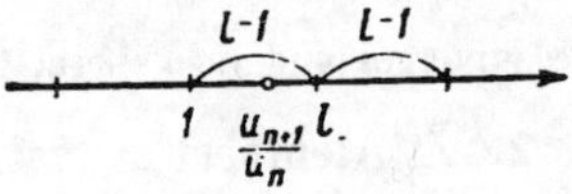

Fig. 110

series (1), after u_{N+1}, are less than the terms of the series (1'). By Theorem 1, Sec. 4.3, and Theorem 1, Sec. 4.1, it follows that the series (1) converges.

(2) Let $l > 1$. Then from the equation $\lim\limits_{n\to\infty} \frac{u_{n+1}}{u_n} = l$ (where $l > 1$) it follows that, after a certain N, that is for $n \geqslant N$, we will have the inequality

$$\frac{u_{n+1}}{u_n} > 1$$

(Fig. 110), or $u_{n+1} > u_n$ for all $n \geqslant N$. But this means that the terms of the series increase after the term $N+1$, and for this reason the general term of the series does not tend to zero. Hence, the series diverges.

Note 1. The series will also diverge when $\lim\limits_{n\to\infty}\frac{u_{n+1}}{u_n}=\infty$. This follows from the fact that if $\lim\limits_{n\to\infty}\frac{u_{n+1}}{u_n}=\infty$, then after a certain $n=N$ we will have the inequality $\frac{u_{n+1}}{u_n}>1$, or $u_{n+1}>u_n$.

Example 1. Test the following series for convergence:

$$1+\frac{1}{1\cdot 2}+\frac{1}{1\cdot 2\cdot 3}+\ldots+\frac{1}{1\cdot 2\cdot 3\cdot\ldots\cdot n}+\ldots$$

Solution. Here,

$$u_n=\frac{1}{1\cdot 2\cdot\ldots\cdot n}=\frac{1}{n!},\qquad u_{n+1}=\frac{1}{1\cdot 2\cdot\ldots\cdot n(n+1)}=\frac{1}{(n+1)!}$$

$$\frac{u_{n+1}}{u_n}=\frac{n!}{(n+1)!}=\frac{1}{n+1}$$

Hence,

$$\lim_{n\to\infty}\frac{u_{n+1}}{u_n}=\lim_{n\to\infty}\frac{1}{n+1}=0<1$$

The series converges.

Example 2. Test for convergence the series

$$\frac{2}{1}+\frac{2^2}{2}+\frac{2^3}{3}+\ldots+\frac{2^n}{n}+\ldots$$

Solution. Here,

$$u_n=\frac{2^n}{n},\quad u_{n+1}=\frac{2^{n+1}}{n+1},\quad \frac{u_{n+1}}{u_n}=2\frac{n}{n+1},\quad \lim_{n\to\infty}\frac{u_{n+1}}{u_n}=\lim_{n\to\infty}2\frac{n}{n+1}=2>1$$

The series diverges and its general term u_n approaches infinity.

Note 2. D'Alembert's test tells us whether a given positive series converges; but it does so only when $\lim\limits_{n\to\infty}\frac{u_{n+1}}{u_n}$ exists and is different from 1. But if this limit does not exist or if it does exist and $\lim\limits_{n\to\infty}\frac{u_{n+1}}{u_n}=1$, then d'Alembert's test does not enable us to tell whether the series converges or diverges, because in this case the series may prove to be both convergent and divergent. Some other test is needed to determine the convergence of such series.

Note 3. If $\lim\limits_{n\to\infty}\frac{u_{n+1}}{u_n}=1$, but the ratio $\frac{u_{n+1}}{u_n}$ for all n (after a certain one) is greater than unity, the series diverges. This follows from the fact that if $\frac{u_{n+1}}{u_n}>1$, then $u_{n+1}>u_n$ and the general term does not approach zero as $n\to\infty$.

To illustrate, let us examine some examples.

Example 3. Test for convergence the series

$$\frac{1}{2}+\frac{2}{3}+\frac{3}{4}+\ldots+\frac{n}{n+1}+\ldots$$

Solution. Here,

$$\lim_{n\to\infty}\frac{u_{n+1}}{u_n}=\lim_{n\to\infty}\frac{\frac{n+1}{n+2}}{\frac{n}{n+1}}=\lim_{n\to\infty}\frac{n^2+2n+1}{n^2+2n}=1$$

In this case the series diverges because $\frac{u_{n+1}}{u_n}>1$ for all n:

$$\frac{u_{n+1}}{u_n}=\frac{n^2+2n+1}{n^2+2n}>1$$

Example 4. Using the d'Alembert test, examine the harmonic series

$$1+\frac{1}{2}+\frac{1}{3}+\ldots+\frac{1}{n}+\ldots$$

We note that $u_n=\frac{1}{n}$, $u_{n+1}=\frac{1}{n+1}$ and, consequently,

$$\lim_{n\to\infty}\frac{u_{n+1}}{u_n}=\lim_{n\to\infty}\frac{n}{n+1}=1$$

Thus, d'Alembert's test does not allow us to determine the convergence or divergence of the given series. But we earlier found out by a different expedient that a harmonic series diverges.

Example 5. Test for convergence the series

$$\frac{1}{1\cdot 2}+\frac{1}{2\cdot 3}+\frac{1}{3\cdot 4}+\ldots+\frac{1}{n(n+1)}+\ldots$$

Solution. Here,

$$u_n=\frac{1}{n(n+1)},\quad u_{n+1}=\frac{1}{(n+1)(n+2)}$$

$$\lim_{n\to\infty}\frac{u_{n+1}}{u_n}=\lim_{n\to\infty}\frac{n(n+1)}{(n+1)(n+2)}=\lim_{n\to\infty}\frac{n}{n+2}=1$$

D'Alembert's test does not permit us to infer that the series converges; but by other reasoning we can establish the fact that this series converges. Noting that

$$\frac{1}{n(n+1)}=\frac{1}{n}-\frac{1}{n+1}$$

we can write the given series in the form

$$\left(\frac{1}{1}-\frac{1}{2}\right)+\left(\frac{1}{2}-\frac{1}{3}\right)+\left(\frac{1}{3}-\frac{1}{4}\right)+\ldots+\left(\frac{1}{n}-\frac{1}{n+1}\right)+\ldots$$

The partial sum of the first n terms, after removing brackets and cancelling, is

$$s_n=1-\frac{1}{n+1}$$

Hence,

$$\lim_{n\to\infty} s_n = \lim_{n\to\infty}\left(1-\frac{1}{n+1}\right)=1$$

that is, the series converges and its sum is 1.

4.5 CAUCHY'S TEST

Theorem (Cauchy's test). *If for a series with positive terms*

$$u_1+u_2+u_3+\ldots+u_n+\ldots \qquad (1)$$

the quantity $\sqrt[n]{u_n}$ *has a finite limit* l *as* $n\to\infty$, *that is,*

$$\lim_{n\to\infty}\sqrt[n]{u_n}=l$$

then: (1) *for* $l<1$, *the series converges*
(2) *for* $l>1$, *the series diverges.*

Proof. (1) Let $l<1$. Consider the number q that satisfies the relation $l<q<1$.

After some $n=N$ we will have the relation

$$\left|\sqrt[n]{u_n}-l\right|<q-l$$

whence it follows that

$$\sqrt[n]{u_n}<q$$

or

$$u_n<q^n$$

for all $n\geqslant N$.

Now consider two series:

$$u_1+u_2+u_3+\ldots+u_N+u_{N+1}+u_{N+2}+\ldots \qquad (1)$$

$$q^N+q^{N+1}+q^{N+2}+\ldots \qquad (1')$$

The series (1′) converges, since its terms form a decreasing geometric progression. The terms of the series (1), after u_N, are less than the terms of the series (1′). Consequently, the series (1) converges.

(2) Let $l>1$. Then, after some $n=N$, we will have

$$\sqrt[n]{u_n}>1$$

or

$$u_n>1$$

But if all the terms of this series, after u_N, exceed 1, then the series diverges, since its general term does not tend to zero.

Example. Test for convergence the series

$$\frac{1}{3}+\left(\frac{2}{5}\right)^2+\left(\frac{3}{7}\right)^3+\ldots+\left(\frac{n}{2n+1}\right)^n+\ldots$$

Solution. Apply the Cauchy test:

$$\lim_{n\to\infty} \sqrt[n]{u_n} = \lim_{n\to\infty} \sqrt[n]{\left(\frac{n}{2n+1}\right)^n} = \lim_{n\to\infty} \frac{n}{2n+1} = \frac{1}{2} < 1$$

The series converges.

Note. As in the d'Alembert test, the case

$$\lim_{n\to\infty} \sqrt[n]{u_n} = l = 1$$

requires further investigation. Among the series that satisfy this condition are convergent and divergent series. Thus, for the harmonic series (which is known to be divergent)

$$\lim_{n\to\infty} \sqrt[n]{u_n} = \lim_{n\to\infty} \sqrt[n]{\frac{1}{n}} = 1$$

To be sure, we shall prove that $\lim_{n\to\infty} \ln \sqrt[n]{\frac{1}{n}} = 0$. Indeed,

$$\lim_{n\to\infty} \ln \sqrt[n]{\frac{1}{n}} = \lim_{n\to\infty} \frac{-\ln n}{n}$$

Here, the numerator and denominator of the fraction approach infinity. Applying l'Hospital's rule, we find

$$\lim_{n\to\infty} \ln \sqrt[n]{\frac{1}{n}} = \lim_{n\to\infty} \frac{-\ln n}{n} = \lim_{n\to\infty} \frac{-\frac{1}{n}}{1} = 0$$

Thus, $\ln \sqrt[n]{\frac{1}{n}} \to 0$, but then $\sqrt[n]{\frac{1}{n}} \to 1$, i.e.,

$$\lim_{n\to\infty} \sqrt[n]{\frac{1}{n}} = 1$$

For the series

$$\frac{1}{1} + \frac{1}{2^2} + \frac{1}{3^2} + \dots + \frac{1}{n^2} + \dots$$

we also have the equality

$$\lim_{n\to\infty} \sqrt[n]{u_n} = \lim_{n\to\infty} \sqrt[n]{\frac{1}{n^2}} = \lim_{n\to\infty} \sqrt[n]{\frac{1}{n}} \sqrt[n]{\frac{1}{n}} = 1$$

but this series converges, since if we suppress the first term, the terms of the remaining series will be less than the corresponding terms of the converging series

$$\frac{1}{1\cdot 2} + \frac{1}{2\cdot 3} + \dots + \frac{1}{n(n+1)} + \dots$$

(see Example 5, Sec. 4.4).

4.6 THE INTEGRAL TEST FOR CONVERGENCE OF A SERIES

Theorem. *Let the terms of the series*

$$u_1+u_2+u_3+\ldots+u_n+\ldots \tag{1}$$

be positive and nonincreasing, that is,

$$u_1 \geqslant u_2 \geqslant u_3 \geqslant \ldots \tag{1'}$$

and let $f(x)$ be a continuous nonincreasing function such that

$$f(1)=u_1, \quad f(2)=u_2, \quad \ldots, \quad f(n)=u_n \tag{2}$$

Then the following assertions hold true:

(1) *if the improper integral*

$$\int_1^\infty f(x)\,dx$$

converges (see Sec. 11.7, Vol. I), *then the series* (1) *converges too;*

(2) *if the given integral diverges, then the series* (1) *diverges as well.*

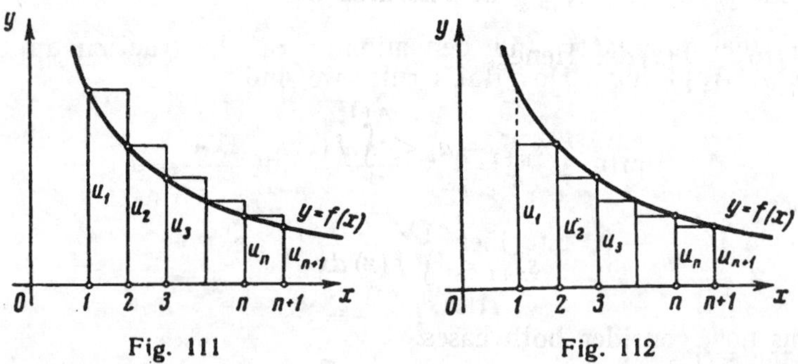

Fig. 111 Fig. 112

Proof. Depict the terms of the series geometrically by plotting on the x-axis the terms $1, 2, 3, \ldots, n, n+1, \ldots$ of the series, and on the y-axis, the corresponding values of the terms of the series $u_1, u_2, \ldots, u_n, \ldots$ (Fig. 111).

In the same coordinate system plot the graph of the continuous nonincreasing function

$$y=f(x)$$

which satisfies condition (2).

An examination of Fig. 111 shows that the first of the constructed rectangles has base equal to 1 and altitude $f(1)=u_1$ The area of this rectangle is thus u_1. The area of the second one is u_2, and so on; finally, the area of the last (nth) of the constructed rectangles is u_n. The sum of the areas of the constructed rectangles is equal to the sum s_n of the first n terms of the series. On the other hand, the step-like figure formed by these rectangles embraces a region bounded by the curve $y=f(x)$ and the straight lines $x=1$, $x=n+1$, $y=0$; the area of this region is equal to $\int_1^{n+1} f(x)\,dx$. Hence,

$$s_n > \int_1^{n+1} f(x)\,dx \tag{3}$$

Let us now consider Fig. 112. Here the first of the constructed rectangles on the left has altitude u_2; and so its area is u_2. The area of the second rectangle is u_3, and so forth. The area of the last of the constructed rectangles is u_{n+1}. Hence, the sum of the areas of all constructed rectangles is equal to the sum of all terms of the series beginning with the second to the $(n+1)$th or $s_{n+1}-u_1$. On the other hand, it is readily seen that the step-like figure formed by these rectangles is contained within the curvilinear figure bounded by the curve $y=f(x)$ and the straight lines $x=1$, $x=n+1$, $y=0$. The area of this curvilinear figure is equal to $\int_1^{n+1} f(x)\,dx$. Hence,

$$s_{n+1}-u_1 < \int_1^{n+1} f(x)\,dx$$

whence

$$s_{n+1} < \int_1^{n+1} f(x)\,dx + u_1 \tag{4}$$

Let us now consider both cases.

1. We assume that the integral $\int_1^{\infty} f(x)\,dx$ converges, that is, has a finite value.

Since

$$\int_1^{n+1} f(x)\,dx < \int_1^{\infty} f(x)\,dx$$

it follows, by virtue of inequality (4), that

$$s_n < s_{n+1} < \int_1^{\infty} f(x)\,dx + u_1$$

Thus, the partial sum s_n remains bounded for all values of n. But it increases with increasing n, since all the terms u_n are positive. Consequently, s_n (as $n \to \infty$) has the finite limit

$$\lim_{n\to\infty} s_n = s$$

and the series converges.

2. Assume, further, that $\int_1^\infty f(x)\,dx = \infty$. This means that $\int_1^{n+1} f(x)\,dx$ increases without bound as n increases. But then, by virtue of inequality (3), s_n likewise increases indefinitely with n; the series diverges.

The proof of the theorem is complete.

Note. This theorem remains valid if inequalities (1′) are fulfilled only from some integer N onwards.

Example. Test for convergence the series

$$\frac{1}{1^p}+\frac{1}{2^p}+\frac{1}{3^p}+\ldots+\frac{1}{n^p}+\ldots$$

Solution. Apply the integral test, putting

$$f(x)=\frac{1}{x^p}$$

This function satisfies all the conditions of the theorem. Consider the integral

$$\int_1^N \frac{dx}{x^p}=\begin{cases}\dfrac{1}{1-p}x^{1-p}\Big|_1^N=\dfrac{1}{1-p}(N^{1-p}-1) & \text{when } p\neq 1\\ \ln x\Big|_1^N=\ln N & \text{when } p=1\end{cases}$$

Allow N to approach infinity and determine whether the improper integral converges in various cases.

It will then be possible to judge about the convergence or divergence of the series for various values of p.

For $p>1$, $\int_1^\infty \frac{dx}{x^p}=\frac{1}{p-1}$, the integral is finite and, hence, the series converges;

for $p<1$, $\int_1^\infty \frac{dx}{x^p}=\infty$, the integral is infinite, and the series diverges;

for $p=1$, $\int_1^\infty \frac{dx}{x}=\infty$, the integral is infinite, and the series diverges.

We note that neither the d'Alembert test nor the Cauchy test, which were considered earlier, decide whether the series is convergent or not, since

$$\lim_{n\to\infty}\frac{u_{n+1}}{u_n}=\lim_{n\to\infty}\left(\frac{n}{n+1}\right)^p=1$$

$$\lim_{n\to\infty}\sqrt[n]{u_n}=\lim_{n\to\infty}\sqrt[n]{\frac{1}{n^p}}=\lim_{n\to\infty}\left(\sqrt[n]{\frac{1}{n}}\right)^p=1^p=1$$

4.7 ALTERNATING SERIES. LEIBNIZ' THEOREM

So far we have been considering series whose terms are all positive. In this section we consider series whose terms have **alternating signs**, that is, series of the form

$$u_1 - u_2 + u_3 - u_4 + \dots$$

where $u_1, u_2, \dots, u_n, \dots$ are positive.

Leibniz' Theorem. *If in the alternating series*

$$u_1 - u_2 + u_3 - u_4 + \dots \quad (u_n > 0) \tag{1}$$

the terms are such that

$$u_1 > u_2 > u_3 > \dots \tag{2}$$

and

$$\lim_{n\to\infty} u_n = 0 \tag{3}$$

then the series (1) *converges, its sum is positive and does not exceed the first term.*

Proof. Consider the sum of the first $n = 2m$ terms of the series (1):

$$s_{2m} = (u_1 - u_2) + (u_3 - u_4) + \dots + (u_{2m-1} - u_{2m})$$

From condition (2) it follows that the expression in each of the brackets is positive. Hence, the sum s_{2m} is positive,

$$s_{2m} > 0$$

and increases with increasing m. Now write this sum as follows:

$$s_{2m} = u_1 - (u_2 - u_3) - (u_4 - u_5) - \dots - (u_{2m-2} - u_{2m-1}) - u_{2m}$$

By virtue of condition (2), each of the parentheses is positive. Therefore, subtracting these parentheses from u_1 we get a number less than u_1, or

$$s_{2m} < u_1$$

We have thus established that s_{2m} increases with increasing m and is bounded above. Whence it follows that s_{2m} has the limit s:

$$\lim_{m\to\infty} s_{2m} = s$$

and

$$0 < s < u_1$$

However, we have not yet proved the convergence of the series; we have only proved that a sequence of "even" partial sums has as its limit the number s. We now prove that "odd" partial sums also approach the limit s.

Consider the sum of the first $n = 2m + 1$ terms of the series (1):

$$s_{2m+1} = s_{2m} + u_{2m+1}$$

Since, by condition (3), $\lim_{m\to\infty} u_{2m+1}=0$, it follows that

$$\lim_{m\to\infty} s_{2m+1}=\lim_{m\to\infty} s_{2m}+\lim_{m\to\infty} u_{2m+1}=\lim_{m\to\infty} s_{2m}=s$$

We have thus proved that $\lim_{n\to\infty} s_n=s$ both for even n and for odd n. Hence, the series (1) converges.

Note 1. The Leibniz theorem holds if inequalities (2) hold true from some N onwards.

Note 2. The Leibniz theorem may be illustrated geometrically as follows. Plot the following partial sums on a number line (Fig. 113)

$$s_1=u_1,\quad s_2=u_1-u_2=s_1-u_2,\quad s_3=s_2+u_3,\quad s_4=s_3-u_4,\quad s_5=s_4+u_5,$$

etc.

The points corresponding to partial sums will approach a certain point s, which depicts the sum of the series. Here, the points corresponding to the even partial sums lie on the left of s, and those corresponding to odd sums, on the right of s.

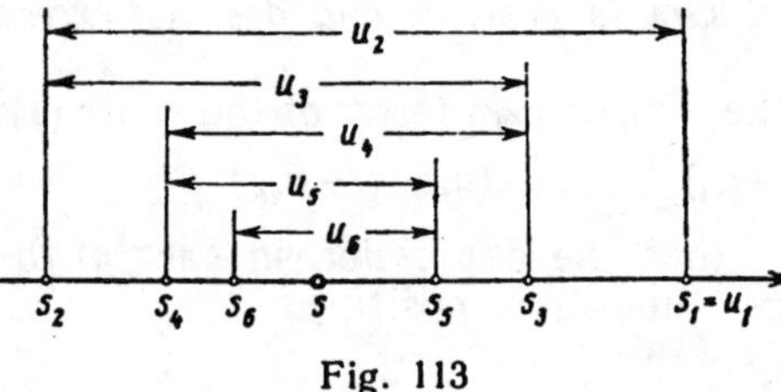

Fig. 113

Note 3. If an alternating series satisfies the statement of the Leibniz theorem, then it is easy to evaluate the error that results if we replace its sum, s, by the partial sum s_n. In this substitution we suppress all terms after u_{n+1}. But these numbers form by themselves an alternating series, whose sum (in absolute value) is less than the first term of this series (that is, less than u_{n+1}). Thus, the error obtained when replacing s by s_n does not exceed (in absolute value) the first of the suppressed terms.

Example 1. The series

$$1-\frac{1}{2}+\frac{1}{3}-\frac{1}{4}+\ldots$$

converges, since (1) $1>\frac{1}{2}>\frac{1}{3}>\ldots$; (2) $\lim_{n\to\infty} u_n=\lim_{n\to\infty}\frac{1}{n}=0.$

The sum of the first n terms of this series

$$s_n=1-\frac{1}{2}+\frac{1}{3}-\frac{1}{4}+\ldots+(-1)^{n+1}\frac{1}{n}$$

differs from the sum s of the series by a quantity less than $\frac{1}{n+1}$.

Example 2. The series

$$1-\frac{1}{2!}+\frac{1}{3!}-\frac{1}{4!}+\ldots$$

converges by virtue of the Leibniz theorem.

4.8 PLUS-AND-MINUS SERIES. ABSOLUTE AND CONDITIONAL CONVERGENCE

We give the name *plus-and-minus* series to a series that has both positive and negative terms.

Obviously, the **alternating** series considered in Sec. 4.7 is a **special case** of plus-and-minus series.*

We shall consider some properties of alternating series.

In contrast to the agreement made in the preceding section we will now assume that the numbers $u_1, u_2, \ldots, u_n \ldots$ can be both positive and negative.

First, let us give an important sufficient condition for the convergence of an alternating series.

Theorem 1. *If the alternating series*

$$u_1+u_2+\ldots+u_n+\ldots \tag{1}$$

is such that a series made up of the absolute values of its terms

$$|u_1|+|u_2|+\ldots+|u_n|+\ldots \tag{2}$$

converges, then the given alternating series also converges.

Proof. Let s_n and σ_n be the sums of the first n terms of the series (1) and (2).

Also, let s_n' be the sum of all the positive terms, and s_n'', the sum of the absolute values of all the negative terms of the first n terms of the given series; then

$$s_n=s_n'-s_n'', \quad \sigma_n=s_n'+s_n''$$

By hypothesis, σ_n has the limit σ; s_n' and s_n'' are positive increasing quantities less than σ. Consequently, they have the limits s' and s''. From the relationship $s_n=s_n'-s_n''$ it follows that s_n also has a limit and that this limit is equal to $s'-s''$, which means that the alternating series (1) converges.

The above-proved theorem enables one to judge about the convergence of certain alternating series. In this case, testing the alternating series for convergence reduces to investigating a series with positive terms.

Consider two examples.

Example 1. Test for convergence the series

$$\frac{\sin\alpha}{1^2}+\frac{\sin 2\alpha}{2^2}+\frac{\sin 3\alpha}{3^2}+\ldots+\frac{\sin n\alpha}{n^2}+\ldots \tag{3}$$

where α is any number.

---

* In this English edition we shall use the term alternating series for both types.—*Tr.*

Solution. Also consider the series

$$\left|\frac{\sin \alpha}{1^2}\right|+\left|\frac{\sin 2\alpha}{2^2}\right|+\left|\frac{\sin 3\alpha}{3^2}\right|+\ldots+\left|\frac{\sin n\alpha}{n^2}\right|+\ldots \tag{4}$$

and

$$\frac{1}{1^2}+\frac{1}{2^2}+\frac{1}{3^2}+\ldots+\frac{1}{n^2}+\ldots \tag{5}$$

The series (5) converges (see Sec. 4.6). The terms of the series (4) do not exceed the corresponding terms of the series (5); hence, the series (4) also converges. But then, in virtue of the theorem just proved, the given series (3) likewise converges.

Example 2. Test for convergence the series

$$\frac{\cos\frac{\pi}{4}}{3}+\frac{\cos\frac{3\pi}{4}}{3^2}+\frac{\cos\frac{5\pi}{4}}{3^3}+\ldots+\frac{\cos\frac{(2n-1)\pi}{4}}{3^n}+\ldots \tag{6}$$

Solution. In addition to this series, consider the series

$$\frac{1}{3}+\frac{1}{3^2}+\frac{1}{3^3}+\ldots+\frac{1}{3^n}+\ldots \tag{7}$$

This series converges because it is a decreasing geometric progression with ratio $\frac{1}{3}$. But then the given series (6) converges, since the absolute values of its terms are less than those of the corresponding terms of the series (7).

We note that the convergence condition that was proved earlier is only a **sufficient** condition for convergence of an alternating series, but not a necessary condition: there are alternating series which converge, but series formed from the absolute values of their terms diverge. In this connection, it is useful to introduce the concepts of absolute and conditional convergence of an alternating series and, on the basis of these concepts, to classify alternating series.

Definition. The alternating series

$$u_1+u_2+u_3+\ldots+u_n+\ldots \tag{1}$$

is called *absolutely convergent* if a series made up of the absolute values of its terms converges:

$$|u_1|+|u_2|+|u_3|+\ldots+|u_n|+\ldots \tag{2}$$

If the alternating series (1) converges, while the series (2) composed of the absolute values of its terms diverges, then the given alternating series (1) is called *a conditionally convergent series.*

Example 3. The alternating series

$$1-\frac{1}{2}+\frac{1}{3}-\frac{1}{4}+\ldots$$

is **conditionally** convergent, since a series composed of the absolute values of its terms is a harmonic series,

$$1+\frac{1}{2}+\frac{1}{3}+\frac{1}{4}+\dots$$

which diverges. The series itself converges (this can be readily verified by Leibniz' test).

Example 4. The alternating series

$$1-\frac{1}{2!}+\frac{1}{3!}-\frac{1}{4!}+\dots$$

is **absolutely** convergent, since a series made up of the absolute values of its terms,

$$1+\frac{1}{2!}+\frac{1}{3!}+\frac{1}{4!}+\dots$$

converges, as established in Sec. 4.4.

Theorem 1 is frequently stated (with the help of the concept of absolute convergence) as follows: *every absolutely convergent series is a convergent series.*

In conclusion, we note (without proof) the following properties of absolutely convergent and conditionally convergent series.

Theorem 2. *If a series converges absolutely, it remains absolutely convergent for any rearrangement of its terms. The sum of the series is independent of the order of its terms.*

This property does not hold for conditionally convergent series.

Theorem 3. *If a series converges conditionally, then no matter what number A is given, the terms of this series can be rearranged in such manner that its sum is exactly equal to A. What is more, it is possible so to rearrange the terms of a conditionally convergent series that the series resulting after the rearrangement is divergent.*

The proof of these theorems is outside the scope of this course, and can be found in more fundamental texts (see, for example, G. M. Fikhtengolts, "Course of Differential and Integral Calculus", 1962, Vol. II, pp. 319-320, in Russian).

To illustrate the fact that the sum of a conditionally convergent series can change upon rearrangement of its terms, consider the following example.

Example 5. The alternating series

$$1-\frac{1}{2}+\frac{1}{3}-\frac{1}{4}+\dots \tag{8}$$

converges conditionally. Denote its sum by s. It is obvious that $s>0$. Rearrange the terms of the series (8) so that two negative terms follow one positive term:

$$\underbrace{1-\frac{1}{2}-\frac{1}{4}}+\underbrace{\frac{1}{3}-\frac{1}{6}-\frac{1}{8}}+\dots+\underbrace{\frac{1}{2k-1}-\frac{1}{4k-2}-\frac{1}{4k}}+\dots \tag{9}$$

We shall prove that the resultant series converges, but that its sum s' is half the sum of the series (8): that is, $\frac{1}{2}s$. Denote by s_n and s_n' the partial sums of the series (8) and (9). Consider the sum of $3k$ terms of the series (9):

$$s_{3k}' = \left(1-\frac{1}{2}-\frac{1}{4}\right)+\left(\frac{1}{3}-\frac{1}{6}-\frac{1}{8}\right)+\ldots+\left(\frac{1}{2k-1}-\frac{1}{4k-2}-\frac{1}{4k}\right)$$
$$=\left(\frac{1}{2}-\frac{1}{4}\right)+\left(\frac{1}{6}-\frac{1}{8}\right)+\ldots+\left(\frac{1}{4k-2}-\frac{1}{4k}\right)$$
$$=\frac{1}{2}\left[\left(1-\frac{1}{2}\right)+\left(\frac{1}{3}-\frac{1}{4}\right)+\ldots+\left(\frac{1}{2k-1}-\frac{1}{2k}\right)\right]$$
$$=\frac{1}{2}\left(1-\frac{1}{2}+\frac{1}{3}-\frac{1}{4}+\ldots+\frac{1}{2k-1}-\frac{1}{2k}\right)=\frac{1}{2}s_{2k}$$

Consequently,

$$\lim_{k\to\infty} s_{3k}' = \lim_{k\to\infty} \frac{1}{2}s_{2k} = \frac{1}{2}s$$

Further,

$$\lim_{k\to\infty} s_{3k+1}' = \lim_{k\to\infty}\left(s_{3k}'+\frac{1}{2k+1}\right)=\frac{1}{2}s$$
$$\lim_{k\to\infty} s_{3k+2}' = \lim_{k\to\infty}\left(s_{3k}'+\frac{1}{2k+1}-\frac{1}{4k+2}\right)=\frac{1}{2}s$$

And we obtain

$$\lim_{n\to\infty} s_n' = s' = \frac{1}{2}s$$

Thus, in this case the sum of the series changed after its terms were rearranged (it diminished by a factor of 2).

4.9 FUNCTIONAL SERIES

A series $u_1+u_2+\ldots+u_n+\ldots$ is called a *functional series* (or series of functions) if its terms are functions of x.

Consider the functional series

$$u_1(x)+u_2(x)+u_3(x)+\ldots+u_n(x)+\ldots \tag{1}$$

Assigning to x definite numerical values, we get different numerical series, which may prove to be convergent or divergent.

The set of all those values of x for which the functional series converges is called the *domain of convergence* of the series.

Obviously, in the domain of convergence of a series its sum is some function of x. Therefore, the sum of a functional series is denoted by $s(x)$.

Example. Consider the functional series

$$1+x+x^2+\ldots+x^n+\ldots$$

This series converges for all values of x in the interval $(-1, 1)$, that is, for all x that satisfy the condition $|x|<1$. For each value of x in the interval $(-1, 1)$, the sum of the series is equal to $\frac{1}{1-x}$ (the sum of a decreasing geometric progression with ratio x). Thus, in the interval $(-1, 1)$ the given series defines the function

$$s(x)=\frac{1}{1-x}$$

which is the sum of the series; that is,

$$\frac{1}{1-x}=1+x+x^2+x^3+\ldots$$

Denote by $s_n(x)$ the sum of the first n terms of the series (1). If this series converges and its sum is equal to $s(x)$, then

$$s(x)=s_n(x)+r_n(x)$$

where $r_n(x)$ is the sum of the series $u_{n+1}(x)+u_{n+2}(x)+\ldots$, i.e.,

$$r_n(x)=u_{n+1}(x)+u_{n+2}(x)+\ldots$$

Here, the quantity $r_n(x)$ is called the *remainder of the series* (1). For all values of x in the domain of convergence of the series we have the relation $\lim\limits_{n\to\infty} s_n(x)=s(x)$; therefore,

$$\lim_{n\to\infty} r_n(x)=\lim_{n\to\infty}[s(x)-s_n(x)]=0$$

which means that the remainder $r_n(x)$ of a convergent series approaches zero as $n\to\infty$.

4.10 DOMINATED SERIES

Definition. The functional series

$$u_1(x)+u_2(x)+u_3(x)+\ldots+u_n(x)+\ldots \tag{1}$$

is called *dominated* in some range of x if there exists a convergent numerical series

$$\alpha_1+\alpha_2+\alpha_3+\ldots+\alpha_n+\ldots \tag{2}$$

with positive terms such that for all values of x from this range the following relations are fulfilled:

$$|u_1(x)|\leqslant\alpha_1,\ |u_2(x)|\leqslant\alpha_2,\ \ldots,\ |u_n(x)|\leqslant\alpha_n,\ \ldots \tag{3}$$

In other words, a series is called *dominated* if each of its terms does not exceed, in absolute value, the corresponding term of some convergent numerical series with positive terms.

For example, the series

$$\frac{\cos x}{1}+\frac{\cos 2x}{2^2}+\frac{\cos 3x}{3^2}+\ldots+\frac{\cos nx}{n^2}+\ldots$$

is a series dominated on the entire x-axis. Indeed, for all values of x, the relation

$$\left|\frac{\cos nx}{n^2}\right| \leqslant \frac{1}{n^2} \quad (n=1,\ 2,\ \ldots)$$

is fulfilled and the series

$$\frac{1}{1}+\frac{1}{2^2}+\frac{1}{3^2}+\ldots$$

as we know, converges.

From the definition it follows straightway that a series dominated in some range converges absolutely at all points of this range (see Sec. 4.8). Also, a dominated series has the following important property.

Theorem. *Let the functional series*

$$u_1(x)+u_2(x)+\ldots+u_n(x)+\ldots$$

be dominated on the interval $[a, b]$. *Let* $s(x)$ *be the sum of this series and* $s_n(x)$ *the sum of the first* n *terms of this series. Then for each arbitrarily small number* $\varepsilon>0$ *there will be a positive integer* N *such that for all* $n \geqslant N$ *the following inequality will be fulfilled,*

$$|s(x)-s_n(x)|<\varepsilon$$

no matter what the x *of the interval* $[a, b]$.

Proof. Denote by σ the sum of the series (2):

$$\sigma=\alpha_1+\alpha_2+\alpha_3+\ldots+\alpha_n+\alpha_{n+1}+\ldots$$

then

$$\sigma=\sigma_n+\varepsilon_n$$

where σ_n is the sum of the first n terms of the series (2), and ε_n is the sum of the remaining terms of this series; that is,

$$\varepsilon_n=\alpha_{n+1}+\alpha_{n+2}+\ldots$$

Since this series converges, it follows that

$$\lim_{n\to\infty}\sigma_n=\sigma$$

and, consequently,

$$\lim_{n\to\infty}\varepsilon_n=0$$

Let us now represent the sum of the functional series (1) in the form

$$s(x)=s_n(x)+r_n(x)$$

where

$$s_n(x)=u_1(x)+\ldots+u_n(x)$$
$$r_n(x)=u_{n+1}(x)+u_{n+2}(x)+u_{n+3}(x)+\ldots$$

From condition (3) it follows that

$$|u_{n+1}(x)| \leqslant \alpha_{n+1}, \quad |u_{n+2}(x)| \leqslant \alpha_{n+2}, \ \ldots$$

and therefore

$$|r_n(x)| \leqslant \varepsilon_n$$

for all x of the range under consideration.

Thus,

$$|s(x) - s_n(x)| < \varepsilon_n$$

for all x of the interval $[a, b]$, and $\varepsilon_n \to 0$ as $n \to \infty$.

Note 1. This result may be represented geometrically as follows. Consider the graph of the function $y = s(x)$. About this curve construct a band of width $2\varepsilon_n$; in other words, construct the curves $y = s(x) + \varepsilon_n$ and $y = s(x) - \varepsilon_n$ (Fig. 114). Then for any ε_n the graph of the function $s_n(x)$ will lie completely in the band under consideration. The graphs of all successive partial sums will likewise lie within this band.

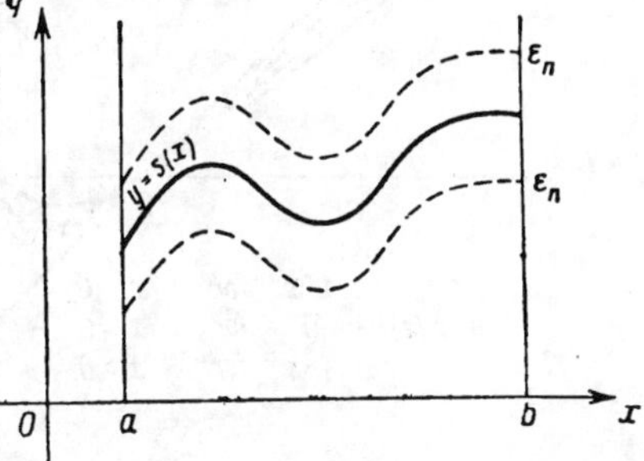

Fig. 114

Note 2. Not every functional series convergent on the interval $[a, b]$ has the property indicated in the foregoing theorem. However, there are nondominated series such that possess this property. A series that possesses this property is called *a uniformly convergent series on the interval* $[a, b]$.

Thus, *the functional series* $u_1(x) + u_2(x) + \ldots + u_n(x) + \ldots$ *is called a uniformly convergent series on the interval* $[a, b]$ *if for any arbitrarily small* $\varepsilon > 0$ *there is an integer* N *such that for all* $n \geqslant N$ *the inequality*

$$|s(x) - s_n(x)| < \varepsilon$$

will be fulfilled for any x *of the interval* $[a, b]$.

From the theorem that has been proved it follows that a dominated series is a series that uniformly converges.

4.11 THE CONTINUITY OF THE SUM OF A SERIES

Let there be a series made up of continuous functions

$$u_1(x) + u_2(x) + \ldots + u_n(x) + \ldots$$

convergent on some interval $[a, b]$.

In Chapter 2, Vol. I, we proved a theorem which stated that the sum of a finite number of continuous functions is a continuous function. This property does not hold for the sum of a series consisting of an infinite number of terms. Some functional series with continuous terms have for the sum a continuous function, while in the case of other functional series with continuous terms, the sum is a discontinuous function.

Example. Consider the series

$$\left(x^{\frac{1}{3}}-x\right)+\left(x^{\frac{1}{5}}-x^{\frac{1}{3}}\right)+\left(x^{\frac{1}{7}}-x^{\frac{1}{5}}\right)+\ldots+\left(x^{\frac{1}{2n+1}}-x^{\frac{1}{2n-1}}\right)+\ldots$$

The terms of this series (each term is bracketed) are continuous functions for all values of x. We shall prove that this series converges and that its sum is a discontinuous function.

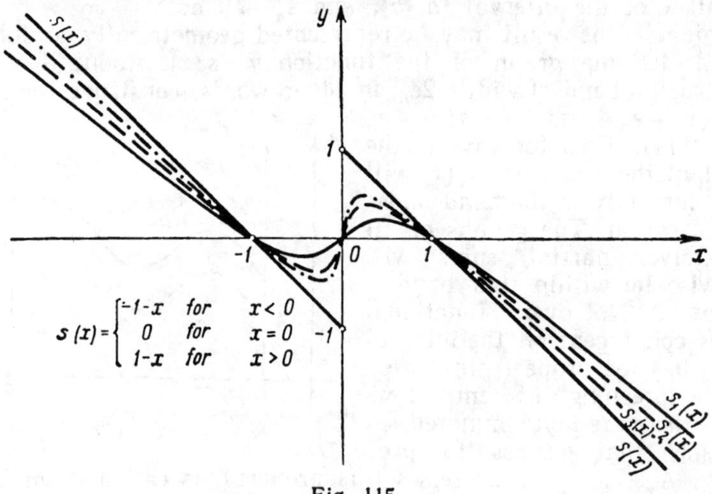

Fig. 115

We find the sum of the first n terms of the series:

$$s_n = x^{\frac{1}{2n+1}} - x$$

Find the sum of the series:

if $x > 0$, then

$$s = \lim_{n\to\infty} s_n(x) = \lim_{n\to\infty}\left(x^{\frac{1}{2n+1}} - x\right) = 1 - x$$

if $x < 0$, then

$$s = \lim_{n\to\infty} s_n(x) = \lim_{n\to\infty}\left(-|x|^{\frac{1}{2n+1}} - x\right) = -1 - x$$

if $x=0$, then $s_n = 0$, and so $s = \lim\limits_{n\to\infty} s_n = 0$. Thus, we have

$$\begin{aligned} s(x) &= -1-x \quad \text{for} \quad x<0 \\ s(x) &= 0 \quad \text{for} \quad x=0 \\ s(x) &= 1-x \quad \text{for} \quad x>0 \end{aligned}$$

And so the sum of the given series is a discontinuous function. Its graph is shown in Fig. 115 along with the graphs of the partial sums $s_1(x)$, $s_2(x)$, and $s_3(x)$.

The following theorem holds true for dominated series.

Theorem. *The sum of a series (of continuous functions) dominated on some interval* $[a, b]$ *is a function continuous on that interval.*

Proof. Let there be a series of continuous functions that is dominated on the interval $[a, b]$:

$$u_1(x)+u_2(x)+u_3(x)+\ldots \tag{1}$$

Let us represent its sum in the form

$$s(x)=s_n(x)+r_n(x)$$

where

$$s_n(x)=u_1(x)+\ldots+u_n(x)$$

and

$$r_n(x)=u_{n+1}(x)+u_{n+2}(x)+\ldots$$

On the interval $[a, b]$ take an arbitrary value of the argument x and give it an increment Δx such that the point $x+\Delta x$ also lies on the interval $[a, b]$.

We introduce the notation

$$\Delta s=s(x+\Delta x)-s(x)$$

$$\Delta s_n=s_n(x+\Delta x)-s_n(x)$$

then

$$\Delta s=\Delta s_n+r_n(x+\Delta x)-r_n(x)$$

from which we have

$$|\Delta s|\leqslant|\Delta s_n|+|r_n(x+\Delta x)|+|r_n(x)| \tag{2}$$

This inequality is true for any integer n.

To prove the continuity of $s(x)$, we have to show that for any preassigned and arbitrarily small $\varepsilon>0$ there will be a number $\delta>0$ such that for all $|\Delta x|<\delta$ we will have $|\Delta s|<\varepsilon$.

Since the given series (1) is dominated, it follows that for any preassigned $\varepsilon>0$ there will be found an integer N such that for all $n\geqslant N$ (and as a particular case, $n=N$) the inequality

$$|r_N(x)|<\frac{\varepsilon}{3} \tag{3}$$

will be fulfilled for any x of the interval $[a, b]$. The value $x+\Delta x$ lies on the interval $[a, b]$ and therefore the following inequality is fulfilled:

$$|r_N(x+\Delta x)|<\frac{\varepsilon}{3} \tag{3'}$$

Further, for the chosen N the partial sum $s_N(x)$ is a continuous function (the sum of a finite number of continuous functions) and, consequently, a positive number δ may be chosen such that for every Δx that satisfies the condition $|\Delta x| < \delta$ the following inequality is fulfilled:

$$|\Delta s_N| < \frac{\varepsilon}{3} \tag{4}$$

By inequalities (2), (3), (3′), and (4), we have

$$|\Delta s| < \frac{\varepsilon}{3} + \frac{\varepsilon}{3} + \frac{\varepsilon}{3} = \varepsilon$$

that is,

$$|\Delta s| < \varepsilon \quad \text{for} \quad |\Delta x| < \delta$$

which means that $s(x)$ is a continuous function at the point x (and, consequently, at any point of the interval $[a, b]$).

Note. From this theorem it follows that if the sum of a series is discontinuous on some interval $[a, b]$, then the series is not dominated on this interval. In particular, the series given in the Example is not dominated on any interval containing the point $x=0$, that is to say, a point of discontinuity of the sum of the series.

We note, finally, that the converse statement is not true: there are series, not dominated on an interval, which, however, converge on that interval to a continuous function. For instance, every series uniformly convergent on the interval $[a, b]$ (even if it is not dominated) has a continuous function for its sum (if, of course, all terms of the series are continuous).

4.12 INTEGRATION AND DIFFERENTIATION OF SERIES

Theorem 1. *Let there be a series of continuous functions*

$$u_1(x) + u_2(x) + \ldots + u_n(x) + \ldots \tag{1}$$

dominated on the interval $[a, b]$ and let $s(x)$ be the sum of the series. Then the integral of $s(x)$ from α to x, which limits belong to the interval $[a, b]$, is equal to the sum of the integrals of the terms of the given series; that is,

$$\int_\alpha^x s(t)\,dt = \int_\alpha^x u_1(t)\,dt + \int_\alpha^x u_2(t)\,dt + \ldots + \int_\alpha^x u_n(t)\,dt + \ldots$$

Proof. The function $s(x)$ may be represented in the form

$$s(x) = s_n(x) + r_n(x)$$

or

$$s(x) = u_1(x) + u_2(x) + \ldots + u_n(x) + r_n(x)$$

Then

$$\int_\alpha^x s(t)\,dt = \int_\alpha^x u_1(t)\,dt + \int_\alpha^x u_2(t)\,dt + \dots + \int_\alpha^x u_n(t)\,dt + \int_\alpha^x r_n(t)\,dt \quad (2)$$

(the integral of the sum of a **finite** number of terms is equal to the sum of the integrals of these terms).

Since the original series (1) is dominated, it follows that for every x we have $|r_n(x)| < \varepsilon_n$, where $\varepsilon_n \to 0$ as $n \to \infty$. Therefore,*

$$\left|\int_\alpha^x r_n(t)\,dt\right| \leqslant \pm\int_\alpha^x |r_n(t)|\,dt < \pm\int_\alpha^x \varepsilon_n\,dt = \pm\varepsilon_n(x-\alpha) \leqslant \varepsilon_n(b-a)$$

Since $\varepsilon_n \to 0$, it follows that

$$\lim_{n\to\infty}\int_\alpha^x r_n(t)\,dt = 0$$

But from equation (2) we have

$$\int_\alpha^x r_n(t)\,dt = \int_\alpha^x s(t)\,dt - \left[\int_\alpha^x u_1(t)\,dt + \dots + \int_\alpha^x u_n(t)\,dt\right]$$

Hence

$$\lim_{n\to\infty}\left\{\int_\alpha^x s(t)\,dt - \left[\int_\alpha^x u_1(t)\,dt + \dots + \int_\alpha^x u_n(t)\,dt\right]\right\} = 0$$

or

$$\lim_{n\to\infty}\left[\int_\alpha^x u_1(t)\,dt + \dots + \int_\alpha^x u_n(t)\,dt\right] = \int_\alpha^x s(t)\,dt \quad (3)$$

The sum in the brackets is a partial sum of the series

$$\int_\alpha^x u_1(t)\,dt + \dots + \int_\alpha^x u_n(t)\,dt + \dots \quad (4)$$

Since the partial sums of this series have a limit, this series converges and its sum, by virtue of equation (3), is equal to $\int_\alpha^x s(t)\,dt$, i.e.,

$$\int_\alpha^x s(t)\,dt = \int_\alpha^x u_1(t)\,dt + \int_\alpha^x u_2(t)\,dt + \dots + \int_\alpha^x u_n(t)\,dt + \dots$$

This is the equation that had to be proved.

---

* In the estimates given below we take the + sign for $\alpha < x$ and the — sign for $x < \alpha$.

Note 1. If a series is not dominated, term-by-term integration of it is not always possible. This is to be understood in the sense that the integral $\int_{\alpha}^{x} s(t)\,dt$ of the sum of the series (1) is not always equal to the sum of the integrals of its terms [that is, to the sum of the series (4)].

Theorem 2. *If a series*

$$u_1(x) + u_2(x) + \ldots + u_n(x) + \ldots \tag{5}$$

made up of functions having continuous derivatives on the interval $[a, b]$ *converges (on this interval) to the sum* $s(x)$ *and the series*

$$u_1'(x) + u_2'(x) + \ldots + u_n'(x) + \ldots \tag{6}$$

made up of the derivatives of its terms is dominated on the same interval, then the sum of the series of derivatives is equal to the derivative of the sum of the original series; that is,

$$s'(x) = u_1'(x) + u_2'(x) + u_3'(x) + \ldots + u_n'(x) + \ldots$$

Proof. Denote by $F(x)$ the sum of the series (6):

$$F(x) = u_1'(x) + u_2'(x) + \ldots + u_n'(x) + \ldots$$

and prove that

$$F(x) = s'(x)$$

Since the series (6) is dominated, it follows, by the preceding theorem, that

$$\int_{\alpha}^{x} F(t)\,dt = \int_{\alpha}^{x} u_1'(t)\,dt + \int_{\alpha}^{x} u_2'(t)\,dt + \ldots + \int_{\alpha}^{x} u_n'(t)\,dt + \ldots$$

Performing the integration, we get

$$\int_{\alpha}^{x} F(t)\,dt = [u_1(x) - u_1(\alpha)]$$

$$+ [u_2(x) - u_2(\alpha)] + \ldots + [u_n(x) - u_n(\alpha)] + \ldots$$

But, by hypothesis,

$$s(x) = u_1(x) + u_2(x) + \ldots + u_n(x) + \ldots$$
$$s(\alpha) = u_1(\alpha) + u_2(\alpha) + \ldots + u_n(\alpha) + \ldots$$

no matter what the numbers x and α on the interval $[a, b]$. Therefore,

$$\int_{\alpha}^{x} F(t)\,dt = s(x) - s(\alpha)$$

Differentiating both sides of this equation with respect to x, we obtain

$$F(x) = s'(x)$$

We have thus proved that when the conditions of the theorem are fulfilled, the derivative of the sum of the series is equal to the sum of the derivatives of the terms of the series.

Note 2. The requirement of dominance (majorization) of a series of derivatives is extremely essential, and if not fulfilled it can make term-by-term differentiation of the series impossible. This is illustrated by a dominated series that does not admit term-by-term differentiation.

Consider the series

$$\frac{\sin 1^4 x}{1^2} + \frac{\sin 2^4 x}{2^2} + \frac{\sin 3^4 x}{3^2} + \ldots + \frac{\sin n^4 x}{n^2} + \ldots$$

This series converges to a continuous function because it is dominated. Indeed, for every x its terms are (in absolute value) less than the terms of the numerical convergent series with positive terms

$$\frac{1}{1} + \frac{1}{2^2} + \frac{1}{3^2} + \ldots + \frac{1}{n^2} + \ldots$$

Write a series composed of the derivatives of the terms of the original series:

$$\cos x + 2^2 \cos 2^4 x + \ldots + n^2 \cos n^4 x + \ldots$$

This series diverges. Thus, for instance, for $x = 0$ it turns into the series

$$1 + 2^2 + 3^2 + \ldots + n^2 + \ldots$$

(It may be shown that it diverges not only for $x = 0$.)

4.13 POWER SERIES. INTERVAL OF CONVERGENCE

Definition 1. *A power series* is a functional series of the form

$$a_0 + a_1 x + a_2 x^2 + \ldots + a_n x^n + \ldots \qquad (1)$$

where $a_0, a_1, a_2, \ldots, a_n, \ldots$ are constants called *coefficients of the series*.

The domain of convergence of a power series is always some interval, which, in a particular case, can degenerate into a point. To convince ourselves of this, let us first prove the following theorem, which is very important for the whole theory of power series.

Theorem 1 (Abel's theorem). (1) *If a power series converges for some nonzero value x_0, then it converges absolutely for any value of x, for which*

$$|x| < |x_0|$$

(2) *if a series diverges for some value x_0', then it diverges for every x for which*

$$|x| > |x_0'|$$

Proof. (1) Since, by assumption, the numerical series

$$a_0 + a_1x_0 + a_2x_0^2 + \dots + a_nx_0^n + \dots \tag{2}$$

converges, it follows that its common term $a_nx_0^n \to 0$ as $n \to \infty$, and this means that there exists a positive number M such that all the terms of the series are less than M in absolute value.

Rewrite the series (1) in the form

$$a_0 + a_1x_0\left(\frac{x}{x_0}\right) + a_2x_0^2\left(\frac{x}{x_0}\right)^2 + \dots + a_nx_0^n\left(\frac{x}{x_0}\right)^n + \dots \tag{3}$$

and consider a series of the absolute values of its terms:

$$|a_0| + |a_1x_0|\left|\frac{x}{x_0}\right| + |a_2x_0^2|\left|\frac{x}{x_0}\right|^2 + \dots + |a_nx_0^n|\left|\frac{x}{x_0}\right|^n + \dots \tag{4}$$

The terms of this series are less than the corresponding terms of the series

$$M + M\left|\frac{x}{x_0}\right| + M\left|\frac{x}{x_0}\right|^2 + \dots + M\left|\frac{x}{x_0}\right|^n + \dots \tag{5}$$

For $|x| < |x_0|$ the latter series is a geometric progression with ratio $\left|\frac{x}{x_0}\right| < 1$ and, consequently, converges. Since the terms of the series (4) are less than the corresponding terms of the series (5), the series (4) also converges, and this means that the series (3) or (1) converges absolutely.

(2) It is now easy to prove the second part of the theorem: let the series (1) diverge at some point x_0'. Then it will diverge at any point x that satisfies the condition $|x| > |x_0'|$. Indeed, if at some point x that satisfies this condition the series converged, then, by virtue of the first part (just proved) of the theorem, it should converge at the point x_0' as well, since $|x_0'| < |x|$. But this contradicts the condition that at the point x_0' the series diverges. Hence the series diverges at the point x as well. The theorem is thus completely proved.

Abel's theorem makes it possible to judge the position of the points of convergence and divergence of a power series. Indeed, if x_0 is a point of convergence, then the entire interval $(-|x_0|, |x_0|)$ is filled with points of absolute convergence. If x_0' is a point of divergence, then the whole infinite half-line to the right of the point $|x_0'|$ and the whole half-line to the left of the point $-|x_0'|$ consist of points of divergence.

From this it may be concluded that there exists a number R such that for $|x| < R$ we have points of absolute convergence and for $|x| > R$, points of divergence.

We thus have the following theorem on the structure of the domain of convergence of a power series:

Theorem 2. *The domain of convergence of a power series is an interval with centre at the coordinate origin.*

Definition 2. The *interval of convergence* of a power series is an interval from $-R$ to $+R$ such that for any point x lying inside this interval, the series converges and converges absolutely, while

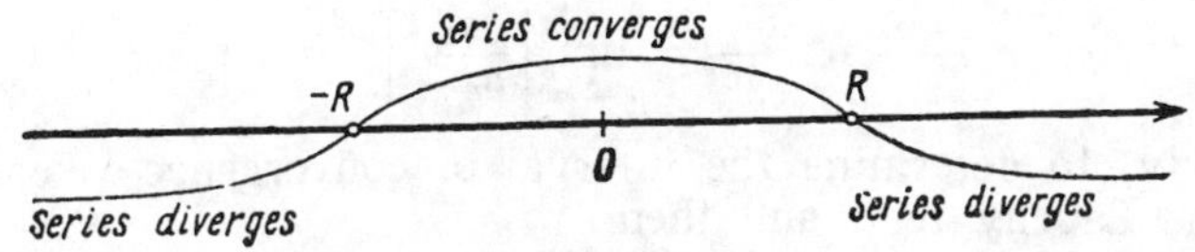

Fig. 116

for points x lying outside it, the series diverges (Fig. 116). The number R is called the *radius of convergence* of the power series.

At the end points of the interval (at $x=R$ and at $x=-R$) the question of the convergence or divergence of a given series is decided separately for each specific series.

We note that in some series the interval of convergence degenerates into a point ($R=0$), while in others it encompasses the entire x-axis ($R=\infty$).

We give a method for determining the radius of convergence of a power series.

Let there be a series

$$a_0+a_1x+a_2x^2+\ldots+a_nx^n+\ldots \tag{1}$$

Consider a series made up of the absolute values of its terms:

$$|a_0|+|a_1||x|+|a_2||x|^2+|a_3||x|^3 \\ +|a_4||x|^4+\ldots+|a_n||x|^n+\ldots \tag{6}$$

To determine the convergence of this series (with positive terms!), apply the d'Alembert test.

Let us assume that there exists a limit:

$$\lim_{n\to\infty}\frac{u_{n+1}}{u_n}=\lim_{n\to\infty}\left|\frac{a_{n+1}x^{n+1}}{a_nx^n}\right|=\lim_{n\to\infty}\left|\frac{a_{n+1}}{a_n}\right||x|=L|x|$$

Then, by the d'Alembert test, the series (6) converges, if $L|x|<1$; that is, if $|x|<\frac{1}{L}$, and diverges if $L|x|>1$, that is, if $|x|>\frac{1}{L}$.

Consequently, series (1) converges absolutely when $|x|<\frac{1}{L}$. But if $|x|>\frac{1}{L}$, then $\lim\limits_{n\to\infty}\frac{u_{n+1}}{u_n}=|x|L>1$ and series (6) diverges, and its general term does not tend to zero. * But then neither does the general term of the given power series (1) tend to zero, and this means that (on the basis of the necessary condition of convergence) this power series diverges $\left(\text{when } |x| > \frac{1}{L}\right)$.

From the foregoing it follows that the interval $\left(-\frac{1}{L}, \frac{1}{L}\right)$ is the interval of convergence of the power series (1), i.e.,

$$R = \frac{1}{L} = \lim_{n \to \infty} \left| \frac{a_n}{a_{n+1}} \right|$$

Similarly, to determine the interval of convergence we can make use of the Cauchy test, and then

$$R = \frac{1}{\lim\limits_{n \to \infty} \sqrt[n]{|a_n|}}$$

Example 1. Determine the interval of convergence of the series

$$1 + x + x^2 + x^3 + \ldots + x^n + \ldots$$

Solution. Applying d'Alembert's test directly, we get

$$\lim_{n \to \infty} \left| \frac{x^{n+1}}{x^n} \right| = |x|$$

Thus, the series converges when $|x| < 1$ and diverges when $|x| > 1$. At the extremities of the interval $(-1, 1)$ it is impossible to investigate the series by means of d'Alembert's test. However, it is immediately apparent that when $x = -1$ and when $x = 1$ the series diverges.

Example 2. Determine the interval of convergence of the series

$$\frac{2x}{1} - \frac{(2x)^2}{2} + \frac{(2x)^3}{3} - \ldots$$

Solution. We apply the d'Alembert test:

$$\lim_{n \to \infty} \left| \frac{\frac{(2x)^{n+1}}{n+1}}{\frac{(2x)^n}{n}} \right| = \lim_{n \to \infty} \left| \frac{n}{n+1} \right| |2x| = |2x|$$

The series converges if $|2x| < 1$, that is, if $|x| < \frac{1}{2}$; when $x = \frac{1}{2}$ the series converges; when $x = -\frac{1}{2}$ the series diverges.

Example 3. Determine the interval of convergence of the series

$$x + \frac{x^2}{2!} + \frac{x^3}{3!} + \ldots + \frac{x^n}{n!} + \ldots$$

---

* It will be recalled that in proving d'Alembert's test (see Sec. 4.4) we found that if $\lim \frac{u_{n+1}}{u_n} > 1$, then the general term of the series increases and, consequently, does not tend to zero.

Solution. Applying the d'Alembert test, we get

$$\lim_{n\to\infty}\left|\frac{u_{n+1}}{u_n}\right|=\lim_{n\to\infty}\left|\frac{x^{n+1}\,n!}{x^n\,(n+1)!}\right|=\lim_{n\to\infty}\left|\frac{x}{n+1}\right|=0<1$$

Since the limit is independent of x and is less than unity, the series converges for all values of x.

Example 4. The series $1+x+(2x)^2+(3x)^3+\ldots+(nx)^n+\ldots$ diverges for all values of x except $x=0$ because $(nx)^n\longrightarrow\infty$ as $n\longrightarrow\infty$ no matter what the x, as long as it is different from zero.

Theorem 3. *The power series*

$$a_0+a_1x+a_2x^2+\ldots+a_nx^n+\ldots \tag{1}$$

is dominated on any interval $[-\rho,\rho]$ *that lies completely inside the interval of convergence.*

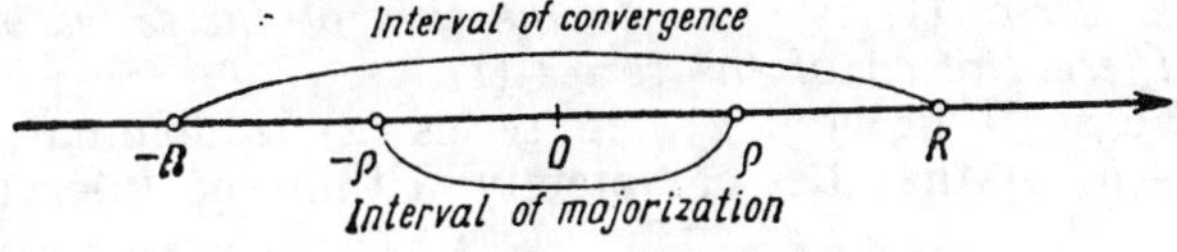

Fig. 117

Proof. It is given that $\rho<R$ (Fig. 117) and therefore the number series (with positive terms)

$$|a_0|+|a_1|\rho+|a_2|\rho^2+\ldots+|a_n|\rho^n \tag{7}$$

converges. But when $|x|<\rho$, the terms of the series (1) do not exceed, in absolute value, the corresponding terms of series (7). Hence, series (1) is dominated on the interval $[-\rho,\rho]$.

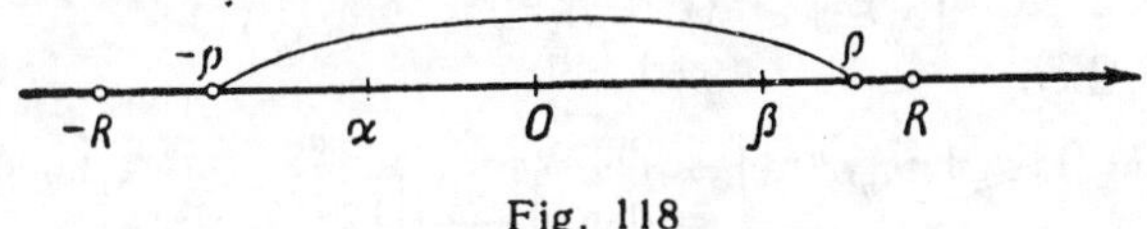

Fig. 118

Corollary 1. *On every interval lying entirely within the interval of convergence, the sum of a power series is a continuous function.* Indeed, the series on this interval is dominated, and its terms are continuous functions of x. Consequently, on the basis of Theorem 1, Sec. 4.11, the sum of this series is a continuous function.

Corollary 2. *If the limits of integration* α, β *lie within the interval of convergence of a power series, then the integral of the sum of the series is equal to the sum of the integrals of the terms of the series*, because the domain of integration may be taken in the interval $[-\rho,\rho]$, where the series is dominated (Fig. 118) (see Theorem 1, Sec. 4.12, on the possibility of term-by-term integration of a dominated series).

4.14 DIFFERENTIATION OF POWER SERIES

Theorem 1. *If a power series*

$$s(x) = a_0 + a_1 x + a_2 x^2 + a_3 x^3 + a_4 x^4 + \ldots + a_n x^n + \ldots \qquad (1)$$

has an interval of convergence $(-R, R)$, *then the series*

$$\varphi(x) = a_1 + 2a_2 x + 3a_3 x^2 + \ldots + n a_n x^{n-1} + \ldots \qquad (2)$$

obtained by termwise differentiation of the series (1) *has the same interval of convergence* $(-R, R)$; *here,*

$$\varphi(x) = s'(x) \quad \text{if } |x| < R$$

i.e., inside the interval of convergence the derivative of the sum of the power series (1) *is equal to the sum of the series obtained by termwise differentiation of the series* (1).

Proof. We shall prove that the series (2) is dominated on any interval $[-\rho, \rho]$ that lies completely within the interval of convergence.

Fig. 119

Take a point ξ such that $\rho < \xi < R$ (Fig. 119). The series (1) converges at this point, hence $\lim\limits_{n \to \infty} a_n \xi^n = 0$; it is therefore possible to indicate a constant number M such that

$$|a_n \xi^n| < M \quad (n = 1, 2, \ldots)$$

If $|x| \leqslant \rho$, then

$$|n a_n x^{n-1}| \leqslant |n a_n \rho^{n-1}| = n |a_n \xi^{n-1}| \left| \frac{\rho}{\xi} \right|^{n-1} < n \frac{M}{\xi} q^{n-1}$$

where

$$q = \frac{\rho}{\xi} < 1$$

Thus the terms of the series (2), when $|x| \leqslant \rho$, are less, in absolute value, than the terms of a positive number series with constant terms:

$$\frac{M}{\xi}(1 + 2q + 3q^2 + \ldots + nq^{n-1} + \ldots)$$

But the latter series converges, as will be evident if we apply the d'Alembert test:

$$\lim_{n \to \infty} \frac{nq^{n-1}}{(n-1) q^{n-2}} = q < 1$$

Hence, the series (2) is dominated on the interval $[-\rho, \rho]$, and by Theorem 2, Sec. 4.12, its sum is a derivative of the sum of the given series on the interval $[-\rho, \rho]$, i.e.,

$$\varphi(x) = s'(x)$$

Since every interior point of the interval $(-R, R)$ may be included in some interval $[-\rho, \rho]$, it follows that the series (2) converges at every interior point of the interval $(-R, R)$.

We shall prove that outside the interval $(-R, R)$ the series (2) diverges. Assume that the series (2) converges when $x_1 > R$. Integrating it termwise in the interval $(0, x_2)$, where $R < x_2 < x_1$, we would find that the series (1) converges at the point x_2, but this contradicts the hypotheses of the theorem. Thus, the interval $(-R, R)$ is the interval of convergence of series (2). And the theorem is proved completely.

Series (2) may again be differentiated term by term, and this may be continued as many times as one pleases. We thus have the conclusion:

Theorem 2. *If a power series converges in an interval $(-R, R)$, its sum is a function which has, inside the interval of convergence, derivatives of any order, each of which is the sum of a series resulting from term-by-term differentiation of the given series an appropriate number of times; here, the interval of convergence of each series obtained by differentiation is the same interval $(-R, R)$.*

4.15 SERIES IN POWERS OF $x-a$

Also called a *power series* is a functional series of the form

$$a_0 + a_1(x-a) + a_2(x-a)^2 + \ldots + a_n(x-a)^n + \ldots \qquad (1)$$

where the constants $a_0, a_1, \ldots, a_n, \ldots$ are likewise termed *coefficients of the series*. This is a power series arranged in powers of the binomial $x-a$.

When $a=0$, we have a power series in powers of x, which, consequently, is a special case of series (1).

To determine the region of convergence of series (1), substitute

$$x-a=X$$

Series (1) then takes on the form

$$a_0 + a_1X + a_2X^2 + \ldots + a_nX^n + \ldots \qquad (2)$$

we thus have a power series in powers of X.

Let the interval $-R < X < R$ be the interval of convergence of the series (2) (Fig. 120, *a*). It thus follows that series (1) will converge for values of x that satisfy the inequality $-R < x-a < R$ or $a-R < x < a+R$. Since series (2) diverges for $|X| > R$, the series (1) will diverge for $|x-a|>R$, that is, it will diverge outside the interval $a-R<x<a+R$ (Fig. 120, *b*).

And so the interval $(a-R, a+R)$ with centre at the point a will be the interval of convergence of series (1). All the properties of a series in powers of x inside the interval of convergence $(-R, +R)$ are retained completely for a series in powers of $x-a$ inside the interval of convergence $(a-R, a+R)$. For example,

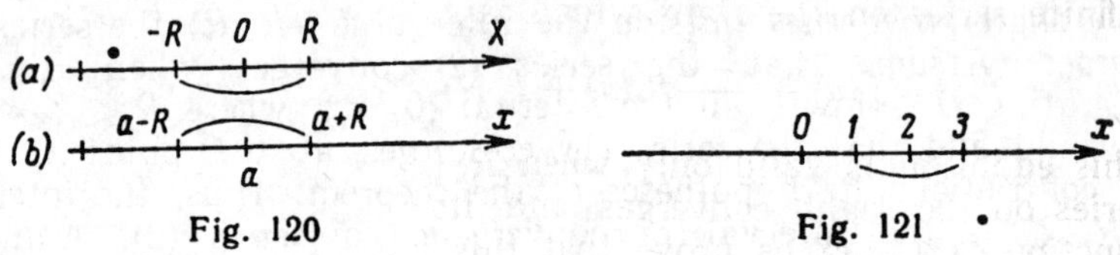

Fig. 120 Fig. 121

after term-by-term integration of the power series (1), if the limits of integration lie within the interval of convergence $(a-R, a+R)$, we get a series whose sum is equal to the corresponding integral of the sum of the given series (1). In the case of termwise differentiation of the power series (1), for all x lying inside the interval of convergence $(a-R, a+R)$ we obtain a series whose sum is equal to the derivative of the sum of the given series (1).

Example. Find the region of convergence of the series

$$(x-2)+(x-2)^2+(x-2)^3+\ldots+(x-2)^n+\ldots$$

Solution. Putting $x-2=X$, we get the series

$$X+X^2+X^3+\ldots+X^n+\ldots$$

This series converges when $-1<X<+1$. Hence, the given series converges for all x that satisfy the inequality $-1<x-2<1$, that is, when $1<x<3$ (Fig. 121).

4.16 TAYLOR'S SERIES AND MACLAURIN'S SERIES

In Sec. 4.6, Vol. I, it was shown that for a function $f(x)$ that has all derivatives up to the $(n+1)$th order inclusive, Taylor's formula holds in the neighbourhood of the point $x=a$ (that is, in some interval containing the point $x=a$):

$$f(x)=f(a)+\frac{x-a}{1}f'(a)$$
$$+\frac{(x-a)^2}{1\cdot 2}f''(a)+\ldots+\frac{(x-a)^n}{n!}f^{(n)}(a)+R_n(x) \qquad (1)$$

where the so-called remainder term $R_n(x)$ is computed from the formula

$$R_n=\frac{(x-a)^{n+1}}{(n+1)!}f^{(n+1)}[a+\theta(x-a)], \quad 0<\theta<1$$

If the function $f(x)$ has derivatives of **all** orders in the neighbourhood of the point $x=a$, then in Taylor's formula the number n may be taken as large as we please. Suppose that in the neighbourhood under consideration the remainder term R_n tends to zero as $n \longrightarrow \infty$:

$$\lim_{n\to\infty} R_n(x)=0$$

Then, passing to the limit in formula (1) as $n \longrightarrow \infty$, we get an infinite series on the right which is called the *Taylor series*:

$$f(x)=f(a)+\frac{x-a}{1}f'(a)+\ldots+\frac{(x-a)^n}{n!}f^{(n)}(a)+\ldots \qquad (2)$$

This equation is valid only when $R_n(x) \longrightarrow 0$ as $n \longrightarrow \infty$. Then the series on the right converges and its sum is equal to the given function $f(x)$. Let us prove that this is indeed the case:

$$f(x)=P_n(x)+R_n(x)$$

where

$$P_n(x)=f(a)+\frac{x-a}{1!}f'(a)+\ldots+\frac{(x-a)^n}{n!}f^{(n)}(a)$$

Since it is given that $\lim\limits_{n\to\infty} R_n(x)=0$, we have

$$f(x)=\lim_{n\to\infty} P_n(x)$$

But $P_n(x)$ is the nth partial sum of the series (2); its limit is equal to the sum of the series on the right side of (2). Hence, (2) is true:

$$f(x)=f(a)+\frac{x-a}{1}f'(a)+\frac{(x-a^2)}{2}f''(a)+\ldots+\frac{(x-a)^n}{n!}f^{(n)}(a)+\ldots$$

From the foregoing it follows that the *Taylor series is a given function* $f(x)$ *only when* $\lim R_n(x)=0$. If $\lim R_n(x)\neq 0$, then the series is not the given function, although it mav converge (to a different function).

If in the Taylor series we put $a=0$, we get a special case of this series known as *Maclaurin's series*:

$$f(x)=f(0)+\frac{x}{1}f'(0)+\frac{x^2}{2!}f''(0)+\ldots+\frac{x^n}{n!}f^{(n)}(0)+\ldots \qquad (3)$$

If for some function we have a formally written Taylor's series, then in order to prove that this series is a given function it is either necessary to prove that the remainder term approaches zero, or to be convinced in some way that this series converges to the given function.

We note that for each of the elementary functions defined in Sec. 1.8, Vol. I, there exists an a and an R such that in the interval $(a-R,\ a+R)$ it may be expanded into a Taylor's series or (if $a=0$) into a Maclaurin's series.

4.17 SERIES EXPANSION OF FUNCTIONS

1. Expanding the function $f(x) = \sin x$ in a Maclaurin's series. In Sec. 4.7, Vol. I, we obtained the formula

$$\sin x = x - \frac{x^3}{3!} + \frac{x^5}{5!} + \ldots + (-1)^{n+1}\frac{x^{2n-1}}{(2n-1)!} + R_{2n}(x)$$

Since it has been proved that $\lim\limits_{n\to\infty} R_{2n}(x) = 0$, it follows, by what was said in the preceding section, that we get an expansion of $\sin x$ in a Maclaurin's series:

$$\sin x = x - \frac{x^3}{3!} + \frac{x^5}{5!} + \ldots + (-1)^{n+1}\frac{x^{2n-1}}{(2n-1)!} + \ldots \qquad (1)$$

Since the remainder term approaches zero for any x, the given series converges and, for its sum, has the function $\sin x$ for any x.

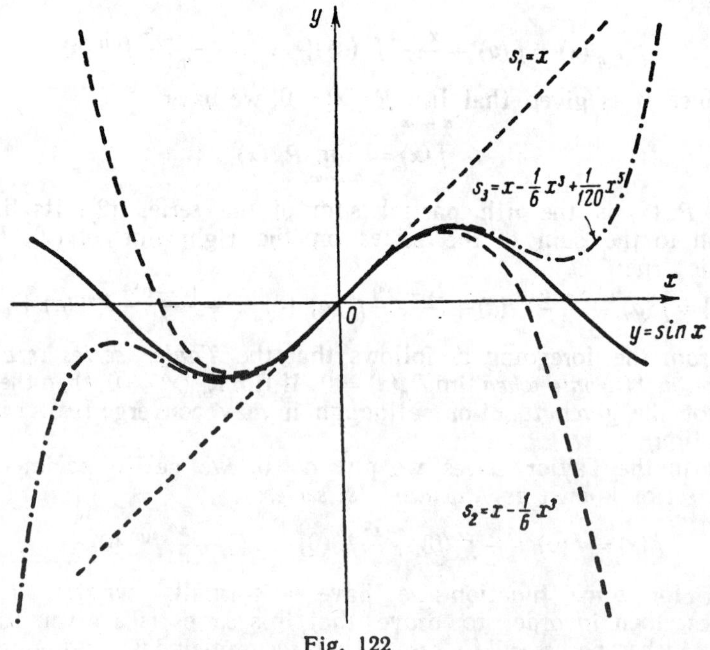

Fig. 122

Fig. 122 shows the graphs of the function $\sin x$ and of the first three partial sums of the series (1).

This series is used to compute the values of $\sin x$ for different values of x.

To illustrate, let us compute $\sin 10°$ to the fifth decimal place.

Since $10^\circ = \frac{\pi}{18}$ (radians) ≈ 0.174533, we have

$$\sin 10^\circ = \frac{\pi}{18} - \frac{1}{3!}\left(\frac{\pi}{18}\right)^3 + \frac{1}{5!}\left(\frac{\pi}{18}\right)^5 - \frac{1}{7!}\left(\frac{\pi}{18}\right)^7 + \dots$$

Confining ourselves to the first two terms, we get the following approximate equation:

$$\sin \frac{\pi}{18} \approx \frac{\pi}{18} - \frac{1}{6}\left(\frac{\pi}{18}\right)^3$$

here, we are in error by δ, which in absolute value is less than the first of the suppressed terms; that is,

$$\delta < \frac{1}{5!}\left(\frac{\pi}{18}\right)^5 < \frac{1}{120}(0.2)^5 < 4 \cdot 10^{-6}$$

If each term in the expression for $\sin \frac{\pi}{18}$ is computed to six decimal places, we get

$$\sin \frac{\pi}{18} = 0.173647$$

We can be sure of the first four decimals.

2. Expanding the function $f(x) = e^x$ in a Maclaurin's series. On the basis of Sec. 4.7, Vol. I, we have

$$e^x = 1 + x + \frac{x^2}{2!} + \frac{x^3}{3!} + \dots + \frac{x^n}{n!} + \dots \tag{2}$$

since it was proved that $\lim_{n \to \infty} R_n(x) = 0$ for any x. Hence, the series converges for all values of x and is the function e^x.

Substituting $x = -x$, in (2), we get

$$e^{-x} = 1 - x + \frac{x^2}{2!} - \frac{x^3}{3!} + \frac{x^4}{4!} - \dots \tag{3}$$

3. Expanding the function $f(x) = \cos x$ in a Maclaurin's series. From Sec. 4.7, Vol. I, we have

$$\cos x = 1 - \frac{x^2}{2!} + \frac{x^4}{4!} - \frac{x^6}{6!} + \dots \tag{4}$$

for all values of x the series converges and represents the function $\cos x$.

4. Expansion, in Maclaurin's series, of the functions

$$f(x) = \sinh x = \frac{e^x - e^{-x}}{2}$$

$$f(x) = \cosh x = \frac{e^x + e^{-x}}{2}$$

These functions are readily expanded by subtracting and adding the series (2) and (3) and dividing by 2.

Thus,

$$\sinh x = x + \frac{x^3}{3!} + \frac{x^5}{5!} + \frac{x^7}{7!} + \ldots \quad (5)$$

$$\cosh x = 1 + \frac{x^2}{2!} + \frac{x^4}{4!} + \frac{x^6}{6!} + \ldots \quad (6)$$

4.18 EULER'S FORMULA

Up till now we have considered only series with real terms and have not dealt with series with complex terms. We shall not give the complete theory of series with complex terms, for this goes beyond the scope of this text. We shall consider only one important example in this field.

In Chapter 7, Vol. I we defined the function e^{x+iy} by the equation

$$e^{x+iy} = e^x (\cos y + i \sin y)$$

When $x = 0$, we get *Euler's formula*:

$$e^{iy} = \cos y + i \sin y$$

If we determine the exponential function e^{iy} with imaginary exponent by means of formula (2), Sec. 4.17, which represents the function e^x in the form of a power series, we will get the very same Euler equation. Indeed, determine e^{iy} by putting the expression iy in place of x in equation (2), Sec. 4.17:

$$e^{iy} = 1 + \frac{iy}{1!} + \frac{(iy)^2}{2!} + \frac{(iy)^3}{3!} + \ldots + \frac{(iy)^n}{n!} + \ldots \quad (1)$$

Taking into account that $i^2 = -1$, $i^3 = -i$, $i^4 = 1$, $i^5 = i$, $i^6 = -1$, and so forth, we transform (1) to the form

$$e^{iy} = 1 + \frac{iy}{1!} - \frac{y^2}{2!} - \frac{iy^3}{3!} + \frac{y^4}{4!} + \frac{iy^5}{5!} - \ldots$$

Separating in this series the reals from the imaginaries, we find

$$e^{iy} = \left(1 - \frac{y^2}{2!} + \frac{y^4}{4!} - \ldots\right) + i\left(\frac{y}{1!} - \frac{y^3}{3!} + \frac{y^5}{5!} - \ldots\right)$$

The parentheses contain power series whose sums are equal to $\cos y$ and $\sin y$, respectively [see formulas (4) and (1) of the preceding section]. Consequently,

$$e^{iy} = \cos y + i \sin y \quad (2)$$

Thus, we have again arrived at *Euler's formula*.

4.19 THE BINOMIAL SERIES

1. Let us expand the following function in a Maclaurin's series:

$$f(x) = (1+x)^m$$

where m is an arbitrary constant number.

Here the evaluation of the remainder term presents certain difficulties and so we shall approach the series expansion of this function somewhat differently.

Noting that the function $f(x) = (1+x)^m$ satisfies the differential equation

$$(1+x) f'(x) = mf(x) \tag{1}$$

and the condition

$$f(0) = 1$$

we find a power series whose sum $s(x)$ satisfies equation (1) and the condition $s(0) = 1$:

$$s(x) = 1 + a_1x + a_2x^2 + \ldots + a_nx^n + \ldots \tag{2}$$

Putting this series into equation (1), we get

$$(1+x)(a_1 + 2a_2x + 3a_3x^2 + \ldots + na_nx^{n-1} + \ldots)$$
$$= m(1 + a_1x + a_2x^2 + \ldots + a_nx^n + \ldots)$$

Equating the coefficients of identical powers of x in different parts of the equation, we find

$$a_1 = m,\ a_1 + 2a_2 = ma_1,\ \ldots,\ na_n + (n+1)a_{n+1} = ma_n,\ \ldots$$

Whence for the coefficients of the series we get the expressions

$$a_0 = 1,\ a_1 = m,\ a_2 = \frac{a_1(m-1)}{2} = \frac{m(m-1)}{2},$$
$$a_3 = \frac{a_2(m-2)}{3} = \frac{m(m-1)(m-2)}{2\cdot 3},\ \ldots$$
$$a_n = \frac{m(m-1)\ldots[m-n+1]}{1\cdot 2\ldots n},\ \ldots$$

These are binomial coefficients.

Putting them into formula (2), we obtain

$$s(x) = 1 + mx + \frac{m(m-1)}{1\cdot 2}x^2 + \ldots$$
$$+ \frac{m(m-1)\ldots[m-(n-1)]}{1\cdot 2\ldots n}x^n + \ldots \tag{3}$$

---

* We took the absolute term equal to unity by virtue of the initial condition $s(0) = 1$.

If m is a positive integer, then beginning with the term containing x^{m+1} all coefficients are equal to zero, and the series is converted into a polynomial. For m fractional or a negative integer, we have an infinite series.

Let us determine the radius of convergence of series (3):

$$u_{n+1}=\frac{m(m-1)\ldots[m-n+1]}{n!}x^n, \qquad u_n=\frac{m(m-1)\ldots[m-n+2]}{(n-1)!}x^{n-1}$$

$$\lim_{n\to\infty}\left|\frac{u_{n+1}}{u_n}\right|=\lim_{n\to\infty}\left|\frac{m(m-1)\ldots(m-n+1)(n-1)!}{m(m-1)\ldots(m-n+2)\,n!}x\right|$$

$$=\lim_{n\to\infty}\left|\frac{m-n+1}{n}\right||x|=|x|$$

Thus, series (3) converges for $|x|<1$.

In the interval $(-1, 1)$, series (3) is a function $s(x)$ that satisfies the differential equation (1) and the condition

$$s(0)=1$$

Since the differential equation (1) and the condition $s(0)=1$ are satisfied by a unique function, it follows that the sum of the series (3) is identically equal to the function $(1+x)^m$, and we obtain the expansion

$$(1+x)^m=1+mx+\frac{m(m-1)}{1\cdot 2}x^2+\frac{m(m-1)(m-2)}{1\cdot 2\cdot 3}x^3+\ldots \qquad (3')$$

For the particular case $m=-1$, we have

$$\frac{1}{1+x}=1-x+x^2-x^3+\ldots \qquad (4)$$

For $m=\frac{1}{2}$ we get

$$\sqrt{1+x}=1+\frac{1}{2}x-\frac{1}{2\cdot 4}x^2+\frac{1\cdot 3}{2\cdot 4\cdot 6}x^3-\frac{1\cdot 3\cdot 5}{2\cdot 4\cdot 6\cdot 8}x^4+\ldots \qquad (5)$$

For $m=-\frac{1}{2}$ we have

$$\frac{1}{\sqrt{1+x}}=1-\frac{1}{2}x+\frac{1\cdot 3}{2\cdot 4}x^2-\frac{1\cdot 3\cdot 5}{2\cdot 4\cdot 6}x^3+\frac{1\cdot 3\cdot 5\cdot 7}{2\cdot 4\cdot 6\cdot 8}x^4-\ldots \qquad (6)$$

2. Let us apply the binomial expansion to the expansion of functions. Expand the function

$$f(x)=\arcsin x$$

in a Maclaurin's series. Putting into equation (6) the expression $-x^2$ in place of x, we get

$$\frac{1}{\sqrt{1-x^2}}=1+\frac{1}{2}x^2+\frac{1\cdot 3}{2\cdot 4}x^4+\frac{1\cdot 3\cdot 5}{2\cdot 4\cdot 6}x^6+\ldots+\frac{1\cdot 3\cdot 5\ldots(2n-1)}{2\cdot 4\cdot 6\ldots 2n}x^{2n}+\ldots$$

By the theorem on integrating power series we have, for $|x| < 1$:

$$\int_0^x \frac{dt}{\sqrt{1-t^2}} = \arcsin x = x + \frac{1}{2}\frac{x^3}{3} + \frac{1\cdot 3}{2\cdot 4}\frac{x^5}{5} + \frac{1\cdot 3\cdot 5}{2\cdot 4\cdot 6}\frac{x^7}{7} + \ldots$$

$$+ \frac{1\cdot 3\cdot 5\ldots(2n-1)}{2\cdot 4\cdot 6\ldots 2n}\frac{x^{2n+1}}{2n+1} + \ldots$$

This series converges in the interval $(-1, 1)$. One could prove that the series converges for $x = \pm 1$ as well as that for these values the sum of the series is likewise equal to $\arcsin x$. Then, setting $x = 1$, we get a formula for computing π:

$$\arcsin 1 = \frac{\pi}{2} = 1 + \frac{1}{2}\cdot\frac{1}{3} + \frac{1\cdot 3}{2\cdot 4}\cdot\frac{1}{5} + \frac{1\cdot 3\cdot 5}{2\cdot 4\cdot 6}\cdot\frac{1}{7} + \ldots$$

4.20 EXPANSION OF THE FUNCTION ln (1 + x) IN A POWER SERIES. COMPUTING LOGARITHMS

Integrating equation (4), Sec. 4.19, from 0 to x (when $|x| < 1$), we obtain

$$\int_0^x \frac{dt}{1+t} = \int_0^x (1 - t + t^2 - t^3 + \ldots)\, dt$$

or

$$\ln(1+x) = x - \frac{x^2}{2} + \frac{x^3}{3} - \frac{x^4}{4} + \ldots + (-1)^{n+1}\frac{x^n}{n} + \ldots \qquad (1)$$

This equation holds true in the interval $(-1, 1)$.

If in this formula x is replaced by $-x$, then we get the series

$$\ln(1-x) = -x - \frac{x^2}{2} - \frac{x^3}{3} - \frac{x^4}{4} - \ldots \qquad (2)$$

which converges in the interval $(-1, 1)$.

Using the series (1) and (2) we can compute the logarithms of numbers lying between zero and two. We note, without proof, that for $x = 1$ the expansion (1) also holds true.

We will now derive a formula for computing the natural logarithms of all integers.

Since in the term-by-term subtraction of two convergent series we get a convergent series (see Sec. 4.1, Theorem 3), then by subtracting equation (2) from equation (1) term by term, we find

$$\ln(1+x) - \ln(1-x) = \ln\frac{1+x}{1-x} = 2\left[x + \frac{x^3}{3} + \frac{x^5}{5} + \ldots\right]$$

Now put $\frac{1+x}{1-x} = \frac{n+1}{n}$; then $x = \frac{1}{2n+1}$. For any $n > 0$ we have

$0<x<1$; therefore

$$\ln\frac{1+x}{1-x}=\ln\frac{n+1}{n}=2\left[\frac{1}{2n+1}+\frac{1}{3(2n+1)^3}+\frac{1}{5(2n+1)^5}+\dots\right]$$

whence

$$\ln(n+1)-\ln n=2\left[\frac{1}{2n+1}+\frac{1}{3(2n+1)^3}+\frac{1}{5(2n+1)^5}+\dots\right] \quad (3)$$

For $n=1$ we then obtain

$$\ln 2=2\left[\frac{1}{1\cdot 3}+\frac{1}{3\cdot 3^3}+\frac{1}{5\cdot 3^5}+\dots\right]$$

To compute $\ln 2$ to a given degree of accuracy δ, one has to compute the partial sum s_p, choosing the number p of its terms such that the sum of the suppressed terms (that is, the error R_p committed when replacing s by s_p) is less than the admissible error δ. To do this, let us evaluate the error R_p:

$$R_p=2\left[\frac{1}{(2p+1)\,3^{2p+1}}+\frac{1}{(2p+3)\,3^{2p+3}}+\frac{1}{(2p+5)\,3^{2p+5}}+\dots\right]$$

Since the numbers $2p+3$, $2p+5$, ... are greater than $2p+1$, it follows that by replacing them by $2p+1$ we increase each fraction. Therefore,

$$R_p<2\left[\frac{1}{(2p+1)\,3^{2p+1}}+\frac{1}{(2p+1)\,3^{2p+3}}+\frac{1}{(2p+1)\,3^{2p+5}}+\dots\right]$$

or

$$R_p<\frac{1}{2p+1}\left[\frac{1}{3^{2p+1}}+\frac{1}{3^{2p+3}}+\frac{1}{3^{2p+5}}+\dots\right]$$

The series in the brackets is a geometric progression with ratio $\frac{1}{9}$. Computing the sum of this progression we find

$$R_p<\frac{2}{2p+1}\,\frac{\frac{1}{3^{2p+1}}}{1-\frac{1}{9}}=\frac{1}{(2p+1)\,3^{2p-1}4} \quad (4)$$

If we now want to compute $\ln 2$ to, for example, nine decimal places, we must choose p such that $R_p<0.000000001$. This can be done by selecting p so that the right side of inequality (4) is less than 0.000000001. By direct choice we find that it is sufficient to take $p=8$. To nine-place accuracy we have

$$\ln 2\approx s_8=2\left[\frac{1}{1\cdot 3}+\frac{1}{3\cdot 3^3}+\frac{1}{5\cdot 3^5}+\frac{1}{7\cdot 3^7}+\frac{1}{9\cdot 3^9}+\frac{1}{11\cdot 3^{11}}\right.$$
$$\left.+\frac{1}{13\cdot 3^{13}}+\frac{1}{15\cdot 3^{15}}\right]=0.693147180$$

Thus, $\ln 2=0.693147180$, correct to nine places.

Assuming $n=2$ in formula (3), we obtain

$$\ln 3 = \ln 2 + 2\left[\frac{1}{5}+\frac{1}{3\cdot 5^3}+\frac{1}{5\cdot 5^5}+\cdots\right] = 1.098612288, \text{ and so forth.}$$

In this way we obtain the **natural** logarithms of any integer.

To get the **common** logarithms of numbers, use the following relation (see Sec. 2.8, Vol. I):

$$\log N = M \ln N$$

where $M=0.434294$. Then, for example, we get

$$\log 2 = 0.434294 \times 0.693147 = 0.30103$$

4.21 SERIES EVALUATION OF DEFINITE INTEGRALS

In Chapters 10 and 11 (Vol. I) it was noted that there exist definite integrals, which, as functions of the upper limit, are not expressible in terms of elementary functions in closed form. It is sometimes convenient to compute such integrals by means of series.

Let us consider several examples.

1. Let it be required to compute the integral

$$\int_0^a e^{-x^2}\,dx$$

Here, the antiderivative of e^{-x^2} is not an elementary function. To evaluate this integral we expand the integrand in a series, replacing x by $-x^2$ in the expansion of e^x [see formula (2), Sec. 4.17]:

$$e^{-x^2} = 1-\frac{x^2}{1!}-\frac{x^4}{2!}-\frac{x^6}{3!}+\cdots+(-1)^n\frac{x^{2n}}{n!}+\cdots$$

Integrating both sides of this equation from 0 to a, we obtain

$$\int_0^a e^{-x^2}\,dx = \left(\frac{x}{1}-\frac{x^3}{1\cdot 3}+\frac{x^5}{2!\cdot 5}-\frac{x^7}{3!\cdot 7}+\cdots\right)\Big|_0^a$$

$$=\frac{a}{1}-\frac{a^3}{1!\cdot 3}+\frac{a^5}{2!\cdot 5}-\frac{a^7}{3!\cdot 7}+\cdots$$

Using this equation, we can calculate the given integral to any degree of accuracy for any a.

2. It is required to evaluate the integral

$$\int_0^a \frac{\sin x}{x}\,dx$$

Expand the integrand in a series: from the equation

$$\sin x = x-\frac{x^3}{3!}+\frac{x^5}{5!}-\frac{x^7}{7!}+\cdots$$

we get

$$\frac{\sin x}{x} = 1 - \frac{x^2}{3!} + \frac{x^4}{5!} - \frac{x^6}{7!} + \dots$$

the latter series converges for all values of x. Integrating term by term, we obtain

$$\int_0^a \frac{\sin x}{x}\, dx = a - \frac{a^3}{3!\cdot 3} + \frac{a^5}{5!\cdot 5} - \frac{a^7}{7!\cdot 7} + \dots$$

The sum of the series is readily computed to any degree of accuracy for any a.

3. Evaluate the elliptic integral

$$\int_0^{\frac{\pi}{2}} \sqrt{1 - k^2 \sin^2 \varphi}\, d\varphi \quad (k < 1)$$

Expand the integrand in a binomial series, putting $m = \frac{1}{2}$, $x = -k^2 \sin^2 \varphi$ [see formula (5), Sec. 4.19]:

$$\sqrt{1 - k^2 \sin^2 \varphi} = 1 - \frac{1}{2} k^2 \sin^2 \varphi - \frac{1}{2}\frac{1}{4} k^4 \sin^4 \varphi - \frac{1}{2}\frac{1}{4}\frac{3}{6} k^6 \sin^6 \varphi - \dots$$

This series converges for all values of φ and admits term-by-term integration because it is dominated on any interval. Therefore,

$$\int_0^{\varphi} \sqrt{1 - k^2 \sin^2 t}\, dt = \varphi - \frac{1}{2} k^2 \int_0^{\varphi} \sin^2 t\, dt - \frac{1}{2}\frac{1}{4} k^4 \int_0^{\varphi} \sin^4 t\, dt - \frac{1}{2}\frac{1}{4}\frac{3}{6} k^6 \int_0^{\varphi} \sin^6 t\, dt - \dots$$

The integrals on the right are computed in elementary fashion. For $\varphi = \frac{\pi}{2}$ we have

$$\int_0^{\frac{\pi}{2}} \sin^{2n} \varphi\, d\varphi = \frac{1\cdot 3 \dots (2n-1)}{2\cdot 4 \dots 2n} \frac{\pi}{2}$$

(see Sec. 11.6, Vol. I) and, hence,

$$\int_0^{\frac{\pi}{2}} \sqrt{1 - k^2 \sin^2 \varphi}\, d\varphi = \frac{\pi}{2}\left[1 - \left(\frac{1}{2}\right)^2 k^2 - \left(\frac{1\cdot 3}{2\cdot 4}\right)^2 \frac{k^4}{3} - \left(\frac{1\cdot 3\cdot 5}{2\cdot 4\cdot 6}\right)^2 \frac{k^6}{5} - \dots\right]$$

4.22 INTEGRATING DIFFERENTIAL EQUATIONS BY MEANS OF SERIES

If the integration of a differential equation does not reduce to quadratures, one resorts to approximate methods of integrating the equation. One of these methods is representing the equation as a Taylor's series; the sum of a finite number of terms of this series will be approximately equal to the desired particular solution.

To take an example, let it be required to find the solution of a second-order differential equation,

$$y'' = F(x, y, y') \tag{1}$$

that satisfies the initial conditions

$$(y)_{x=x_0} = y_0, \quad (y')_{x=x_0} = y'_0 \tag{2}$$

Suppose that the solution $y = f(x)$ exists and may be given in the form of a Taylor's series (we will not discuss the conditions under which this occurs):

$$y = f(x) = f(x_0) + \frac{x - x_0}{1} f'(x_0) + \frac{(x - x_0)^2}{1 \cdot 2} f''(x_0) + \ldots \tag{3}$$

We have to find $f(x_0)$, $f'(x_0)$, $f''(x_0)$, ..., i.e., the values of the derivatives of the particular solution when $x = x_0$. But this can be done by means of equation (1) and conditions (2).

Indeed, from conditions (2) it follows that

$$f(x_0) = y_0, \quad f'(x_0) = y'_0$$

from equation (1) we have

$$f''(x_0) = (y'')_{x=x_0} = F(x, y_0, y'_0)$$

Differentiating both sides of (1) with respect to x, we get

$$y''' = F'_x(x, y, y') + F'_y(x, y, y')\, y' + F'_{y'}(x, y, y')\, y'' \tag{4}$$

and substituting the value $x = x_0$ into the right side, we find

$$f'''(x_0) = (y''')_{x=x_0}$$

Differentiating the relationship (4) once again, we find

$$f^{\mathrm{IV}}(x_0) = (y^{\mathrm{IV}})_{x=x_0}$$

and so on.

We put these values of the derivatives into (3). For those values of x for which this series converges, this series represents the solution of the equation.

Example 1. Find the solution of the equation

$$y'' = -yx^2$$

which satisfies the initial conditions

$$(y)_{x=0} = 1, \quad (y')_{x=0} = 0$$

Solution. We have

$$f(0)=y_0=1,\quad f'(0)=y_0'=0$$

From the given equation we find $(y'')_{x=0}=f''(0)=0$; further,

$$y'''=-y'x^2-2xy,\qquad (y''')_{x=0}=f'''(0)=0,$$
$$y^{IV}=-y''x^2-4xy'-2y,\quad (y^{IV})_{x=0}=-2$$

and, generally, differentiating k times both sides of the equation by the Leibniz formula, we find (Sec. 3.22, Vol. I)

$$y^{(k+2)}=-y^{(k)}x^2-2kxy^{(k-1)}-k(k-1)\,y^{(k-2)}$$

Putting $x=0$, we have

$$y_0^{(k+2)}=-k(k-1)\,y_0^{(k-2)}$$

or, setting $k+2=n$,

$$y_0^{(n)}=-(n-3)(n-2)\,y_0^{(n-4)}$$

Whence

$$y_0^{IV}=-1\cdot 2,\ y_0^{(8)}=-5\cdot 6y_0^{IV}=(-1)^2(1\cdot 2)(5\cdot 6)$$
$$y_0^{(12)}=-9\cdot 10y_0^{(8)}=(-1)^3(1\cdot 2)(5\cdot 6)(9\cdot 10)$$
$$\cdots\cdots\cdots\cdots\cdots$$
$$y_0^{(4k)}=(-1)^k(1\cdot 2)(5\cdot 6)(9\cdot 10)\ldots[(4k-3)(4k-2)]$$
$$\cdots\cdots\cdots\cdots\cdots$$

In addition,

$$y_0^{(5)}=0,\quad y_0^{(9)}=0,\quad \ldots,\quad y_0^{(4k+1)}=0,\ \ldots,$$
$$y_0^{(6)}=0,\quad y_0^{(10)}=0,\quad \ldots,\quad y_0^{(4k+2)}=0,\ \ldots,$$
$$y_0^{(7)}=0,\quad y_0^{(11)}=0,\quad \ldots,\quad y_0^{(4k+3)}=0,\ \ldots$$

Thus, only those derivatives whose order is a multiple of four do not become zero.

Putting the values of the derivatives that we have found into a Maclaurin series, we get the solution of the equation

$$y=1-\frac{x^4}{4!}1\cdot 2+\frac{x^8}{8!}(1\cdot 2)(5\cdot 6)-\frac{x^{12}}{12!}(1\cdot 2)(5\cdot 6)(9\cdot 10)+\ldots$$
$$+(-1)^k\frac{x^{4k}}{(4k)!}(1\cdot 2)(5\cdot 6)\ldots[(4k-3)(4k-2)]+\ldots$$

By means of d'Alembert's test we can verify that this series converges for all values of x; hence, it is the solution of the equation.

If the equation is linear, it is more convenient to seek the coefficients of expansion of the particular solution by the method of undetermined coefficients. To do this, we put the series

$$y=a_0+a_1x+a_2x^2+\ldots+a_nx^n+\ldots$$

into the differential equation and equate the coefficients of identical powers of x on different sides of the equation.

Example 2. Find the solution of the equation

$$y''=2xy'+4y$$

that satisfies the initial conditions

$$(y)_{x=0}=0, \ (y')_{x=0}=1$$

Solution. We set

$$y=a_0+a_1x+a_2x^2+a_3x^3+\ldots+a_nx^n+\ldots$$

On the basis of the initial conditions we find

$$a_0=0, \ a_1=1$$

Hence,

$$y=x+a_2x^2+a_3x^3+\ldots+a_nx^n+\ldots$$
$$y'=1+2a_2x+3a_3x^2+\ldots+na_nx^{n-1}+\ldots$$
$$y''=2a_2+3\cdot 2a_3x+\ldots+n(n-1)a_nx^{n-2}+\ldots$$

Putting these expressions into the given equation and equating the coefficients of identical powers of x, we obtain

$$2a_2=0, \qquad \text{whence } a_2=0$$
$$3\cdot 2a_3=2+4, \qquad \text{whence } a_3=1$$
$$4\cdot 3a_4=4a_2+4a_2, \qquad \text{whence } a_4=0$$
$$\ldots\ldots\ldots\ldots\ldots$$
$$n(n-1)a_n=(n-2)2a_{n-2}+4a_{n-2}, \quad \text{whence } a_n=\frac{2a_{n-2}}{n-1}$$
$$\ldots\ldots\ldots\ldots\ldots$$

Consequently,

$$a_5=\frac{2\cdot 1}{4}=\frac{1}{2!}, \quad a_7=\frac{2\cdot\frac{1}{2}}{6}=\frac{1}{3!}, \quad a_9=\frac{1}{4!}, \ \ldots,$$

$$a_{2k+1}=\frac{2\cdot\frac{1}{(k-1)!}}{2k}=\frac{1}{k!}, \ \ldots,$$

$$a_4=0; \ a_6=0, \ \ldots, \ a_{2k}=0, \ \ldots$$

Substituting the coefficients which we have found, we get the desired solution:

$$y=x+\frac{x^3}{1}+\frac{x^5}{2!}+\frac{x^7}{3!}+\ldots+\frac{x^{2k+1}}{k!}+\ldots$$

The series thus obtained converges for all values of x.

It will be noted that this particular solution may be expressed in terms of elementary functions: taking x outside the brackets we get (inside the brackets) an expansion of the function e^{x^2}. Hence,

$$y=xe^{x^2}$$

4.23 BESSEL'S EQUATION

Bessel's equation is a differential equation of the form

$$x^2y''+xy'+(x^2-p^2)y=0 \quad (p=\text{const}) \tag{1}$$

The solution of this equation (and also of certain other equations with variable coefficients) should be sought not in the form of a power series, but in the form of a product of some power of x by a power series:

$$y = x^r \sum_{k=0}^{\infty} a_k x^k \tag{2}$$

The coefficient a_0 may be considered nonzero due to the indefiniteness of the exponent r.

We rewrite the expression (2) in the form

$$y = \sum_{k=0}^{\infty} a_k x^{r+k}$$

and find its derivatives:

$$y' = \sum_{k=0}^{\infty} (r+k) a_k x^{r+k-1}$$

$$y'' = \sum_{k=0}^{\infty} (r+k)(r+k-1) a_k x^{r+k-2}$$

Put these expressions into equation (1):

$$x^2 \sum_{k=0}^{\infty} (r+k)(r+k-1) a_k x^{r+k-2}$$

$$+ x \sum_{k=0}^{\infty} (r+k) a_k x^{r+k-1} + (x^2 - p^2) \sum_{k=0}^{\infty} a_k x^{r+k} = 0$$

Equating to zero the coefficients of x to the powers r, $r+1$, $r+2$, ..., $r+k$, we get a system of equations:

$$\left.\begin{array}{ll}
[r(r-1)+r-p^2]\, a_0 = 0 \text{ or} & (r^2-p^2)\, a_0 = 0 \\
[(r+1)r+(r+1)-p^2]\, a_1 = 0 \text{ or} & [(r+1)^2-p^2]\, a_1 = 0 \\
[(r+2)(r+1)+(r+2)-p^2]\, a_2 + a_0 = 0 \text{ or} & [(r+2)^2-p^2]\, a_2 + a_0 = 0 \\
\cdots\cdots\cdots & \cdots\cdots\cdots \\
[(r+k)(r+k-1)+(r+k)-p^2]\, a_k + a_{k-2} = 0 \text{ or} & \\
 & [(r+k)^2-p^2]\, a_k + a_{k-2} = 0 \\
\cdots\cdots\cdots & \cdots\cdots\cdots
\end{array}\right\} \tag{3}$$

Let us consider the equation

$$[(r+k)^2 - p^2]\, a_k + a_{k-2} = 0 \tag{3'}$$

It may be rewritten as follows:

$$[(r+k-p)(r+k+p)]\, a_k + a_{k-2} = 0$$

It is given that $a_0 \neq 0$; hence,

$$r^2 - p^2 = 0$$

therefore, $r_1 = p$ or $r_2 = -p$.

Let us first consider the solution for $r_1 = p > 0$.

From the system of equations (3) we determine all the coefficients a_1, a_2, ... in succession; a_0 remains arbitrary. For instance, put $a_0 = 1$. Then

$$a_k = -\frac{a_{k-2}}{k(2p+k)}$$

Assigning various values to k, we find

$$\left.\begin{aligned} &a_1 = 0, \quad a_3 = 0 \text{ and, generally, } a_{2m+1} = 0, \\ &a_2 = -\frac{1}{2(2p+2)}, \quad a_4 = \frac{1}{2\cdot 4(2p+2)(2p+4)}, \ldots, \\ &a_{2\nu} = (-1)^{\nu+1}\frac{1}{2\cdot 4\cdot 6\ldots 2\nu(2p+2)(2p+4)\ldots(2p+2\nu)}, \ \ldots \end{aligned}\right\} \tag{4}$$

Putting the coefficients found into (2), we obtain

$$y_1 = x^p\left[1 - \frac{x^2}{2(2p+2)} + \frac{x^4}{2\cdot 4(2p+2)(2p+4)} - \frac{x^6}{2\cdot 4\cdot 6(2p+2)(2p+4)(2p+6)} + \ldots\right] \tag{5}$$

All the coefficients $a_{2\nu}$ will be determined, since for every k the coefficient of a_k in (3),

$$(r_1+k)^2 - p^2$$

will be different from zero.

Thus, y_1 is a particular solution of equation (1).

Let us further establish the conditions under which all the coefficients a_k will be determined for the second root $r_2 = -p$ as well. This will occur if for any even integral positive k the following inequalities are fulfilled:

$$(r_2+k)^2 - p^2 \neq 0 \tag{6}$$

or

$$r_2 + k \neq p$$

But $p = r_1$; hence,

$$r_2 + k \neq r_1$$

Thus, condition (6) is in this case equivalent to the following

$$r_1 - r_2 \neq k$$

where k is a positive even integer. But

$$r_1 = p, \ r_2 = -p$$

hence

$$r_1 - r_2 = 2p$$

Thus, if p is not equal to an integer, it is possible to write a second particular solution that is obtained from expression (5) by substituting $-p$ for p:

$$y_2=x^{-p}\left[1-\frac{x^2}{2(-2p+2)}+\frac{x^4}{2\cdot 4(-2p+2)(-2p+4)}\right.$$
$$\left.-\frac{x^6}{2\cdot 4\cdot 6(-2p+2)(-2p+4)(-2p+6)}+\dots\right] \qquad (5')$$

The power series (5) and (5′) converge for all values of x; this is readily found by d'Alembert's test. It is likewise obvious that y_1 and y_2 are linearly independent.*

The solution y_1 multiplied by a certain constant is called a *Bessel function of the first kind of order p* and is designated by the symbol J_p. The solution y_2 is denoted by the symbol J_{-p}.

Thus, for p not equal to an integer, the general solution of equation (1) has the form

$$y=C_1J_p+C_2J_{-p}$$

For instance, when $p=\frac{1}{2}$ the series (5) will have the form

$$x^{\frac{1}{2}}\left[1-\frac{x^2}{2\cdot 3}+\frac{x^4}{2\cdot 4\cdot 3\cdot 5}-\frac{x^6}{2\cdot 4\cdot 6\cdot 3\cdot 5\cdot 7}+\dots\right]$$
$$=\frac{1}{\sqrt{x}}\left[x-\frac{x^3}{3!}+\frac{x^5}{5!}-\frac{x^7}{7!}+\dots\right]$$

This solution multiplied by the constant factor $\sqrt{\frac{2}{\pi}}$ is called Bessel's function $J_{\frac{1}{2}}$; we note that the brackets contain a series whose sum is equal to $\sin x$. Hence,

$$J_{\frac{1}{2}}(x)=\sqrt{\frac{2}{\pi x}}\sin x$$

In exactly the same way, using formula (5′), we obtain

$$J_{-\frac{1}{2}}(x)=\sqrt{\frac{2}{\pi x}}\cos x$$

---

* The linear independence of functions is verified as follows. Consider the relation

$$\frac{y_2}{y_1}=x^{-2p}\frac{1-\frac{x^2}{2(-2p+2)}+\frac{x^4}{2\cdot 4(-2p+2)(-2p+4)}-\dots}{1-\frac{x^2}{2(2p+2)}+\frac{x^4}{2\cdot 4(2p+2)(2p+4)}-\dots}$$

This relation is not constant, since it approaches infinity as $x\to 0$. Hence the functions y_1 and y_2 are linearly independent.

The complete integral of (1) for $p=\frac{1}{2}$ is

$$y=C_1J_{\frac{1}{2}}(x)+C_2J_{-\frac{1}{2}}(x)$$

Now let p be an integer which we shall denote by $n\ (n\geqslant 0)$. The solution (5) will in this case be meaningful and is the first particular solution of (1).

But the solution (5′) will not be meaningful because one of the factors of the denominator will become zero upon expansion.

For positive integral $p=n$ the Bessel function J_n is determined by the series (5) multiplied into the constant factor $\frac{1}{2^n n!}$ (when $n=0$ we multiply by 1):

$$J_n(x)=\frac{x^n}{2^n n!}\left[1-\frac{x^2}{2(2n+2)}+\frac{x^4}{2\cdot 4(2n+2)(2n+4)}-\frac{x^6}{2\cdot 4\cdot 6(2n+2)(2n+4)(2n+6)}+\ldots\right]$$

or

$$J_n(x)=\sum_{\nu=0}^{\infty}\frac{(-1)^{\nu}}{\nu!\,(n+\nu)!}\left(\frac{x}{2}\right)^{n+2\nu} \tag{7}$$

It may be shown that the second particular solution should in this case be sought in the form

$$K_n(x)=J_n(x)\ln x+x^{-n}\sum_{k=0}^{\infty}b_kx^k$$

Putting this expression into (1), we determine the coefficients b_k.

The function $K_n(x)$, with the coefficients thus determined, multiplied by a certain constant is called *Bessel's function of the second kind of order n*.

This is the second solution of (1), which, together with the first one, form a linearly independent system.

The general integral will be of the form

$$y=C_1J_n(x)+C_2K_n(x) \tag{8}$$

We note that

$$\lim_{x\to 0}K_n(x)=\infty$$

Hence, if we want to consider the final solutions for $x=0$, then we must put $C_2=0$ into formula (8).

Example. Find the solution of Bessel's equation, for $p=0$,

$$y''+\frac{1}{x}y'+y=0$$

that satisfies the initial conditions: for $x=0$,

$$y=2, \; y'=0$$

Solution. From (7) we find one particular solution:

$$J_0(x)=\sum_{\nu=0}^{\infty}\frac{(-1)^{\nu}}{(\nu!)^2}\left(\frac{x}{2}\right)^{2\nu}=1-\frac{1}{(1!)^2}\left(\frac{x}{2}\right)^2+\frac{1}{(2!)^2}\left(\frac{x}{2}\right)^4-\frac{1}{(3!)^2}\left(\frac{x}{2}\right)^6+\ldots$$

Using this solution, we can write a solution that satisfies the given initial conditions, namely:

$$y=2J_0(x)$$

Note. If we had to find the general integral of this given equation we would seek the second particular solution in the form

$$K_0(x)=J_0(x)\ln x+\sum_{k=0}^{\infty}b_k x^k$$

Without giving all the computations, we indicate that the second particular solution, which we denote by $K_0(x)$, is of the form

$$K_0(x)=J_0(x)\ln x+\frac{x^2}{2^2}-\frac{1}{(2!)^2}\left(\frac{x}{2}\right)^4\left(1+\frac{1}{2}\right)+\frac{1}{(3!)^2}\left(\frac{x}{2}\right)^6\left(1+\frac{1}{2}+\frac{1}{3}\right)-\ldots$$

This function multiplied by some constant factor is called *Bessel's function of the second kind of order zero.*

4.24 SERIES WITH COMPLEX TERMS

Let us consider a sequence of complex numbers:

$$z_1, \; z_2, \; \ldots, \; z_n, \; \ldots$$

where

$$z_n=a_n+ib_n \quad (n=1, \; 2, \; \ldots)$$

Definition 1. A complex number $z_0=a+ib$ is the *limit of a sequence* of complex numbers $z_n=a_n+ib_n$ if

$$\lim_{n\to\infty}|z_n-z_0|=0 \tag{1}$$

Let us write condition (1) in expanded form:

$$z_n-z_0=(a_n+ib_n)-(a+ib)=(a_n-a)+i(b_n-b)$$

$$\lim_{n\to\infty}|z_n-z_0|=\lim_{n\to\infty}\sqrt{(a_n-a)^2+(b_n-b)^2}=0 \tag{2}$$

From (2) it follows that condition (1) will hold provided the following conditions are valid:

$$\lim_{n\to\infty}a_n=a, \qquad \lim_{n\to\infty}b_n=b \tag{3}$$

Let us form a series of complex numbers

$$w_1+w_2+\ldots+w_n+\ldots \tag{4}$$

where

$$w_n = u_n + iv_n \quad (n = 1, 2, \ldots)$$

We consider a sum of n terms of the series (4), which we denote by s_n:

$$s_n = w_1 + w_2 + \ldots + w_n \tag{5}$$

where s_n is a complex number:

$$s_n = \left(\sum_{k=1}^{n} u_k\right) + i\left(\sum_{k=1}^{n} v_k\right) \tag{6}$$

Definition 2. If the limit

$$\lim_{n \to \infty} s_n = s = A + iB$$

exists, then series (4) is termed a *convergent series* and s is called the *sum* of the series:

$$s = \sum_{k=1}^{\infty} w_k = A + iB \tag{7}$$

From condition (6), on the basis of (3), follow the equations

$$A = \lim_{n \to \infty} \sum_{k=1}^{n} u_k, \quad B = \lim_{n \to \infty} \sum_{k=1}^{n} v_k \tag{8}$$

If there is no $\lim s_n$, then series (4) is a *divergent series*.

An effective theorem for testing the convergence of series (4) is the following.

Theorem 1. *If a series made up of the absolute values (moduli) of the terms of series* (4) *converges,*

$$|w_1| + |w_2| + \ldots + |w_n| + \ldots, \quad \textit{where } |w_n| = \sqrt{u_n^2 + v_n^2} \tag{9}$$

then series (4) *converges as well.*

Proof. From the convergence of series (9) and from the conditions

$$|u_n| \leqslant \sqrt{u_n^2 + v_n^2} = |w_n|, \quad |v_n| \leqslant \sqrt{u_n^2 + v_n^2} = |w_n|$$

follow equations (8) (on the basis of the appropriate theorem on the absolute convergence of series with real terms) and, hence, equation (7).

This theorem enables us, when testing the convergence of a series with complex terms, to apply all the sufficient criteria of convergence of series with positive terms.

4.25 POWER SERIES IN A COMPLEX VARIABLE

We now consider power series with complex terms.

Definition 1. A series

$$c_0 + c_1 z + c_2 z^2 + \ldots + c_n z^n + \ldots \tag{1}$$

where $z = x + iy$ is a complex variable, x and y are real numbers, and c_n are constant complex or real numbers, is called a *power series in a complex variable*. The theory of complex power series is similar to the theory of power series with real terms.

Definition 2. The collection of values of z in the complex plane for which the series (1) converges is called the *domain of convergence* of the power series (1) (for each concrete value of z, the series (1) becomes a numerical series with complex terms of type (4), Sec. 4.24).

Definition 3. The series (1) is *absolutely convergent* if the series composed of the moduli of the terms of (1) converges:

$$|c_0|+|c_1z|+|c_2z^2|+\ldots+|c_nz^n|+\ldots \quad (2)$$

We give without proof the following theorem.

Theorem 1. *The domain of convergence of a power series with complex terms* (1) *is a circle in the complex plane with centre at the origin. This circle is called the circle of convergence. The series* (1) *converges absolutely at points lying inside the circle of convergence.*

The radius R of the circle of convergence is termed the *radius of convergence* of the power series. If R is the radius of convergence of the power series (1), then we write that the series converges in the domain

$$|z|<R$$

(The question of the convergence of the series at the boundary points $|z|=R$ requires a supplementary investigation; this is similar to the problem of the convergence of a power series in a real variable at the end points of an interval.) Note that the radius R of convergence can take on any value from $R=0$ to $R=\infty$ depending on the nature of the coefficients c_n. In the former case, the series converges only at the point $z=0$, in the latter, it converges for any value of z.

We write the equation

$$f(z)=c_0+c_1z+c_2z^2+\ldots+c_nz^n+\ldots \quad (3)$$

If z assumes distinct values within the circle of convergence, the function $f(z)$ will assume distinct values. Thus, every power series in a complex variable defines an appropriate function of the complex variable within the circle of convergence. This function is called an *analytic function* of a complex variable The following are some examples of functions of a complex variable **defined by power series in a complex variable.**

1. $$e^z=1+z+\frac{z^2}{2!}+\frac{z^3}{3!}+\ldots+\frac{z^n}{n!}+\ldots \quad (4)$$

This is the *exponential function of a complex variable.* If $y=0$, then formula (4) becomes formula (2), Sec. 4.17. If $x=0$, then we get equation (1), Sec. 4.18.

2. $\sin z = z - \frac{z^3}{3!} + \frac{z^5}{5!} - \frac{z^7}{7!} + \ldots$ (5)

This is the *sine of a complex variable*. For $y=0$, formula (5) turns into formula (1) of Sec. 4.17.

3. $\cos z = 1 - \frac{z^2}{2!} + \frac{z^4}{4!} - \frac{z^6}{6!} + \ldots$ (6)

This is the *cosine of a complex variable*. If in formula (4) we put (iz) in place of z on the right and on the left, then (like what was done in Sec. 4.18) we get

$$e^{iz} = \cos z + i \sin z \tag{7}$$

This is *Euler's formula for a complex* z. If z is a real number, then this formula coincides with formula (2) of Sec. 4.18.

4. $\sinh z = z + \frac{z^3}{3!} + \frac{z^5}{5!} + \ldots$ (8)

5. $\cosh z = 1 + \frac{z^2}{2!} + \frac{z^4}{4!} + \frac{z^6}{6!} + \ldots$ (9)

These last two formulas are similar to formulas (5) and (6) of Sec. 4.17 and coincide with the latter when $z=x$ is a real number.

By (4), (5), (6), (8), and (9) and via addition, subtraction of the series and replacement of z by (iz), we get the following equations:

$$e^z = \cosh z + \sinh z, \tag{10}$$

$$e^{-z} = \cosh z - \sinh z, \tag{11}$$

$$\cosh z = \frac{e^z + e^{-z}}{2}, \qquad \sinh z = \frac{e^z - e^{-z}}{2}, \tag{12}$$

$$\cos iz = \frac{e^{-z} + e^z}{2}, \qquad \sin iz = \frac{e^{-z} - e^z}{2i} \tag{13}$$

Note that the series (4), (5), (6), (8), (9) converge for all values of z. This is clearly seen on the basis of Theorem 1 and of Sec. 4.24. As in the case of power series of a real variable, we can consider series of a complex variable in powers of $(z-z_0)$, where z_0 is a complex number, c_n are complex or real constants

$$c_0 + c_1(z-z_0) + c_2(z-z_0)^2 + \ldots + c_n(z-z_0)^n + \ldots \tag{14}$$

The series (14) is reduced to (1) by the substitution $z - z_0 = z^*$. All the properties and theorems that are valid for series of type (1) carry over to series of type (14), only the centre of the circle of convergence of (14) lies at point z_0 instead of at the origin.

If R is the radius of convergence of series (14), then we write: the series converges in the domain

$$|z - z_0| < R$$

4.26. THE SOLUTION OF FIRST-ORDER DIFFERENTIAL EQUATIONS BY THE METHOD OF SUCCESSIVE APPROXIMATIONS (METHOD OF ITERATION)

In Secs. 1.32, 1.33, and 1.34 we considered the approximate integration of differential equations and systems of differential equations by difference methods. Here we will consider a different method of approximate integration of a differential equation. It will be noted that this is at the same time proof of the theorem of the existence of a solution of a differential equation (see Sec. 1.2). We will need the theory of series in this case.

Let it be required to find the solution to the differential equation

$$\frac{dy}{dx}=f(x, y) \tag{1}$$

that satisfies the initial condition

$$y=y_0 \text{ for } x=x_0 \tag{2}$$

Integrating the terms of equation (1) from x_0 to x and taking into account that $y|_{x=x_0}=y_0$, we get

$$y=\int_{x_0}^{x} f(x, y)\,dx+y_0 \tag{3}$$

In this equation, the desired function y is under the integral sign and so the equation is called an *integral equation*.

The function $y=y(x)$ which satisfies (1) and the initial conditions (2) satisfies equation (3). It is clear that the function $y=y(x)$ which satisfies (3) satisfies equation (1) and the initial conditions (2).

Let us first consider a method of obtaining approximate solutions to equation (1) for the initial conditions (2).

We will take y_0 for the *zeroth* approximation of the solution. Substituting y_0 into the integrand on the right of (3) in place of y, we get

$$y_1(x)=\int_{x_0}^{x} f(x, y_0)\,dx+y_0 \tag{4}$$

This is the *first* approximation of the differential equation (1) that satisfies the initial conditions (2).

Substituting the first approximation $y_1(x)$ into the integrand in (3), we get

$$y_2(x)=\int_{x_0}^{x} f[x, y_1(x)]\,dx+y_0 \tag{5}$$

This is the *second* approximation. We continue the process:

$$\left.\begin{array}{l} y_3(x)=\int\limits_{x_0}^{x} f[x,\ y_2(x)]\,dx+y_0, \\ \cdots\cdots\cdots\cdots\cdots \\ y_n(x)=\int\limits_{x_0}^{x} f[x,\ y_{n-1}(x)]\,dx+y_0, \\ \cdots\cdots\cdots\cdots\cdots \end{array}\right\} \quad (6)$$

We will indicate below what approximation is to be taken so as to satisfy our requirements of accuracy.

Example. Approximate the equation $y'=x+y^2$ which satisfies the initial conditions

$$y_0=1 \text{ for } x=0$$

Solution. By formula (4) we have

$$y_1=\int\limits_0^x (x+1)\,dx+1=\frac{x^2}{2}+x+1$$

$$y_2=\int\limits_0^x \left[x+\left(\frac{x^2}{2}+x+1\right)^2\right] dx+1=\frac{x^5}{20}+\frac{x^4}{4}+2\frac{x^3}{3}+3\frac{x^2}{2}+x+1$$

and so forth.

4.27. PROOF OF THE EXISTENCE OF A SOLUTION OF A DIFFERENTIAL EQUATION. ERROR ESTIMATION IN APPROXIMATE SOLUTIONS

We prove the following theorem.

Theorem. *Given the differential equation*

$$\frac{dy}{dx}=f(x,\ y) \quad (1)$$

and the initial conditions

$$y=y_0 \text{ for } x=x_0 \quad (2)$$

Let $f(x, y)$ and $f'_y(x, y)$ be continuous in a closed domain D:

$$D\{x_0-a\leqslant x\leqslant x_0+a,\ y_0-b\leqslant y\leqslant y_0+b\} \text{ (Fig. 123)} \quad (3)$$

Then, in some interval

$$x_0-l<x<x_0+l \quad (4)$$

there exists a solution to equation (1) which satisfies the initial conditions (2); and the solution is unique. The number l will be defined below.

Proof. Note that from the fact that $f(x, y)$ and $f'_y(x, y)$ are continuous in a closed domain D it follows that there exist constants $M>0$ and $N>0$ such that for all points of the domain the following relations hold:

$$|f(x, y)| \leqslant M, \quad (5)$$

$$|f'_y(x, y)| \leqslant N \quad (6)$$

(This property is similar to the property indicated in Sec. 2.10, Vol. I.) The number l in (4) is the smaller of the numbers a and $\frac{b}{M}$, that is,

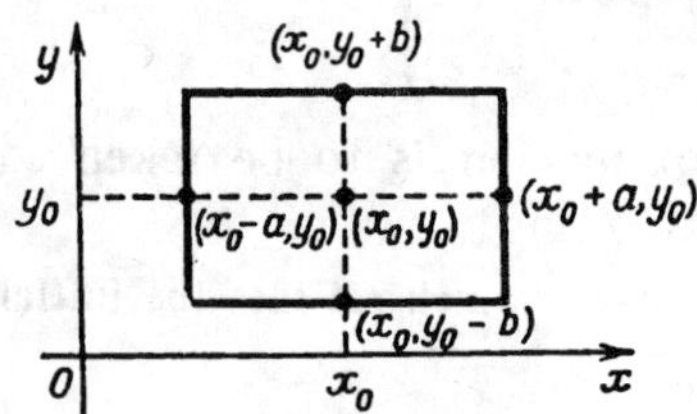

Fig. 123

$$l=\min\left(a, \frac{b}{M}\right) \quad (7)$$

Let us apply the Lagrange theorem to the function $f(x, y)$ for two arbitrary points $A_1(x, y_1)$ and $A_2(x, y_2)$ belonging to D:

$$f(x, y_2)-f(x, y_1)=f'_y(x, \eta)(y_2-y_1)$$

where $y_1<\eta<y_2$, hence, $|f'_y(x, \eta)| \leqslant N$. Thus, for any two points the following inequality is true:*

$$|f(x, y_2)-f(x, y_1)| \leqslant N|y_2-y_1| \quad (8)$$

Let us return to equation (4) of Sec. 4.26. From this equation, taking into account (5), (4), (7), we get

$$|y_1-y_0|=\left|\int_{x_0}^{x} f(x, y_0)\,dx\right| \leqslant \int_{x_0}^{x} M\,dx=M|x-x_0| \leqslant Ml \leqslant b \quad (9)$$

Thus, the function $y=y_1(x)$ defined by (4), Sec. 4.26, on the interval (4) does not go outside domain D.

Now we take up equation (5) of Sec. 4.26. The arguments of the function $f[x, y_1(x)]$ do not leave D. Hence, we can write

$$|y_2-y_0|=\left|\int_{x_0}^{x} f[x, y_1(x)]\,dx\right| \leqslant M|x-x_0| \leqslant Ml \leqslant b \quad (10)$$

---

* Note that if for some function $F(y)$ the following condition is satisfied:

$$|F(y_2)-F(y_1)| \leqslant K|y_2-y_1|$$

where y_2 and y_1 are any points in the domain and K is a constant, then this condition is called the *Lipschitz condition*. Thus, having established the relation (8), we demonstrated that if a function $f(x, y)$ has a derivative $\frac{\partial f}{\partial y}$ bounded in some domain, then it satisfies the Lipschitz condition in that domain. The converse may not prove to be true.

By the method of complete induction we can prove that for arbitrary n

$$|y_n - y_0| \leqslant b \tag{11}$$

if x lies in the interval (4).

We will now prove that there exists a limit

$$\lim_{n \to \infty} y_n(x) = y(x) \tag{12}$$

and the function $y(x)$ satisfies the differential equation (1) and the initial conditions (2). To prove this, consider the series

$$y_0 + (y_1 - y_0) + (y_2 - y_1) + \ldots + (y_{n-1} - y_{n-2}) + (y_n - y_{n-1}) + \ldots \tag{13}$$

with general term $u_n = y_n - y_{n-1}$ with $u_0 = y_0$. The sum of $n+1$ terms of this series is clearly equal to

$$S_{n+1} = \sum_{i=0}^{n} u_i = y_n \tag{14}$$

Let us estimate the terms of series (13) in absolute value:

$$|y_1 - y_0| = \left| \int_{x_0}^{x} f(x,\ y_0)\, dx \right| \leqslant M\,|x - x_0| \tag{15}$$

On the basis of (4), (5), Sec. 4.26, and (6) we find

$$|y_2 - y_1| = \left| \int_{x_0}^{x} [f(x,\ y_1) - f(x,\ y_0)]\, dx \right| = \left| \int_{x_0}^{x} f'_y(x,\ \eta_1)(y_2 - y_1)\, dx \right|$$

$$\leqslant \pm N \int_{x_0}^{x} M\,|x - x_0|\, dx = N\frac{M}{2}|x - x_0|^2$$

(the plus sign is taken if $x_0 < x$, and the minus sign if $x_0 > x$). Thus

$$|y_2 - y_1| \leqslant M\frac{N}{1\cdot 2}|x - x_0|^2 \tag{16}$$

Similarly, taking into account (16),

$$|y_3 - y_2| = \left| \int_{x_0}^{x} [f(x,\ y_2) - f(x,\ y_1)]\, dx \right|$$

$$= \left| \int_{x_0}^{x} f'_y(x, \eta_2)(y_2 - y_1)\, dx \right|$$

$$\leqslant \pm N \int_{x_0}^{x} \frac{NM}{2}|x - x_0|^2\, dx = M\frac{N^2}{1\cdot 2\cdot 3}|x - x_0|^3 \tag{17}$$

Continuing in the same fashion, we have

$$|y_n - y_{n-1}| \leqslant M \frac{N^{n-1}}{n!} |x - x_0|^n \tag{18}$$

Thus, for the interval $|x - x_0| < l$ the series (13) of functions is dominated. The corresponding numerical series with positive terms, which exceed in absolute value the corresponding terms of series (13), is

$$y_0 + Ml + M\frac{Nl^2}{2!} + M\frac{N^2 l^3}{3!} + \ldots + M\frac{N^{n-1} l^n}{n!} + \ldots \tag{19}$$

with general term $v_n = M\frac{N^{n-1} l^n}{n!}$. This series converges, as may readily be seen by applying d'Alembert's test:

$$\lim_{n\to\infty} \frac{v_n}{v_{n-1}} = \lim_{n\to\infty} \frac{M\frac{N^{n-1} l^n}{n!}}{M\frac{N^{n-2} l^{n-1}}{(n-1)!}} = \lim_{n\to\infty} \frac{Nl}{n} = 0 < 1$$

Thus, the series (13) is dominated and hence converges. Since its terms are continuous functions, it converges to a continuous function $y(x)$. Thus

$$\lim_{n\to\infty} s_{n-1} = \lim_{n\to\infty} y_n = y(x) \tag{20}$$

where $y(x)$ is a continuous function. This function satisfies the initial condition since for all n

$$y_n(x_0) = y_0$$

We will prove that the function $y(x)$ thus obtained satisfies equation (1). We again write down the last equation of (6), Sec. 4.26:

$$y_n = y_0 + \int_{x_0}^{x} f(x, y_{n-1})\, dx \tag{21}$$

We will prove that

$$\lim_{n\to\infty} \int_{x_0}^{x} f[x, y_{n-1}(x)]\, dx = \int_{x_0}^{x} f(x, y)\, dx \tag{22}$$

where $y(x)$ is defined by equation (20).

First note the following. Since the series (13) is dominated, it follows from (20) that for any $\varepsilon > 0$ there is an n such that

$$|y - y_n| < \varepsilon \tag{23}$$

Having regard for (23) over the entire interval (4), we can write

$$\left| \int_{x_0}^{x} f(x, y)\, dx - \int_{x_0}^{x} f(x, y_n)\, dx \right| \leqslant \pm \int_{x_0}^{x} |f(x, y) - f(x, y_n)|\, dx$$

$$\leqslant \pm \int_{x_0}^{x} N |y - y_n|\, dx \leqslant N\varepsilon |x - x_0|$$

But $\lim\limits_{n\to\infty} \varepsilon = 0$. Hence

$$\lim_{n\to\infty}\left|\int_{x_0}^{x} f(x,\ y)\,dx - \int_{x_0}^{x} f(x,\ y_n)\,dx\right| = 0$$

From this equation follows equation (22).

Now passing to the limit in both members of (21) as $n \to \infty$, we find that $y(x)$, defined by (20), satisfies the equation

$$y = y_0 + \int_{x_0}^{x} f(x,\ y)\,dx \tag{24}$$

From this it follows, as was pointed out above, that the function $y(x)$ thus found satisfies the differential equation (1) and the initial conditions (2).

Note 1. Using other methods of proof it is possible to assert that there exists a solution of equation (1) that satisfies the conditions (2) if the function $f(x,\ y)$ is continuous in a domain D (Peano's theorem).

Note 2. Using a device similar to the one employed to obtain relation (18), we can demonstrate that the error resulting from replacing the solution $y(x)$ by its nth approximation y_n is given by the formula

$$|y - y_n| \leqslant \frac{N^n M}{(n+1)!}\,|x - x_0|^{n+1} \leqslant \frac{MN^n\, l^{n+1}}{(n+1)!} \tag{25}$$

Example. Let us apply this estimate with respect to the fifth approximation y_5 to the solution of the equation

$$y' = x + y^2$$

given the initial conditions $y_0 = 1$ when $x = 0$.

Suppose the domain D is

$$D\{-1/2 \leqslant x \leqslant 1/2,\quad -1 \leqslant y \leqslant 1\}$$

that is, $a = 1/2$, $b = 1$. Then $M = 3/2$, $N = 2$. We then determine

$$l = \min\left(a,\ \frac{b}{M}\right) = \min\left(\frac{1}{2},\ \frac{2}{3}\right) = \frac{1}{2}$$

By formula (25) we get

$$|y - y_5| \leqslant \frac{2^5 \cdot \frac{3}{2} \cdot \left(\frac{1}{2}\right)^6}{6!} = \frac{1}{960}$$

Note that the estimate (25) is rather rough. In the case at hand, we could show by other methods that the error is tens of times less.

4.28 THE UNIQUENESS THEOREM OF THE SOLUTION OF A DIFFERENTIAL EQUATION

We will prove the following theorem.

Theorem. *If a function $f(x, y)$ is continuous and has a continuous derivative $\frac{\partial f}{\partial y}$ (defined in Sec. 4.27) in a domain D then the solution of the differential equation*

$$\frac{dy}{dx}=f(x, y) \tag{1}$$

is unique for the initial conditions

$$y=y_0 \quad when \quad x=x_0 \tag{2}$$

That is, through the point (x_0, y_0) there passes a unique integral curve of equation (1).

Proof. Assume there are two solutions to equation (1) satisfying the conditions (2), that is, two curves emanating from the point $A(x_0, y_0)$: $y(x)$ and $z(x)$. Consequently, both functions satisfy equation (24) of Sec. 4.27:

$$y=y_0+\int_{x_0}^{x} f(x, y)\,dx, \qquad z=y_0+\int_{x_0}^{x} f(x, z)=dx$$

Consider the difference

$$y(x)-z(x)=\int_{x_0}^{x}[f(x, y)-f(x, z)]\,dx \tag{3}$$

Transform the difference in the integrand by Lagrange's formula taking into account the inequality (6) of Sec. 4.27:

$$f(x, y)-f(x, z)=\frac{\partial f(x, \eta)}{y}(y-z) \tag{4}$$

From this equation we get

$$|f(x, y)-f(x, z)|\leqslant N|y-z| \tag{5}$$

On the basis of (3), with regard for (5), we can write the inequality

$$|y-z|=\left|\int_{x_0}^{x}\frac{\partial f}{\partial y}\cdot(y-z)\,dx\right|\leqslant\pm\int_{x_0}^{x}N|y-z|\,dx \tag{6}$$

Let us consider a value of x such that $|x-x_0|<\frac{1}{N}$. For the sake of definiteness, we assume that $x_0<x$; for the case of $x<x_0$ the proof is analogous.

Suppose, on the interval $x-x_0<\frac{1}{N}$, $|y-z|$ assumes a maximum value for $x=x^*$ and let it be equal to λ. Then inequality (6) assumes, for the point x^*, the form

$$\lambda \leqslant N \int_{x_0}^{x^*} \lambda \, dx = N\lambda (x^* - x_0) < N\lambda \frac{1}{N} < \lambda$$

or

$$\lambda < \lambda$$

With the assumption that there exist two distinct solutions we arrived at a contradiction. Hence, the solution is unique.

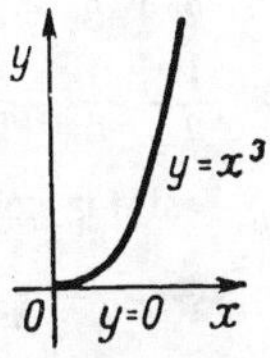

Note 1. It may be demonstrated that the solution is unique for less strict restrictions on the function $f(x, y)$. See, for example, I. G. Petrovsky's "Lectures on the Theory of Ordinary Differential Equations".

Note 2. If a function $f(x, y)$ has an unbounded partial derivative $\frac{\partial f}{\partial y}$ in the domain, then there may be several solutions satisfying the equation (1) and the initial conditions (2).

Fig. 124

Indeed, consider the equation

$$y' = 3x \sqrt[3]{y} \tag{7}$$

with initial conditions

$$y = 0 \quad \text{for} \quad x = 0 \tag{8}$$

Here $\frac{\partial f}{\partial y} = xy^{-2/3} \rightarrow \infty$ as $y \rightarrow 0$. In this case there are two solutions to equation (7) that satisfy the initial conditions (8):

$$y = 0, \quad y = x^3$$

Direct substitution of the functions into the equation convinces us that they are solutions of (7). Two integral curves pass through the coordinate origin (Fig. 124).

Exercises on Chapter 4

Write the first few terms of the series on the basis of the given general term:

1. $u_n = \frac{1}{n(n+1)}$. **2.** $u_n = \frac{n^3}{n+1}$. **3.** $u_n = \frac{(n!)^2}{(2n)!}$. **4.** $u_n = (-1)^{n+1} \frac{x^n}{n^k}$.

5. $u_n = \sqrt[3]{n^3+1} - \sqrt{n^2+1}$.

Test the following series for convergence:

6. $\frac{1}{2} + \frac{2}{2^2} + \frac{3}{2^3} + \ldots + \frac{n}{2^n} \ldots$. *Ans.* Converges.

7. $\frac{1}{\sqrt{10}} + \frac{1}{\sqrt{20}} + \frac{1}{\sqrt{30}} + \ldots + \frac{1}{\sqrt{10n}} + \ldots$. *Ans.* Diverges.

8. $2+\frac{3}{2}+\frac{4}{3}+\ldots+\frac{n+1}{n}+\ldots$. *Ans.* Diverges.

9. $\frac{1}{\sqrt[3]{7}}+\frac{1}{\sqrt[3]{8}}+\ldots+\frac{1}{\sqrt[3]{n+6}}+\ldots$. *Ans.* Diverges.

10. $\frac{1}{2}+\left(\frac{2}{3}\right)^4+\left(\frac{3}{4}\right)^9+\ldots+\left(\frac{n}{n+1}\right)^{n^2}+\ldots$. *Ans.* Converges.

11. $\frac{1}{2}+\frac{2}{5}+\frac{3}{10}+\frac{4}{17}+\ldots\frac{n}{n^2+1}+\ldots$. *Ans.* Diverges.

12. $\frac{1}{2}+\frac{1}{5}+\frac{1}{10}+\frac{1}{17}+\ldots+\frac{1}{1+n^2}+\ldots$. *Ans.* Converges.

Test for convergence the following series with given general terms:

13. $u_n=\frac{1}{n^3}$. *Ans.* Converges. 14. $u_n=\frac{1}{\sqrt[3]{n^2}}$. *Ans.* Diverges. 15. $u_n=\frac{2}{5n+1}$. *Ans.* Diverges. 16. $u_n=\frac{1+n}{3+n^2}$. *Ans.* Diverges.

17. $u_n=\frac{1}{n^2+2n+3}$. *Ans.* Converges. 18. $u_{n-1}=\frac{1}{n\ln n}$. *Ans.* Diverges. 19. Prove the inequality

$$1+\frac{1}{2}+\frac{1}{3}+\ldots+\frac{1}{n}>\ln(n+1)>\frac{1}{2}+\frac{1}{3}+\ldots+\frac{1}{n+1}.$$

20. Is the Leibniz theorem applicable to the series

$$\frac{1}{\sqrt{2}-1}-\frac{1}{\sqrt{2}+1}+\frac{1}{\sqrt{3}-1}-\frac{1}{\sqrt{3}+1}+\ldots+\frac{1}{\sqrt{n}-1}-\frac{1}{\sqrt{n}+1}+\ldots?$$

Ans. It is not applicable because the terms of the series do not decrease monotonically in absolute value. The series diverges.

How many first terms must be taken in the series so that their sum should not differ by more than 10^{-6} from the sum of the corresponding series:

21. $\frac{1}{2}-\frac{1}{2^2}+\frac{1}{2^3}-\frac{1}{2^4}+\ldots+(-1)^{n+1}\frac{1}{2^n}+\ldots$. *Ans.* $n=20$.

22. $\frac{1}{2}-\frac{1}{3}+\frac{1}{4}-\frac{1}{5}+\ldots(-1)^n\frac{1}{n}+\ldots$. *Ans.* $n=10^6$.

23. $\frac{1}{2^2}-\frac{1}{3^2}+\frac{1}{4^2}-\frac{1}{5^2}+\ldots+(-1)^n\frac{1}{n^2}+\ldots$. *Ans.* $n=10^3$.

24. $\frac{1}{2}-\frac{1}{2\cdot3}+\frac{1}{2\cdot3\cdot4}-\frac{1}{2\cdot3\cdot4\cdot5}+\ldots+(-1)^n\frac{1}{n!}+\ldots$. *Ans.* $n=10$.

Find out which of the following series converges absolutely:

25. $1-\frac{1}{3^2}+\frac{1}{5^2}-\frac{1}{7^2}+\frac{1}{9^2}+\ldots+(-1)^{n+1}\frac{1}{(2n-1)^2}+\ldots$. *Ans.* Converges absolutely. 26. $\frac{1}{2}-\frac{1}{2}\cdot\frac{1}{2^2}+\frac{1}{3}\cdot\frac{1}{2^3}-\ldots+(-1)^{n+1}\frac{1}{n}\cdot\frac{1}{2^n}+\ldots$. *Ans.* Converges absolutely.

27. $\frac{1}{\ln 2}-\frac{1}{\ln 3}+\frac{1}{\ln 4}-\frac{1}{\ln 5}+\ldots+(-1)^n\frac{1}{\ln n}+\ldots$. *Ans.* Converges conditionally. 28. $-1+\frac{1}{\sqrt[5]{2}}-\frac{1}{\sqrt[5]{3}}+\frac{1}{\sqrt[5]{4}}+\ldots+(-1)^n\frac{1}{\sqrt[5]{n}}+\ldots$ *Ans.* Converges conditionally.

Find the sum of the series:

29. $\frac{1}{1\cdot 2\cdot 3}+\frac{1}{2\cdot 3\cdot 4}+\ldots+\frac{1}{n(n+1)(n+2)}+\ldots$ *Ans.* $\frac{1}{4}$.

For what values of x do the following series converge:

30. $1+\frac{x}{2}+\frac{x^2}{4}+\ldots+\frac{x^n}{2^n}+\ldots$ *Ans.* $-2<x<2$. 31. $x-\frac{x^2}{2^2}+\frac{x^3}{3^2}-\frac{x^4}{4^2}+\ldots+(-1)^{n+1}\frac{x^n}{n^2}+\ldots$ *Ans.* $-1\leqslant x\leqslant 1$. 32. $3x+3^4x^4+3^9x^9+\ldots+3^{n^2}x^{n^2}+\ldots$ *Ans.* $|x|<\frac{1}{3}$. 33. $1+\frac{100x}{1\cdot 3}+\frac{10\,000x^2}{1\cdot 3\cdot 5}+\frac{1\,000\,000x^3}{1\cdot 3\cdot 5\cdot 7}+\ldots$ *Ans.* $-\infty<x<\infty$.

34. $\sin x+2\sin\frac{x}{3}+4\sin\frac{x}{9}+\ldots+2^n\sin\frac{x}{3^n}+\ldots$ *Ans.* $-\infty<x<\infty$.

35. $\frac{x}{1+\sqrt{1}}+\frac{x^2}{2+\sqrt{2}}+\ldots+\frac{x^n}{n+\sqrt{n}}+\ldots$ *Ans.* $-1\leqslant x<1$..

36. $x+\frac{2^k}{2!}x^2+\frac{3^k}{3!}x^3+\ldots+\frac{n^k}{n!}x^n+\ldots$ *Ans.* $-\infty<x<\infty$.

37. $x+\frac{2!}{2^2}x^2+\frac{3!}{3^3}x^3+\ldots+\frac{n!}{n^n}x^n+\ldots$ *Ans.* $-e<x<e$.

38. $x+\frac{2^2}{4!}x^2+\frac{(1\cdot 2\cdot 3)^2}{6!}x^3+\ldots+\frac{(n!)^2}{(2n)!}x^n+\ldots$ *Ans.* $-4<x<4$.

39. Find the sum of the series $x+2x^2+\ldots+nx^n+\ldots(|x|<1)$. *Ans.* $\frac{x}{(1-x)^2}$

Determine which of the following series are dominated on the indicated intervals:

40. $1+\frac{x}{1^2}+\frac{x^2}{2^2}+\ldots+\frac{x^n}{n^2}+\ldots(0\leqslant x\leqslant 1)$. *Ans.* Dominated. 41. $1+\frac{x}{1}+\frac{x^2}{2}+\frac{x^3}{3}+\ldots+\frac{x^n}{n}+\ldots(0\leqslant x\leqslant 1)$. *Ans.* Not dominated. 42. $\frac{\sin x}{1^2}+\frac{\sin 2x}{2^2}+\frac{\sin 3x}{3^2}+\ldots+\frac{\sin nx}{n^2}+\ldots[0, 2\pi]$. *Ans.* Dominated.

Expanding Functions in Series

43. Expand $\frac{1}{10+x}$ in powers of x and determine the interval of convergence. *Ans.* The series converges for $-10<x<10$.

44. Expand $\cos x$ in powers of $\left(x-\frac{\pi}{4}\right)$. *Ans.* $\frac{1}{\sqrt{2}}-\frac{1}{\sqrt{2}}\left(x-\frac{\pi}{4}\right)-\frac{1}{2\sqrt{2}}\left(x-\frac{\pi}{4}\right)^2+\frac{1}{6\sqrt{2}}\left(x-\frac{\pi}{4}\right)^3+\ldots$

45. Expand e^{-x} in powers of x. *Ans.* $1-x+\frac{x^2}{2!}-\frac{x^3}{3!}+\ldots$

46. Expand e^x in powers of $(x-2)$. *Ans.* $e^2+e^2(x-2)+\frac{e^2}{2!}(x-2)^2+\frac{e^2}{3!}(x-2)^3+\ldots$

47. Expand x^3-2x^2+5x-7 in powers of $(x-1)$. *Ans.* $-3+4(x-1)+(x-1)^2+(x-1)^3$.

48. Expand the polynomial $x^{10}+2x^9-3x^7-6x^6+3x^4+6x^3-x-2$ in a Taylor's series in powers of $(x-1)$; check to see that this polynomial has the

number 1 for a triple root. *Ans.* $f(x)=81(x-1)^3+270(x-1)^4+405(x-1)^5+$ $+351(x-1)^6+189(x-1)^7+63(x-1)^8+12(x-1)^9+(x-1)^{10}$.

49. Expand $\cos(x+a)$ in powers of x. *Ans.* $\cos a-x\sin a-\frac{x^2}{2!}\cos a+$ $+\frac{x^3}{3!}\sin a+\frac{x^4}{4!}\cos a-\ldots$.

50. Expand $\ln x$ in powers of $(x-1)$. *Ans.* $(x-1)-\frac{1}{2}(x-1)^2+\frac{1}{3}(x-1)^3-$ $-\frac{1}{4}(x-1)^4+\ldots$.

51. Expand e^x in a series of powers of $(x+2)$. *Ans.* $e^{-2}\left[1+\sum_{n=1}^{\infty}\frac{(x+2)^n}{n!}\right]$.

52. Expand $\cos^2 x$ in a series of powers of $\left(x-\frac{\pi}{4}\right)$.

Ans. $\frac{1}{2}+\sum_{n=1}^{\infty}(-1)^n\frac{4^{n-1}\left(x-\frac{\pi}{4}\right)^{2n-1}}{(2n-1)!}$ $(|x|<\infty)$.

53. Expand $\frac{1}{x^2}$ in a series of powers of $(x+1)$. *Ans.* $\sum_{n=0}^{\infty}(n+1)(x+1)^n$ $(-2<x<0)$.

54. Expand $\tan x$ in a series of powers of $\left(x-\frac{\pi}{4}\right)$. *Ans.* $1+2\left(x-\frac{\pi}{4}\right)+$ $+\left(x-\frac{\pi}{4}\right)^2+\ldots$.

Write the first four terms of the series expansion, in powers of x, of the following functions:

55. $\tan x$. *Ans.* $x+\frac{x^3}{3}+\frac{2x^5}{15}+\frac{17x^7}{315}+\ldots$.

56. $e^{\cos x}$. *Ans.* $e\left(1-\frac{x^2}{2!}+\frac{4x^4}{4!}-\frac{31x^6}{720}-\ldots\right)$.

57. $e^{\arctan x}$. *Ans.* $1+x+\frac{x^2}{2}-\frac{x^3}{6}-\frac{7x^4}{24}+\ldots$.

58. $\ln(1+e^x)$. *Ans.* $\ln 2+\frac{x}{2}+\frac{x^2}{8}-\frac{x^4}{192}+\ldots$.

59. $e^{\sin x}$. *Ans.* $1+x+\frac{x^2}{2}-\frac{x^4}{8}+\ldots$.

60. $(1+x)^x$. *Ans.* $1+x^2-\frac{x^3}{2}+\frac{5}{6}x^4-\ldots$.

61. $\sec x$. *Ans.* $1+\frac{x^2}{2}+\frac{5x^4}{24}+\ldots$.

62. $\ln\cos x$. *Ans.* $-\frac{x^2}{2}-\frac{x^4}{12}-\frac{x^6}{45}-\ldots$.

63. Expand $\sin kx$ in powers of x. *Ans.* $kx-\frac{(kx)^3}{3!}+\frac{(kx)^5}{5!}-\frac{(kx)^7}{7!}+\ldots$.

64. Expand $\sin^2 x$ in powers of x and determine the interval of convergence. *Ans.* $\frac{2x^2}{2!}-\frac{2^3x^4}{4!}+\frac{2^5x^6}{6!}-\ldots+(-1)^{n-1}\frac{2^{2n-1}x^{2n}}{(2n)!}+\ldots$. The series converges for all values of x.

65. Expand $\frac{1}{1+x^2}$ in a power series of x. *Ans.* $1-x^2+x^4-x^6+\ldots$.

66. Expand $\arctan x$ in a power series of x.

Hint. Take advantage of the formula $\arctan x=\int\limits_0^x \frac{dx}{1+x^2}$. *Ans.* $x-\frac{x^3}{3}+\frac{x^5}{5}-\frac{x^7}{7}+\ldots (-1\leqslant x\leqslant 1)$.

67. Expand $\frac{1}{(1+x)^2}$ in a series of powers of x. *Ans.* $1-2x+3x^2-4x^3+\ldots$ $-1<x<1)$.

Using the formulas for expansion of the functions e^x, $\sin x$, $\cos x$, $\ln(1+x)$ and $(1+x)^m$ into power series and applying various procedures, expand the following functions in power series and determine the intervals of convergence:

68. $\sinh x$. *Ans.* $x+\frac{x^3}{3!}+\frac{x^5}{5!}+\ldots (-\infty<x<\infty)$.

69. $\cosh x$. *Ans.* $1+\frac{x^2}{2!}+\frac{x^4}{4!}+\ldots (-\infty<x<\infty)$.

70. $\cos^2 x$. *Ans.* $1+\frac{1}{2}\sum\limits_{n=1}^{\infty}\frac{(-1)^n(2x)^{2n}}{(2n)!} (-\infty<x<\infty)$.

71. $(1+x)\ln(1+x)$. *Ans.* $x+\sum\limits_{n=2}^{\infty}(-1)^n\frac{x^n}{(n-1)n} \quad (|x|\leqslant 1)$.

72. $(1+x)e^{-x}$. *Ans.* $1+\sum\limits_{n=2}^{\infty}(-1)^{n-1}\frac{n-1}{n!}x^n \quad (-\infty<x<\infty)$.

73. $\frac{1}{4-x^4}$. *Ans.* $\sum\limits_{n=0}^{\infty}\frac{x^{4n}}{4^{n+1}} (|x|<\sqrt{2})$. 74. $\frac{e^x-1}{x}$. *Ans.* $1+\frac{x}{2!}+\frac{x^2}{3!}+\ldots+\frac{x^{n-1}}{n!}+\ldots(-\infty<x<\infty)$ 75. $\frac{1}{(1-x)^2}$. *Ans.* $\sum\limits_{n=0}^{\infty}(n+1)x^n \quad (|x|<1)$.

76. $e^x\sin x$. *Ans.* $x+x^2+\frac{2x^3}{3!}-\frac{4x^5}{5!}+\ldots+\sqrt{2^n}\sin\frac{n\pi}{4}\frac{x^n}{n!}+\ldots \quad (-\infty<x<\infty)$.

77. $\ln(x+\sqrt{1+x^2})$. *Ans.* $x-\frac{1}{2}\frac{x^3}{3}+\frac{1\cdot 3}{2\cdot 4}\frac{x^5}{5}+\ldots+(-1)^{n+1}\frac{1\cdot 3\ldots(2n-1)}{2^n\cdot n!}\frac{x^{2n+1}}{2n+1}+\ldots \quad (-1\leqslant x\leqslant 1)$.

78. $\int\limits_0^x\frac{\ln(1+x)}{x}dx$. *Ans.* $\sum\limits_{n=1}^{\infty}(-1)^{n+1}\frac{x^n}{n^2} \quad (|x|\leqslant 1)$.

79. $\int\limits_0^x\frac{\arctan x}{x}dx$. *Ans.* $\sum\limits_{n=0}^{\infty}(-1)^n\frac{x^{2n+1}}{(2n+1)^2} \quad (-1\leqslant x\leqslant 1)$.

80. $\int\frac{\cos x}{x}dx$. *Ans.* $C+\ln|x|+\sum\limits_{n=1}^{\infty}(-1)^n\frac{x^{2n}}{(2n)(2n)!} \quad (-\infty<x<0$ and $0<x<\infty)$.

81. $\int_0^x \frac{dx}{1-x^9}$. *Ans.* $\sum_{n=1}^{\infty} \frac{x^{9n-8}}{9n-8}$.

82. Prove the equations

$$\sin(a+x) = \sin a \cos x + \cos a \sin x$$
$$\cos(a+x) = \cos a \cos x - \sin a \sin x$$

by expanding the left sides in powers of x.

Utilizing appropriate series, compute:

83. $\cos 10°$ to four decimals. *Ans.* 0.9848. **84.** $\sin 1°$ to four decimals. *Ans.* 0.0175. **85.** $\sin 18°$ to three decimals. *Ans.* 0.309. **86.** $\sin \frac{\pi}{4}$ to four decimals. *Ans.* 0.7071. **87.** $\arctan \frac{1}{5}$ to four decimals. *Ans.* 0.1973. **88.** $\ln 5$ to three decimals. *Ans.* 1.609. **89.** $\log_{10} 5$ to three decimals. *Ans.* 0.699. **90.** $\arcsin 1$ to within 0.0001. *Ans.* 1.5708. **91.** $\sqrt{e}$ to within 0.0001. *Ans.* 1.6487. **92.** $\log e$ to within 0.00001. *Ans.* 0.43429. **93.** $\cos 1$ to within 0.00001. *Ans.* 0.5403.

Using a Maclaurin series expansion of the function $f(x) = \sqrt[n]{a^n + x}$, compute to within 0.001:

94. $\sqrt[3]{30}$. *Ans.* 3.107. **95.** $\sqrt{70}$. *Ans.* 4.121. **96.** $\sqrt[3]{500}$. *Ans.* 7.937. **97.** $\sqrt[5]{250}$. *Ans.* 3.017. **98.** $\sqrt{84}$. *Ans.* 9.165. **99.** $\sqrt[3]{2}$. *Ans.* 1.2598.

Expanding the integrand in a series, compute the integrals:

100. $\int_0^1 \frac{\sin x}{x} dx$ to five decimal places. *Ans.* 0.94608. **101.** $\int_0^1 e^{-x^2} dx$ to four decimals. *Ans.* 0.7468. **102.** $\int_0^{\frac{\pi}{4}} \sin(x^2) dx$ to four decimals. *Ans.* 0.1571. **103.** $\int_0^{\frac{1}{2}} e^{\sqrt{x}} dx$ to two decimals. *Ans.* 0.81. **104.** $\int_0^{0.5} \frac{\arctan x}{x} dx$ to three decimal places. *Ans.* 0.487. **105.** $\int_0^1 \cos \sqrt{x}\, dx$ to within 0.001. *Ans.* 0.764. **106.** $\int_0^{\frac{1}{4}} \ln(1+\sqrt{x})\, dx$ to within 0.001. *Ans.* 0.071. **107.** $\int_0^1 e^{-\frac{x^2}{4}} dx$ to within 0.0001. *Ans.* 0.9226. **108.** $\int_0^{\frac{1}{5}} \frac{\sin x}{\sqrt{1-x}} dx$ to within 0.0001. *Ans.* 0.0214.

109. $\int_0^{0.5} \frac{dx}{1+x^4}$ to within 0.001. *Ans.* 0.494. **110.** $\int_0^1 \frac{\ln(1+x)}{x} dx$. *Ans.* $\frac{\pi^2}{12}$.

Note. When solving this exercise and the two following ones it is well to bear in mind the equations:

$$\sum_{n=1}^{\infty}\frac{1}{n^2}=\frac{\pi^2}{6},\quad \sum_{n=1}^{\infty}\frac{(-1)^{n-1}}{n^2}=\frac{\pi^2}{12},\quad \sum_{n=1}^{\infty}\frac{1}{(2n-1)^2}=\frac{\pi^2}{8}$$

which will be established in Sec. 5.2.

111. $\int\limits_0^1 \frac{\ln(1-x)}{x}\,dx$. *Ans.* $\frac{\pi^2}{6}$. **112.** $\int\limits_0^1 \ln\frac{1+x}{1-x}\frac{dx}{x}$. *Ans.* $\frac{\pi^2}{4}$.

Integrating Differential Equations by Means of Series

113. Find the solution of the equation $y''=xy$ that satisfies the initial conditions for $x=0$, $y=1$, $y'=0$.

Hint. Look for the solution in the form of a series. *Ans.* $1+\frac{x^3}{2\cdot 3}+\frac{x^6}{2\cdot 3\cdot 5\cdot 6}+ +\ldots+\frac{x^{3k}}{2\cdot 3\cdot 5\cdot 6\ldots(3k-1)\,3k}+\ldots$.

114. Find the solution of the equation $y''+xy'+y=0$ that satisfies the initial conditions for $x=0$, $y=0$, $y'=1$. *Ans.* $x-\frac{x^3}{3}+\frac{x^5}{1\cdot 3\cdot 5}-\ldots+\frac{(-1)^{n+1}x^{2n-1}}{1\cdot 3\cdot 5\ldots(2n-1)}$.

115. Find the general solution of the equation

$$x^2y''+xy'+\left(x^2-\frac{1}{4}\right)y=0$$

Hint. Seek the solution in the form $y=x^p(A_0+A_1x+A_2x^2+\ldots)$.

Ans. $C_1x^{\frac{1}{2}}\left[1-\frac{x^2}{3!}+\frac{x^4}{5!}-\frac{x^6}{7!}+\ldots\right]+C_2x^{-\frac{1}{2}}\left[1-\frac{x^2}{2!}+\frac{x^4}{4!}-\ldots\right]= =C_1\frac{\sin x}{\sqrt{x}}+C_2\frac{\cos x}{\sqrt{x}}$.

116. Find the solution of the equation $xy''+y'+xy=0$ that satisfies the initial conditions for $x=0$, $y=1$, $y'=0$. *Ans.* $1-\frac{x^2}{2^2}+\frac{x^4}{(1\cdot 2)^2 2^4}-\frac{x^6}{(1\cdot 2\cdot 3)^2 2^6}+ +\ldots+(-1)^k\frac{x^{2k}}{(k!)^2 2^{2k}}+\ldots$.

Hint. The last two differential equations are particular cases of the Bessel equation

$$x^2y''+xy'+(x^2-n^2)\,y=0$$

for $n=\frac{1}{2}$ and $n=0$.

117. Find the general solution of the equation

$$4xy''+2y'+y=0$$

Hint. Seek the solution in the form of a series $x^p(a_0+a_1x+a_2x^2+\ldots)$.
Ans. $C_1\cos\sqrt{x}+C_2\sin\sqrt{x}$.

118. Find the solution of the equation $(1-x^2)\,y''-xy'=0$ that satisfies the initial conditions $y=0$, $y'=1$ when $x=0$. *Ans.* $x+\frac{1}{2}\frac{x^3}{3}+\frac{1}{2}\frac{3}{4}\frac{x^5}{5}+ +\frac{1}{2}\frac{3}{4}\frac{5}{6}\frac{x^7}{7}+\ldots$.

119. Find the solution of the equation $(1+x^2)y''+2xy'=0$ that satisfies the initial conditions $y=0$, $y'=1$ when $x=0$. *Ans.* $x-\frac{x^3}{3}+\frac{x^5}{5}-\frac{x^7}{7}+\ldots$.

120. Find the solution of the equation $y''=xyy'$ that satisfies the initial conditions $y=1$, $y'=1$ when $x=0$. *Ans.* $1+x+\frac{x^3}{3!}+\frac{2x^4}{4!}+\frac{3x^5}{5!}+\ldots$.

121. Find the solution of the equation $(1-x)y'=1+x-y$ that satisfies the initial conditions $y=0$ when $x=0$, and indicate the interval of convergence of the series obtained. *Ans.* $x+\frac{x^2}{1\cdot 2}+\frac{x^3}{2\cdot 3}+\frac{x^4}{3\cdot 4}+\ldots$ $(-1\leqslant x\leqslant 1)$.

122. Find the solution of the equation $xy''+y=0$ that satisfies the initial conditions $y=0$, $y'=1$ when $x=0$ and indicate the interval of convergence. *Ans.* $x-\frac{x^2}{(1!)^2\,2}+\frac{x^3}{(2!)^2\,3}-\frac{x^4}{(3!)^2\,4}+\ldots+(-1)^{n+1}\frac{x^n}{[(n-1)!]^2\,n}+\ldots$ $(-\infty<x<\infty)$.

123. Find the solution of the equation $y''+\frac{2}{x}y'+y=0$ that satisfies the initial conditions $y=1$, $y'=1$ when $x=0$. *Ans.* $\frac{\sin x}{x}$.

124. Find the solution of the equation $y''+\frac{1}{x}y'+y=0$ that satisfies the initial conditions $y=1$, $y'=0$ when $x=0$, and indicate the interval of convergence of the series obtained. *Ans.* $1-\frac{x^2}{2^2}+\frac{x^4}{2^2\cdot 4^2}-\frac{x^6}{2^2\cdot 4^2\cdot 6^2}+\ldots+(-1)^n\frac{x^{2n}}{2^{2n}(n!)^2}+\ldots$ $(|x|<\infty)$.

Find the first three terms of the power-series expansion of the solutions of the following differential equations for the given initial conditions:

125. $y'=x^2+y^2$; for $x=0$, $y=1$. *Ans.* $1+x+x^2+\frac{4x^3}{3}+\ldots$.

126. $y''=e^y+x$; for $x=0$, $y=1$, $y'=0$. *Ans.* $1+\frac{ex^2}{2}+\frac{x^3}{6}+\ldots$.

127. $y'=\sin y-\sin x$; for $x=0$, $y=0$. *Ans.* $-\frac{x^2}{2}-\frac{x^3}{6}-\ldots$.

Find several terms of the series expansion of solutions of differential equations under the indicated initial conditions:

128. $y''=yy'-x^2$ when $x=0$, $y=1$ and $y'=1$. *Ans.* $1+x+\frac{x^2}{2!}+\frac{2x^3}{3!}+\frac{3x^4}{4!}+\frac{14x^5}{5!}+\ldots$. **129.** $y'=y^2+x^3$ when $x=0$ and $y=\frac{1}{2}$. *Ans.* $\frac{1}{2}+\frac{1}{4}x+\frac{1}{8}x^2+\frac{1}{16}x^3+\frac{9}{32}x^4+\ldots$. **130.** $y'=x^2-y^2$ when $x=0$ and $y=0$. *Ans.* $\frac{1}{3}x^3-\frac{1}{7\cdot 9}x^7+\frac{2}{7\cdot 11\cdot 27}x^{11}-\ldots$. **131.** $y'=x^2y^2-1$ when $x=0$ and $y=1$. *Ans.* $1-x+\frac{x^3}{3}-\frac{x^4}{2}+\frac{x^5}{5}-\ldots$. **132.** $y'=e^y+xy$ when $x=0$ and $y=0$. *Ans.* $x+\frac{x^2}{2}+\frac{2x^3}{3}+\frac{11x^4}{2\cdot 3\cdot 4}+\ldots$.

CHAPTER 5

FOURIER SERIES

5.1 DEFINITION. STATEMENT OF THE PROBLEM

A functional series of the form

$$\frac{a_0}{2}+a_1\cos x+b_1\sin x+a_2\cos 2x+b_2\sin 2x+\ldots$$

or, more compactly, a series of the form

$$\frac{a_0}{2}+\sum_{n=1}^{\infty}(a_n\cos nx+b_n\sin nx) \qquad (1)$$

is called a *trigonometric series*. The constants a_0, a_n and b_n $(n=1, 2, \ldots)$ are called *coefficients of the trigonometric series*.

If series (1) converges, then its sum is a periodic function $f(x)$ with period 2π, since $\sin nx$ and $\cos nx$ are periodic functions with period 2π.

Thus,

$$f(x)=f(x+2\pi)$$

Let us pose the following problem.

Given a function $f(x)$ which is periodic and has a period 2π, *Under what conditions for $f(x)$ is it possible to find a trigonometric series convergent to the given function?*

That is the problem that we shall solve in this chapter.

Determining the coefficients of a series from Fourier's formulas. Let the periodic function $f(x)$ with period 2π be such that it may be represented as a trigonometric series convergent to a given function in the interval $(-\pi, \pi)$; i.e., that it is the sum of this series:

$$f(x)=\frac{a_0}{2}+\sum_{n=1}^{\infty}(a_n\cos nx+b_n\sin nx) \qquad (2)$$

Suppose that the integral of the function on the left-hand side of this equation is equal to the sum of the integrals of the terms of the series (2). This will be the case, for example, if we assume that the numerical series made up of the coefficients of the given trigonometric series converges absolutely; that is, that the following positive number series converges:

$$\left|\frac{a_0}{2}\right|+|a_1|+|b_1|+|a_2|+|b_2|+\ldots+|a_n|+|b_n|+\ldots \qquad (3)$$

Then series (1) is dominated and, consequently, it may be integrated termwise in the interval from $-\pi$ to π. Let us take advantage of this for computing the coefficient a_0.

Integrate both sides of (2) from $-\pi$ to $+\pi$:

$$\int_{-\pi}^{\pi} f(x)\,dx=\int_{-\pi}^{\pi}\frac{a_0}{2}\,dx+\sum_{n=1}^{\infty}\left(\int_{-\pi}^{\pi} a_n\cos nx\,dx+\int_{-\pi}^{\pi} b_n\sin nx\,dx\right)$$

Evaluate separately each integral on the right side:

$$\int_{-\pi}^{\pi}\frac{a_0}{2}\,dx=\pi a_0$$

$$\int_{-\pi}^{\pi} a_n\cos nx\,dx=a_n\int_{-\pi}^{\pi}\cos nx\,dx=\frac{a_n\sin nx}{n}\Bigg|_{-\pi}^{\pi}=0$$

$$\int_{-\pi}^{\pi} b_n\sin nx\,dx=b_n\int_{-\pi}^{\pi}\sin nx\,dx=-b_n\frac{\cos nx}{n}\Bigg|_{-\pi}^{\pi}=0$$

Consequently,

$$\int_{-\pi}^{\pi} f(x)\,dx=\pi a_0$$

whence

$$a_0=\frac{1}{\pi}\int_{-\pi}^{\pi} f(x)\,dx \qquad (4)$$

To calculate the other coefficients of the series we shall need certain definite integrals, which we will consider first.

If n and k are integers, then we have the following equations: if $n\neq k$, then

$$\left.\begin{aligned}\int_{-\pi}^{\pi}\cos nx\cos kx\,dx=0\\ \int_{-\pi}^{\pi}\cos nx\sin kx\,dx=0\\ \int_{-\pi}^{\pi}\sin nx\sin kx\,dx=0\end{aligned}\right\} \qquad (\mathrm{I})$$

but if $n=k$, then

$$\left.\begin{aligned}&\int_{-\pi}^{\pi}\cos^2 kx\,dx=\pi\\&\int_{-\pi}^{\pi}\sin kx\cos kx\,dx=0\\&\int_{-\pi}^{\pi}\sin^2 kx\,dx=\pi\end{aligned}\right\}\qquad\text{(II)}$$

To take an example, evaluate the first integral of group (I). Since

$$\cos nx\cos kx=\frac{1}{2}[\cos(n+k)x+\cos(n-k)x]$$

it follows that

$$\int_{-\pi}^{\pi}\cos nx\cos kx\,dx=\frac{1}{2}\int_{-\pi}^{\pi}\cos(n+k)x\,dx+\frac{1}{2}\int_{-\pi}^{\pi}\cos(n-k)x\,dx=0$$

The other formulas of (I) * are obtained in similar fashion. The integrals of group (II) are computed directly (see Ch. 10, Vol. I).

Now we can compute the coefficients a_k and b_k of series (2).

To find the coefficient a_k for some definite value $k\neq 0$, multiply both sides of (2) by $\cos kx$:

$$f(x)\cos kx=\frac{a_0}{2}\cos kx+\sum_{n=1}^{\infty}(a_n\cos nx\cos kx+b_n\sin nx\cos kx)\qquad(2')$$

The resulting series on the right is dominated, since its terms do not exceed (in absolute value) the terms of the convergent positive series (3). We can therefore integrate it termwise on any interval.

Integrate (2′) from $-\pi$ to π:

$$\int_{-\pi}^{\pi}f(x)\cos kx\,dx=\frac{a_0}{2}\int_{-\pi}^{\pi}\cos kx\,dx$$
$$+\sum_{n=1}^{\infty}\left(a_n\int_{-\pi}^{\pi}\cos nx\cos kx\,dx+b_n\int_{-\pi}^{\pi}\sin nx\cos kx\,dx\right)$$

Taking into account formulas (II) and (I), we see that all the integrals on the right are equal to zero, with the exception of the

---

* By means of the formulas

$$\cos nx\sin kx={}^1/_2[\sin(n+k)x-\sin(n-k)x]$$
$$\sin nx\sin kx={}^1/_2[-\cos(n+k)x+\cos(n-k)x]$$

integral with coefficient a_k. Hence,

$$\int_{-\pi}^{\pi} f(x)\cos kx\,dx = a_k \int_{-\pi}^{\pi} \cos^2 kx\,dx = a_k\pi$$

whence

$$a_k = \frac{1}{\pi}\int_{-\pi}^{\pi} f(x)\cos kx\,dx \qquad (5)$$

Multiplying both sides of (2) by $\sin kx$ and again integrating from $-\pi$ to π, we find

$$\int_{-\pi}^{\pi} f(x)\sin kx\,dx = b_k \int_{-\pi}^{\pi} \sin^2 kx\,dx = b_k\pi,$$

whence

$$b_k = \frac{1}{\pi}\int_{-\pi}^{\pi} f(x)\sin kx\,dx \qquad (6)$$

The coefficients determined from formulas (4), (5) and (6) are called *Fourier coefficients* of the function $f(x)$, and the trigonometric series (1) with such coefficients is called a *Fourier series* of the function $f(x)$.

Let us now revert to the question posed at the beginning of this section: What properties must a function have so that the Fourier series constructed for it should converge and so that the sum of the constructed Fourier series should equal the values of the given function at corresponding points? We shall here state a theorem that will yield sufficient conditions for representing a function $f(x)$ by a Fourier series.

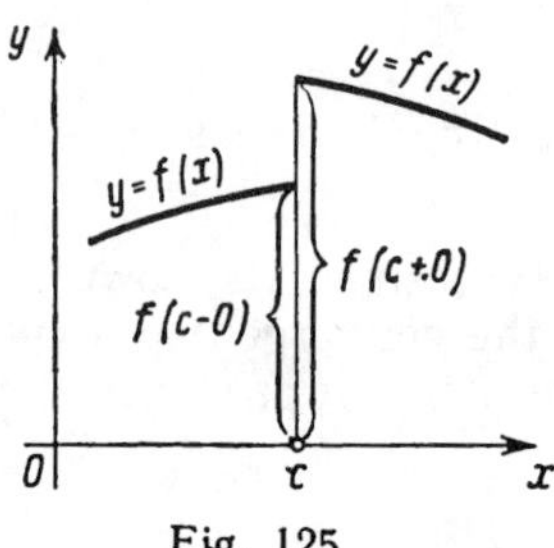

Fig. 125

Definition. A function $f(x)$ is called **piecewise monotonic** on the interval $[a, b]$ if this interval may be divided by a finite number of points $x_1, x_2, \ldots, x_{n-1}$ into subintervals $(a, x_1), (x_1, x_2), \ldots, (x_{n-1}, b)$ such that the function is monotonic (that is, either nonincreasing or nondecreasing) on each of the subintervals.

From the definition it follows that if the function $f(x)$ is piecewise monotonic and bounded on the interval $[a, b]$, then it can have only discontinuities of the first kind. Indeed, if $x=c$ is a point of discontinuity of the function $f(x)$, then, by virtue of the monotonicity of the function, there exist the limits

$$\lim_{x\to c-0} f(x) = f(c-0), \qquad \lim_{x\to c+0} f(x) = f(c+0)$$

i.e., the point c is a discontinuity of the first kind (Fig. 125) We now state the following theorem.

Theorem. *If a periodic function $f(x)$ with period 2π is piecewise monotonic and bounded on the interval $[-\pi, \pi]$, then the Fourier series constructed for this function converges at all points. The sum of the resultant series $s(x)$ is equal to the value of $f(x)$ at the points of its continuity. At the discontinuities of $f(x)$, the sum of the series is equal to the arithmetic mean of the limits of $f(x)$ on the right and on the left; that is, if $x=c$ is a discontinuity of the function $f(x)$, then*

$$s(x)_{x=c}=\frac{f(c-0)+f(c+0)}{2}$$

From this theorem it follows that the class of functions that may be represented by Fourier series is rather broad. That is why Fourier series have found extensive applications in various divisions of mathematics. Particularly effective use is made of Fourier series in mathematical physics and its applications to specific problems of mechanics and physics (see Ch. 6).

We give this theorem without proof. In Secs. 5.8-5.10 we will prove another sufficient condition for the expandability of a function in a Fourier series, which condition in a certain sense deals with a narrower class of functions.

5.2 EXPANSIONS OF FUNCTIONS IN FOURIER SERIES

The following are some instances of the expansion of functions of Fourier series.

Example 1. A periodic function $f(x)$ with period 2π is defined as follows:

$$f(x)=x, \quad -\pi < x \leqslant \pi$$

This function is piecewise monotonic and bounded (Fig. 126). Hence, it admits expansion in a Fourier series.

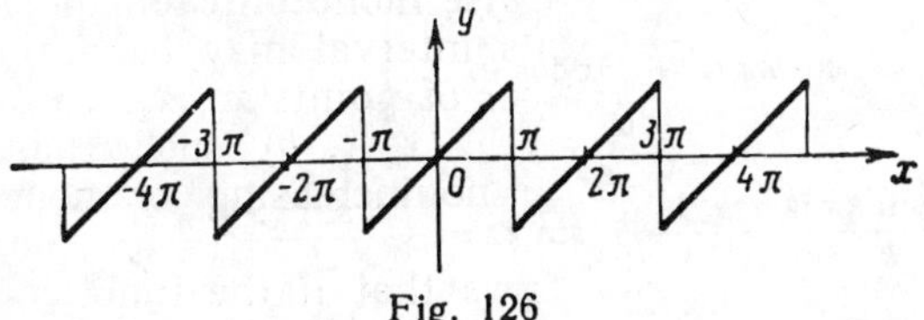

Fig. 126

By formula (4), Sec. 5.1, we find

$$a_0=\frac{1}{\pi}\int_{-\pi}^{\pi} x\,dx=\frac{1}{\pi}\frac{x^2}{2}\Big|_{-\pi}^{\pi}=0.$$

Applying formula (5), Sec. 5.1, and integrating by parts, we find

$$a_k=\frac{1}{\pi}\int_{-\pi}^{\pi} x\cos kx\,dx=\frac{1}{\pi}\left[x\frac{\sin kx}{k}\Big|_{-\pi}^{\pi}-\frac{1}{k}\int_{-\pi}^{\pi}\sin kx\,dx\right]=0$$

By formula (6), Sec. 5.1, we have

$$b_k=\frac{1}{\pi}\int_{-\pi}^{\pi} x\sin kx\,dx=\frac{1}{\pi}\left[-x\frac{\cos kx}{k}\Big|_{-\pi}^{\pi}+\frac{1}{k}\int_{-\pi}^{\pi}\cos kx\,dx\right]=(-1)^{k+1}\frac{2}{k}$$

Thus, we get the series

$$f(x)=2\left[\frac{\sin x}{1}-\frac{\sin 2x}{2}+\frac{\sin 3x}{3}-\ldots(-1)^{k+1}\frac{\sin kx}{k}+\ldots\right]$$

This equation occurs at all points except at points of discontinuity. At each discontinuity, the sum of the series is equal to the arithmetical mean of its limits on the right and left, which is zero.

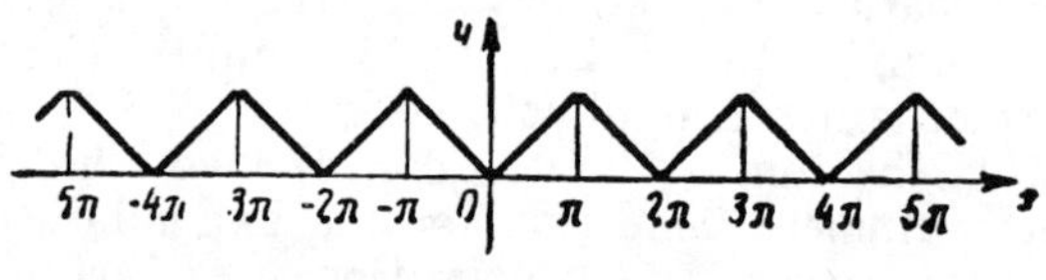

Fig. 127

Example 2. A periodic function $f(x)$ with period 2π is defined as follows:

$$f(x)=-x \quad \text{for} \quad -\pi\leqslant x\leqslant 0$$
$$f(x)=x \quad \text{for} \quad 0<x\leqslant \pi$$

[or $f(x)=|x|$] (Fig. 127). This function is also piecewise monotonic and bounded on the interval $-\pi\leqslant x\leqslant\pi$.

Let us determine its Fourier coefficients:

$$a_0=\frac{1}{\pi}\int_{-\pi}^{\pi} f(x)\,dx=\frac{1}{\pi}\left[\int_{-\pi}^{0}(-x)\,dx+\int_{0}^{\pi}x\,dx\right]=\pi$$

$$a_k=\frac{1}{\pi}\left[\int_{-\pi}^{0}(-x)\cos kx\,dx+\int_{0}^{\pi}x\cos kx\,dx\right]$$

$$=\frac{1}{\pi}\left[-\frac{x\sin kx}{k}\Big|_{-\pi}^{0}+\frac{1}{k}\int_{-\pi}^{0}\sin kx\,dx+\frac{x\sin kx}{k}\Big|_{0}^{\pi}-\frac{1}{k}\int_{0}^{\pi}\sin kx\,dx\right]$$

$$=\frac{1}{\pi k}\left[-\frac{\cos kx}{k}\Big|_{-\pi}^{0}+\frac{\cos kx}{k}\Big|_{0}^{\pi}\right]=\frac{2}{\pi k^2}(\cos k\pi-1)=\begin{cases}0 & \text{for } k \text{ even}\\ -\dfrac{4}{\pi k^2} & \text{for } k \text{ odd}\end{cases}$$

$$b_k=\frac{1}{\pi}\left[\int_{-\pi}^{0}(-x)\sin kx\,dx+\int_{0}^{\pi}x\sin kx\,dx\right]=0$$

We thus obtain the series

$$f(x)=\frac{\pi}{2}-\frac{4}{\pi}\left[\frac{\cos x}{1^2}+\frac{\cos 3x}{3^2}+\frac{\cos 5x}{5^2}+\ldots+\frac{\cos(2p+1)x}{(2p+1)^2}+\ldots\right]$$

This series converges at all points, and its sum is equal to the given function.

Example 3. A periodic function $f(x)$ with period 2π is defined as follows:

$$f(x)=-1 \quad \text{for} \quad -\pi<x<0$$
$$f(x)=1 \quad \text{for} \quad 0\leqslant x\leqslant\pi$$

This function (Fig. 128) is piecewise monotonic and bounded on the interval $-\pi\leqslant x\leqslant\pi$.

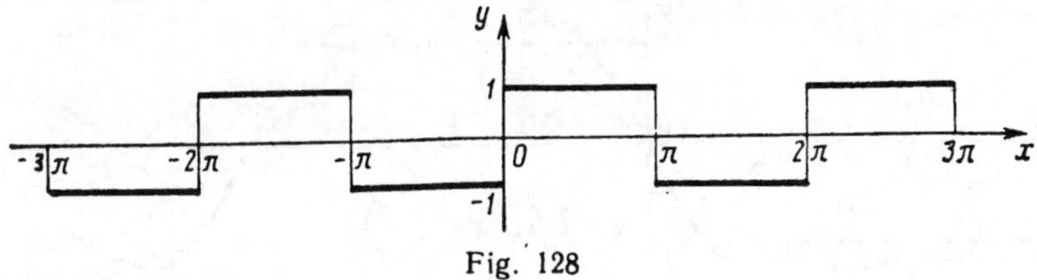

Fig. 128

Let us compute its Fourier coefficients:

$$a_0=\frac{1}{\pi}\int_{-\pi}^{\pi} f(x)\,dx=\frac{1}{\pi}\left[\int_{-\pi}^{0}(-1)\,dx+\int_{0}^{\pi}dx\right]=0$$

$$a_k=\frac{1}{\pi}\left[\int_{-\pi}^{0}(-1)\cos kx\,dx+\int_{0}^{\pi}\cos kx\,dx\right]=-1\cdot\frac{\sin kx}{k\pi}\Big|_{-\pi}^{0}+\frac{\sin kx}{k\pi}\Big|_{0}^{\pi}=0$$

$$b_k=\frac{1}{\pi}\left[\int_{-\pi}^{0}(-1)\sin kx\,dx+\int_{0}^{\pi}\sin kx\,dx\right]=\frac{1}{\pi}\left[\frac{\cos kx}{k}\Big|_{-\pi}^{0}-\frac{\cos kx}{k}\Big|_{0}^{\pi}\right]$$

$$=\frac{2}{\pi k}[1-\cos\pi k]=\begin{cases}0 & \text{for } k \text{ even}\\ \dfrac{4}{\pi k} & \text{for } k \text{ odd}\end{cases}$$

Consequently, for the function at hand the Fourier series has the form

$$f(x)=\frac{4}{\pi}\left[\frac{\sin x}{1}+\frac{\sin 3x}{3}+\frac{\sin 5x}{5}+\ldots+\frac{\sin(2p+1)x}{2p+1}+\ldots\right]$$

This equation holds at all points with the exception of discontinuities.

Fig. 129 illustrates how the partial sums s_n of the series represent more and more accurately the function $f(x)$ as $n\to\infty$.

Example 4. A periodic function $f(x)$ with period 2π is defined as follows:

$$f(x)=x^2, \quad -\pi\leqslant x\leqslant\pi \qquad \text{(Fig. 130)}$$

Determine its Fourier coefficients.

$$a_0=\frac{1}{\pi}\int_{-\pi}^{\pi}x^2\,dx=\frac{1}{\pi}\,\frac{x^3}{3}\Big|_{-\pi}^{\pi}=\frac{2\pi^2}{3}$$

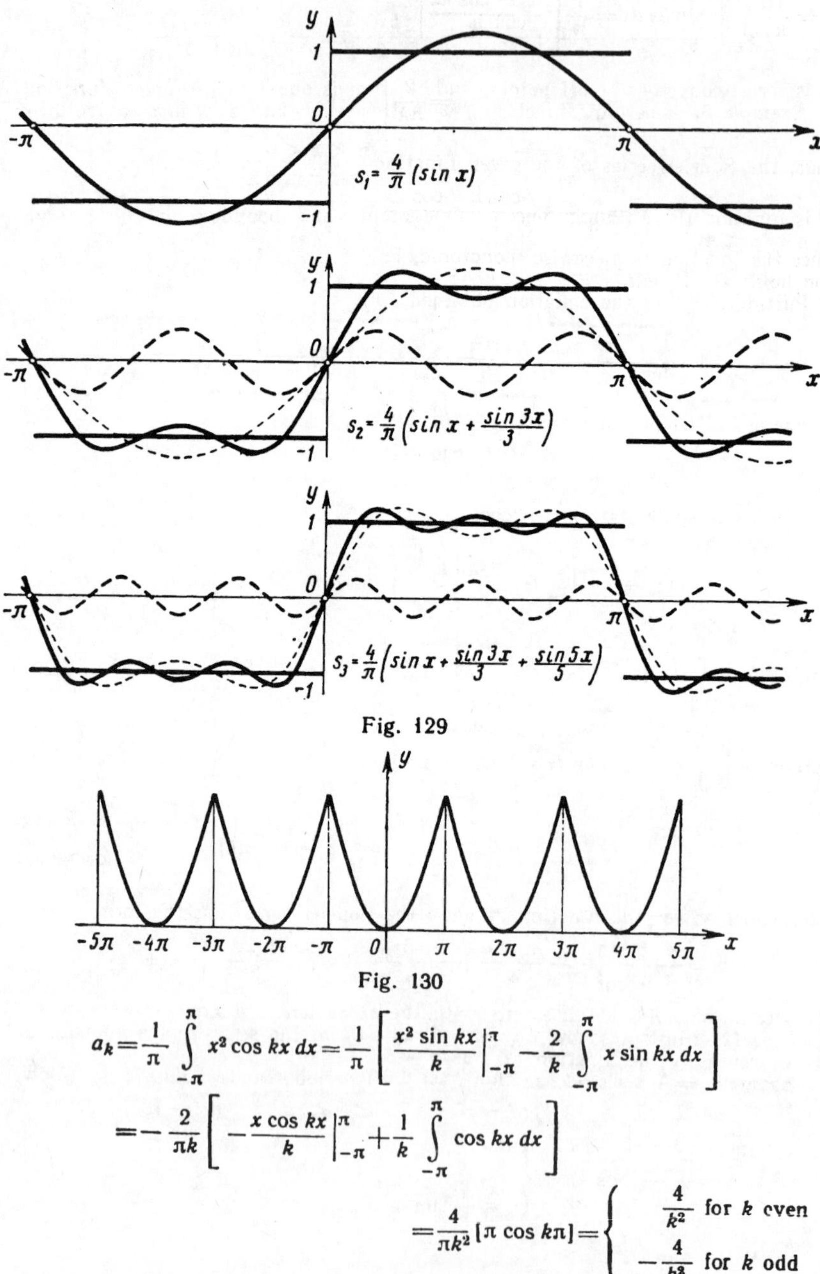

Fig. 129

Fig. 130

$$a_k = \frac{1}{\pi}\int_{-\pi}^{\pi} x^2 \cos kx\,dx = \frac{1}{\pi}\left[\frac{x^2 \sin kx}{k}\Big|_{-\pi}^{\pi} - \frac{2}{k}\int_{-\pi}^{\pi} x \sin kx\,dx\right]$$

$$= -\frac{2}{\pi k}\left[-\frac{x\cos kx}{k}\Big|_{-\pi}^{\pi} + \frac{1}{k}\int_{-\pi}^{\pi} \cos kx\,dx\right]$$

$$= \frac{4}{\pi k^2}[\pi \cos k\pi] = \begin{cases} \frac{4}{k^2} & \text{for } k \text{ even} \\ -\frac{4}{k^2} & \text{for } k \text{ odd} \end{cases}$$

$$b_k=\frac{1}{\pi}\int\limits_{-\pi}^{\pi} x^2 \sin kx\,dx=\frac{1}{\pi}\left[-\frac{x^2\cos kx}{k}\Big|_{-\pi}^{\pi}+\frac{2}{k}\int\limits_{-\pi}^{\pi} x\cos kx\,dx\right]$$

$$=\frac{2}{\pi k}\left[\frac{x\sin kx}{k}\Big|_{-\pi}^{\pi}-\frac{1}{k}\int\limits_{-\pi}^{\pi}\sin kx\,dx\right]=0$$

Thus, the Fourier series of the given function has the form

$$x^2=\frac{\pi^2}{3}-4\left(\frac{\cos x}{1}-\frac{\cos 2x}{2^2}+\frac{\cos 3x}{3^2}-\ldots\right)$$

Since the function is piecewise monotonic, bounded and continuous, this equation holds at all points.

Putting $x=\pi$ in the equation obtained, we get

$$\frac{\pi^2}{6}=\sum_{n=1}^{\infty}\frac{1}{n^2}$$

Example 5. A periodic function $f(x)$ with period 2π is defined as follows:

$$f(x)=0 \quad \text{for} \quad -\pi<x\leqslant 0$$
$$f(x)=x \quad \text{for} \quad 0<x\leqslant\pi \qquad \text{(Fig. 131)}$$

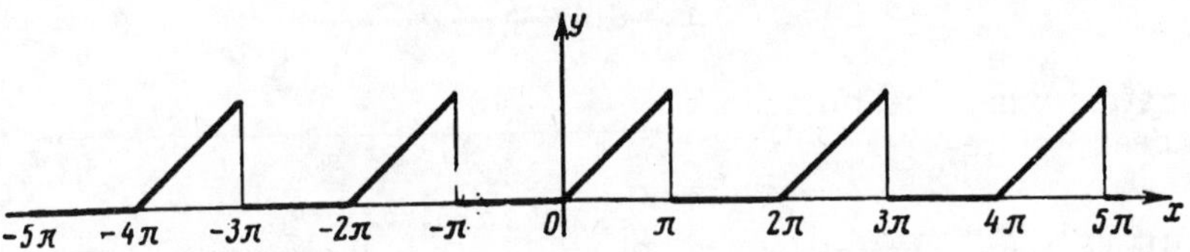

Fig. 131

Determine the Fourier coefficients:

$$a_0=\frac{1}{\pi}\int\limits_{-\pi}^{\pi} f(x)\,dx=\frac{1}{\pi}\left[\int\limits_{-\pi}^{0}0\cdot dx+\int\limits_{0}^{\pi}x\,dx\right]=\frac{1}{\pi}\,\frac{\pi^2}{2}=\frac{\pi}{2}$$

$$a_k=\frac{1}{\pi}\int\limits_{0}^{\pi}x\cos kx\,dx=\frac{1}{\pi}\left[\frac{x\sin kx}{k}\Big|_{0}^{\pi}-\frac{1}{k}\int\limits_{0}^{\pi}\sin kx\,dx\right]$$

$$=\frac{1}{\pi k}\,\frac{\cos kx}{k}\Big|_{0}^{\pi}=\begin{cases}-\dfrac{2}{\pi k^2} & \text{for } k \text{ odd}\\ 0 & \text{for } k \text{ even}\end{cases}$$

$$b_k=\frac{1}{\pi}\int\limits_{0}^{\pi}x\sin kx\,dx=\frac{1}{\pi}\left[-\frac{x\cos kx}{k}\Big|_{0}^{\pi}+\frac{1}{k}\int\limits_{0}^{\pi}\cos kx\,dx\right]$$

$$=-\frac{\pi}{\pi k}\cos k\pi=\begin{cases}\dfrac{1}{k} & \text{for } k \text{ odd}\\ -\dfrac{1}{k} & \text{for } k \text{ even}\end{cases}$$

The Fourier series will thus have the form

$$f(x)=\frac{\pi}{4}-\frac{2}{\pi}\left(\frac{\cos x}{1^2}+\frac{\cos 3x}{3^2}+\frac{\cos 5x}{5^2}+\ldots\right)+\left(\frac{\sin x}{1}-\frac{\sin 2x}{2}+\frac{\sin 3x}{3}-\ldots\right)$$

At the discontinuities of the function $f(x)$, the sum of the series is equal to the arithmetic mean of its limits on the right and left $\left(\text{in this case, to the number } \frac{\pi}{2}\right)$.

Putting $x=0$ in the equation obtained, we get

$$\frac{\pi^2}{8}=\sum_{n=1}^{\infty}\frac{1}{(2n-1)^2}$$

5.3 A REMARK ON THE EXPANSION OF A PERIODIC FUNCTION IN A FOURIER SERIES

We note the following property of a periodic function $\psi(x)$ with period 2π:

$$\int_{-\pi}^{\pi}\psi(x)\,dx=\int_{\lambda}^{\lambda+2\pi}\psi(x)\,dx$$

no matter what the number λ.

Indeed, since

$$\psi(\xi-2\pi)=\psi(\xi)$$

it follows that, putting $x=\xi-2\pi$, we can write (for all c and d):

$$\int_{c}^{d}\psi(x)\,dx=\int_{c+2\pi}^{d+2\pi}\psi(\xi-2\pi)\,d\xi=\int_{c+2\pi}^{d+2\pi}\psi(\xi)\,d\xi=\int_{c+2\pi}^{d+2\pi}\psi(x)\,dx$$

In particular, taking $c=-\pi$, $d=\lambda$, we get

$$\int_{-\pi}^{\lambda}\psi(x)\,dx=\int_{\pi}^{\lambda+2\pi}\psi(x)\,dx$$

therefore,

$$\int_{\lambda}^{\lambda+2\pi}\psi(x)\,dx=\int_{\lambda}^{-\pi}\psi(x)\,dx+\int_{-\pi}^{\pi}\psi(x)\,dx+\int_{\pi}^{\lambda+2\pi}\psi(x)\,dx$$

$$=\int_{\lambda}^{-\pi}\psi(x)\,dx+\int_{-\pi}^{\pi}\psi(x)\,dx+\int_{-\pi}^{\lambda}\psi(x)\,dx=\int_{-\pi}^{\pi}\psi(x)\,dx$$

This property means that the *integral of a periodic function* $\psi(x)$ *over any interval whose length is equal to the period always has the same value*. This fact is readily illustrated geometrically: the cross-hatched areas in Fig. 132 are equal.

From the property that has been proved it follows that when computing Fourier coefficients we can replace the interval of integration $(-\pi, \pi)$ by the interval of integration $(\lambda, \lambda+2\pi)$, that is, we can put

$$\left.\begin{array}{l} a_0=\frac{1}{\pi}\int\limits_{\lambda}^{\lambda+2\pi} f(x)\,dx, \quad a_n=\frac{1}{\pi}\int\limits_{\lambda}^{\lambda+2\pi} f(x)\cos nx\,dx \\ b_n=\frac{1}{\pi}\int\limits_{\lambda}^{\lambda+2\pi} f(x)\sin nx\,dx \end{array}\right\} \quad (1)$$

where λ is any number.

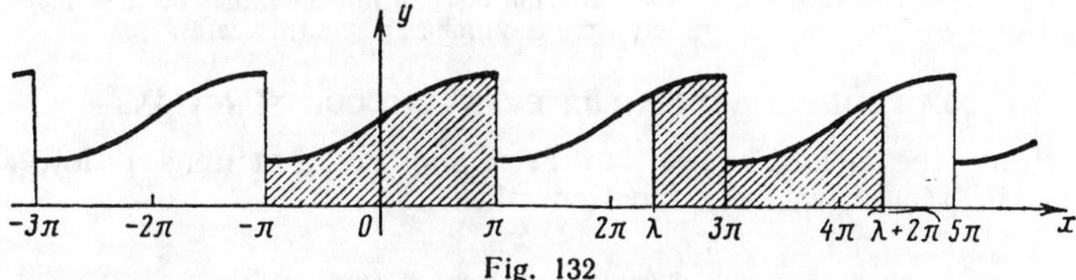

Fig. 132

This follows from the fact that the function $f(x)$ is, by hypothesis, periodic with period 2π; hence, both the functions $f(x)\cos nx$ and $f(x)\sin nx$ are periodic functions with period 2π. We now illustrate how this property simplifies the process of finding coefficients in certain cases.

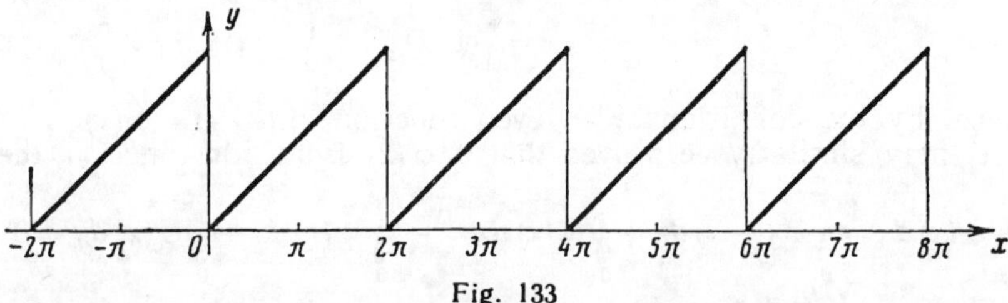

Fig. 133

Example. Let it be required to expand in a Fourier series the function $f(x)$ with period 2π, which is given on the interval $0 \leqslant x \leqslant 2\pi$ by the equation

$$f(x)=x$$

The graph of $f(x)$ is shown in Fig. 133. On the interval $[-\pi, \pi]$ this function is represented by two formulas: $f(x)=x+2\pi$ on the interval $[-\pi, 0]$ and $f(x)=x$ on the interval $[0, \pi]$. Yet, on $[0, 2\pi]$ it is far more simply represented by a single formula $f(x)=x$. Therefore, to expand this function in a Fourier series it is better to make use of formulas (1), setting $\lambda=0$:

$$a_0=\frac{1}{\pi}\int\limits_0^{2\pi} f(x)\,dx=\frac{1}{\pi}\int\limits_0^{2\pi} x\,dx=2\pi$$

$$a_n=\frac{1}{\pi}\int_0^{2\pi} f(x)\cos nx\,dx=\frac{1}{\pi}\int_0^{2\pi} x\cos nx\,dx=\frac{1}{\pi}\left[\frac{x\sin nx}{n}+\frac{\cos nx}{n^2}\right]_0^{2\pi}=0$$

$$b_n=\frac{1}{\pi}\int_0^{2\pi} f(x)\sin nx\,dx=\frac{1}{\pi}\int_0^{2\pi} x\sin nx\,dx=\frac{1}{\pi}\left[-\frac{x\cos nx}{n}+\frac{\sin nx}{n^2}\right]_0^{2\pi}=-\frac{2}{n}$$

Consequently,

$$f(x)=\pi-2\sin x-\frac{2}{2}\sin 2x-\frac{2}{3}\sin 3x-\frac{2}{4}\sin 4x-\frac{2}{5}\sin 5x-\ldots$$

This series yields the given function at all points with the exception of points of discontinuity (i.e., except the points $x=0, 2\pi, 4\pi, \ldots$). At these points the sum of the series is equal to the half sum of the limiting values of the function $f(x)$ on the right and on the left (to the number π, in this case).

5.4 FOURIER SERIES FOR EVEN AND ODD FUNCTIONS

From the definitions of an even and odd functions it follows that if $\psi(x)$ is an **even** function, then

$$\int_{-\pi}^{\pi}\psi(x)\,dx=2\int_0^{\pi}\psi(x)\,dx$$

Indeed,

$$\int_{-\pi}^{\pi}\psi(x)\,dx=\int_{-\pi}^{0}\psi(x)\,dx+\int_0^{\pi}\psi(x)\,dx=\int_0^{\pi}\psi(-x)\,dx+\int_0^{\pi}\psi(x)\,dx$$

$$=\int_0^{\pi}\psi(x)\,dx+\int_0^{\pi}\psi(x)\,dx=2\int_0^{\pi}\psi(x)\,dx$$

since, by the definition of an even function, $\psi(-x)=\psi(x)$.

It may similarly be proved that if $\varphi(x)$ is an **odd** function, then

$$\int_{-\pi}^{\pi}\varphi(x)\,dx=\int_0^{\pi}\varphi(-x)\,dx+\int_0^{\pi}\varphi(x)\,dx=-\int_0^{\pi}\varphi(x)\,dx+\int_0^{\pi}\varphi(x)\,dx=0$$

If an **odd** function $f(x)$ is expanded in a Fourier series, then the product $f(x)\cos kx$ is also an odd function, while $f(x)\sin kx$ is an even function; hence,

$$\left.\begin{aligned} a_0&=\frac{1}{\pi}\int_{-\pi}^{\pi} f(x)\,dx=0\\ a_k&=\frac{1}{\pi}\int_{-\pi}^{\pi} f(x)\cos kx\,dx=0\\ b_k&=\frac{1}{\pi}\int_{-\pi}^{\pi} f(x)\sin kx\,dx=\frac{2}{\pi}\int_0^{\pi} f(x)\sin kx\,dx \end{aligned}\right\}\qquad(1)$$

Thus the Fourier series of an **odd** function contains **"only sines"** (see Example 1, Sec. 5.2).

If an **even** function is expanded in a Fourier series, the product $f(x)\sin kx$ is an odd function, while $f(x)\cos kx$ is an even function and, hence,

$$\left.\begin{aligned} a_0 &= \frac{2}{\pi}\int_0^{\pi} f(x)\,dx \\ a_k &= \frac{2}{\pi}\int_0^{\pi} f(x)\cos kx\,dx \\ b_k &= \frac{1}{\pi}\int_{-\pi}^{\pi} f(x)\sin kx\,dx = 0 \end{aligned}\right\} \tag{2}$$

Thus, the Fourier series of an **even** function contains **"only cosines"** (see Example 2, Sec. 5.2).

The formulas obtained permit simplifying computations when seeking Fourier coefficients in cases when the given function is even or odd. It is obvious that not every periodic function is even or odd (see Example 5, Sec. 5.2).

Example. Let it be required to expand in a Fourier series the even function $f(x)$ which has a period of 2π and on the interval $[0, \pi]$ is given by the equation

$$y = x$$

We have already expanded this function in a Fourier series in Example 2, Sec. 5.2 (Fig. 127). Let us again compute the Fourier series of this function, taking advantage of the fact that the given function is even.

By virtue of formulas (2) $b_k = 0$ for any k;

$$a_0 = \frac{2}{\pi}\int_0^{\pi} x\,dx = \pi, \quad a_k = \frac{2}{\pi}\int_0^{\pi} x\cos kx\,dx$$

$$= \frac{2}{\pi}\left[\frac{x\sin kx}{k} + \frac{\cos kx}{k^2}\right]_0^{\pi} = \frac{2}{\pi k^2}[(-1)^k - 1] = \begin{cases} 0 & \text{for } k \text{ even} \\ -\dfrac{4}{\pi k^2} & \text{for } k \text{ odd} \end{cases}$$

We obtained the same coefficients as in Example 2, Sec. 5.2, but this time by a short cut.

5.5 THE FOURIER SERIES FOR A FUNCTION WITH PERIOD 2l

Let $f(x)$ be a periodic function with period $2l$, generally speaking, different from 2π. Expand it in a Fourier series.

Make the substitution

$$x = \frac{lt}{\pi}$$

Then the function $f\left(\frac{lt}{\pi}\right)$ will be a periodic function of t with period 2π.

It may be expanded in a Fourier series on the interval $-\pi \leqslant \leqslant x \leqslant \pi$:

$$f\left(\frac{l}{\pi}t\right)=\frac{a_0}{2}+\sum_{k=1}^{\infty}(a_k \cos kt+b_k \sin kt) \tag{1}$$

where

$$a_0=\frac{1}{\pi}\int_{-\pi}^{\pi} f\left(\frac{l}{\pi}t\right)dt, \quad a_k=\frac{1}{\pi}\int_{-\pi}^{\pi} f\left(\frac{l}{\pi}t\right)\cos kt\, dt$$

$$b_k=\frac{1}{\pi}\int_{-\pi}^{\pi} f\left(\frac{l}{\pi}t\right)\sin kt\, dt$$

Now let us return to the original variable x:

$$x=\frac{l}{\pi}t, \quad t=x\frac{\pi}{l}, \quad dt=\frac{\pi}{l}dx$$

We will then have

$$\left.\begin{aligned} a_0&=\frac{1}{l}\int_{-l}^{l} f(x)\,dx, \quad a_k=\frac{1}{l}\int_{-l}^{l} f(x)\cos k\frac{\pi}{l}x\,dx \\ b_k&=\frac{1}{l}\int_{-l}^{l} f(x)\sin k\frac{\pi}{l}x\,dx \end{aligned}\right\} \tag{2}$$

Formula (1) takes the form

$$f(x)=\frac{a_0}{2}+\sum_{k=1}^{\infty}\left(a_k \cos\frac{k\pi}{l}x+b_k \sin\frac{k\pi}{l}x\right) \tag{3}$$

where the coefficients a_0, a_k, b_k are computed from formulas (2). This is the Fourier series for a periodic function with period $2l$.

We note that all the theorems that hold for Fourier series of periodic functions with period 2π hold also for Fourier series of periodic functions with some other period $2l$. In particular, the sufficient condition for expansion of a function in a Fourier series (see end of Sec. 5.1) holds true, as do also the remark on the possibility of computing coefficients of the series by integrating over any interval whose length is equal to the period (see Sec. 5.3), and the remark on the possibility of simplifying computation of coefficients of the series if the function is even or odd (Sec. 5.4).

Example. Expand in a Fourier series the periodic function $f(x)$ with period $2l$ which on the interval $[-l, l]$ is given by the equation $f(x)=|x|$ (Fig. 134).

Solution. Since the function at hand is even, it follows that

$$b_k = 0, \quad a_0 = \frac{2}{l}\int_0^l x\,dx = l$$

$$a_k = \frac{2}{l}\int_0^l x\cos\frac{k\pi x}{l}\,dx = \frac{2l}{\pi^2}\int_0^\pi x\cos kx\,dx = \begin{cases} 0 & \text{for } k \text{ even} \\ -\dfrac{4l}{\pi^2 k^2} & \text{for } k \text{ odd} \end{cases}$$

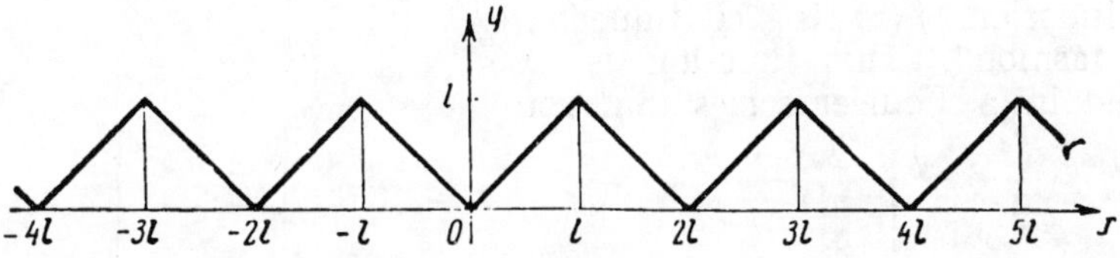

Fig. 134

Hence, the expansion is of the form

$$|x| = \frac{l}{2} - \frac{4l}{\pi^2}\left[\frac{\cos\frac{\pi}{l}x}{1} + \frac{\cos\frac{3\pi}{l}x}{3^2} + \ldots + \frac{\cos\frac{(2p+1)\pi}{l}x}{(2p+1)^2} + \ldots\right]$$

5.6 ON THE EXPANSION OF A NONPERIODIC FUNCTION IN A FOURIER SERIES

Let there be given, on some interval $[a, b]$, a piecewise monotonic function $f(x)$ (Fig. 135). We shall show that this function $f(x)$ may be represented in the form of a sum of a Fourier series

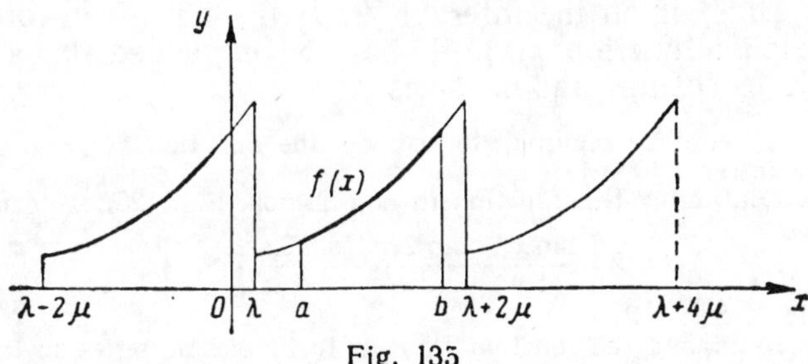

Fig. 135

at the points of its discontinuity. To do this, let us consider an arbitrary periodic piecewise monotonic function $f_1(x)$ with period $2\mu \geqslant b-a$, which coincides with the function $f(x)$ on the interval $[a, b]$. [We have redefined the function $f(x)$.]

Expand $f_1(x)$ in a Fourier series. At all points of the interval $[a, b]$ (with the exception of points of discontinuity) the sum of this series coincides with the given function $f(x)$; in other words, we have expanded the function $f(x)$ in a Fourier series on the interval $[a, b]$.

Let us now consider the following important case. Let a function $f(x)$ be given on the interval $[0, l]$. Redefining this function in arbitrary fashion on the interval $[-l, 0]$ (retaining piecewise monotonicity), we can expand it in a Fourier series. In particular, if we redefine this function so that when $-l \leqslant x < 0$, $f(x) = f(-x)$, we will get an even function (Fig. 136). [In this case we say that the function $f(x)$ is "continued in even fashion".] This function is expanded in a Fourier series that contains only cosines. Thus, we have expanded in cosines the function $f(x)$ given on the interval $[0, l]$.

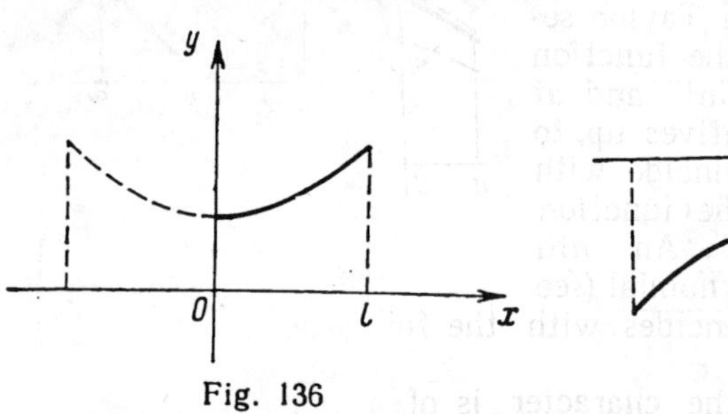

Fig. 136

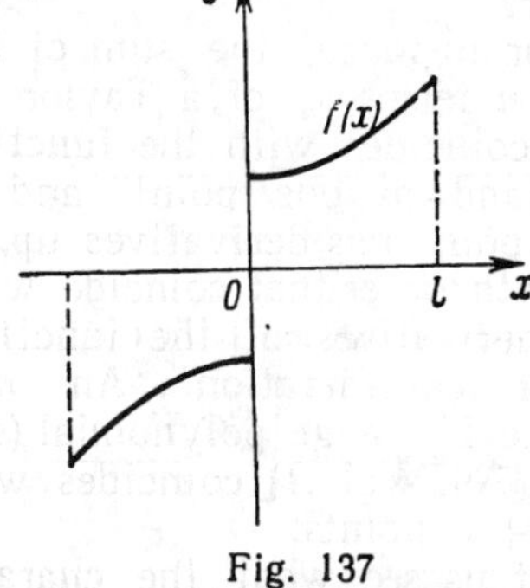

Fig. 137

But if we redefine the function $f(x)$ when $-l \leqslant x < 0$ as follows: $f(x) = -f(-x)$, then we get an odd function which may be expanded in sines (Fig. 137). [The function $f(x)$ is "continued in odd fashion".] Thus, if on the interval $[0, l]$ there is given some piecewise monotonic function $f(x)$, it may be expanded in a Fourier series both in cosines and in sines.

Example 1. Let it be required to expand the function $f(x) = x$ in the sine series on the interval $[0, \pi]$.

Solution. Continuing this function in odd fashion (Fig. 126), we get the series

$$x = 2\left[\frac{\sin x}{1} - \frac{\sin 2x}{2} + \frac{\sin 3x}{3} - \ldots\right]$$

(see Example 1, Sec. 5.2).

Example 2. Expand the function $f(x) = x$ in the cosine series on the interval $[0, \pi]$.

Solution. Continuing this function in even fashion, we get

$$f(x) = |x|, \quad -\pi < x \leqslant \pi$$

(Fig. 127). Expanding it in a series we find

$$f(x) = \frac{\pi}{2} - \frac{4}{\pi}\left[\frac{\cos x}{1} + \frac{\cos 3x}{3^2} + \frac{\cos 5x}{5^2} + \ldots\right]$$

(see Example 2, Sec. 5.2). And so on the interval $[0, \pi]$ we have the equation

$$x = \frac{\pi}{2} - \frac{4}{\pi}\left[\frac{\cos x}{1} + \frac{\cos 3x}{3^2} + \frac{\cos 5x}{5^2} + \ldots\right]$$

5.7 MEAN APPROXIMATION OF A GIVEN FUNCTION BY A TRIGONOMETRIC POLYNOMIAL

Representing a function by an infinite series (Fourier's, Taylor's and so forth) has the following meaning in practice: the **finite** sum obtained in terminating the series with the nth term is an **approximate expression** of the function being expanded. This approximate expression may be made as accurate as desired by choosing a sufficiently large value of n. However, the character of the approximate representation may differ.

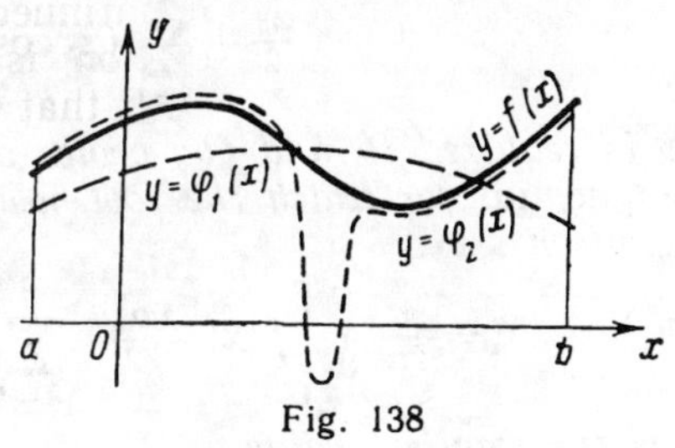

Fig. 138

For instance, the sum of the first n terms s_n of a Taylor series coincides with the function at hand at one point, and at this point has derivatives up to the nth order that coincide with the derivatives of the function under consideration. An nth degree Lagrange polynomial (see Sec. 7.9, Vol. I) coincides with the function under consideration at $n+1$ points.

Let us see what the character is of an approximate representation of a periodic function $f(x)$ by trigonometric polynomials of the form

$$s_n(x) = \frac{a_0}{2} + \sum_{k=1}^{n} a_k \cos kx + b_k \sin kx$$

where $a_0, a_1, b_1, a_2, b_2, \ldots, a_n, b_n$ are Fourier coefficients; that is, by the sum of the first n terms of a Fourier series. We first make several remarks.

Suppose we regard some function $y=f(x)$ on the interval $[a, b]$ and want to evaluate the error when replacing this function by another function $\varphi(x)$. For the measure of error we can, for instance, take $\max |f(x)-\varphi(x)|$ on the interval $[a, b]$, which is the so-called *maximum deviation* of $\varphi(x)$ from $f(x)$. But it is sometimes more natural to take for the measure of error the so-called *root-mean-square deviation* δ, which is defined by the equation

$$\delta^2 = \frac{1}{(b-a)} \int_a^b [f(x)-\varphi(x)]^2 \, dx$$

Fig. 138 illustrates the difference between the root-mean-square deviation and the maximum deviation.

Let the solid line depict the function $y=f(x)$, the dashed lines the approximations $\varphi_1(x)$ and $\varphi_2(x)$. The maximum deviation of the curve $y=\varphi_1(x)$ is less than that of the curve $y=\varphi_2(x)$, but the root-mean-square deviation of the first curve is greater than that of the second because the curve $y=\varphi_2(x)$ is considerably different from the curve $y=f(x)$ only on a narrow section and for this reason characterizes the curve $y=f(x)$ better than the first.

Now let us return to our problem.

Let there be given a periodic function $f(x)$ with period 2π. From among all the trigonometric polynomials of order n

$$\frac{\alpha_0}{2}+\sum_{k=1}^{n}(\alpha_k\cos kx+\beta_k\sin kx)$$

it is required to find (by choice of the coefficients α_k and β_k) that polynomial for which the root-mean-square deviation defined by the equation

$$\delta_n^2=\frac{1}{2\pi}\int_{-\pi}^{\pi}\left[f(x)-\frac{\alpha_0}{2}-\sum_{k=1}^{n}(\alpha_k\cos kx+\beta_k\sin kx)\right]^2dx$$

has the smallest value.

The problem reduces to finding the minimum of the function of $2n+1$ variables $\alpha_0, \alpha_1, \ldots, \alpha_n, \beta_1, \beta_2, \ldots, \beta_n$.

Expanding the square under the integral sign and integrating termwise, we get

$$\delta_n^2=\frac{1}{2\pi}\int_{-\pi}^{\pi}\left\{f^2(x)-2f(x)\left[\frac{\alpha_0}{2}+\sum_{k=1}^{n}(\alpha_k\cos kx+\beta_k\sin kx)\right]\right.$$
$$\left.+\left[\frac{\alpha_0}{2}+\sum_{k=1}^{n}(\alpha_k\cos kx+\beta_k\sin kx)\right]^2\right\}dx$$
$$=\frac{1}{2\pi}\int_{-\pi}^{\pi}f^2(x)\,dx-\frac{\alpha_0}{2\pi}\int_{-\pi}^{\pi}f(x)\,dx-\frac{1}{\pi}\sum_{k=1}^{n}\alpha_k\int_{-\pi}^{\pi}f(x)\cos kx\,dx$$
$$-\frac{1}{\pi}\sum_{k=1}^{n}\beta_k\int_{-\pi}^{\pi}f(x)\sin kx\,dx+\frac{1}{2\pi}\frac{\alpha_0^2}{4}\int_{-\pi}^{\pi}dx+\frac{1}{2\pi}\sum_{k=1}^{n}\alpha_k^2\int_{-\pi}^{\pi}\cos^2 kx\,dx$$
$$+\frac{1}{2\pi}\sum_{k=1}^{n}\beta_k^2\int_{-\pi}^{\pi}\sin^2 kx\,dx+\frac{1}{2\pi}\alpha_0\sum_{k=1}^{n}\alpha_k\int_{-\pi}^{\pi}\cos kx\,dx$$
$$+\frac{1}{2\pi}\alpha_0\sum_{k=1}^{n}\beta_k\int_{-\pi}^{\pi}\sin kx\,dx+\frac{1}{\pi}\sum_{\substack{k=1\\k\neq j}}^{n}\sum_{j=1}^{n}\alpha_k\alpha_j\int_{-\pi}^{\pi}\cos kx\cos jx\,dx$$
$$+\frac{1}{\pi}\sum_{k=1}^{n}\sum_{j=1}^{n}\alpha_k\beta_j\int_{-\pi}^{\pi}\cos kx\sin jx\,dx+\frac{1}{\pi}\sum_{\substack{k=1\\k\neq j}}^{n}\sum_{j=1}^{n}\beta_k\beta_j\int_{-\pi}^{\pi}\sin kx\sin jx\,dx$$

We note that

$$\frac{1}{\pi}\int_{-\pi}^{\pi} f(x)\,dx=a_0,\quad \frac{1}{\pi}\int_{-\pi}^{\pi} f(x)\cos kx\,dx=a_k$$

$$\frac{1}{\pi}\int_{-\pi}^{\pi} f(x)\sin kx\,dx=b_k$$

are the Fourier coefficients of the function $f(x)$.

Further, by formulas (I) and (II), Sec. 5.1, we have: for $k=j$

$$\int_{-\pi}^{\pi}\cos^2 kx\,dx=\pi,\quad \int_{-\pi}^{\pi}\sin^2 kx\,dx=\pi$$

for arbitrary k and j

$$\int_{-\pi}^{\pi}\sin kx\cos jx\,dx=0$$

and for $k\neq j$

$$\int_{-\pi}^{\pi}\cos kx\cos jx\,dx=0.\quad \int_{-\pi}^{\pi}\sin kx\sin jx\,dx=0$$

Thus, we obtain

$$\delta_n^2=\frac{1}{2\pi}\int_{-\pi}^{\pi} f^2(x)\,dx-\frac{\alpha_0 a_0}{2}-\sum_{k=1}^{n}(\alpha_k a_k+\beta_k b_k)+\frac{\alpha_0^2}{4}+\frac{1}{2}\sum_{k=1}^{n}(\alpha_k^2+\beta_k^2)$$

Adding and subtracting the sum

$$\frac{a_0^2}{4}+\frac{1}{2}\sum_{k=1}^{n}(a_k^2+b_k^2)$$

we will have

$$\delta_n^2=\frac{1}{2\pi}\int_{-\pi}^{\pi} f^2(x)\,dx-\frac{a_0^2}{4}-\frac{1}{2}\sum_{k=1}^{n}(a_k^2+b_k^2)+\frac{1}{4}(\alpha_0-a_0)^2$$

$$+\frac{1}{2}\sum_{k=1}^{n}[(\alpha_k-a_k)^2+(\beta_k-b_k)^2] \qquad (1)$$

The first three terms of this sum are independent of the choice of coefficients $\alpha_0, \alpha_1, \ldots, \alpha_n, \beta_1, \ldots, \beta_n$. The remaining terms

$$\frac{1}{4}(\alpha_0-a_0)^2+\frac{1}{2}\sum_{k=1}^{n}[(\alpha_k-a_k)^2+(\beta_k-b_k)^2]$$

are nonnegative. Their sum reaches the least value (equal to zero) if we put $\alpha_0 = a_0$, $\alpha_1 = a_1$, ..., $\alpha_n = a_n$, $\beta_1 = b_1$, ..., $\beta_n = b_n$. With this choice of coefficients $\alpha_0, \alpha_1, \ldots, \alpha_n, \beta_1, \ldots, \beta_n$ the trigonometric polynomial

$$\frac{\alpha_0}{2} + \sum_{k=1}^{n} (\alpha_k \cos kx + \beta_k \sin kx)$$

will least of all differ from the function $f(x)$ in the sense that in such a choice of coefficients the square deviation δ_n^2 will be the least.

We have thus proved the theorem:

Of all trigonometric polynomials of order n, that polynomial has the least root-mean-square deviation from the function f(x), the coefficients of which polynomial are the Fourier coefficients of the function f(x).

The least square deviation is

$$\delta_n^2 = \frac{1}{2\pi} \int_{-\pi}^{\pi} f^2(x)\, dx - \frac{a_0^2}{4} - \frac{1}{2} \sum_{k=1}^{n} (a_k^2 + b_k^2) \tag{2}$$

Since $\delta_n^2 \geqslant 0$, it follows that for any n we have

$$\frac{1}{2\pi} \int_{-\pi}^{\pi} f^2(x)\, dx \geqslant \frac{a_0^2}{4} + \frac{1}{2} \sum_{k=1}^{n} (a_k^2 + b_k^2)$$

Hence, the series on the right converges (as $n \to \infty$), and we can write

$$\frac{1}{\pi} \int_{-\pi}^{\pi} f^2(x)\, dx \geqslant \frac{a_0^2}{2} + \sum_{k=1}^{\infty} (a_k^2 + b_k^2) \tag{3}$$

This relation is called *Bessel's inequality*.

We note without proof that for any bounded and piecewise monotonic function the root-mean-square deviation obtained upon replacing the given function by the nth partial sum of the Fourier series tends to zero as $n \to \infty$, that is, $\delta_n^2 \to 0$ as $n \to \infty$. But then from formula (2) there follows the equation

$$\frac{a_0^2}{2} + \sum_{k=1}^{\infty} (a_k^2 + b_k^2) = \frac{1}{\pi} \int_{-\pi}^{\pi} f^2(x)\, dx \tag{3'}$$

which is called the *Lyapunov-Parseval equation*. We note that this equation has been proved for a broader class of functions than that which we here consider.

From what has been proved it follows that for a function which satisfies the Lyapunov-Parseval equation (in particular, for any bounded piecewise monotonic function), the corresponding Fourier series yields a root-mean-square deviation equal to zero.

Note. Let us establish a property of Fourier coefficients that will be needed in the future. We first introduce a definition.

A function $f(x)$ is called *piecewise continuous* on an interval $[a, b]$ if it has a definite number of discontinuities of the first kind on this interval (or is everywhere continuous).

We shall prove the following proposition.

If a function $f(x)$ is piecewise continuous on the interval $[-\pi, \pi]$, then its Fourier coefficients approach zero as $n \to \infty$; that is,

$$\lim_{n\to\infty} a_n = 0, \quad \lim_{n\to\infty} b_n = 0 \tag{4}$$

Proof. If the function $f(x)$ is piecewise continuous on the interval $[-\pi, \pi]$, then the function $f^2(x)$ too is piecewise continuous on this interval. Then $\int\limits_{-\pi}^{\pi} f^2(x)\,dx$ exists and is a finite number.* In this case, from the Bessel inequality (3) it follows that the series $\sum\limits_{n=1}^{\infty}(a_n^2+b_n^2)$ converges. But if the series converges then its general term approaches zero; in this case, $\lim\limits_{n\to 0}(a_n^2+b_n^2)=0$. Whence we get equations (4) directly. Thus, the following equations are valid for a piecewise continuous and bounded function:

$$\lim_{n\to\infty}\int_{-\pi}^{\pi} f(x)\cos nx\,dx = 0$$

$$\lim_{n\to\infty}\int_{-\pi}^{\pi} f(x)\sin nx\,dx = 0$$

If a function $f(x)$ is periodic with period 2π, then the latter equations may be written as follows (for any a):

$$\lim_{n\to\infty}\int_{a}^{a+2\pi} f(x)\cos nx\,dx = 0, \quad \lim_{n\to\infty}\int_{a}^{a+2\pi} f(x)\sin nx\,dx = 0$$

We note that these equations continue to hold if in the integrals we take any arbitrary interval of integration $[a, b]$, which is to say that the integrals

$$\int_a^b f(x)\cos nx\,dx \quad \text{and} \quad \int_a^b f(x)\sin nx\,dx$$

* This integral may be presented as the sum of definite integrals of continuous functions over the subintervals into which the interval $[-\pi, \pi]$ is subdivided.

approach zero when n increases without bound if $f(x)$ is a bounded and piecewise continuous function.

Indeed, taking $b-a<2\pi$ for definiteness, we consider the auxiliary function $\varphi(x)$ with period 2π defined as follows:

$$\varphi(x)=f(x) \qquad \text{when } a\leqslant x\leqslant b$$
$$\varphi(x)=0 \qquad \text{when } b<x\leqslant a+2\pi$$

Then

$$\int_a^b f(x)\cos nx\,dx=\int_a^{a+2\pi}\varphi(x)\cos nx\,dx$$
$$\int_a^b f(x)\sin nx\,dx=\int_a^{a+2\pi}\varphi(x)\sin nx\,dx$$

Since $\varphi(x)$ is a bounded and piecewise continuous function, the integrals on the right approach zero as $n\to\infty$. Hence, the integrals on the left approach zero as well. Thus, the proposition is proved; that is,

$$\lim_{n\to\infty}\int_a^b f(x)\cos nx\,dx=0, \qquad \lim_{n\to\infty}\int_a^b f(x)\sin nx\,dx=0 \tag{5}$$

for any numbers a and b and any function $f(x)$ piecewise continuous and bounded on $[a, b]$.

5.8 THE DIRICHLET INTEGRAL

In this section we shall derive a formula that expresses the nth partial sum of a Fourier series in terms of a certain integral. This formula will be needed in the subsequent sections.

Consider the nth partial sum of a Fourier series for the periodic function $f(x)$ with period 2π:

$$s_n(x)=\frac{a_0}{2}+\sum(a_k\cos kx+b_k\sin kx)$$

where

$$a_k=\frac{1}{\pi}\int_{-\pi}^{\pi}f(t)\cos kt\,dt, \qquad b_k=\frac{1}{\pi}\int_{-\pi}^{\pi}f(t)\sin kt\,dt$$

Putting these expressions into the formula for $s_n(x)$, we obtain

$$s_n(x)=\frac{1}{2\pi}\int_{-\pi}^{\pi}f(t)\,dt$$
$$+\sum_{k=1}^{n}\left[\frac{\cos kx}{\pi}\int_{-\pi}^{\pi}f(t)\cos kt\,dt+\frac{\sin kx}{\pi}\int_{-\pi}^{\pi}f(t)\sin kt\,dt\right]$$

or bringing $\cos kx$ and $\sin kx$ under the integral sign (which is possible since $\cos kx$ and $\sin kx$ are independent of the variable of integration and, hence, can be regarded as constants), we get

$$s_n(x) = \frac{1}{2\pi} \int_{-\pi}^{\pi} f(t)\,dt$$

$$+ \frac{1}{\pi} \sum_{k=1}^{n} \left[\int_{-\pi}^{\pi} f(t) \cos kx \cos kt\,dt + \int_{-\pi}^{\pi} f(t) \sin kx \sin kt\,dt \right]$$

Now taking $\frac{1}{\pi}$ outside the brackets and replacing the sum of integrals by the integral of the sum, we obtain

$$s_n(x) = \frac{1}{\pi} \int_{-\pi}^{\pi} \left\{ \frac{f(t)}{2} + \sum_{k=1}^{n} [f(t) \cos kx \cos kt + f(t) \sin kx \sin kt] \right\} dt$$

or

$$s_n(x) = \frac{1}{\pi} \int_{-\pi}^{\pi} f(t) \left[\frac{1}{2} + \sum_{k=1}^{n} (\cos kt \cos kx + \sin kt \sin kx) \right] dt$$

$$= \frac{1}{\pi} \int_{-\pi}^{\pi} f(t) \left[\frac{1}{2} + \sum_{k=1}^{n} \cos k(t-x) \right] dt \qquad (1)$$

Transform the expression in the brackets. Let

$$\sigma_n(z) = \frac{1}{2} + \cos z + \cos 2z + \ldots + \cos nz$$

then

$$2\sigma_n(z) \cos z = \cos z + 2 \cos z \cos z + 2 \cos z \cos 2z + \ldots$$
$$+ 2 \cos z \cos nz = \cos z + (1 + \cos 2z) + (\cos z + \cos 3z)$$
$$+ (\cos 2z + \cos 4z) + \ldots + [\cos (n-1) z + \cos (n+1) z]$$
$$= 1 + 2 \cos z + 2 \cos 2z + \ldots + 2 \cos (n-1) z + \cos nz + \cos (n+1) z$$

or

$$2\sigma_n(z) \cos z = 2\sigma_n(z) - \cos nz + \cos (n+1) z$$

$$\sigma_n(z) = \frac{\cos nz - \cos (n+1) z}{2(1-\cos z)}$$

But

$$\cos nz - \cos (n+1) z = 2 \sin (2n+1) \frac{z}{2} \sin \frac{z}{2}$$

$$1 - \cos z = 2 \sin^2 \frac{z}{2}$$

Hence,

$$\sigma_n(z)=\frac{\sin(2n+1)\frac{z}{2}}{2\sin\frac{z}{2}}$$

Thus, equation (1) may be rewritten as

$$s_n(x)=\frac{1}{\pi}\int_{-\pi}^{\pi} f(t)\frac{\sin(2n+1)\frac{t-x}{2}}{2\sin\frac{t-x}{2}}\,dt$$

Since the integrand is periodic (with period 2π), it follows that the integral retains its value on any interval of integration of length 2π. We can therefore write

$$s_n(x)=\frac{1}{\pi}\int_{x-\pi}^{x+\pi} f(t)\frac{\sin(2n+1)\frac{t-x}{2}}{2\sin\frac{t-x}{2}}\,dt$$

Introducing a new variable α, we put

$$t-x=\alpha,\quad t=x+\alpha$$

Then we get the formula

$$s_n(x)=\frac{1}{\pi}\int_{-\pi}^{\pi} f(x+\alpha)\frac{\sin(2n+1)\frac{\alpha}{2}}{2\sin\frac{\alpha}{2}}\,d\alpha \tag{2}$$

The integral on the right is *Dirichlet's integral*.

In this formula put $f(x)\equiv 1$; then $a_0=2$, $a_k=0$, $b_k=0$ when $k>0$; hence, $s_n(x)=1$ for any n and we get the identity

$$1=\frac{1}{\pi}\int_{-\pi}^{\pi}\frac{\sin(2n+1)\frac{\alpha}{2}}{2\sin\frac{\alpha}{2}}\,d\alpha \tag{3}$$

which we will need later on.

5.9 THE CONVERGENCE OF A FOURIER SERIES AT A GIVEN POINT

Assume that the function $f(x)$ is piecewise continuous on the interval $[-\pi, \pi]$.

Multiplying both sides of the identity (3) of the preceding section by $f(x)$ and bringing $f(x)$ under the integral sign, we get the equation

$$f(x)=\frac{1}{\pi}\int_{-\pi}^{\pi} f(x)\frac{\sin(2n+1)\frac{\alpha}{2}}{2\sin\frac{\alpha}{2}}\,d\alpha$$

Subtract the terms of the latter equation from the corresponding terms of (2) of the preceding section; we get

$$s_n(x)-f(x)=\frac{1}{\pi}\int_{-\pi}^{\pi}[f(x+\alpha)-f(x)]\frac{\sin(2n+1)\frac{\alpha}{2}}{2\sin\frac{\alpha}{2}}\,d\alpha$$

Thus, the convergence of a Fourier series to the value of a function $f(x)$ at a given point depends on whether the integral on the right approaches zero as $n\to\infty$.

Let us break up this integral into two integrals:

$$s_n(x)-f(x)=\frac{1}{\pi}\int_{-\pi}^{\pi}[f(x+\alpha)-f(x)]\frac{\cos\frac{\alpha}{2}}{2\sin\frac{\alpha}{2}}\sin n\alpha\,d\alpha$$

$$+\frac{1}{2}\frac{1}{\pi}\int_{-\pi}^{\pi}[f(x+\alpha)-f(x)]\cos n\alpha\,d\alpha$$

taking advantage of the fact that $\sin(2n+1)\frac{\alpha}{2}=\sin n\alpha\cos\frac{\alpha}{2}+\cos n\alpha\sin\frac{\alpha}{2}$. Break up the first of the integrals on the right of the latter equation into three integrals:

$$s_n(x)-f(x)=\frac{1}{\pi}\int_{-\delta}^{\delta}[f(x+\alpha)-f(x)]\frac{\cos\frac{\alpha}{2}}{2\sin\frac{\alpha}{2}}\sin n\alpha\,d\alpha$$

$$+\frac{1}{\pi}\int_{-\pi}^{-\delta}[f(x+\alpha)-f(x)]\frac{\cos\frac{\alpha}{2}}{2\sin\frac{\alpha}{2}}\sin n\alpha\,d\alpha$$

$$+\frac{1}{\pi}\int_{\delta}^{\pi}[f(x+\alpha)-f(x)]\frac{\cos\frac{\alpha}{2}}{2\sin\frac{\alpha}{2}}\sin n\alpha\,d\alpha$$

$$+\frac{1}{\pi}\int_{-\pi}^{\pi}[f(x+\alpha)-f(x)]\frac{1}{2}\cos n\alpha\,d\alpha$$

Put $\Phi_1(\alpha)=\frac{f(x+\alpha)-f(x)}{2}$. Since $f(x)$ is a bounded piecewise continuous function, it follows that $\Phi_1(\alpha)$ is also a bounded and piecewise continuous periodic function of α. Hence, the latter integral approaches zero as $n\to\infty$, since it is a Fourier coefficient of this function. The function

$$\Phi_2(\alpha)=[f(x+\alpha)-f(x)]\frac{\cos\frac{\alpha}{2}}{2\sin\frac{\alpha}{2}}$$

is bounded when $-\pi\leqslant\alpha<-\delta$ and $\delta\leqslant\alpha\leqslant\pi$ and

$$|\Phi_2(\alpha)|\leqslant[M+M]\frac{1}{2\sin\frac{\delta}{2}}$$

where M is the upper limit of the quantity $|f(x)|$. Also, the function $\Phi_2(\alpha)$ is piecewise continuous. Hence, by formulas (5) of Sec. 5.7, the second and third integrals approach zero as $n\to\infty$.

We can thus write

$$\lim_{n\to\infty}[s_n(x)-f(x)]=\lim_{n\to\infty}\frac{1}{\pi}\int_{-\delta}^{\delta}[f(x+\alpha)-f(x)]\frac{\cos\frac{\alpha}{2}}{2\sin\frac{\alpha}{2}}\sin n\alpha\,d\alpha \qquad (1)$$

In the expression on the right, the integration is performed over the interval $-\delta\leqslant\alpha\leqslant\delta$; consequently, the integral is dependent on the values of the function $f(x)$ only in the interval from $x-\delta$ to $x+\delta$. An important proposition thus follows from the last equation: *the convergence of a Fourier series at a given point x depends only on the behaviour of the function $f(x)$ in an arbitrarily small neighbourhood of this point.*

Therein lies the so-called *principle of localization in the study of Fourier series. If two functions $f_1(x)$ and $f_2(x)$ coincide in the neighbourhood of some point x, then their Fourier series simultaneously either converge or diverge at this point.*

5.10 CERTAIN SUFFICIENT CONDITIONS FOR THE CONVERGENCE OF A FOURIER SERIES

In the preceding section it was shown that if a function $f(x)$ is piecewise continuous in the interval $[-\pi, \pi]$, then the convergence of a Fourier series at a given point x_0 to a value of the function $f(x_0)$ depends on the behaviour of the function in a certain arbitrarily small neighbourhood $[x_0-\delta,\ x_0+\delta]$ with centre at the point x_0.

Let us now prove that if *in the neighbourhood of the point* x_0 *the function* $f(x)$ *is such that there exist finite limits*

$$\lim_{\alpha\to-0}\frac{f(x_0+\alpha)-f(x_0)}{\alpha}=k_1 \quad (1)$$

$$\lim_{\alpha\to+0}\frac{f(x_0+\alpha)-f(x_0)}{\alpha}=k_2 \quad (2)$$

while the function is continuous at the very point x_0 (Fig. 139), *then the Fourier series converges at this point to a corresponding value of the function* $f(x)$ *.

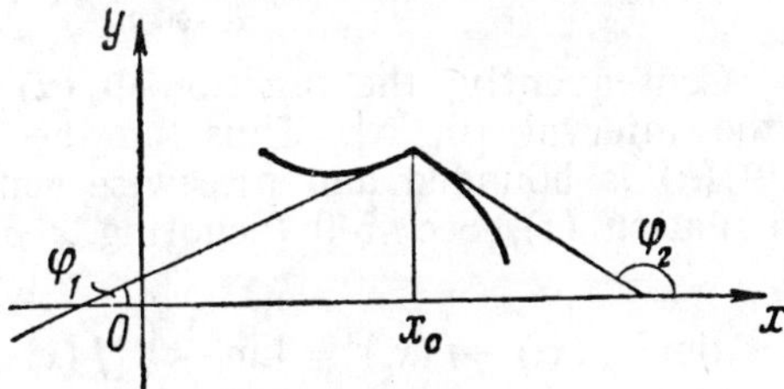

Fig. 139

Proof. Let us consider the function $\Phi_2(\alpha)$ defined in the preceding section:

$$\Phi_2(\alpha)=[f(x_0+\alpha)-f(x_0)]\frac{\cos\frac{\alpha}{2}}{2\sin\frac{\alpha}{2}}$$

since the function $f(x)$ is piecewise continuous on the interval $[-\pi,\ \pi]$ and is continuous at the point x_0, it is therefore continuous in some neighbourhood $[x_0-\delta,\ x_0+\delta]$ of the point x_0. For this reason, the function $\Phi_2(\alpha)$ is continuous at all points where $\alpha\neq0$ and $|\alpha|\leqslant\delta$. When $\alpha=0$ the function $\Phi_2(\alpha)$ is not defined.

Let us find $\lim\limits_{\alpha\to0-0}\Phi_2(\alpha)$ and $\lim\limits_{\alpha\to0+0}\Phi_2(\alpha)$, making use of conditions (1) and (2):

$$\lim_{\alpha\to0-0}\Phi_2(\alpha)=\lim_{\alpha\to0-0}[f(x_0+\alpha)-f(x_0)]\frac{\cos\frac{\alpha}{2}}{2\sin\frac{\alpha}{2}}$$

$$=\lim_{\alpha\to0-0}\frac{f(x_0+\alpha)-f(x_0)}{\alpha}\,\frac{\frac{\alpha}{2}}{\sin\frac{\alpha}{2}}\cos\frac{\alpha}{2}$$

$$=\lim_{\alpha\to0-0}\frac{f(x_0+\alpha)-f(x_0)}{\alpha}\lim_{\alpha\to0-0}\frac{\frac{\alpha}{2}}{\sin\frac{\alpha}{2}}\lim_{\alpha\to0-0}\cos\frac{\alpha}{2}=k_1\cdot1\cdot1=k_1$$

* If conditions (1) and (2) are fulfilled, then we say that the function $f(x)$ has, at the point x, a derivative on the right and a derivative on the left. Fig. 139 shows a function where $k_1=\tan\varphi_1$, $k_2=\tan\varphi_2$, $k_1\neq k_2$. If $k_1=k_2$, that is, if the derivatives on the right and left are equal, then the function will be differentiable at the given point.

Thus, if we redefine the function $\Phi_2(\alpha)$ by putting $\Phi_2(0)=k_1$, then it will be continuous on the interval $[-\delta, 0]$, and, hence, bounded as well. Similarly we prove that

$$\lim_{\alpha\to 0+0} \Phi_2(\alpha) = k_2$$

Consequently, the function $\Phi_2(\alpha)$ is bounded and continuous on the interval $[0, \delta]$. Thus, on the interval $[-\delta, \delta]$ the function $\Phi_2(\alpha)$ is bounded and piecewise continuous. Now let us return to equation (1), Sec. 5.9 (denoting x by x_0),

$$\lim_{n\to\infty} [s_n(x_0) - f(x_0)] = \lim_{n\to\infty} \frac{1}{\pi} \int_{-\delta}^{\delta} [f(x_0+\alpha) - f(x_0)] \frac{\cos\frac{\alpha}{2}}{2\sin\frac{\alpha}{2}} \sin n\alpha \, d\alpha$$

or

$$\lim_{n\to\infty} [s_n(x_0) - f(x_0)] = \lim_{n\to\infty} \frac{1}{\pi} \int_{-\delta}^{\delta} \Phi_2(\alpha) \sin n\alpha \, d\alpha$$

From formulas (5) of Sec. 5.7 we conclude that the limit on the right is equal to zero, and therefore

$$\lim_{n\to\infty} [s_n(x_0) - f(x_0)] = 0$$

or

$$\lim_{n\to\infty} s_n(x_0) = f(x_0)$$

The theorem is proved.

This theorem differs from the theorem stated in Sec. 5.1 in that in the latter case it was required, for convergence of the Fourier series at a point x_0 to the value of the function $f(x_0)$, that the point x_0 should be a point of continuity on the interval $[-\pi, \pi]$, whereas the function should be piecewise monotonic; here, however, it is required that the function at the point x_0 should be a point of continuity and that the conditions (1) and (2) be fulfilled, while throughout the interval $[-\pi, \pi]$ the function should be piecewise continuous and bounded. It is obvious that these conditions are different.

Note 1. If a piecewise continuous function is differentiable at the point x_0, it is obvious that conditions (1) and (2) are fulfilled. Here, $k_1 = k_2$. Hence, at points where the function $f(x)$ is differentiable, the Fourier series converges to the value of the function at the corresponding point.

Note 2. (a) The function considered in Example 2, Sec. 5.2 (Fig. 127), satisfies conditions (1) and (2) at the points $0, \pm 2\pi, \pm 4\pi, \ldots$. At all the other points it is differentiable. Consequently, a Fourier series constructed for it converges at each point to the value of this function.

(b) The function considered in Example 4, Sec. 5.2 (Fig. 130); satisfies conditions (1) and (2) at the points $\pm\pi$, $\pm 3\pi$, $\pm 5\pi$. It is differentiable at all the other points. It is represented by a Fourier series at each point.

(c) The function considered in Example 1, Sec. 5.2 (Fig. 126), is discontinuous at the points $\pm\pi$, $\pm 3\pi$, $\pm 5\pi$. At all other points it is differentiable. Hence, at all points, with the exception of points of discontinuity, the Fourier series corresponding to it converges to the value of the function at the corresponding points. At the discontinuities, the sum of the Fourier series is equal to the arithmetic mean limit of the function on the right and on the left (in this case, zero).

5.11 PRACTICAL HARMONIC ANALYSIS

The theory of expanding functions in Fourier series is called *harmonic analysis*. We shall now make several remarks about approximate computation of the coefficients of a Fourier series, that is to say, about practical harmonic analysis.

As we know, the Fourier coefficients of a function $f(x)$ with period 2π are defined by the formulas

$$a_0 = \frac{1}{\pi}\int_{-\pi}^{\pi} f(x)\,dx, \qquad a_k = \frac{1}{\pi}\int_{-\pi}^{\pi} f(x)\cos kx\,dx,$$

$$b_k = \frac{1}{\pi}\int_{-\pi}^{\pi} f(x)\sin kx\,dx$$

In many practical cases, the function $f(x)$ is represented either in tabular form (when the functional relation is obtained by experiment) or in the form of a curve which is plotted by some kind of instrument. In these cases the Fourier coefficients are calculated by means of approximate methods of integration (see Sec. 11.8, Vol. I).

Let us consider the interval $-\pi \leqslant x \leqslant \pi$ of length 2π. This can always be done by proper choice of scale on the x-axis.

Divide the interval $[-\pi, \pi]$ into n equal parts by the points

$$-\pi = x_0,\ x_1,\ x_2,\ \ldots,\ x_n = \pi$$

Then the subinterval will be

$$\Delta x = \frac{2\pi}{n}$$

We denote the values of the function $f(x)$ at the points x_0, x_1, x_2, ..., x_n (respectively) by

$$y_0,\ y_1,\ y_2,\ \ldots,\ y_n$$

These values are determined either from a table or from the graph of the given function (by measuring the corresponding ordinates).

Then, taking advantage, for example, of the rectangular formula [see formula (1′), Sec. 11.8, Vol. I], we determine the Fourier coefficients:

$$a_0=\frac{2}{n}\sum_{i=1}^{n} y_i,\quad a_k=\frac{2}{n}\sum_{i=1}^{n} y_i\cos kx_i,\quad b_k=\frac{2}{n}\sum_{i=1}^{n} y_i\sin kx_i$$

Diagrams have been devised that simplify computation of Fourier coefficients. We cannot deal here with the details but we can note that there are instruments (Fourier analyzers) which permit approximating the values of Fourier coefficients from the graph of the function.

5.12 THE FOURIER SERIES IN COMPLEX FORM

Suppose we have a Fourier series for a periodic function $f(x)$ with period 2π:

$$f(x)=\frac{a_0}{2}+\sum_{n=1}^{\infty} a_n\cos nx+b_n\sin nx \tag{1}$$

We express $\cos nx$ and $\sin nx$ in terms of exponential functions [see formulas (3), Sec. 7.5, Vol. I]:

$$\cos y=\frac{e^{iy}+e^{-iy}}{2},\qquad \sin y=\frac{e^{iy}-e^{-iy}}{2i}$$

Thus,

$$\cos nx=\frac{e^{inx}+e^{-inx}}{2},\qquad \sin nx=\frac{e^{inx}-e^{-inx}}{2i}=-i\,\frac{e^{inx}-e^{-inx}}{2}$$

Putting these values of $\cos nx$ and $\sin nx$ into formula (1) and performing appropriate manipulations, we get

$$f(x)=\frac{a_0}{2}+\sum_{n=1}^{\infty} a_n\frac{e^{inx}+e^{-inx}}{2}-ib_n\frac{e^{inx}-e^{-inx}}{2}$$

$$=\frac{a_0}{2}+\sum_{n=1}^{\infty}\left(\frac{a_n-ib_n}{2}e^{inx}+\frac{a_n+ib_n}{2}e^{-inx}\right) \tag{2}$$

Introducing the notation

$$\frac{a_0}{2}=c_0,\qquad \frac{a_n-ib_n}{2}=c_n,\qquad \frac{a_n+ib_n}{2}=c_{-n} \tag{3}$$

we see that (2) becomes

$$f(x)=c_0+\sum_{n=1}^{\infty}(c_ne^{inx}+c_{-n}e^{-inx})$$

This equation can be written more compactly as

$$f(x)=\sum_{n=-\infty}^{\infty} c_n e^{inx} \tag{4}$$

that is the *complex form of the Fourier series.*

Let us express the coefficients c_n and c_{-n} in terms of integrals. Using formulas (4), (5), and (6) of Sec. 5.1, we can write formulas (3) as follows:

$$c_n=\frac{1}{2\pi}\left[\int_{-\pi}^{\pi} f(x)\cos nx\,dx-i\int_{-\pi}^{\pi} f(x)\sin nx\,dx\right]$$

$$=\frac{1}{2\pi}\int_{-\pi}^{\pi} f(x)(\cos nx-i\sin nx)\,dx=\frac{1}{2\pi}\int_{-\pi}^{\pi} f(x)e^{-inx}dx$$

Thus

$$c_n=\frac{1}{2\pi}\int_{-\pi}^{\pi} f(x)e^{-inx}dx \tag{5'}$$

Similarly

$$c_{-n}=\frac{1}{2\pi}\int_{-\pi}^{\pi} f(x)e^{inx}dx \tag{5''}$$

The expression for c_0 and formulas (5′) and (5″) can be combined into a single formula

$$c_n=\frac{1}{2\pi}\int_{-\pi}^{\pi} f(x)e^{inx}dx \quad (n=0,\ \pm 1,\ \pm 2,\ \pm 3,\ \ldots) \tag{6}$$

c_n and c_{-n} are termed the complex Fourier coefficients of the function $f(x)$.

If the function $f(x)$ is periodic with period $2l$, then the Fourier series of $f(x)$ will be

$$f(x)=\frac{a_0}{2}+\sum_{n=1}^{\infty} a_n\cos\frac{n\pi}{l}x+b_n\sin\frac{n\pi}{l}x \tag{7}$$

[see formula (3), Sec. 5.5].

In this case, clearly, the Fourier series in complex form will be expressed by the formula

$$f(x)=\sum_{n=-\infty}^{\infty} c_n e^{i\frac{n\pi}{l}x} \tag{8}$$

instead of formula (4).

The coefficients c_n of the series are expressed by the formulas

$$c_n=\frac{1}{2l}\int_{-l}^{l} f(x)\,e^{-i\frac{n\pi}{l}x}\,dx \quad (n=0,\ \pm 1,\ \pm 2,\ \ldots) \tag{9}$$

The terminology in electrical and radio engineering is as follows: the expressions $e^{i\frac{n\pi}{l}x}$ are called *harmonics*, the numbers $\alpha_n=\frac{n\pi}{l}$ $(n=0,\ \pm 1,\ \pm 2,\ \ldots)$ are called the *wave numbers* of the function

$$f(x)=\sum_{n=-\infty}^{\infty} c_n e^{i\alpha_n x} \tag{10}$$

The set of wave numbers forms what is known as a *spectrum*. Plotting these numbers on a number axis, we get a collection of separate points. This collection of points is called a discrete collection, and the associated spectrum is termed a *discrete spectrum*. The coefficients c_n defined by formulas (9) are called the *complex amplitude*. In some engineering texts, the collection of moduli of the amplitudes, $|c_n|$, is also called the *spectrum* of the function $f(x)$.

5.13 FOURIER INTEGRAL

Let a function $f(x)$ be defined in an infinite interval $(-\infty, \infty)$ and absolutely integrable over it; that is, there exists an integral

$$\int_{-\infty}^{\infty} |f(x)|\,dx=Q \tag{1}$$

Further, let the function $f(x)$ be such that it is expandable into a Fourier series in any interval $(-l, +l)$:

$$f(x)=\frac{a_0}{2}+\sum_{k=1}^{\infty} a_k \cos\frac{k\pi}{l}x+b_k \sin\frac{k\pi}{l}x \tag{2}$$

where

$$a_k=\frac{1}{l}\int_{-l}^{l} f(t)\cos\frac{k\pi}{l}t\,dt,\quad b_k=\frac{1}{l}\int_{-l}^{l} f(t)\sin\frac{k\pi}{l}t\,dt \tag{3}$$

Putting into series (2) the expressions of the coefficients a_k and b_k from formulas (3), we can write

$$f(x)=\frac{1}{2l}\int_{-l}^{l} f(t)\,dt+\frac{1}{l}\sum_{k=1}^{\infty}\left(\int_{-l}^{l} f(t)\cos\frac{k\pi}{l}t\,dt\right)\cos\frac{k\pi}{l}x$$

$$+\frac{1}{l}\sum_{k=1}^{\infty}\left(\int_{-l}^{l} f(t)\sin\frac{k\pi}{l}t\,dt\right)\sin\frac{k\pi}{l}x$$

$$=\frac{1}{2l}\int_{-l}^{l} f(t)\,dt+\frac{1}{l}\sum_{k=1}^{\infty}\int_{-l}^{l} f(t)\left[\cos\frac{k\pi}{l}t\cos\frac{k\pi}{l}x+\sin\frac{k\pi}{l}t\sin\frac{k\pi}{l}x\right]dt$$

or

$$f(x)=\frac{1}{2l}\int_{-l}^{l} f(t)\,dt+\frac{1}{l}\sum_{k=1}^{\infty}\int_{-l}^{l} f(t)\cos\frac{k\pi(t-x)}{l}\,dt. \tag{4}$$

Let us investigate what form expansion (4) will take when passing to the limit as $l\to\infty$.

We introduce the following notation:

$$\alpha_1=\frac{\pi}{l},\quad \alpha_2=\frac{2\pi}{l},\quad \ldots,\quad \alpha_k=\frac{k\pi}{l},\quad \ldots \text{ and } \Delta\alpha_k=\frac{\pi}{l} \tag{5}$$

Substituting into (4), we get

$$f(x)=\frac{l}{2l}\int_{-l}^{l} f(t)\,dt+\frac{1}{\pi}\sum_{k=1}^{\infty}\left(\int_{-l}^{l} f(t)\cos\alpha_k(t-x)\,dt\right)\Delta\alpha_k \tag{6}$$

As $l\to\infty$, the first term on the right approaches zero. Indeed,

$$\left|\frac{1}{2l}\int_{-l}^{l} f(t)\,dt\right|\leqslant\frac{1}{2l}\int_{-l}^{l}|f(t)|\,dt<\frac{1}{2l}\int_{-\infty}^{\infty}|f(t)|\,dt=\frac{1}{2l}Q\to 0$$

For any fixed l, the expression in the parentheses is a function of α_k [see formulas (5)], which takes on values from $\frac{\pi}{l}$ to ∞. We will show, without proof, that if the function $f(x)$ is piecewise monotonic on every finite interval, is bounded on an infinite interval and satisfies condition (1), then as $l\to+\infty$ formula (6) takes the form

$$f(x)=\frac{1}{\pi}\int_{0}^{\infty}\left(\int_{-\infty}^{\infty} f(t)\cos\alpha(t-x)\,dt\right)d\alpha \tag{7}$$

The expression on the right is known as the *Fourier integral* of the function $f(x)$. Equation (7) occurs for all points where the function is continuous. At points of discontinuity we have the equation

$$\frac{1}{\pi}\int_{0}^{\infty}\left(\int_{-\infty}^{\infty} f(t)\cos\alpha(t-x)\,dt\right)dx=\frac{f(x+0)+f(x-0)}{2} \tag{7'}$$

Let us transform the integral on the right of (7) by expanding $\cos\alpha(t-x)$:

$$\cos\alpha(t-x)=\cos\alpha t\cos\alpha x+\sin\alpha t\sin\alpha x$$

Putting this expression into formula (7) and taking $\cos\alpha x$ and $\sin\alpha x$ outside the integral signs, where the integration is performed with respect to the variable t, we get

$$f(x)=\frac{1}{\pi}\int_0^\infty\left(\int_{-\infty}^{\infty}f(t)\cos\alpha t\,dt\right)\cos\alpha x\,d\alpha$$
$$+\frac{1}{\pi}\int_0^\infty\left(\int_{-\infty}^{\infty}f(t)\sin\alpha t\,dt\right)\sin\alpha x\,d\alpha \tag{8}$$

Each of the integrals in brackets with respect to t exists, since the function $f(t)$ is absolutely integrable in the interval $(-\infty, \infty)$, and therefore the functions $f(t)\cos\alpha t$ and $f(t)\sin\alpha t$ are also absolutely integrable.

Let us consider particular cases of formula (8).

1. Let $f(x)$ be even. Then $f(t)\cos\alpha t$ is an even function, while $f(t)\sin\alpha t$ is odd and we have

$$\int_{-\infty}^{\infty}f(t)\cos\alpha t\,dt=2\int_0^\infty f(t)\cos\alpha t\,dt$$

$$\int_{-\infty}^{\infty}f(t)\sin\alpha t\,dt=0$$

Formula (8) in this case takes the form

$$f(x)=\frac{2}{\pi}\int_0^\infty\left(\int_0^\infty f(t)\cos\alpha t\,dt\right)\cos\alpha x\,d\alpha \tag{9}$$

2. Let $f(x)$ be odd. Analyzing the character of the integrals in formula (8) in this case, we obtain

$$f(x)=\frac{2}{\pi}\int_0^\infty\left(\int_0^\infty f(t)\sin\alpha t\,dt\right)\sin\alpha x\,d\alpha \tag{10}$$

If $f(x)$ is defined only in the interval $(0, \infty)$, then for $x>0$ it may be represented by either formula (9) or (10). In the first case we redefine it in the interval $(-\infty, 0)$ in even fashion; in the latter case, in odd fashion.

Let it be noted once again that at the points of discontinuity one should write the following expression in place of $f(x)$ in the left-hand members of (9) and (10):

$$\frac{f(x+0)+f(x-0)}{2}$$

Let us return to formula (8). The integrals in brackets are functions of α. We introduce the following notation:

$$A(\alpha)=\frac{1}{\pi}\int_{-\infty}^{\infty} f(t)\cos\alpha t\,dt$$

$$B(\alpha)=\frac{1}{\pi}\int_{-\infty}^{\infty} f(t)\sin\alpha t\,dt$$

Then formula (8) may be rewritten as follows:

$$f(x)=\int_{0}^{\infty}[A(\alpha)\cos\alpha x+B(\alpha)\sin\alpha x]\,d\alpha \tag{11}$$

We say that formula (11) yields an expansion of the function $f(x)$ into harmonics with a frequency α that varies continuously from 0 to ∞. The law of distribution of amplitudes and initial phases as dependent upon the frequency α is expressed in terms of the functions $A(\alpha)$ and $B(\alpha)$.

Let us return to formula (9). We set

$$F(\alpha)=\sqrt{\frac{2}{\pi}}\int_{0}^{\infty} f(t)\cos\alpha t\,dt \tag{12}$$

then formula (9) takes the form

$$f(x)=\sqrt{\frac{2}{\pi}}\int_{0}^{\infty} F(\alpha)\cos\alpha x\,d\alpha \tag{13}$$

The function $F(\alpha)$ is called the *Fourier cosine transform* of the function $f(x)$.

If in (12) we consider $F(\alpha)$ as given and $f(t)$ as the unknown function, then it is an *integral equation* of the function $f(t)$. Formula (13) gives the solution of this equation.

On the basis of formula (10) we can write the following equations:

$$\Phi(\alpha)=\sqrt{\frac{2}{\pi}}\int_{0}^{\infty} f(t)\sin\alpha t\,dt \tag{14}$$

$$f(x)=\sqrt{\frac{2}{\pi}}\int_{0}^{\infty} \Phi(\alpha)\sin\alpha x\,d\alpha \tag{15}$$

The function $\Phi(\alpha)$ is called the *Fourier sine transform*.

Example. Let

$$f(x) = e^{-\beta x} \quad (\beta > 0, \ x \geqslant 0)$$

From (12) we determine the Fourier cosine transform:

$$F(\alpha) = \sqrt{\frac{2}{\pi}} \int_0^\infty e^{-\beta t} \cos \alpha t \, dt = \sqrt{\frac{2}{\pi}} \frac{\beta}{\beta^2 + \alpha^2}$$

From (14) we determine the Fourier sine transform:

$$\Phi(\alpha) = \sqrt{\frac{2}{\pi}} \int_0^\infty e^{-\beta t} \sin \alpha t \, dt = \sqrt{\frac{2}{\pi}} \frac{\alpha}{\beta^2 + \alpha^2}$$

From formulas (13) and (15) we find the reciprocal relationships:

$$\frac{2\beta}{\pi} \int_0^\infty \frac{\cos \alpha x}{\beta^2 + \alpha^2} \, d\alpha = e^{-\beta x} \quad (x \geqslant 0)$$

$$\frac{2}{\pi} \int_0^\infty \frac{\alpha \sin \alpha x}{\beta^2 + \alpha^2} \, d\alpha = e^{-\beta x} \quad (x > 0)$$

5.14 THE FOURIER INTEGRAL IN COMPLEX FORM

In the Fourier integral [formula (7), Sec. 5.12], the brackets contain an even function of α; hence, it is defined for negative values of α as well. On the basis of the foregoing, formula (7) can be rewritten as follows:

$$f(x) = \frac{1}{2\pi} \int_{-\infty}^{\infty} \left(\int_{-\infty}^{\infty} f(t) \cos \alpha (t - x) \, dt \right) d\alpha \qquad (1)$$

Let us now consider the following expression, which is identically equal to zero:

$$\int_{-M}^{M} \left(\int_{-\infty}^{\infty} f(t) \sin \alpha (t - x) \, dt \right) d\alpha = 0$$

The expression on the left is identically equal to zero because the function of α in the brackets is an odd function, and the integral of an odd function from $-M$ to $+M$ is equal to zero. It is obvious that

$$\lim_{M \to 0} \int_{-M}^{M} \left(\int_{-\infty}^{\infty} f(t) \sin \alpha (t - x) \, dt \right) d\alpha = 0$$

or

$$\int_{-\infty}^{\infty} \left(\int_{-\infty}^{\infty} f(t) \sin \alpha (t - x) \, dt \right) d\alpha = 0 \qquad (2)$$

Note. It is necessary to point to the following. A convergent integral with infinite limits is defined as follows:

$$\int_{-\infty}^{\infty} \varphi(\alpha)\, d\alpha = \int_{-\infty}^{c} \varphi(\alpha)\, d\alpha + \int_{c}^{\infty} \varphi(\alpha)\, d\alpha$$

$$= \lim_{M \to \infty} \int_{-M}^{c} \varphi(\alpha)\, d\alpha + \lim_{M \to \infty} \int_{c}^{M} \varphi(\alpha)\, d\alpha \qquad (*)$$

provided that each of the limits to the right exists (see Sec. 11.7, Vol. I). But in equation (2) we wrote

$$\int_{-\infty}^{\infty} \varphi(\alpha)\, d\alpha = \lim_{M \to \infty} \int_{-M}^{M} \varphi(\alpha)\, d\alpha \qquad (**)$$

Obviously, it may happen that the limit (**) exists, while the limits on the right side of equation (*) do not exist. The expression on the right of (**) is called the *principal value of the integral*. Thus, in equation (2) we consider the principal value of the improper (outer) integral. The subsequent integrals of this section will be written in this sense.

Let us multiply the terms of (2) by $-\frac{i}{2\pi}$ and add them to the corresponding terms of (1); we then get

$$f(x) = \frac{1}{2\pi} \int_{-\infty}^{\infty} \left[\int_{-\infty}^{\infty} f(t) (\cos \alpha (t - x) - i \sin \alpha (t - x))\, dt \right] d\alpha$$

or

$$f(x) = \frac{1}{2\pi} \int_{-\infty}^{\infty} \left[\int_{-\infty}^{\infty} f(t)\, e^{-i\alpha(t-x)}\, dt \right] d\alpha \qquad (3)$$

The right member in formula (3) is called the *Fourier integral in complex form of the function* $f(x)$.

Let us rewrite formula (3) thus:

$$f(x) = \int_{-\infty}^{\infty} \left[\frac{1}{2\pi} \int_{-\infty}^{\infty} f(t) e^{-i\alpha t} dt \right] e^{i\alpha x} d\alpha \qquad (4)$$

or, compactly

$$f(x) = \int_{-\infty}^{\infty} C(\alpha)\, e^{i\alpha x}\, d\alpha \qquad (5)$$

where

$$C(\alpha) = \frac{1}{2\pi} \int_{-\infty}^{\infty} f(t)\, e^{-i\alpha t}\, dt \qquad (6)$$

Formula (5) is similar to formula (10) of Sec. 5.12; α is also called the *wave number*, but here it assumes all values from $-\infty$ to $+\infty$ and the spectrum of wave numbers is termed a *continuous spectrum*. There are other similarities between (5) and (10), Sec. 5.12, whereas in formula (10) of Sec. 5.12 the wave number α_n is associated with the complex amplitude c_n, in formula (5) the wave numbers lying in interval $(\alpha_1, \alpha_1+\Delta\alpha)$ are associated with the complex amplitude $C(\alpha_1)$. The function $C(\alpha)$ is called the *spectral density* or the *spectral function*. (The term density is used here in same meaning as in Sec. 2.8, where we discussed the density of distribution over a two-dimensional domain.)

Equation (4) is usually written as two equations:

$$F^*(\alpha)=\frac{1}{\sqrt{2\pi}}\int_{-\infty}^{\infty} f(t)\,e^{-i\alpha t}\,dt \tag{7}$$

$$f(x)=\frac{1}{\sqrt{2\pi}}\int_{-\infty}^{\infty} F^*(\alpha)\,e^{i\alpha x}\,d\alpha \tag{8}$$

The function $F^*(\alpha)$ defined by formula (7) is called the *Fourier transform* of the function $f(x)$. The function $f(x)$ defined by formula (8) is called the *inverse Fourier transform* of the function $F^*(\alpha)$ (the transforms differ in the sign in front of i). The function $F^*(\alpha)$ differs from the function $C(\alpha)$ by the constant factor $\frac{1}{\sqrt{2\pi}}$.

From the transformations (7) and (8) follow the transformations (12), (14), (13) and (15), Sec. 5.13 (to within the constant factor 1/2). The transformations (12) and (14) are obtained if we substitute into (7)

$$e^{-i\alpha t}=\cos\alpha t-i\sin\alpha t,\qquad F^*(\alpha)=F(\alpha)-i\Phi(\alpha)$$

and equate the real and imaginary parts. In similar fashion it is possible to obtain transformations (13) and (15) from the transformation (8).

Note that in Chapter 7 ("Operational Calculus and Certain of Its Applications") we will make use of transformations that are similar to the Fourier transformations.

5.15 FOURIER SERIES EXPANSION WITH RESPECT TO AN ORTHOGONAL SYSTEM OF FUNCTIONS

Definition 1. An infinite sequence (system) of functions

$$\varphi_1(x),\quad \varphi_2(x),\quad \ldots,\quad \varphi_n(x),\quad \ldots \tag{1}$$

is said to be *orthogonal on the interval* $[a, b]$ if for any $n\neq k$

the equation

$$\int_a^b \varphi_n(x)\,\varphi_k(x)\,dx = 0 \tag{2}$$

holds true. It is assumed here that

$$\int_a^b [\varphi_n(x)]^2\,dx \neq 0$$

Example 1. The system of functions

$$1,\ \cos x,\ \sin x,\ \cos 2x,\ \sin 2x,\ \ldots,\ \cos nx,\ \sin nx,\ \ldots \tag{3}$$

is orthogonal on the interval $[-\pi, \pi]$. This follows from the equations (I) and (II) of Sec. 5.1.

Example 2. The system of functions

$$1,\ \cos\frac{\pi}{l}x,\ \sin\frac{\pi}{l}x,\ \cos 2\frac{\pi}{l}x,\ \sin 2\frac{\pi}{l}x,\ \ldots,\ \cos\frac{n\pi x}{l},\ \sin\frac{n\pi x}{l},\ \ldots \tag{3'}$$

is orthogonal on the interval $[-l, l]$. This can readily be verified.

Example 3. The system of functions

$$1,\ \cos x,\ \cos 2x,\ \cos 3x,\ \ldots,\ \cos nx,\ \ldots \tag{4}$$

is orthogonal on the interval $[0, \pi]$.

Example 4. The system of functions

$$\sin x,\ \sin 2x,\ \ldots,\ \sin nx,\ \ldots \tag{5}$$

is orthogonal on the interval $[0, \pi]$.

Below we give other systems of orthogonal functions.

Suppose a function $f(x)$, defined on the interval $[a, b]$, is such that it can be represented by a series relative to the functions of the orthogonal system (1), which converges to the given function on $[a, b]$:

$$f(x) = \sum_{n=0}^{\infty} c_n \varphi_n(x) \tag{6}$$

We determine the coefficients c_n. Suppose that the series obtained by multiplying the series (6) by any function $\varphi_k(x)$ admits termwise integration.

We multiply both members of (6) by $\varphi_k(x)$ and integrate from a to b. Taking into account equation (2), we get

$$\int_a^b f(x)\,\varphi_k(x)\,dx = c_k \int_a^b \varphi_k^2(x)\,dx$$

whence

$$c_k = \frac{\int_a^b f(x)\,\varphi_k(x)\,dx}{\int_a^b \varphi_k^2(x)\,dx} \tag{7}$$

The coefficients c_k computed from formulas (7) are called the *Fourier coefficients* of the function $f(x)$ relative to the system of orthogonal functions (1). Series (6) is called the *Fourier series* relative to the system of functions (1).

Definition 2. The orthogonal system of functions

$$\varphi_1(x), \quad \varphi_2(x), \quad \ldots, \quad \varphi_n(x), \quad \ldots$$

is called a *complete* system if for any quadratically integrable function $f(x)$, that is, such that

$$\int_a^b f^2(x)\,dx < \infty$$

the following equation holds true:

$$\lim_{n\to\infty} \int_a^b \left[f(x) - \sum_{i=0}^{n} c_i \varphi_i(x) \right]^2 dx = 0 \tag{8}$$

By virtue of the definitions of Sec. 5.7, equation (8) may also be interpreted as follows. The root-mean-square deviation of the sum $\sum_{i=0}^{n} c_i \varphi_i(x)$ from the function $f(x)$ tends to zero as $n \to \infty$.

If equation (8) holds, then we say that the Fourier series (6) converges to the function $f(x)$ *in the mean*.

It is quite obvious that convergence in the mean does not imply convergence at every point of the interval $[a, b]$.

We note without proof that the trigonometric systems indicated in examples 1 to 4 are complete over the appropriate intervals.

Extensive use is made of the *system of Bessel functions*

$$J_n(\lambda_1 x), \quad J_n(\lambda_2 x), \quad \ldots, \quad J_n(\lambda_i x), \quad \ldots \tag{9}$$

which were considered in Sec. 4.23. Here, $\lambda_1, \lambda_2, \ldots, \lambda_i, \ldots$ are the *roots of the Bessel function*, that is, numbers such that satisfy the relation

$$J_n(\lambda_i) = 0 \quad (i = 1, 2, \ldots)$$

We will now point out, without proof, that the system of functions

$$\sqrt{x}\, J_n(\lambda_1 x), \quad \sqrt{x}\, J_n(\lambda_2 x), \quad \ldots, \quad \sqrt{x}\, J_n(\lambda_i x), \quad \ldots \tag{10}$$

is orthogonal on the interval [0, 1]:

$$\int_0^1 xJ_n(\lambda_k x)\,J_n(\lambda_j x)\,dx=0^* \quad (k\neq j) \tag{11}$$

Systems of orthogonal Legendre polynomials also find applications. They are defined as follows:

$$P_0(x)=1,\quad P_n(x)=\frac{1}{2^n n!}\frac{d^n[(x^2-1)^n]}{dx^n}\quad (n=1,\ 2,\ \ldots)$$

They satisfy the equations

$$(x^2-1)\,y''+2xy'-n\,(n+1)\,y=0$$

Other systems of orthogonal polynomials are also used.

5.16 THE CONCEPT OF A LINEAR FUNCTION SPACE. EXPANSION OF FUNCTIONS IN FOURIER SERIES COMPARED WITH DECOMPOSITION OF VECTORS

In analytic geometry, a vector in three-dimensional space is defined as

$$\boldsymbol{A}=A_x\boldsymbol{i}+A_y\boldsymbol{j}+A_z\boldsymbol{k}$$

where $\boldsymbol{i}$, $\boldsymbol{j}$, $\boldsymbol{k}$ are unit, mutually perpendicular vectors taken along the coordinate axes. The vectors $\boldsymbol{i}$, $\boldsymbol{j}$, $\boldsymbol{k}$ will from now on be denoted by $\boldsymbol{e}_1$, $\boldsymbol{e}_2$, $\boldsymbol{e}_3$.

In similar fashion we can define a vector in n-dimensional space:

$$\boldsymbol{A}=\sum_{i=1}^{n}A_i\boldsymbol{e}_i$$

A set of vectors of the form $\boldsymbol{A}$ will be called an *n-dimensional Euclidean space* and will be symbolized by E_n. The vectors $\boldsymbol{A}$ will be called the elements or points of the n-dimensional Euclidean space. (It is also possible to consider vectors in an infinite-dimensional space.) Let us examine the properties of the space E_n.

Suppose we have two vectors in the space E_n:

$$\boldsymbol{A}=\sum_{i=1}^{n}A_i\boldsymbol{e}_i \quad \text{and} \quad \boldsymbol{B}=\sum_{i=1}^{n}B_i\boldsymbol{e}_i$$

---

* If for the functions $\varphi_k(x)$, $\varphi_j(x)$ the relation

$$\int_a^b \rho(x)\,\varphi_k(x)\,\varphi_j(x)\,dx=0 \quad (j\neq k)$$

holds, then we say that the functions $\varphi_i(x)$ are orthogonal with weight $\rho(x)$. Hence, the functions $J_n(\lambda_i x)$ (for $k\neq j$) are orthogonal with weight x.

If C_1 and C_2 are real numbers, then by analogy with three-dimensional space

$$C_1A+C_2B \tag{1}$$

is a vector in E_n.

The *scalar product* (also called dot product or inner product) of two vectors $\boldsymbol{A}$ and $\boldsymbol{B}$ is given by the expression

$$(\boldsymbol{AB})=\sum_{i=1}^{n} A_iB_i \tag{2}$$

The vectors $\boldsymbol{e}_1, \boldsymbol{e}_2, \ldots, \boldsymbol{e}_n$ lie in the space E_n and formula (2) holds true.

Hence we get

$$(\boldsymbol{e}_i\boldsymbol{e}_j)=0 \quad (i\neq j) \tag{2'}$$

and

$$(\boldsymbol{e}_i\boldsymbol{e}_i)=1 \quad (i=j)$$

Vectors whose scalar product is equal to zero are called *orthogonal vectors*. Consequently, the vectors $\boldsymbol{e}_1, \boldsymbol{e}_2, \ldots, \boldsymbol{e}_n$ are orthogonal.

As in the case of three-dimensional space, we can easily establish the following properties of a scalar product:

$$\left.\begin{aligned}(\boldsymbol{AB})&=(\boldsymbol{BA})\\ (\boldsymbol{A}+\boldsymbol{B},\ \boldsymbol{C})&=(\boldsymbol{AC})+(\boldsymbol{BC})\\ (\lambda\boldsymbol{A},\ \boldsymbol{B})&=\lambda(\boldsymbol{AB})\end{aligned}\right\} \tag{3}$$

The *length* or *modulus of a vector* $\boldsymbol{A}$ is defined as in three-dimensional space:

$$|\boldsymbol{A}|=\sqrt{(\boldsymbol{AA})}=\sqrt{\sum_{i=1}^{n} A_i^2} \tag{4}$$

The length of the difference between two vectors is defined naturally as follows:

$$|\boldsymbol{A}-\boldsymbol{B}|=\sqrt{\sum_{i=1}^{n}(A_i-B_i)^2} \tag{5}$$

In particular,

$$|\boldsymbol{A}-\boldsymbol{A}|=\sqrt{\sum_{i=1}^{n}(A_i-A_i)^2}=0$$

The *angle* φ between two vectors is defined as

$$\cos\varphi=\frac{(\boldsymbol{AB})}{|\boldsymbol{A}|\,|\boldsymbol{B}|} \tag{6}$$

Let us consider the collection of all piecewise monotonic bounded functions on the interval $[a, b]$. * We denote this collection by Φ and call it the *space of functions* Φ. The functions of this space will be called *elements* or *points* of the space Φ. It is possible to establish operations on the functions of the space Φ that are similar to the operations we performed on the vectors of the space E_n.

If C_1 and C_2 are any real numbers and $f_1(x)$, $f_2(x)$ are elements of the space Φ, then

$$C_1 f_1(x) + C_2 f_2(x) \tag{7}$$

is an element of Φ.

If $f(x)$ and $\varphi(x)$ are two functions in Φ, then the *scalar product of the functions* $f(x)$ and $\varphi(x)$ is given by the expression

$$(f, \varphi) = \int_a^b f(x) \cdot \varphi(x)\, dx \tag{8}$$

This expression is similar to expression (2). It is easy to verify that the scalar product (8) has properties similar to the properties (3) for vectors:

$$\left.\begin{aligned} (f, \varphi) &= (\varphi, f) \\ (f_1 + f_2, \varphi) &= (f_1, \varphi) + (f_2, \varphi) \\ (\lambda f, \varphi) &= \lambda (f, \varphi) \end{aligned}\right\} \tag{9}$$

Like the definition of the modulus of a vector by formula (4), we have the so-called *norm* of an element $f(x)$ of the space Φ:

$$\| f \| = \sqrt{(f, f)} = \sqrt{\int_a^b [f(x)]^2\, dx} \tag{10}$$

The *distance between elements* $f(x)$ *and* $\varphi(x)$ of space Φ is defined by the expression

$$\| f - \varphi \| = \sqrt{\int_a^b [f(x) - \varphi(x)]^2\, dx} \tag{11}$$

which is similar to formula (5).

Expression (11), which gives the distance between elements of the space, is called the *metric* of the space. To within the factor $\sqrt{b-a}$, it coincides with the root-mean-square deviation δ defined in Sec. 5.7.

---

* A class of functions of this kind is considered in the theorem of Sec. 5.1. We could also consider a broader class of functions for which all the assertions of Sec. 5.1 are preserved.

Clearly, if $f(x) \equiv \varphi(x)$, that is, $f(x)$ and $\varphi(x)$ coincide at all points of the interval $[a, b]$, then $\|f-\varphi\|=0$. But if $\|f-\varphi\|=0$, then $f(x)=\varphi(x)$ at all, except a finite number, points of $[a, b]$.* But in this case we also say that the elements of the space Φ are identical.

A space of piecewise monotonic bounded functions, in which are defined the operations (7), (8) and the metric is given by equation (11), is called a *linear function space with quadratic metric*. The elements of the space Φ are called *points* of the space or *vectors*.

Now let us consider a sequence of functions

$$\varphi_1(x), \quad \varphi_2(x), \quad \ldots, \quad \varphi_k(x), \quad \ldots \tag{12}$$

lying in the space Φ.

The sequence of functions (12) is said to be *orthogonal on the interval* $[a, b]$ if for arbitrary i, j $(i \neq j)$ the equations

$$(\varphi_i, \varphi_j) = \int_a^b \varphi_i(x)\,\varphi_j(x)\,dx = 0 \tag{13}$$

hold true.

On the basis of equations (I), Sec. 5.1, it follows that, for instance, the system of functions

$$1, \cos x, \sin x, \cos 2x, \sin 2x, \cos 3x, \sin 3x, \ldots$$

is orthogonal on the interval $[-\pi, \pi]$.

We will now show that expanding a function in a Fourier series of orthogonal functions is similar to the decomposition of a vector into orthogonal vectors. Suppose we have a vector

$$\boldsymbol{A} = A_1\boldsymbol{e}_1 + A_2\boldsymbol{e}_2 + \ldots + A_k\boldsymbol{e}_k + \ldots + A_n\boldsymbol{e}_n \tag{14}$$

We assume that the vectors $\boldsymbol{e}_1, \boldsymbol{e}_2, \ldots, \boldsymbol{e}_n$ are orthogonal, that is, if $i \neq j$, then

$$(\boldsymbol{e}_i\boldsymbol{e}_j) = 0 \tag{15}$$

To determine the projection A_k, we form the scalar products of each member of (14) by the vector $\boldsymbol{e}_k$. By properties (2) and (3) we get

$$(\boldsymbol{A}\boldsymbol{e}_k) = A_1(\boldsymbol{e}_1\boldsymbol{e}_k) + A_2(\boldsymbol{e}_2\boldsymbol{e}_k) + \ldots + A_k(\boldsymbol{e}_k\boldsymbol{e}_k) + \ldots + A_n(\boldsymbol{e}_n\boldsymbol{e}_k)$$

Taking account of (15) we obtain

$$(\boldsymbol{A}\boldsymbol{e}_k) = A_k(\boldsymbol{e}_k\boldsymbol{e}_k)$$

whence

$$A_k = \frac{(\boldsymbol{A}\boldsymbol{e}_k)}{(\boldsymbol{e}_k\boldsymbol{e}_k)} \quad (k = 1, 2, \ldots, n) \tag{16}$$

* There can also be an infinite number of points, where $f(x) \neq \varphi(x)$.

Further assume that the function $f(x)$ has been expanded relative to a system of orthogonal functions:

$$f(x)=\sum_{k=1}^{\infty} a_k\varphi_k(x) \tag{17}$$

Forming scalar products of each member of (17) by $\varphi_k(x)$ and taking into account (9) and (13), we get *

$$(f, \varphi_k)=a_k(\varphi_k, \varphi_k)$$

whence

$$a_k=\frac{(f, \varphi_k)}{(\varphi_k, \varphi_k)}=\frac{\int_a^b f(x)\varphi(x)\,dx}{\int_a^b [\varphi_k(x)]^2\,dx} \tag{18}$$

Formula (18) is similar to formula (16).

Now set

$$s_n=\sum_{k=1}^{n} a_k\varphi_k(x) \tag{19}$$

$$\delta_n=\|f-s_n\| \quad (n=1, 2, \ldots) \tag{20}$$

If

$$\lim_{n\to\infty}\delta_n=0$$

then the system of orthogonal functions (12) is *complete* on the interval $[a, b]$.

The Fourier series (17) converges to the function $f(x)$ *in the mean*.

Exercises on Chapter 5

1. Expand the following function in a Fourier series in the interval $(-\pi, \pi)$:

$$f(x)=2x \text{ for } 0\leqslant x\leqslant\pi$$
$$f(x)=x \text{ for } -\pi<x\leqslant 0$$

Ans. $\frac{1}{4}\pi-\frac{2}{\pi}\left(\frac{\cos x}{1^2}+\frac{\cos 3x}{3^2}+\frac{\cos 5x}{5^2}+\ldots\right)+3\left(\frac{\sin x}{1}-\frac{\sin 2x}{2}+\frac{\sin 3x}{3}-\ldots\right)$

2. Taking advantage of the expansion of the function $f(x)=1$ in the interval $(0, \pi)$ in the sines of multiple arcs, calculate the sum of the series $1-\frac{1}{3}+\frac{1}{5}-\frac{1}{7}+\ldots$. *Ans.* $\frac{\pi}{4}$.

---

* We presume that the series obtained in the process converge and that term-by-term integration is legitimate.

3. Utilizing the expansion of the function $f(x)=x^2$ in a Fourier series, compute the sum of the series $\frac{1}{1^2}-\frac{1}{2^2}+\frac{1}{3^2}-\frac{1}{4^2}+\ldots$. *Ans.* $\frac{\pi^2}{12}$.

4. Expand the function $f(x)=\frac{\pi^2}{12}-\frac{x^2}{4}$ in a Fourier series in the interval $(-\pi, \pi)$. *Ans.* $\cos x-\frac{\cos 2x}{2^2}+\frac{\cos 3x}{3^2}-\frac{\cos 4x}{4^2}+\ldots$.

5. Expand the following function in a Fourier series in the interval $(-\pi, \pi)$

$$f(x)=-\frac{(\pi+x)}{2} \quad \text{for } -\pi\leqslant x<0$$

$$f(x)=\frac{1}{2}(\pi-x) \quad \text{for } 0\leqslant x<\pi$$

Ans. $\sin x+\frac{1}{2}\sin 2x+\frac{1}{3}\sin 3x+\ldots$.

6. Expand in a Fourier series, in the interval $(-\pi, \pi)$, the function

$$f(x)=-x \quad \text{for } -\pi<x\leqslant 0$$
$$f(x)=0 \quad \text{for } 0<x\leqslant\pi$$

Ans. $\frac{\pi}{4}-\frac{2}{\pi}\sum_{m=0}^{\infty}\frac{\cos(2m+1)x}{(2m+1)^2}-\sum_{n=1}^{\infty}(-1)^{n-1}\frac{\sin nx}{n}$.

7. Expand in a Fourier series, in the interval $(-\pi, \pi)$, the function

$$f(x)=1 \quad \text{for } -\pi<x\leqslant 0$$
$$f(x)=-2 \quad \text{for } 0<x\leqslant\pi$$

Ans. $-\frac{1}{2}-\frac{6}{\pi}\sum_{n=0}^{\infty}\frac{\sin(2n+1)x}{2n+1}$.

8. Expand the function $f(x)=x^2$, in the interval $(0, \pi)$, in a series of sines.

Ans. $\frac{2}{\pi}\sum_{n=1}^{\infty}(-1)^{n+1}\left\{\frac{\pi^2}{n}+\frac{2}{n^3}[(-1)^n-1]\right\}\sin nx$.

9. Expand the function $y=\cos 2x$ in a series of sines in the interval $(0, \pi)$.

Ans. $-\frac{4}{\pi}\left[\frac{\sin x}{2^2-1}+\frac{3\sin 3x}{2^2-3^2}+\frac{5\sin 5x}{2^2-5^2}+\ldots\right]$.

10. Expand the function $y=\sin x$ in a series of cosines in the interval $(0, \pi)$.

Ans. $\frac{4}{\pi}\left[\frac{1}{2}+\frac{\cos 2x}{1-2^2}+\frac{\cos 4x}{1-4^2}+\ldots\right]$.

11. Expand the function $y=e^x$ in a Fourier series in the interval $(-l, l)$.

Ans. $\frac{e^l-e^{-l}}{2l}+l(e^l-e^{-l})\sum_{n=1}^{\infty}\frac{(-1)^n\sin\frac{n\pi x}{l}}{l^2+n^2\pi^2}$

$$+\pi(e^l-e^{-l})\sum_{n=1}^{\infty}\frac{(-1)^{n-1}n\sin\frac{n\pi x}{l}}{l^2+n^2\pi^2}.$$

12. Expand the function $f(x)=2x$ in a series of sines in the interval $(0, 1)$.

Ans. $\frac{4}{\pi}\sum_{n=1}^{\infty}(-1)^{n+1}\frac{\sin n\pi x}{n^2}$.

13. Expand the function $f(x)=x$ in a series of sines in the interval $(0, l)$.

Ans. $\frac{2l}{\pi}\sum_{n=1}^{\infty}(-1)^{n+1}\frac{\sin\frac{n\pi x}{l}}{n}$.

14. Expand the function

$$f(x)=\begin{cases} x & \text{for } 0<x\leqslant 1 \\ 2-x & \text{for } 1<x<2 \end{cases}$$

in the interval $(0, 2)$: (a) in a series of sines; (b) in a series of cosines.

Ans. (a) $\frac{8}{\pi^2}\sum_{n=0}^{\infty}(-1)^n\frac{\sin\frac{(2n+1)\pi x}{2}}{(2n+1)^2}$;

(b) $\frac{1}{2}-\frac{4}{\pi^2}\sum_{n=0}^{\infty}\frac{\cos(2n+1)\pi x}{(2n+1)^2}$.

CHAPTER 6

EQUATIONS OF MATHEMATICAL PHYSICS

6.1 BASIC TYPES OF EQUATIONS OF MATHEMATICAL PHYSICS

The basic equations of mathematical physics (for the case of functions of two independent variables) are the following second-order partial differential equations.

I. Wave equation:

$$\frac{\partial^2 u}{\partial t^2}=a^2\frac{\partial^2 u}{\partial x^2} \tag{1}$$

This equation is invoked in the study of processes of transversal vibrations of a string, the longitudinal vibrations of rods, electric oscillations in wires, the torsional oscillations of shafts, oscillations in gases and so forth. This equation is the simplest of the class of *equations of hyperbolic type.*

II. Fourier equation for heat conduction:

$$\frac{\partial u}{\partial t}=a^2\frac{\partial^2 u}{\partial x^2} \tag{2}$$

This equation is invoked in the study of processes of the propagation of heat, the filtration of liquids and gases in a porous medium (for example, the filtration of oil and gas in subterranean sandstones), some problems in probability theory, etc. This equation is the simplest of the class of *equations of parabolic type.*

III. Laplace equation:

$$\frac{\partial^2 u}{\partial x^2}+\frac{\partial^2 u}{\partial y^2}=0 \tag{3}$$

This equation is invoked in the study of problems dealing with electric and magnetic fields, stationary thermal state, problems in hydrodynamics, diffusion, and so on. This equation is the simplest in the class of *equations of elliptic type.*

In equations (1), (2), and (3), the unknown function u depends on two variables. Also considered are appropriate equations of functions with a larger number of variables. Thus, the wave equation in three independent variables is of the form

$$\frac{\partial^2 u}{\partial t^2}=a^2\left(\frac{\partial^2 u}{\partial x^2}+\frac{\partial^2 u}{\partial y^2}\right) \tag{1'}$$

the heat-conduction equation in three independent variables is of the form

$$\frac{\partial u}{\partial t}=a^2\left(\frac{\partial^2 u}{\partial x^2}+\frac{\partial^2 u}{\partial y^2}\right) \qquad (2')$$

the Laplace equation in three independent variables has the form

$$\frac{\partial^2 u}{\partial x^2}+\frac{\partial^2 u}{\partial y^2}+\frac{\partial^2 u}{\partial z^2}=0 \qquad (3')$$

6.2 DERIVING THE EQUATION OF THE VIBRATING STRING. FORMULATING THE BOUNDARY-VALUE PROBLEM. DERIVING EQUATIONS OF ELECTRIC OSCILLATIONS IN WIRES

In mathematical physics a string is understood to be a flexible and elastic thread. The tensions that arise in a string at any instant of time are directed along a tangent to its profile. Let a string of length l be, at the initial instant, directed along a segment of the x-axis from 0 to l. Assume that the ends of the string are fixed at the points $x=0$ and $x=l$. If the string is deflected from its original position and then let loose; or if without deflecting the string we impart to its points a certain velocity at the initial time, or if we deflect the string and impart a velocity to its points, then the points of the string will perform certain motions; we say that the string is set into vibration. The problem is to determine the shape of the string at any instant of time and to determine the law of motion of every point of the string as a function of time.

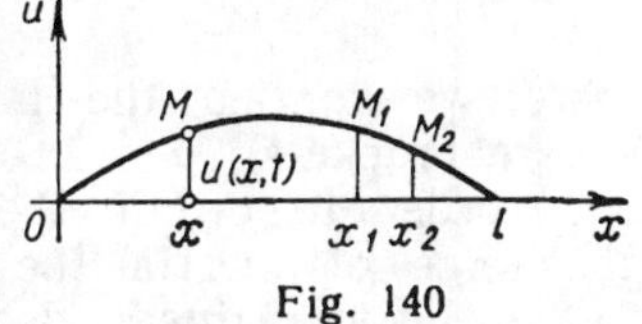

Fig. 140

Let us consider small deflections of the points of the string from the initial position. We may suppose that the motion of the points of the string is perpendicular to the x-axis and in a single plane. On this assumption, the process of vibration of the string is described by a single function $u(x, t)$, which yields the amount that a point of the string with abscissa x has moved at time t (Fig. 140).

Since we consider small deflections of the string in the xu-plane, we shall assume that the length of an element of string $\smile M_1M_2$ is equal to its projection on the x-axis, that is,* $\smile M_1M_2=$

* This assumption is equivalent to neglecting $u_x'^2$ as compared with 1. Indeed,

$$\smile M_1M_2=\int_{x_1}^{x_2}\sqrt{1+u_x'^2}\,dx=\int_{x_1}^{x_2}\left(1+\frac{1}{2}u_x'^2-\dots\right)dx\approx\int_{x_1}^{x_2}dx=x_2-x_1$$

$=x_2-x_1$. We also assume that the tension of the string at all points is the same; we denote it by T.

Consider an element of the string MM' (Fig. 141). Forces T act at the ends of this element along tangents to the string. Let the tangents form with the x-axis angles φ and $\varphi+\Delta\varphi$. Then the projection on the u-axis of forces acting on the element MM' will be equal to $T\sin(\varphi+\Delta\varphi)-T\sin\varphi$. Since the angle φ is small, we can put $\tan\varphi\approx\sin\varphi$, and we will have

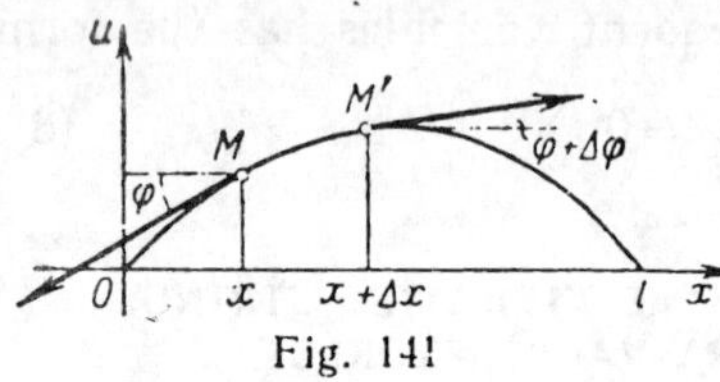

Fig. 141

$$T\sin(\varphi+\Delta\varphi)-T\sin\varphi$$
$$\approx T\tan(\varphi+\Delta\varphi)-T\tan\varphi=T\left[\frac{\partial u(x+\Delta x, t)}{\partial x}-\frac{\partial u(x, t)}{\partial x}\right]$$
$$=T\,\frac{\partial^2 u(x+\theta\Delta x, t)}{\partial x^2}\Delta x\approx T\,\frac{\partial^2 u(x, t)}{\partial x^2}\Delta x$$
$$0<\theta<1$$

(here, we applied the Lagrange theorem to the expression in the square brackets).

In order to obtain the equation of motion, we must equate to the force of inertia the external forces applied to the element. Let ρ be the linear density of the string. Then the mass of an element of string will be $\rho\Delta x$. The acceleration of the element is $\frac{\partial^2 u}{\partial t^2}$. Hence, by d'Alembert's principle we will have

$$\rho\Delta x\frac{\partial^2 u}{\partial t^2}=T\,\frac{\partial^2 u}{\partial x^2}\Delta x$$

Cancelling out Δx and denoting $\frac{T}{\rho}=a^2$, we get the equation of motion:

$$\frac{\partial^2 u}{\partial t^2}=a^2\frac{\partial^2 u}{\partial x^2} \tag{1}$$

This is the *wave equation*, the equation of the vibrating string. Equation (1) by itself is not sufficient for a complete definition of the motion of a string. The desired function $u(x, t)$ must also satisfy *boundary conditions* that indicate what occurs at the ends of the string ($x=0$ and $x=l$) and *initial conditions*, which describe the state of the string at the initial time ($t=0$). The boundary and initial conditions are referred to collectively as *boundary-value conditions*.

For example, as we assumed, let the ends of the string at $x=0$ and $x=l$ be fixed. Then for any t the following equations must hold:

$$u(0, t) = 0 \tag{2}$$

$$u(l, t) = 0 \tag{2'}$$

These equations are the *boundary conditions* for our problem.

At $t = 0$ the string has a definite shape, that which we gave it. Let this shape be defined by a function $f(x)$. We should then have

$$u(x, 0) = u|_{t=0} = f(x) \tag{3}$$

Furthermore, at the initial instant the velocity at each point of the string must be given; it is defined by the function $\varphi(x)$. Thus, we should have

$$\left.\frac{\partial u}{\partial t}\right|_{t=0} = \varphi(x) \tag{3'}$$

The conditions (3) and (3′) are the *initial conditions.*

Note. For a special case we may have $f(x) \equiv 0$ or $\varphi(x) \equiv 0$. But if $f(x) \equiv 0$ and $\varphi(x) \equiv 0$, then the string will be in a state of rest; hence, $u(x, t) \equiv 0$.

As has already been pointed out, the problem of electric oscillations in wires likewise leads to equation (1). Let us show this to be the case. The electric current in a wire is characterized by the quantity $i(x, t)$ and the voltage $v(x, t)$ which are dependent on the coordinate x of a point of the wire and on the time t. Regarding an element of wire Δx, we can write that the voltage drop on the element Δx is equal to $v(x, t) - v(x + \Delta x, t) \approx$ $\approx -\frac{\partial v}{\partial x}\Delta x$. This voltage drop consists of the resistance drop, equal to $iR\Delta x$, and the inductive drop, equal to $\frac{\partial i}{\partial t}L\Delta x$. Thus,

$$-\frac{\partial v}{\partial x}\Delta x = iR\Delta x + \frac{\partial i}{\partial t}L\Delta x \tag{4}$$

where R and L are the resistance and the inductance reckoned per unit length of wire. The minus sign indicates that the current flow is in a direction opposite to the build-up of v. Cancelling out Δx, we get the equation

$$\frac{\partial v}{\partial x} + iR + L\frac{\partial i}{\partial t} = 0 \tag{5}$$

Further, the difference between the current leaving element Δx and entering it during time Δt will be

$$[i(x, t) - i(x + \Delta x, t)]\,\Delta t \approx -\frac{\partial i}{\partial x}\Delta x\,\Delta t$$

It is taken up in charging the element $\left(\text{this is equal to } C\Delta x \frac{\partial v}{\partial t}\Delta t\right)$

and in leakage through the lateral surface of the wire due to imperfect insulation, equal to $Av\Delta x\Delta t$ (here A is the leakage coefficient). Equating these expressions and cancelling out $\Delta x\Delta t$, we get the equation

$$\frac{\partial i}{\partial x}+C\frac{\partial v}{\partial t}+Av=0 \tag{6}$$

Equations (5) and (6) are generally called *telegraph equations*.

From the system of equations (5) and (6) we can obtain an equation that contains only the desired function $i(x, t)$, and an equation containing only the desired function $v(x, t)$. Differentiate the terms of equation (6) with respect to x; differentiate the terms of (5) with respect to t and multiply them by C. Subtracting, we get

$$\frac{\partial^2 i}{\partial x^2}+A\frac{\partial v}{\partial x}-CR\frac{\partial i}{\partial x}-CL\frac{\partial^2 i}{\partial t^2}=0$$

Substituting into this equation the expression $\frac{\partial v}{\partial x}$ from (5), we get

$$\frac{\partial^2 i}{\partial x^2}+A\left(-iR-L\frac{\partial i}{\partial t}\right)-CR\frac{\partial i}{\partial x}-CL\frac{\partial^2 i}{\partial t^2}=0$$

or

$$\frac{\partial^2 i}{\partial x^2}=CL\frac{\partial^2 i}{\partial t^2}+(CR+AL)\frac{\partial i}{\partial t}+ARi \tag{7}$$

Similarly, we obtain an equation for determining $v(x, t)$:

$$\frac{\partial^2 v}{\partial x^2}=CL\frac{\partial^2 v}{\partial t^2}+(CR+AL)\frac{\partial v}{\partial t}+ARv \tag{8}$$

If we neglect the leakage through the insulation ($A=0$) and the resistance ($R=0$), then equations (7) and (8) pass into the wave equations

$$a^2\frac{\partial^2 i}{\partial x^2}=\frac{\partial^2 i}{\partial t^2}, \quad a^2\frac{\partial^2 v}{\partial x^2}=\frac{\partial^2 v}{\partial t^2}$$

where $a^2=\frac{1}{CL}$. The physical conditions dictate the formulation of the boundary and initial conditions of the problem.

6.3 SOLUTION OF THE EQUATION OF THE VIBRATING STRING BY THE METHOD OF SEPARATION OF VARIABLES (THE FOURIER METHOD)

The method of separation of variables (or the Fourier method), which we shall now discuss, is typical of the solution of many problems in mathematical physics. Let it be required to find the solution of the equation

$$\frac{\partial^2 u}{\partial t^2}=a^2\frac{\partial^2 u}{\partial x^2} \tag{1}$$

which satisfies the boundary-value conditions

$$u(0, t) = 0 \tag{2}$$

$$u(l, t) = 0 \tag{3}$$

$$u(x, 0) = f(x) \tag{4}$$

$$\left.\frac{\partial u}{\partial t}\right|_{t=0} = \varphi(x) \tag{5}$$

We shall seek a particular solution (not identically equal to zero) of equation (1) that satisfies the boundary conditions (2) and (3), in the form of a product of two functions $X(x)$ and $T(t)$, of which the former is dependent only on x, and the latter, only on t:

$$u(x, t) = X(x)\,T(t) \tag{6}$$

Substituting into equation (1), we get $X(x)\,T''(t) = a^2X''(x)\,T(t)$, and dividing the terms of the equation by a^2XT,

$$\frac{T''}{a^2T} = \frac{X''}{X} \tag{7}$$

The left member of this equation is a function that does not depend on x, the right member is a function that does not depend on t. Equation (7) is possible only when the left and right members are not dependent either on x or on t, that is, are equal to a constant number. We denote it by $-\lambda$, where $\lambda > 0$ (later on we will consider the case $\lambda < 0$). Thus,

$$\frac{T''}{a^2T} = \frac{X''}{X} = -\lambda$$

From these equations we get two equations:

$$X'' + \lambda X = 0 \tag{8}$$

$$T'' + a^2\lambda T = 0 \tag{9}$$

The general solutions of these equations are (see Sec. 1.21):

$$X(x) = A\cos\sqrt{\lambda}x + B\sin\sqrt{\lambda}x \tag{10}$$

$$T(t) = C\cos a\sqrt{\lambda}t + D\sin a\sqrt{\lambda}t \tag{11}$$

where A, B, C and D are arbitrary constants.

Substituting the expressions $X(x)$ and $T(t)$ into (6), we get

$$u(x, t) = (A\cos\sqrt{\lambda}x + B\sin\sqrt{\lambda}x)\,(C\cos a\sqrt{\lambda}t + D\sin a\sqrt{\lambda}t)$$

Now choose the constants A and B so that the conditions (2) and (3) are satisfied. Since $T(t) \not\equiv 0$ [otherwise we would have $u(x, t) \equiv 0$, which contradicts the hypothesis], the function $X(x)$ must satisfy the conditions (2) and (3); that is, we must have $X(0) = 0$, $X(l) = 0$. Putting the values $x = 0$ and $x = l$ into (10), we obtain, on the basis of (2) and (3),

$$0 = A \cdot 1 + B \cdot 0$$
$$0 = A \cos \sqrt{\lambda} l + B \sin \sqrt{\lambda} l$$

From the first equation we find $A = 0$. From the second it follows that

$$B \sin \sqrt{\lambda} l = 0$$

$B \neq 0$, since otherwise we would have $X \equiv 0$ and $u \equiv 0$, which contradicts the hypothesis. Consequently, we must have

$$\sin \sqrt{\lambda} l = 0$$

whence

$$\sqrt{\lambda} = \frac{n\pi}{l} \quad (n = 1, 2, \ldots) \tag{12}$$

(we do not take the value $n = 0$, since then we would have $X \equiv 0$ and $u \equiv 0$). And so we have

$$X = B \sin \frac{n\pi}{l} x \tag{13}$$

These values of λ are called *eigenvalues* of the given boundary-value problem. The functions $X(x)$ corresponding to them are called *eigenfunctions*.

Note. If in place of $-\lambda$ we took the expression $+\lambda = k^2$, then equation (8) would take the form

$$X'' - k^2 X = 0$$

The general solution of this equation is

$$X = Ae^{kx} + Be^{-kx}$$

A nonzero solution in this form cannot satisfy the boundary conditions (2) and (3).

Knowing $\sqrt{\lambda}$ we can [utilizing (11)] write

$$T(t) = C \cos \frac{an\pi}{l} t + D \sin \frac{an\pi}{l} t \quad (n = 1, 2, \ldots) \tag{14}$$

For each value of n, hence for every λ, we put the expressions (13) and (14) into (6) and obtain a solution of equation (1) that satisfies the boundary conditions (2) and (3). We denote this solution by $u_n(x, t)$:

$$u_n(x, t) = \sin \frac{n\pi}{l} x \left(C_n \cos \frac{an\pi}{l} t + D_n \sin \frac{an\pi}{l} t \right) \tag{15}$$

For each value of n we can take the constants C and D and thus write C_n and D_n (the constant B is included in C_n and D_n). Since equation (1) is linear and homogeneous, the sum of the solutions is also a solution, and therefore the function represented by the series

$$u(x, t) = \sum_{n=1}^{\infty} u_n(x, t)$$

or

$$u(x, t) = \sum_{n=1}^{\infty} \left(C_n \cos \frac{an\pi}{l} t + D_n \sin \frac{an\pi}{l} t \right) \sin \frac{n\pi}{l} x \qquad (16)$$

will likewise be a solution of the differential equation (1), which will satisfy the boundary conditions (2) and (3). Series (16) will obviously be a solution of equation (1) only if the coefficients C_n and D_n are such that this series converges and that the series resulting from a double term-by-term differentiation with respect to x and to t converge as well.

The solution (16) should also satisfy the initial conditions (4) and (5). We shall try to do this by choosing the constants C_n and D_n. Substituting into (16) $t = 0$, we get [see condition (4)]:

$$f(x) = \sum_{n=1}^{\infty} C_n \sin \frac{n\pi}{l} x \qquad (17)$$

If the function $f(x)$ is such that in the interval $(0, l)$ it may be expanded in a Fourier series (see Sec. 5.1), the condition (17) will be fulfilled if we put

$$C_n = \frac{2}{l} \int_0^l f(x) \sin \frac{n\pi}{l} x \, dx \qquad (18)$$

We then differentiate the terms of (16) with respect to t and substitute $t = 0$. From condition (5) we get the equation

$$\varphi(x) = \sum_{n=1}^{\infty} D_n \frac{an\pi}{l} \sin \frac{n\pi}{l} x$$

We define the Fourier coefficients of this series:

$$D_n \frac{an\pi}{l} = \frac{2}{l} \int_0^l \varphi(x) \sin \frac{n\pi}{l} x \, dx$$

or

$$D_n = \frac{2}{an\pi} \int_0^l \varphi(x) \sin \frac{n\pi}{l} x \, dx \qquad (19)$$

Thus, we have proved that the series (16), where the coefficients C_n and D_n are defined by formulas (18) and (19) [if it admits double termwise differentiation], is a function $u(x, t)$, which is the solution of equation (1) and satisfies the boundary and initial conditions (2) to (5).

Note. Solving the problem at hand for the wave equation by a different method, we can prove that the series (16) is a solution even when it does not admit termwise differentiation. In this case the function $f(x)$ must be twice differentiable and $\varphi(x)$ must be once differentiable.*

6.4 THE EQUATION OF HEAT CONDUCTION IN A ROD. FORMULATION OF THE BOUNDARY-VALUE PROBLEM

Let us consider a homogeneous rod of length l. We assume that the lateral surface of the rod is impenetrable to heat transfer and that the temperature is the same at all points of any cross-sectional area of the rod. Let us study the process of propagation of heat in the rod.

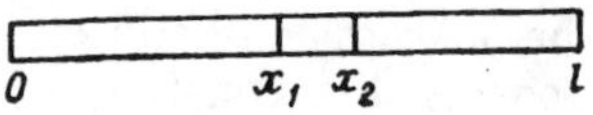

Fig. 142

We place the x-axis so that one end of the rod coincides with the point $x=0$, the other with the point $x=l$ (Fig. 142). Let $u(x, t)$ be the temperature in the cross section of the rod with abscissa x at time t. Experiment tells us that the rate of propagation of heat (that is, the quantity of heat passing through a cross section with abscissa x in unit time) is given by the formula

$$q=-k\frac{\partial u}{\partial x}S \tag{1}$$

where S is the cross-sectional area of the rod and k is the coefficient of thermal conductivity.**

Let us examine an element of rod contained between cross sections with abscissas x_1 and x_2 $(x_2-x_1=\Delta x)$. The quantity of heat passing through the cross section with abscissa x_1 during time Δt will be equal to

$$\Delta Q_1=-k\frac{\partial u}{\partial x}\bigg|_{x=x_1} S\Delta t \tag{2}$$

---

* These conditions are dealt with in detail in *Equations of Mathematical Physics* by A. N. Tikhonov and A. A. Samarsky, Gostekhizdat, 1954 (in Russian).

** The rate of heat transfer, or the rate of heat flux, is determined by

$$q=\lim_{\Delta t\to 0}\frac{\Delta Q}{\Delta t}$$

where ΔQ is the quantity of heat that has passed through a cross section S during a time Δt.

and the same for the cross section with abscissa x_2:

$$\Delta Q_2 = -k \left.\frac{\partial u}{\partial x}\right|_{x=x_2} S\Delta t \tag{3}$$

The influx of heat $\Delta Q_1 - \Delta Q_2$ into the rod element during time Δt will be

$$\Delta Q_1 - \Delta Q_2 = \left[-k \left.\frac{\partial u}{\partial x}\right|_{x=x_1} S\Delta t\right] - \left[-k \left.\frac{\partial u}{\partial x}\right|_{x=x_2} S\Delta t\right] \approx k \frac{\partial^2 u}{\partial x^2} \Delta x S \Delta t \tag{4}$$

(we applied the Lagrange theorem to the difference $\left.\frac{\partial u}{\partial x}\right|_{x=x_2} - \left.\frac{\partial u}{\partial x}\right|_{x=x_1}$). This influx of heat during time Δt was spent in raising the temperature of the rod element by Δu:

$$\Delta Q_1 - \Delta Q_2 = c\rho \Delta x S \Delta u$$

or

$$\Delta Q_1 - \Delta Q_2 \approx c\rho \Delta x S \frac{\partial u}{\partial t} \Delta t \tag{5}$$

where c is the heat capacity of the substance of the rod and ρ is the density of the substance ($\rho \Delta x S$ is the mass of an element of rod).

Equating expressions (4) and (5) of one and the same quantity of heat $\Delta Q_1 - \Delta Q_2$, we get

$$k \frac{\partial^2 u}{\partial x^2} \Delta x S \Delta t = c\rho \Delta x S \frac{\partial u}{\partial t} \Delta t$$

or

$$\frac{\partial u}{\partial t} = \frac{k}{c\rho} \frac{\partial^2 u}{\partial x^2}$$

Denoting $\frac{k}{c\rho} = a^2$, we finally get

$$\frac{\partial u}{\partial t} = a^2 \frac{\partial^2 u}{\partial x^2} \tag{6}$$

This is the equation for the propagation of heat (*the equation of heat conduction*) in a homogeneous rod.

For the solution of equation (6) to be definite, the function $u(x, t)$ must satisfy the boundary-value conditions corresponding to the physical conditions of the problem. For the solution of equation (6), the boundary-value conditions may differ. The conditions which correspond to the so-called ***first boundary-value problem*** for $0 \leqslant t \leqslant T$ are as follows:

$$u(x, 0) = \varphi(x) \tag{7}$$

$$u(0, t) = \psi_1(t) \tag{8}$$

$$u(l, t) = \psi_2(t) \tag{9}$$

Physically, condition (7) *(the initial condition)* corresponds to the fact that for $t=0$ a temperature is given in various cross sections of the rod equal to $\varphi(x)$. Conditions (8) and (9) *(the boundary conditions)* correspond to the fact that at the ends of the rod, $x=0$ and $x=l$, a temperature is maintained equal to $\psi_1(t)$ and $\psi_2(t)$, respectively.

It is proved that the equation (6) has only one solution in the region $0 \leqslant x \leqslant l$, $0 \leqslant t \leqslant T$, which satisfies the conditions (7), (8), and (9).

6.5 HEAT TRANSFER IN SPACE

Let us further consider the process of heat transfer in three-dimensional space. Let $u(x, y, z, t)$ be the temperature at a point with coordinates (x, y, z) at time t. Experiment states that the rate of heat passage through an area Δs, that is, the amount of heat passing through in unit time is governed by the formula [similar to formula (1) of the preceding section]

$$\Delta Q = -k \frac{\partial u}{\partial n} \Delta s \tag{1}$$

where k is the coefficient of thermal conductivity of the medium under consideration, which we regard as homogeneous and isotropic, $\boldsymbol{n}$ is the unit vector directed normally to the area Δs in the direction of flow of the heat. Taking advantage of Sec. 8.14, Vol. I, we can write

$$\frac{\partial u}{\partial n} = \frac{\partial u}{\partial x} \cos\alpha + \frac{\partial u}{\partial y} \cos\beta + \frac{\partial u}{\partial z} \cos\gamma$$

where $\cos\alpha$, $\cos\beta$, $\cos\gamma$ are the direction cosines of the vector $\boldsymbol{n}$, or

$$\frac{\partial u}{\partial n} = \boldsymbol{n} \operatorname{grad} u$$

Substituting the expression $\frac{\partial u}{\partial n}$ into formula (1), we get

$$\Delta Q = -k\boldsymbol{n} \operatorname{grad} u \, \Delta s$$

The amount of heat flowing in time Δt through the elementary area Δs will be

$$\Delta Q \, \Delta t = -k\boldsymbol{n} \operatorname{grad} u \, \Delta t \, \Delta s$$

Now let us return to the problem posed at the beginning of the section. In the medium at hand we pick out a small volume V bounded by the surface S. The amount of heat flowing through the surface S will be

$$Q = -\Delta t \iint_S k\boldsymbol{n} \operatorname{grad} u \, ds \tag{2}$$

where $\boldsymbol{n}$ is the unit vector directed along the external normal to the surface S. It is obvious that formula (2) yields the amount of heat entering the volume V (or leaving the volume V) during time Δt. The amount of heat entering V is spent in raising the temperature of the substance of this volume.

Let us consider an elementary volume Δv. Let its temperature rise by Δu in time Δt. Obviously, the amount of heat expended on raising the temperature of the element Δv will be

$$c\Delta v\rho\,\Delta u \approx c\Delta v\rho\,\frac{\partial u}{\partial t}\,\Delta t$$

where c is the heat capacity of the substance and ρ is the density. The total amount of heat consumed in raising the temperature in the volume V during time Δt will be

$$\Delta t\iiint\limits_V c\rho\,\frac{\partial u}{\partial t}\,dv$$

But this is the heat that has entered the volume V during the time Δt; it is defined by formula (2). Thus, we have the equation

$$-\Delta t\iint\limits_S k\boldsymbol{n}\operatorname{grad} u\,ds=\Delta t\iiint\limits_V c\rho\,\frac{\partial u}{\partial t}\,dv$$

Cancelling out Δt, we get

$$-\iint\limits_S k\boldsymbol{n}\operatorname{grad} u\,ds=\iiint\limits_V c\rho\,\frac{\partial u}{\partial t}\,dv \qquad (3)$$

The surface integral on the left-hand side of this equation we transform by the Ostrogradsky formula (see Sec. 3.8), assuming $\boldsymbol{F}=k\operatorname{grad} u$:

$$\iint\limits_S (k\operatorname{grad} u)\,\boldsymbol{n}\,ds=\iiint\limits_V \operatorname{div}(k\operatorname{grad} u)\,dv$$

Replacing the double integral on the left of (3) by a triple integral, we get

$$-\iiint\limits_V \operatorname{div}(k\operatorname{grad} u)\,dv=\iiint\limits_V c\rho\,\frac{\partial u}{\partial t}\,dv$$

or

$$\iiint\limits_V \left[\operatorname{div}(k\operatorname{grad} u)+c\rho\,\frac{\partial u}{\partial t}\right]dv=0 \qquad (4)$$

Applying the mean-value theorem to the triple integral on the left (see Sec. 2.12), we get

$$\left[\operatorname{div}(k\operatorname{grad} u)+c\rho\,\frac{\partial u}{\partial t}\right]_{x=x_1,\ y=y_1,\ z=z_1}=0 \qquad (5)$$

where the point $P(x_1, y_1, z_1)$ is some point in the volume V.

Since we can pick out an arbitrary volume V in three-dimensional space where heat is being transferred, and since we assume that the integrand in (4) is continuous, (5) will hold true at each point of the space. Thus,

$$c\rho \frac{\partial u}{\partial t} = -\operatorname{div}(k \operatorname{grad} u) \tag{6}$$

But

$$k \operatorname{grad} u = k \frac{\partial u}{\partial x} \boldsymbol{i} + k \frac{\partial u}{\partial y} \boldsymbol{j} + k \frac{\partial u}{\partial z} \boldsymbol{k}$$

and (see Sec. 3.8)

$$\operatorname{div}(k \operatorname{grad} u) = \frac{\partial}{\partial x}\left(k \frac{\partial u}{\partial x}\right) + \frac{\partial}{\partial y}\left(k \frac{\partial u}{\partial y}\right) + \frac{\partial}{\partial z}\left(k \frac{\partial u}{\partial z}\right)$$

Substituting into (6), we obtain

$$-c\rho \frac{\partial u}{\partial t} = \frac{\partial}{\partial x}\left(k \frac{\partial u}{\partial x}\right) + \frac{\partial}{\partial y}\left(k \frac{\partial u}{\partial y}\right) + \frac{\partial}{\partial z}\left(k \frac{\partial u}{\partial z}\right) \tag{7}$$

If k is a constant, then

$$\operatorname{div}(k \operatorname{grad} u) = k \operatorname{div}(\operatorname{grad} u) = k\left(\frac{\partial^2 u}{\partial x^2} + \frac{\partial^2 u}{\partial y^2} + \frac{\partial^2 u}{\partial z^2}\right)$$

and equation (6) then yields

$$-c\rho \frac{\partial u}{\partial t} = k\left(\frac{\partial^2 u}{\partial x^2} + \frac{\partial^2 u}{\partial y^2} + \frac{\partial^2 u}{dz^2}\right)$$

or, putting $-\frac{k}{c\rho} = a^2$,

$$\frac{\partial u}{\partial t} = a^2\left(\frac{\partial^2 u}{\partial x^2} + \frac{\partial^2 u}{\partial y^2} + \frac{\partial^2 u}{\partial z^2}\right) \tag{8}$$

Equation (8) is compactly written as

$$\frac{\partial u}{\partial t} = a^2 \Delta u$$

where $\Delta u = \frac{\partial^2 u}{\partial x^2} + \frac{\partial^2 u}{\partial y^2} + \frac{\partial^2 u}{\partial z^2}$ is the Laplacian operator. Equation (8) is the *equation of heat conduction in space.* To find its unique solution that corresponds to the problem posed here, it is necessary to specify the boundary-value conditions.

Let there be a body Ω with a surface σ. In this body we consider the process of heat transfer. At the initial time the temperature of the body is specified, which means that the solution is known for $t = 0$ (*the initial condition*):

$$u(x, y, z, 0) = \varphi(x, y, z) \tag{9}$$

In addition to that we must know the temperature at any point M of the surface σ of the body at any time t (*the boundary condi-*

tion):

$$u(M, t) = \psi(M, t) \tag{10}$$

(Other boundary conditions are possible too.)

If the desired function $u(x, y, z, t)$ is independent of z, which corresponds to the temperature being independent of z, we obtain the equation

$$\frac{\partial u}{\partial t} = a^2 \left(\frac{\partial^2 u}{\partial x^2} + \frac{\partial^2 u}{\partial y^2} \right) \tag{11}$$

which is the *equation of heat transfer in a plane*.

If we consider heat transfer in a flat domain D with boundary C, then the boundary-value conditions, like (9) and (10), are formulated as follows:

$$u(x, y, 0) = \varphi(x, y)$$
$$u(M, t) = \psi(M, t)$$

where φ and ψ are specified functions and M is a point on the boundary C.

But if the function u does not depend either on z or on y, then we get the equation

$$\frac{\partial u}{\partial t} = a^2 \frac{\partial^2 u}{\partial x^2}$$

which is the *equation of heat transfer in a rod*.

6.6 SOLUTION OF THE FIRST BOUNDARY-VALUE PROBLEM FOR THE HEAT-CONDUCTION EQUATION BY THE METHOD OF FINITE DIFFERENCES

As in the case of ordinary differential equations, when we solve partial differential equations by the method of finite differences, the derivatives are replaced by appropriate differences (see Fig. 143):

$$\frac{\partial u(x, t)}{\partial x} \approx \frac{u(x+h, t) - u(x, t)}{h} \tag{1}$$

$$\frac{\partial^2 u(x, t)}{\partial x^2} \approx \frac{1}{h} \left[\frac{u(x+h, t) - u(x, t)}{h} - \frac{u(x, t) - u(x-h, t)}{h} \right]$$

or

$$\frac{\partial^2 u(x, t)}{\partial x^2} \approx \frac{u(x+h, t) - 2u(x, t) + u(x-h, t)}{h^2} \tag{2}$$

similarly,

$$\frac{\partial u(x, t)}{\partial t} \approx \frac{u(x, t+l) - u(x, t)}{l} \tag{3}$$

The first boundary-value problem for the heat-conduction equation is stated (see Sec. 6.4) as follows. It is required to find the solution of the equation

$$\frac{\partial u}{\partial t}=a^2\frac{\partial^2 u}{\partial x^2} \tag{4}$$

that satisfies the boundary-value conditions

$$u(x,\ 0)=\varphi(x),\qquad 0\leqslant x\leqslant L \tag{5}$$

$$u(0,\ t)=\psi_1(t),\qquad 0\leqslant t\leqslant T \tag{6}$$

$$u(l,\ t)=\psi_2(t),\qquad 0\leqslant t\leqslant T \tag{7}$$

that is, we have to find the solution $u(x,\ t)$ in a rectangle bounded by the straight lines $t=0$, $x=0$, $x=L$, $t=T$, if the values of the desired function are given on three of its sides: $t=0$,

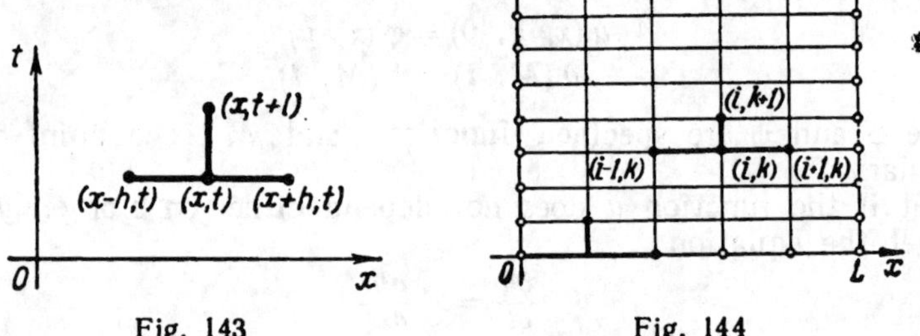

Fig. 143 Fig. 144

$x=0$, $x=L$ (Fig. 144). We cover our region with a grid formed by the straight lines

$$x=ih,\qquad i=1,\ 2,\ \ldots$$

$$t=kl,\qquad k=1,\ 2,\ \ldots$$

and approximate the values at the lattice points of the grid, that is, at the points of intersection of these lines. Introducing the notation $u(ih,\ kl)=u_{i,\,k}$, we write [in place of equation (4)] a corresponding difference equation for the point $(ih,\ kl)$. In accord with (3) and (2), we get

$$\frac{u_{i,\,k+1}-u_{i,\,k}}{l}=a^2\frac{u_{i+1,\,k}-2u_{i,\,k}+u_{i-1,\,k}}{h^2} \tag{8}$$

We determine $u_{i,\,k+1}$:

$$u_{i,\,k+1}=\left(1-\frac{2a^2l}{h^2}\right)u_{i,\,k}+a^2\frac{l}{h^2}\left(u_{i+1,\,k}+u_{i-1,\,k}\right) \tag{9}$$

From (9) it follows that if we know three values in the kth row: $u_{i,\,k}$, $u_{i+1,\,k}$, $u_{i-1,\,k}$, we can determine the value $u_{i,\,k+1}$ in the $(k+1)$th row. We know all the values on the straight line $t=0$ [see formula (5)]. By formula (9) determine the values at all the interior points of the segment $t=l$. We know the values at the end points of this segment by virtue of (6) and (7). In this way, row by row, we determine the values of the desired solution at all lattice points of the grid.

It may be proved that from formula (9) we can obtain an approximate value of the solution not for an arbitrary relationship between the steps h and l, but only for $l \leqslant \frac{h^2}{2a^2}$. Formula (9) is greatly simplified if the step length l along the t-axis is chosen so that

$$1-\frac{2a^2 l}{h^2}=0$$

or

$$l=\frac{h^2}{2a^2}$$

In this case, (9) takes the form

$$u_{i,\,k+1}=\frac{1}{2}(u_{i+1,\,k}+u_{i-1,\,k}) \tag{10}$$

This formula is particularly convenient for computations (Fig. 145). This method gives the solution at the lattice points of the grid. Solutions between these points may be obtained, for example, by extrapolation, by drawing a plane through every three points in the space (x, t, u). Let us denote by $u_h(x, t)$ a solution obtained by formula (10) and this extrapolation. It can be proved that

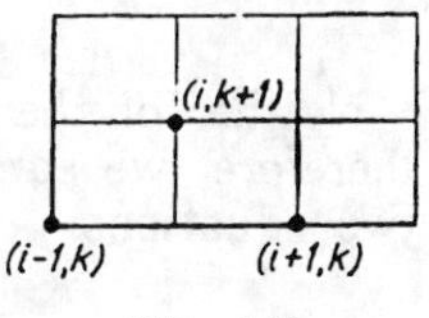

Fig. 145

$$\lim_{h \to 0} u_h(x, t) = u(x, t)$$

where $u(x, t)$ is the solution of our problem. It can also be proved* that

$$|u_h(x, t)-u(x, t)| < Mh^2$$

where M is a constant independent of h.

6.7 HEAT TRANSFER IN AN UNBOUNDED ROD

Let the temperature be given at various sections of an unbounded rod at an initial instant of time. It is required to determine the temperature distribution in the rod at subsequent instants of time. (Physical problems reduce to that of heat transfer in an unbounded rod when the rod is so long that the temperature in the

* This question is dealt with in more detail in D. Yu. Panov's *Handbook of Numerical Solutions of Partial Differential Equations*, Gostekhizdat, 1951; Lothar Collatz, *Numerische Behandlung von Differentialgluchungen*, 2nd edition, Springer, 1955.

interior points of the rod at the instants of time under consideration is but slightly dependent on the conditions at the ends of the rod.)

If the rod coincides with the x-axis, the problem is stated mathematically as follows. Find the solution to the equation

$$\frac{\partial u}{\partial t}=a^2\frac{\partial^2 u}{\partial x^2} \tag{1}$$

in the region $-\infty<x<\infty$, $0<t$, which satisfies the initial condition

$$u(x,\ 0)=\varphi(x) \tag{2}$$

To find the solution, we apply the method of separation of variables (see Sec. 6.3); that is, we shall seek a particular solution of equation (1) in the form of a product of two functions:

$$u(x,\ t)=X(x)\,T(t) \tag{3}$$

Putting this into equation (1) we have $X(x)\,T'(t)=a^2X''(x)\,T(t)$ or

$$\frac{T'}{a^2T}=\frac{X''}{X}=-\lambda^2 \tag{4}$$

Neither of these relations can be dependent either on x or on t; therefore, we equate them to a constant,* $-\lambda^2$. From (4) we get two equations:

$$T'+a^2\lambda^2T=0 \tag{5}$$

$$X''+\lambda^2X=0 \tag{6}$$

Solving them we find

$$T=Ce^{-a^2\lambda^2t}$$

$$X=A\cos\lambda x+B\sin\lambda x$$

Substituting into (3), we obtain

$$u_\lambda(x,\ t)=e^{-a^2\lambda^2t}\left[A(\lambda)\cos\lambda x+B(\lambda)\sin\lambda x\right] \tag{7}$$

[the constant C is included in $A(\lambda)$ and in $B(\lambda)$].

For each value of λ we obtain a solution of the form (7). For each value of λ the arbitrary constants A and B have definite values. We can therefore consider A and B functions of λ. The sum of the solutions of form (7) is likewise a solution [since equation (1) is linear]:

$$\sum_\lambda e^{-a^2\lambda^2t}\left[A(\lambda)\cos\lambda x+B(\lambda)\sin\lambda x\right]$$

---

* Since from the meaning of the problem $T(t)$ must be bounded for any t if $\varphi(x)$ is bounded, it follows that $\frac{T'}{T}$ must be negative. And so we write $-\lambda^2$.

Integrating expression (7) with respect to the parameter λ between 0 and ∞, we also get a solution

$$u(x, t) = \int_0^\infty e^{-a^2\lambda^2 t} [A(\lambda) \cos \lambda x + B(\lambda) \sin \lambda x] \, d\lambda \tag{8}$$

if $A(\lambda)$ and $B(\lambda)$ are such that this integral, its derivative with respect to t and the second derivative with respect to x exist and are obtained by differentiation of the integral with respect to t and x. We choose $A(\lambda)$ and $B(\lambda)$ such that the solution $u(x, t)$ satisfies the condition (2). Putting $t=0$ in (8), we get [on the basis of condition (2)]:

$$u(x, 0) = \varphi(x) = \int_0^\infty [A(\lambda) \cos \lambda x + B(\lambda) \sin \lambda x] \, d\lambda \tag{9}$$

Suppose that the function $\varphi(x)$ is such that it may be represented by the Fourier integral (see Sec. 5.13):

$$\varphi(x) = \frac{1}{\pi} \int_0^\infty \left(\int_{-\infty}^\infty \varphi(\alpha) \cos \lambda (\alpha - x) \, d\alpha \right) d\lambda$$

or

$$\varphi(x) = \frac{1}{\pi} \int_0^\infty \left[\left(\int_{-\infty}^\infty \varphi(\alpha) \cos \lambda\alpha \, d\alpha \right) \cos \lambda x + \left(\int_{-\infty}^\infty \varphi(\alpha) \sin \lambda\alpha \, d\alpha \right) \sin \lambda x \right] d\lambda \tag{10}$$

Comparing the right sides of (9) and (10), we get

$$\left. \begin{aligned} A(\lambda) &= \frac{1}{\pi} \int_{-\infty}^\infty \varphi(\alpha) \cos \lambda\alpha \, d\alpha \\ B(\lambda) &= \frac{1}{\pi} \int_{-\infty}^\infty \varphi(\alpha) \sin \lambda\alpha \, d\alpha \end{aligned} \right\} \tag{11}$$

Putting the expressions thus found of $A(\lambda)$ and $B(\lambda)$ into (8), we obtain

$$u(x, t) = \frac{1}{\pi} \int_0^\infty e^{-a^2\lambda^2 t} \left[\left(\int_{-\infty}^\infty \varphi(\alpha) \cos \lambda\alpha \, d\alpha \right) \cos \lambda x + \left(\int_{-\infty}^{+\infty} \varphi(\alpha) \sin \lambda\alpha \, d\alpha \right) \sin \lambda x \right] d\lambda$$

$$=\frac{1}{\pi}\int_0^\infty e^{-a^2\lambda^2 t}\left[\int_{-\infty}^{\infty}\varphi(\alpha)(\cos\lambda\alpha\cos\lambda x+\sin\lambda\alpha\sin\lambda x)\,d\alpha\right]d\lambda$$

$$=\frac{1}{\pi}\int_0^\infty e^{-a^2\lambda^2 t}\left(\int_{-\infty}^{\infty}\varphi(\alpha)\cos\lambda(\alpha-x)\,d\alpha\right)d\lambda$$

or, interchanging the order of integration, we finally get

$$u(x, t)=\frac{1}{\pi}\int_{-\infty}^{\infty}\left[\varphi(\alpha)\left(\int_0^\infty e^{-a^2\lambda^2 t}\cos\lambda(\alpha-x)\,d\lambda\right)\right]d\alpha \qquad (12)$$

This is the solution of the problem.

Let us transform formula (12). Compute the integral in the parentheses:

$$\int_0^\infty e^{-a^2\lambda^2 t}\cos\lambda(\alpha-x)\,d\lambda=\frac{1}{a\sqrt{t}}\int_0^\infty e^{-z^2}\cos\beta z\,dz \qquad (13)$$

The integral is transformed by substitution:

$$a\lambda\sqrt{t}=z, \quad \frac{\alpha-x}{a\sqrt{t}}=\beta \qquad (14)$$

We denote

$$K(\beta)=\int_0^\infty e^{-z^2}\cos\beta z\,dz \qquad (15)$$

Differentiating,* we get

$$K'(\beta)=-\int_0^\infty e^{-z^2}z\sin\beta z\,dz$$

Integrating by parts, we find

$$K'(\beta)=\frac{1}{2}\left[e^{-z^2}\sin\beta z\right]_0^\infty-\frac{\beta}{2}\int_0^\infty e^{-z^2}\cos\beta z\,dz$$

$$K'(\beta)=-\frac{\beta}{2}K(\beta)$$

Integrating this differential equation, we obtain

$$K(\beta)=Ce^{-\frac{\beta^2}{4}} \qquad (16)$$

* Differentiation here is easily justified.

Determine the constant C. From (15) it follows that

$$K(0)=\int_0^\infty e^{-z^2}\,dz=\frac{\sqrt{\pi}}{2}$$

(see Sec. 2.5). Hence, in (16) we must have

$$C=\frac{\sqrt{\pi}}{2}$$

And so

$$K(\beta)=\frac{\sqrt{\pi}}{2}e^{-\frac{\beta^2}{4}} \tag{17}$$

Put the value (17) of the integral (15) into (13):

$$\int_0^\infty e^{-a^2\lambda^2 t}\cos\lambda(\alpha-x)\,d\lambda=\frac{1}{a\sqrt{t}}\,\frac{\sqrt{\pi}}{2}\,e^{-\frac{\beta^2}{4}}$$

In place of β we substitute its expression (14) and finally get the value of the integral (13):

$$\int_0^\infty e^{-a^2\lambda^2 t}\cos\lambda(\alpha-x)\,d\lambda=\frac{1}{2a}\sqrt{\frac{\pi}{t}}\,e^{-\frac{(\alpha-x)^2}{4a^2t}} \tag{18}$$

Putting this expression of the integral into the solution (12), we finally get

$$u(x,\ t)=\frac{1}{2a\sqrt{\pi t}}\int_{-\infty}^{\infty}\varphi(\alpha)\,e^{-\frac{(\alpha-x)^2}{4a^2t}}\,d\alpha \tag{19}$$

This formula, called the *Poisson integral*, is the solution to the problem of heat transfer in an unbounded rod.

Note. It may be proved that the function $u(x,\ t)$, defined by integral (19), is a solution of equation (1) and satisfies condition (2) if the function $\varphi(x)$ is bounded on an infinite interval $(-\infty,\ \infty)$.

Let us establish the physical meaning of formula (19). We consider the function

$$\varphi^*(x)=\begin{cases}0 & \text{for } -\infty<x<x_0\\ \varphi(x) & \text{for } x_0\leqslant x\leqslant x_0+\Delta x\\ 0 & \text{for } x_0+\Delta x<x<\infty\end{cases} \tag{20}$$

Then the function

$$u^*(x,\ t)=\frac{1}{2a\sqrt{\pi t}}\int_{-\infty}^{\infty}\varphi^*(\alpha)\,e^{-\frac{(\alpha-x)^2}{4a^2t}}\,d\alpha \tag{21}$$

is the solution to equation (1), which solution takes on the value $\varphi^*(x)$ when $t=0$. Taking (20) into consideration, we can write

$$u^*(x, t)=\frac{1}{2a\sqrt{\pi t}}\int_{x_0}^{x_0+\Delta x}\varphi(\alpha)\,e^{-\frac{(\alpha-x)^2}{4a^2t}}\,d\alpha$$

Applying the mean-value theorem to the latter integral, we get

$$u^*(x, t)=\frac{\varphi(\xi)\,\Delta x}{2a\sqrt{\pi t}}\,e^{-\frac{(\xi-x)^2}{4a^2t}}, \qquad x_0<\xi<x_0+\Delta x \tag{22}$$

Formula (22) gives the value of temperature at a point in the rod at any time if for $t=0$ the temperature in the rod is everywhere $u^*=0$, with the exception of the interval $[x_0, x_0+\Delta x]$ where it is $\varphi(x)$. The sum of temperatures of form (22) is what yields the solution of (19). It will be noted that if ρ is the linear density of the rod, c the heat capacity of the material, then the quantity of heat in the element $[x_0, x_0+\Delta x]$ for $t=0$ will be

$$\Delta Q\approx\varphi(\xi)\,\Delta x\,\rho c \tag{23}$$

Let us now consider the function

$$\frac{1}{2a\sqrt{\pi t}}\,e^{-\frac{(\xi-x)^2}{4a^2t}} \tag{24}$$

Comparing it with the right side of (22) and taking into account (23), we may say that it yields the temperature at any point of the rod at any instant of time t if for $t=0$ there was an instantaneous heat source with amount of heat $Q=c\rho$ in the cross section ξ (the limiting case as $\Delta x\to 0$).

6.8 PROBLEMS THAT REDUCE TO INVESTIGATING SOLUTIONS OF THE LAPLACE EQUATION. STATING BOUNDARY-VALUE PROBLEMS

In this section we shall consider certain problems that reduce to the solution of the *Laplace equation:*

$$\frac{\partial^2 u}{\partial x^2}+\frac{\partial^2 u}{\partial y^2}+\frac{\partial^2 u}{\partial z^2}=0 \tag{1}$$

As already pointed out, the left side of equation (1) is symbolized by

$$\frac{\partial^2 u}{\partial x^2}+\frac{\partial^2 u}{\partial y^2}+\frac{\partial^2 u}{\partial z^2}\equiv\Delta u$$

where Δ is called the *Laplacian operator*. The functions u which satisfy the Laplace equation are called *harmonic functions*.

I. A stationary (steady-state) distribution of temperature in a homogeneous body. Let there be a homogeneous body Ω bounded by a surface σ. In Sec. 6.5 it was shown that the temperature at various points of the body satisfies the equation

$$\frac{\partial u}{\partial t}=a^2\left(\frac{\partial^2 u}{\partial x^2}+\frac{\partial^2 u}{\partial y^2}+\frac{\partial^2 u}{\partial z^2}\right)$$

If the process is steady-state, that is, if the temperature is not dependent on the time, but only on the coordinates of the points of the body, then $\frac{\partial u}{\partial t}=0$ and, consequently, the temperature satisfies the Laplace equation

$$\frac{\partial^2 u}{\partial x^2}+\frac{\partial^2 u}{\partial y^2}+\frac{\partial^2 u}{\partial z^2}=0 \tag{1}$$

To determine the temperature in the body uniquely from this equation, one has to know the temperature on the surface σ. Thus, for equation (1), the boundary-value problem is formulated as follows.

Find the function $u(x, y, z)$ that satisfies equation (1) inside the volume Ω and that takes on specified values at each point M of the surface σ:

$$u|_\sigma=\psi(M) \tag{2}$$

This problem is called the *Dirichlet problem* or the *first boundary-value problem* of equation (1).

If the temperature on the surface of the body is not known, but the heat flux at every point of the surface is, which is proportional to $\frac{\partial u}{\partial n}$ (see Sec. 6.5), then in place of the boundary-value condition (2) on the surface σ we will have the condition

$$\left.\frac{\partial u}{\partial n}\right|_\sigma=\psi^*(M) \tag{3}$$

The problem of finding the solution to (1) that satisfies the boundary-value condition (3) is called the *Neumann problem* or the *second boundary-value problem*.

If we consider the temperature distribution in a two-dimensional domain D bounded by a contour C, then the function u will depend on two variables x and y and will satisfy the equation

$$\frac{\partial^2 u}{\partial x^2}+\frac{\partial^2 u}{\partial y^2}=0 \tag{4}$$

which is called the Laplace equation in a plane. The boundary-value conditions (2) and (3) must be fulfilled on the contour C.

II. The potential flow of a fluid. Equation of continuity. Let there be a flow of liquid inside a volume Ω bounded by a surface σ

(in a particular case, Ω may also be unbounded). Let ρ be the density of the liquid. We denote the velocity of the liquid by

$$v = v_x\boldsymbol{i} + v_y\boldsymbol{j} + v_z\boldsymbol{k} \tag{5}$$

where v_x, v_y, v_z are the projections of the vector $\boldsymbol{v}$ on the coordinate axes. In the body Ω pick out a small volume ω, bounded by the surface S. The following quantity of liquid will pass through each element Δs of the surface S in a time Δt:

$$\Delta Q = \rho\,\boldsymbol{vn}\,\Delta s\,\Delta t$$

where $\boldsymbol{n}$ is the unit vector directed along the outer normal to the surface S. The total amount of liquid Q entering the volume ω (or flowing out of the volume ω) is expressed by the integral

$$Q = \Delta t \iint\limits_S \rho\,\boldsymbol{vn}\,ds \tag{6}$$

(see Secs. 3.5 and 3.6). The amount of liquid in the volume ω at time t was

$$\iiint\limits_\omega \rho\,d\omega$$

During time Δt the amount of liquid will change (due to changes in density) by the amount

$$Q = \iiint\limits_\omega \Delta\rho\,d\omega \approx \Delta t \iiint\limits_\omega \frac{\partial\rho}{\partial t}\,d\omega \tag{7}$$

Assuming that there are no sources in the volume ω, we conclude that this change is brought about by an influx of liquid to an amount that is determined by equation (6). Equating the right sides of (6) and (7) and cancelling out Δt, we get

$$\iint\limits_S \rho\,\boldsymbol{vn}\,ds = + \iiint\limits_\omega \frac{\partial\rho}{\partial t}\,d\omega \tag{8}$$

We transform the iterated integral on the left by Ostrogradsky's formula (Sec. 3.8). Then (8) will assume the form

$$\iiint\limits_\omega \operatorname{div}(\rho\boldsymbol{v})\,d\omega = \iiint\limits_\omega \frac{\partial\rho}{\partial t}\,d\omega$$

or

$$\iiint\limits_\omega \left(\frac{\partial\rho}{\partial t} - \operatorname{div}(\rho\boldsymbol{v})\right) d\omega = 0$$

Since the volume ω is arbitrary and the integrand is continuous, we obtain

$$\frac{\partial\rho}{\partial t} - \operatorname{div}(\rho\boldsymbol{v}) = 0 \tag{9}$$

or

$$\frac{\partial \rho}{\partial t} - \frac{\partial}{\partial x}(\rho v_x) - \frac{\partial}{\partial y}(\rho v_y) - \frac{\partial}{\partial z}(\rho v_z) = 0 \qquad (9')$$

This is the *equation of continuity of a compressible liquid.*

Note. In certain problems, for instance when considering the movement of oil or gas in a subterranean porous medium to a well, it may be taken that

$$\boldsymbol{v} = -\frac{k}{\rho}\operatorname{grad} p$$

where p is the pressure and k is the coefficient of permeability and

$$\frac{\partial \rho}{\partial t} \approx \lambda \frac{\partial p}{\partial t}$$

$\lambda = \text{const}$. Substituting into the equation of continuity (9), we get

$$\lambda \frac{\partial p}{\partial t} + \operatorname{div}(k \operatorname{grad} p) = 0$$

or

$$-\lambda \frac{\partial p}{\partial t} = \frac{\partial}{\partial x}\left(k \frac{\partial p}{\partial x}\right) + \frac{\partial}{\partial y}\left(k \frac{\partial p}{\partial y}\right) + \frac{\partial}{\partial z}\left(k \frac{\partial p}{\partial z}\right) \qquad (10)$$

If k is a constant, then this equation takes on the form

$$\frac{\partial p}{\partial t} = -\frac{k}{\lambda}\left(\frac{\partial^2 p}{\partial x^2} + \frac{\partial^2 p}{\partial y^2} + \frac{\partial^2 p}{\partial z^2}\right) \qquad (11)$$

and we arrive at the heat-conduction equation.

Let us return to equation (9). If the liquid is incompressible, then $\rho = \text{const}$, $\frac{\partial \rho}{\partial t} = 0$, and (9) becomes

$$\operatorname{div} \boldsymbol{v} = 0 \qquad (12)$$

If the motion is potential, that is, if the vector $\boldsymbol{v}$ is the gradient of some function φ:

$$\boldsymbol{v} = \operatorname{grad} \varphi$$

then equation (12) takes the form

$$\operatorname{div}(\operatorname{grad} \varphi) = 0$$

or

$$\frac{\partial^2 \varphi}{\partial x^2} + \frac{\partial^2 \varphi}{\partial y^2} + \frac{\partial^2 \varphi}{\partial z^2} = 0 \qquad (13)$$

that is, the potential function of the velocity φ must satisfy the Laplace equation.

In many problems, as, for example, those dealing with filtration, we can put

$$\boldsymbol{v} = -k_1 \operatorname{grad} p$$

where p is the pressure and k_1 is a constant; we then get the Laplace equation for determining the pressure:

$$\frac{\partial^2 p}{\partial x^2} + \frac{\partial^2 p}{\partial y^2} + \frac{\partial^2 p}{\partial z^2} = 0 \qquad (13')$$

The boundary-value conditions for equation (13) or (13′) may be the following:

1. On the surface σ are specified the values of the desired function p, pressure [condition (2)]. This is the Dirichlet problem.
2. On the surface σ are specified the values of the normal derivative $\frac{\partial p}{\partial n}$; the flow through the surface is specified [condition (3)]. This is the Neumann problem.
3. On parts of the surface σ are specified the values of the desired function p, pressure, and on parts of the surface are specified the values of the normal derivative $\frac{\partial p}{\partial n}$, the flux through the surface. This is the Dirichlet-Neumann problem.

If the motion is two-dimensional-parallel—that is, the function φ (or p) does not depend on z—then we get the Laplace equation in a two-dimensional domain D with boundary C:

$$\frac{\partial^2 \varphi}{\partial x^2} + \frac{\partial^2 \varphi}{\partial y^2} = 0 \qquad (14)$$

Boundary-value conditions of type (2), the Dirichlet problem, or of type (3), the Neumann problem, are specified on the contour C.

III. The potential of a steady-state electric current. Let a homogeneous medium fill some volume V, and let an electric current pass through it whose density at each point is given by the vector $\boldsymbol{J}(x, y, z) = J_x\boldsymbol{i} + J_y\boldsymbol{j} + J_z\boldsymbol{k}$. Suppose that the current density is independent of the time t. Further assume that there are no current sources in the volume under consideration. Thus, the flux of the vector $\boldsymbol{J}$ through any closed surface S lying inside the volume V will be equal to zero:

$$\iint_S \boldsymbol{J}\boldsymbol{n}\, ds = 0$$

where $\boldsymbol{n}$ is a unit vector directed along the outer normal to the surface. From Ostrogradsky's formula we conclude that

$$\operatorname{div} \boldsymbol{J} = 0 \qquad (15)$$

The electric force $\boldsymbol{E}$ in the conducting medium at hand is, on the basis of Ohm's generalized law,

$$\boldsymbol{E}=\frac{\boldsymbol{J}}{\lambda} \tag{16}$$

or

$$\boldsymbol{J}=\lambda \boldsymbol{E}$$

where λ is the conductivity of the medium, which we shall consider constant.

From the general electromagnetic-field equations it follows that if the process is stationary, then the vector field $\boldsymbol{E}$ is irrotational, that is, rot $\boldsymbol{E}=0$. Then, like the case we had when considering the velocity field of a liquid, the vector field is potential (see Sec. 3.9). There is a function φ such that

$$\boldsymbol{E}=\operatorname{grad} \varphi \tag{17}$$

From (16) we get

$$\boldsymbol{J}=\lambda \operatorname{grad} \varphi \tag{18}$$

From (15) and (18) we have

$$\lambda \operatorname{div}(\operatorname{grad} \varphi)=0$$

or

$$\frac{\partial^2 \varphi}{\partial x^2}+\frac{\partial^2 \varphi}{\partial y^2}+\frac{\partial^2 \varphi}{\partial z^2}=0 \tag{19}$$

We thus have the Laplace equation.

Solving this equation for appropriate boundary-value conditions, we find the function φ, and from formulas (18) and (17) we find the current $\boldsymbol{J}$ and the electric force $\boldsymbol{E}$.

6.9 THE LAPLACE EQUATION IN CYLINDRICAL COORDINATES. SOLUTION OF THE DIRICHLET PROBLEM FOR AN ANNULUS WITH CONSTANT VALUES OF THE DESIRED FUNCTION ON THE INNER AND OUTER CIRCUMFERENCES

Let $u(x, y, z)$ be a harmonic function of three variables. Then

$$\frac{\partial^2 u}{\partial x^2}+\frac{\partial^2 u}{\partial y^2}+\frac{\partial^2 u}{\partial z^2}=0 \tag{1}$$

We introduce the cylindrical coordinates (r, φ, z):

$$x=r \cos \varphi, \quad x=r \sin \varphi, \quad z=z$$

whence

$$r=\sqrt{x^2+y^2}, \quad \varphi=\arctan \frac{y}{x}, \quad z=z \tag{2}$$

Replacing the independent variables x, y, and z by r, φ, and z, we arrive at the function u^*:

$$u(x, y, z) = u^*(r, \varphi, z)$$

Let us find the equation that will be satisfied by $u^*(r, \varphi, z)$ as a function of the arguments r, φ, and z; we have

$$\frac{\partial u}{\partial x} = \frac{\partial u^*}{\partial r}\frac{\partial r}{\partial x} + \frac{\partial u^*}{\partial \varphi}\frac{\partial \varphi}{\partial x}$$

$$\frac{\partial^2 u}{\partial x^2} = \frac{\partial^2 u^*}{\partial r^2}\left(\frac{\partial r}{\partial x}\right)^2 + \frac{\partial u^*}{\partial r}\frac{\partial^2 r}{\partial x^2} + 2\frac{\partial^2 u^*}{\partial r\,\partial \varphi}\frac{\partial r}{\partial x}\frac{\partial \varphi}{\partial x} + \frac{\partial^2 u^*}{\partial \varphi^2}\left(\frac{\partial \varphi}{\partial x}\right)^2 + \frac{\partial u^*}{\partial \varphi}\frac{\partial^2 \varphi}{\partial x^2} \quad (3)$$

similarly,

$$\frac{\partial^2 u}{\partial y^2} = \frac{\partial^2 u^*}{\partial r^2}\left(\frac{\partial r}{\partial y}\right)^2 + \frac{\partial u^*}{\partial r}\frac{\partial^2 r}{\partial y^2} + 2\frac{\partial^2 u^*}{\partial r\,\partial \varphi}\frac{\partial r}{\partial y}\frac{\partial \varphi}{\partial y} + \frac{\partial^2 u^*}{\partial \varphi^2}\left(\frac{\partial \varphi}{\partial y}\right)^2 + \frac{\partial u^*}{\partial \varphi}\frac{\partial^2 \varphi}{\partial y^2} \quad (4)$$

besides,

$$\frac{\partial^2 u}{\partial z^2} = \frac{\partial^2 u^*}{\partial z^2} \quad (5)$$

We find the expressions for

$$\frac{\partial r}{\partial x}, \frac{\partial r}{\partial y}, \frac{\partial^2 r}{\partial x^2}, \frac{\partial^2 r}{\partial y^2}, \frac{\partial \varphi}{\partial x}, \frac{\partial \varphi}{\partial y}, \frac{\partial^2 \varphi}{\partial x^2}, \frac{\partial^2 \varphi}{\partial y^2}$$

from equations (2). Adding the right sides of (3), (4) and (5), and equating the sum to zero [since the sum of the left-hand sides of these equations is zero by virtue of (1)], we get

$$\frac{\partial^2 u^*}{\partial r^2} + \frac{1}{r}\frac{\partial u^*}{\partial r} + \frac{1}{r^2}\frac{\partial^2 u^*}{\partial \varphi^2} + \frac{\partial^2 u^*}{\partial z^2} = 0 \quad (6)$$

This is the *Laplace equation in cylindrical coordinates*.

If the function u is independent of z and is dependent on x and y, then the function u^*, dependent only on r and φ, satisfies the equation

$$\frac{\partial^2 u^*}{\partial r^2} + \frac{1}{r}\frac{\partial u^*}{\partial r} + \frac{1}{r^2}\frac{\partial^2 u^*}{\partial \varphi^2} = 0 \quad (7)$$

where r and φ are polar coordinates in a plane.

Now let us find the solution to Laplace's equation in the domain D (annulus) bounded by the circles C_1, $x^2 + y^2 = R_1^2$, and C_2, $x^2 + y^2 = R_2^2$, with the following boundary values imposed:

$$u|_{C_1} = u_1 \quad (8)$$

$$u|_{C_2} = u_2 \quad (9)$$

where u_1 and u_2 are constants.

We will solve the problem in polar coordinates. Obviously, it is desirable to seek a solution that is independent of φ. Equation

(7) in this case takes the form

$$\frac{\partial^2 u}{\partial r^2}+\frac{1}{r}\frac{\partial u}{\partial r}=0$$

Integrating this equation we find

$$u=A\ln r+B \tag{10}$$

We determine A and B from conditions (8) and (9):

$$u_1=A\ln R_1+B$$
$$u_2=A\ln R_2+B$$

Whence we find

$$A=\frac{u_2-u_1}{\ln\frac{R_2}{R_1}},\qquad B=u_1-(u_2-u_1)\frac{\ln R_1}{\ln\frac{R_2}{R_1}}=\frac{u_1\ln R_2-u_2\ln R_1}{\ln\frac{R_2}{R_1}}$$

Substituting the values of A and B thus found into (10), we finally get

$$u=u_1+\frac{\ln\frac{r}{R_1}}{\ln\frac{R_2}{R_1}}(u_2-u_1)=\frac{u_2\ln\frac{r}{R_1}-u_1\ln\frac{r}{R_2}}{\ln\frac{R_2}{R_1}} \tag{11}$$

Note. We have actually solved the following problem. Find the function u that satisfies the Laplace equation in a region bounded by the surfaces (in cylindrical coordinates)

$$r=R_1,\quad r=R_2,\quad z=0,\quad z=H$$

and that satisfies the following boundary conditions:

$$u\big|_{r=R_1}=u_1,\quad u\big|_{r=R_2}=u_2$$
$$\left.\frac{\partial u}{\partial z}\right|_{z=0}=0,\quad \left.\frac{\partial u}{\partial z}\right|_{z=H}=0$$

(the Dirichlet-Neumann problem). It is obvious that the desired solution does not depend either on z or on φ and is given by formula (11).

6.10 THE SOLUTION OF DIRICHLET'S PROBLEM FOR A CIRCLE

In an xy-plane, let there be a circle of radius R with centre at the origin and let a certain function $f(\varphi)$, where φ is the polar angle, be given on its circumference. It is required to find the function $u(r, \varphi)$ continuous in the circle (including the boundary) and satisfying (inside the circle) the Laplace equation

$$\frac{\partial^2 u}{\partial x^2}+\frac{\partial^2 u}{\partial y^2}=0 \tag{1}$$

and, on the circumference, assuming the specified values

$$u|_{r=R}=f(\varphi) \tag{2}$$

We shall solve the problem in polar coordinates. Rewrite equation (1) in these coordinates:

$$\frac{\partial^2 u}{\partial r^2}+\frac{1}{r}\frac{\partial u}{\partial r}+\frac{1}{r^2}\frac{\partial^2 u}{\partial \varphi^2}=0$$

or

$$r^2\frac{\partial^2 u}{\partial r^2}+r\frac{\partial u}{\partial r}+\frac{\partial^2 u}{\partial \varphi^2}=0 \tag{1'}$$

We shall seek the solution by the method of separation of variables, placing

$$u=\Phi(\varphi)R(r) \tag{3}$$

Substituting into equation (1'), we get

$$r^2\Phi(\varphi)R''(r)+r\Phi(\varphi)R'(r)+\Phi''(\varphi)R(r)=0$$

or

$$\frac{\Phi''(\varphi)}{\Phi(\varphi)}=-\frac{r^2R''(r)+rR'(r)}{R(r)}=-k^2 \tag{4}$$

Since the left side of this equation is independent of r and the right is independent of φ, it follows that they are equal to a constant which we denote by $-k^2$. Thus, equation (4) yields two equations:

$$\Phi''(\varphi)+k^2\Phi(\varphi)=0 \tag{5}$$

$$r^2R''(r)+rR'(r)-k^2R(r)=0 \tag{5'}$$

The general solution of (5) will be

$$\Phi=A\cos k\varphi+B\sin k\varphi \tag{6}$$

We seek the solution of (5') in the form $R(r)=r^m$. Substituting $R(r)=r^m$ into (5'), we get

$$r^2m(m-1)r^{m-2}+rmr^{m-1}-k^2r^m=0$$

or

$$m^2-k^2=0$$

Thus, there are two particular linearly independent solutions r^k and r^{-k}. The general solution of equation (5') is

$$R=Cr^k+Dr^{-k} \tag{7}$$

We substitute expressions (6) and (7) into (3):

$$u_k=(A_k\cos k\varphi+B_k\sin k\varphi)(C_kr^k+D_kr^{-k}) \tag{8}$$

Function (8) will be the solution of (1') for any value of k different from zero. If $k=0$, then equations (5) and (5′) take the form

$$\Phi''=0,\ rR''(r)+R'(r)=0$$

and, consequently,

$$u_0=(A_0+B_0\varphi)(C_0+D_0\ln r) \qquad (8')$$

The solution must be a periodic function of φ, since for one and the same value of r for φ and $\varphi+2\pi$ we must have the same solution, because one and the same point of the circle is considered. It is therefore obvious that in formula (8′) we must have $B_0=0$. To continue, we seek a solution that is continuous and finite in the circle. Hence, in the centre of the circle the solution must be finite for $r=0$, and for that reason we must have $D_0=0$ in (8′) and $D_k=0$ in (8).

Thus, the right side of (8′) becomes the product A_0C_0, which we denote by $A_0/2$. Thus,

$$u_0=\frac{A_0}{2} \qquad (8'')$$

We shall form the solution to our problem as a sum of solutions of the form (8), since a sum of solutions is a solution. The sum must be a periodic function of φ. This will be the case if each term is a periodic function of φ. For this, k must take on integral values. [We note that if we equated the sides of (4) to the number $+k^2$, we would not obtain a periodic solution.] We shall confine ourselves only to positive values:

$$k=1,\ 2,\ \ldots,\ n,\ \ldots$$

because the constants A, B, C, D are arbitrary and therefore the negative values of k do not yield new particular solutions.

Thus,

$$u(r,\ \varphi)=\frac{A_0}{2}+\sum_{n=1}^{\infty}(A_n\cos n\varphi+B_n\sin n\varphi)\,r^n \qquad (9)$$

(the constant C_n is included in A_n and B_n). Let us now choose arbitrary constants A_n and B_n so as to satisfy the boundary-value condition (2). Putting into (9) $r=R$, we get, from condition (2),

$$f(\varphi)=\frac{A_0}{2}+\sum_{n=1}^{\infty}(A_n\cos n\varphi+B_n\sin n\varphi)\,R^n \qquad (10)$$

For (10) to be valid, it is necessary that the function $f(\varphi)$ be expandable in a Fourier series in the interval $(-\pi,\ \pi)$ and that A_nR^n and B_nR^n be its Fourier coefficients. Hence, A_n and B_n must be defined by the formulas

$$\left.\begin{aligned} A_n &= \frac{1}{\pi R^n}\int_{-\pi}^{\pi} f(t)\cos nt\,dt \\ B_n &= \frac{1}{\pi R^n}\int_{-\pi}^{\pi} f(t)\sin nt\,dt \end{aligned}\right\} \tag{11}$$

Thus, the series (9) with coefficients defined by formulas (11) will be a solution of our problem if it admits termwise iterated differentiation with respect to r and φ (but we have not proved this). Let us transform formula (9). Putting, in place of A_n and B_n, their expressions (11) and performing the trigonometric transformations, we get

$$\begin{aligned} u(r,\varphi) &= \frac{1}{2\pi}\int_{-\pi}^{\pi} f(t)\,dt + \frac{1}{\pi}\sum_{n=1}^{\infty}\int_{-\pi}^{\pi} f(t)\cos n(t-\varphi)\,dt\left(\frac{r}{R}\right)^n \\ &= \frac{1}{2\pi}\int_{-\pi}^{\pi} f(t)\left[1+2\sum_{n=1}^{\infty}\left(\frac{r}{R}\right)^n\cos n(t-\varphi)\right]dt \end{aligned} \tag{12}$$

Let us transform the expression in the square brackets:*

$$\begin{aligned} 1+2\sum_{n=1}^{\infty}\left(\frac{r}{R}\right)^n\cos n(t-\varphi) &= 1+\sum_{n=1}^{\infty}\left(\frac{r}{R}\right)^n\left[e^{in(t-\varphi)}+e^{-in(t-\varphi)}\right] \\ &= 1+\sum_{n=1}^{\infty}\left[\left(\frac{r}{R}e^{i(t-\varphi)}\right)^n+\left(\frac{r}{R}e^{-i(t-\varphi)}\right)^n\right] \\ &= 1+\frac{\frac{r}{R}e^{i(t-\varphi)}}{1-\frac{r}{R}e^{i(t-\varphi)}}+\frac{\frac{r}{R}e^{-i(t-\varphi)}}{1-\frac{r}{R}e^{-i(t-\varphi)}} \\ &= \frac{1-\left(\frac{r}{R}\right)^2}{1-2\frac{r}{R}\cos(t-\varphi)+\left(\frac{r}{R}\right)^2}=\frac{R^2-r^2}{R^2-2Rr\cos(t-\varphi)+r^2} \end{aligned} \tag{13}$$

Replacing the expression in square brackets in (12) by expression (13), we get

$$u(r,\varphi)=\frac{1}{2\pi}\int_{-\pi}^{\pi} f(t)\,\frac{R^2-r^2}{R^2-2rR\cos(t-\varphi)+r^2}\,dt \tag{14}$$

* In the derivation we determine the sum of an infinite geometric progression whose ratio is a complex number, the modulus of which is less than unity. This formula of the sum of a geometric progression is derived in the same way as in the case of real numbers. It is also necessary to take into account the definition of the limit of a complex function of a real argument. Here, the argument is n (see Sec. 7.4, Vol. I).

Formula (14) is called *Poisson's integral*. By an analysis of this formula it is possible to prove that if the function $f(\varphi)$ is continuous, then the function $u(r, \varphi)$ defined by the integral (14) also satisfies equation (1') and $u(r, \varphi) \to f(\varphi)$ as $r \to R$. That is, it is a solution of the Dirichlet problem for a circle.

6.11 SOLUTION OF THE DIRICHLET PROBLEM BY THE METHOD OF FINITE DIFFERENCES

In an xy-plane, let there be given a domain D bounded by a contour C. Let there be given a continuous function f on the contour C. It is required to find an approximate solution to the Laplace equation

$$\frac{\partial^2 u}{\partial x^2}+\frac{\partial^2 u}{\partial y^2}=0 \tag{1}$$

that satisfies the boundary condition

$$u|_C=f \tag{2}$$

We draw two families of straight lines:

$$x=ih \text{ and } y=kh \tag{3}$$

where h is the given number, and i and k assume successive integral values. We shall say that D is covered with a *grid*. We call the points of intersection of the straight lines *lattice points of the grid*.

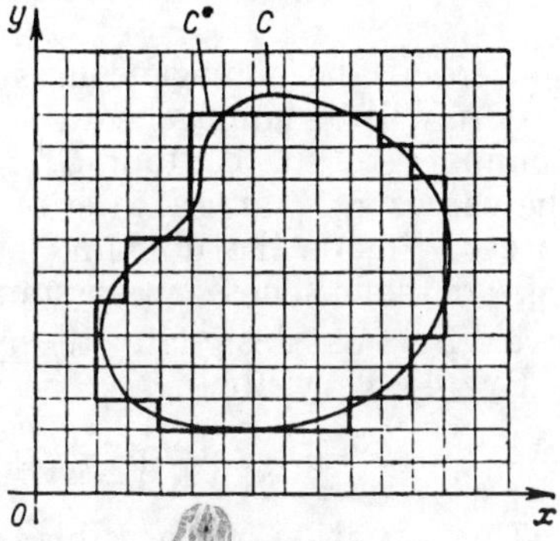

We denote by $u_{i,k}$ the approximate value of the desired function at the point $x=ih$, $y=kh$; that is, $u(ih, kh)=u_{i,k}$. We approximate D by the grid domain D^*, which consists of all the squares that lie completely in D and of some that are crossed by the boundary C (these may be disregarded). Here, the contour C is approximated by the contour C^*, which consists of segments of straight lines of type (3). In each lattice point lying on the contour C^* we specify a value f^*, which is equal to the value of the function f at the closest point of the contour C (Fig. 146).

Fig. 146

The values of the desired function will be considered only at the lattice points of the grid. As has already been pointed out in Sec. 6.6 the derivatives in this approximate method are replaced by finite differences:

$$\left.\frac{\partial^2 u}{\partial x^2}\right|_{x=ih,\ y=kh}=\frac{u_{i+1,\,k}-2u_{i,\,k}+u_{i-1,\,k}}{h^2}$$

$$\left.\frac{\partial^2 u}{\partial y^2}\right|_{x=ih,\ y=kh}=\frac{u_{i,\,k+1}-2u_{i,\,k}+u_{i,\,k-1}}{h^2}$$

The differential equation (1) is replaced by a *difference equation* (after cancelling out h^2):

$$u_{i+1,\,k}-2u_{i,\,k}+u_{i-1,\,k}+u_{i,\,k+1}-2u_{i,\,k}+u_{i,\,k-1}=0$$

or (Fig. 147)

$$u_{i,\,k}=\frac{1}{4}(u_{i+1,\,k}+u_{i,\,k+1}+u_{i-1,\,k}+u_{i,\,k-})\tag{4}$$

For each lattice point of the grid lying inside D^* (and not lying on the boundary C^*), we form an equation (4). If the point ($x=ih$, $y=kh$) is adjacent to the point of the contour C^*, then the right side of (4) will contain known values of f^*. Thus, we obtain a nonhomogeneous system of N equations in N unknowns, where N is the number of lattice points of the grid lying inside D^*.

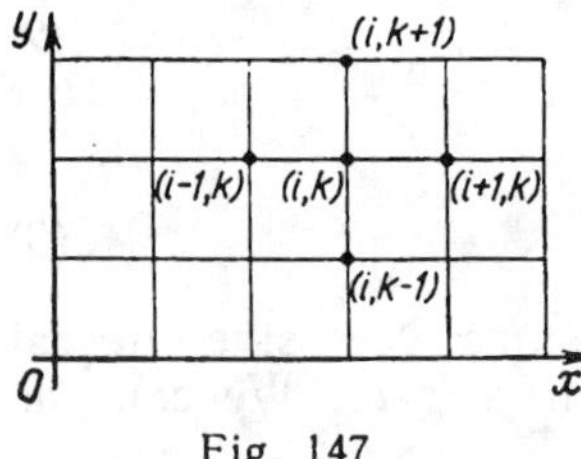

Fig. 147

We shall prove that the system (4) has one, and only one, solution. This is a system of N linear equations in N unknowns. It has a unique solution if the determinant of the system is not zero. The determinant of the system is nonzero if the homogeneous system has only a trivial solution. The system will be homogeneous if $f^*=0$ at the lattice points on the boundary of the contour C^*. We shall prove that in this case all the values $u_{i,\,k}$ at all interior lattice points of the grid are equal to zero. Inside the domain, let $u_{i,\,k}$ be different from zero. For the sake of definiteness, we assume that the greatest of them is positive. Let us designate it by $\bar{u}_{i,\,k}>0$.

By (4) we write

$$\bar{u}_{i,\,k}=\frac{1}{4}(u_{i+1,k}+u_{i,k+1}+u_{i-1,\,k}+u_{i,\,k-1})\tag{4'}$$

This equation is possible only if all the values of u on the right are equal to the greatest $\bar{u}_{i,\,k}$. We now have five points at which the values of the desired function are $\bar{u}_{i,\,k}$. If none of these points is a boundary point, then, taking one of them and writing for it the equation (4), we will prove that at certain other points the value of the desired function will be equal to $\bar{u}_{i,\,k}$. Continuing in this fashion, we will reach the boundary and will prove that at the boundary point the value of the function will be equal to $\bar{u}_{i,k}$, which is contrary to the fact that $f^*=0$ at boundary points. Assuming that inside the domain there is the least negative value, we will prove that on the boundary the value of the function is negative, which contradicts the hypothesis.

And so system (4) has a solution and it is a unique solution. *The values* $u_{i,k}$ *defined from the system* (4) *are approximate values of the solution of the Dirichlet problem formulated above.* It was proved that if the solution of the Dirichlet problem for a given domain D and a given function f exists [we denote it by $u(x, y)$] and if $u_{i,k}$ is a solution of (4), then we have the relation

$$|u(x, y)-u_{i,k}|<Ah^2 \tag{5}$$

where A is a constant independent of h.

Note. It is sometimes justifiable (though this has not been rigorously proved) to use the following procedure for estimating the error of the approximate solution. Let $u_{i,k}^{(2h)}$ be an approximate solution for a step $2h$, $u_{i,k}^{(h)}$ an approximate solution for a step h, and let $E_h(x, y)$ be the error of the solution $u_{i,k}^{(h)}$. Then we have the approximate equation

$$E_h(x, y)\approx\frac{1}{3}(u_{i,k}^{(2h)}-u_{i,k}^{(h)})$$

at the common lattice points of the grids. Thus, in order to dete mine the error of the approximate solution for a step h, it necessary to find the solution for a step $2h$. One third of the diff rence of these approximate solutions is the error estimate .. the solution for a step (mesh-length) of h. This remark also refers to the solution of the heat-conduction equation by the finite-difference method.

Exercises on Chapter 6

1. Derive an equation of torsional vibrations of a homogeneous cylindrical rod. **Hint**. The torque in a cross section of the rod with abscissa x is determined by the formula $M=GI\frac{\partial\theta}{\partial x}$, where $\theta(x, t)$ is the angle of torque of a cross section with abscissa x at time t, G is the shear modulus, and I is the polar moment of inertia of a cross section of the rod.

Ans. $\frac{\partial^2\theta}{\partial t^2}=a^2\frac{\partial^2\theta}{\partial x^2}$, where $a^2=\frac{GI}{k}$, and k is the moment of inertia of unit length of the rod.

2. Find a solution of the equation $\frac{\partial^2\theta}{\partial t^2}=a^2\frac{\partial^2\theta}{\partial x^2}$ that satisfies the conditions $\theta(0, t)=0$, $\theta(l, t)=0$, $\theta(x, 0)=\varphi(x)$, $\frac{\partial\theta(x, 0)}{\partial t}=0$, where

$$\varphi(x)=\frac{2\theta_0 x}{l} \quad \text{for} \quad 0\leqslant x\leqslant\frac{l}{2}$$

$$\varphi(x)=-\frac{2\theta_0 x}{l}+2\theta_0 \quad \text{for} \quad \frac{l}{2}\leqslant x\leqslant l$$

Give a mechanical interpretation of the problem.

Ans. $\theta(x, t)=\frac{8\theta_0}{\pi^2}\sum_{k=0}^{\infty}\frac{(-1)^k}{(2k+1)^2}\sin\frac{(2k+1)\pi x}{l}\cos\frac{(2k+1)\pi at}{l}$.

3. Derive an equation of longitudinal vibrations of a homogeneous cylindrical rod.

Hint. If $u(x, t)$ is the translation of a cross section of the rod with abscissa x at time t, then the tensile stress T in a cross section x is defined by the formula $T=ES\frac{\partial u}{\partial x}$, where E is the elasticity modulus of the material and S is the cross-sectional area of the rod.

Ans. $\frac{\partial^2 u}{\partial t^2}=a^2\frac{\partial^2 u}{\partial x^2}$, where $a^2=\frac{E}{\rho}$, and ρ is the density of the rod material.

4. A homogeneous rod of length $2l$ was shortened by 2λ under the action of forces applied to its ends. At $t=0$ it is free of forces acting externally. Determine the displacement $u(x, t)$ of a cross section of the rod with abscissa x at time t (the mid-point of the axis of the rod has abscissa $x=0$).

Ans. $u(x, t)=\frac{8\lambda}{\pi^2}\sum_{k=0}^{\infty}\frac{(-1)^{k+1}}{(2k+1)^2}\sin\frac{(2k+1)\pi x}{2l}\cos\frac{(2k+1)\pi at}{2l}$

5. One end of a rod of length l is fixed, the other end is acted upon by a tensile force P. Find the longitudinal vibrations of the rod if the force P does not operate when $t=0$. *Ans.* $\frac{8Pl}{ES\pi^2}\sum_{n=0}^{\infty}\frac{(-1)^n}{(2n+1)^2}\sin\frac{(2n+1)\pi x}{2l}\cos\frac{(2n+1)\pi at}{2l}$

(E and S as in Problem 3).

6. Find a solution to the equation $\frac{\partial^2 u}{\partial t^2}=a^2\frac{\partial^2 u}{\partial x^2}$ that satisfies the conditions

$$u(0, t)=0, \quad u(l, t)=A\sin\omega t$$
$$u(x, 0)=0, \quad \frac{\partial u(x, 0)}{\partial t}=0$$

Give a mechanical interpretation of the problem.

Ans. $u(x, t)=\frac{A\sin\frac{\omega}{a}x\sin\omega t}{\sin\frac{\omega}{a}l}+\frac{2A\omega a}{l}\sum_{n=1}^{\infty}\frac{(-1)^{n-1}}{\omega^2-\left(\frac{n\pi a}{l}\right)^2}\sin\frac{n\pi at}{l}\sin\frac{n\pi x}{l}$.

Hint. Seek the solution in the form of a sum of two solutions:

$$u=v+w, \quad \text{where} \quad w=\frac{A\sin\frac{\omega}{a}x\sin\omega t}{\sin\frac{\omega}{a}l}$$

is the solution that satisfies the conditions

$$v(0, t)=0, \quad v(l, t)=0$$
$$v(x, 0)=-w(x, 0), \quad \frac{\partial v(x, 0)}{\partial t}=-\frac{\partial w(x, 0)}{\partial t}$$

$\left(\text{It is assumed that } \sin\frac{\omega}{a}l\neq 0.\right)$

7. Find a solution to the equation $\frac{\partial u}{\partial t}=a^2\frac{\partial^2 u}{\partial x^2}$ that satisfies the conditions

$$u(0, t)=0, \quad u(l, t)=0, \quad t>0$$

$$u(x, 0)=\begin{cases} x \text{ when } 0\leqslant x\leqslant \frac{l}{2} \\ l-x \text{ when } \frac{l}{2}<x<l \end{cases}$$

Ans. $u(x, t)=\frac{4l}{\pi^2}\sum_{n=0}^{\infty}\frac{(-1)^n}{(2n+1)^2}e^{-\frac{(2n+1)^2\pi^2a^2t}{l^2}}\sin\frac{(2n+1)\pi x}{l}$.

Hint. Solve the problem by the method of separation of variables.

8. Find a solution to the equation $\frac{\partial u}{\partial t}=a^2\frac{\partial^2 u}{\partial x^2}$ that satisfies the conditions

$$u(0, t)=u(l, t)=0, \qquad u(x, 0)=\frac{x(l-x)}{l^2}$$

Ans. $u(x, t)=\frac{8}{\pi^3}\sum_{n=0}^{\infty}\frac{1}{(2n+1)^3}e^{-\frac{(2n+1)^2\pi^2a^2t}{l^2}}\sin\frac{(2n+1)\pi x}{l}$.

9. Find a solution to the equation $\frac{\partial u}{\partial t}=a^2\frac{\partial^2 u}{\partial x^2}$ that satisfies the conditions

$$\left.\frac{\partial u}{\partial x}\right|_{x=0}=0, \ u(l, t)=u_0, \quad u(x, 0)=\varphi(x)$$

Point out the physical meaning of the problem.

Ans. $u(x, t)=u_0+\sum_{n=0}^{\infty}A_n e^{-a^2\lambda_n^2 t}\cos\frac{(2n+1)\pi}{2l}x$, where

$$A_n=\frac{2}{l}\int_0^l\varphi(x)\cos\times\frac{(2n+1)\pi x}{2l}dx-\frac{(-1)^n 4u_0}{\pi(2n+1)}.$$

Hint. Seek the solution in the form $u=u_0+v(x, t)$.

10. Find a solution to the equation $\frac{\partial u}{\partial t}=a^2\frac{\partial^2 u}{\partial x^2}$ that satisfies the conditions

$$u(0, t)=0, \qquad \left.\frac{\partial u}{\partial x}\right|_{x=l}=-\left.Hu\right|_{x=l}, \qquad u(x, 0)=\varphi(x)$$

Point out the physical meaning of the problem.

Ans. $$u(x, t)=\sum_{n=1}^{\infty}A_n\frac{p^2+\mu_n^2}{p(p+1)+\mu_n^2}e^{-\frac{\mu_n^2a^2t^2}{l^2}}\sin\frac{\mu_n x}{l},$$ where

$A_n=\frac{2}{l}\int_0^l\varphi(x)\sin\frac{\mu_n x}{l}dx$, $p=Hl$, $\mu_1, \mu_2, \ldots, \mu_n$ are positive roots of the equation $\tan\mu=-\frac{\mu}{p}$.

Hint. At the end of the rod (when $x=l$) a heat exchange occurs with the environment which has a temperature of zero.

11. Find [by formula (10), Sec. 6.6, putting $h=0.2$] an approximate solution to the equation $\frac{\partial u}{\partial t}=2\frac{\partial^2 u}{\partial x^2}$ that satisfies the conditions

$$u(x, 0)=x\left(\frac{3}{2}-x\right), \quad u(0, t)=0, \quad u(1, t)=\frac{1}{2}, \quad 0\leqslant t\leqslant 4l$$

12. Find a solution to the Laplace equation $\frac{\partial^2 u}{\partial x^2}+\frac{\partial^2 u}{\partial y^2}=0$, in a strip $0\leqslant x\leqslant a$, $0\leqslant y<\infty$ that satisfies the conditions

$$u(0, y)=0, \quad u(a, y)=0, \quad u(x, 0)=A\left(1-\frac{x}{a}\right), \quad u(x, \infty)=0$$

Ans. $u(x, t)=\frac{2A}{\pi}\sum_{n=1}^{\infty}\frac{1}{n}e^{-\frac{n\pi}{a}y}\sin\frac{n\pi x}{a}$.

Hint. Use the method of the separation of variables.

13. Find a solution to the Laplace equation $\frac{\partial^2 u}{\partial x^2}+\frac{\partial^2 u}{\partial y^2}=0$ in the rectangle $0\leqslant x\leqslant a$, $0\leqslant y\leqslant b$ that satisfies the conditions

$$u(x, 0)=0, \quad u(x, b)=0, \quad u(0, y)=Ay(b-y), \quad u(a, y)=0.$$

Ans. $u(x, t)=\frac{8Ab^2}{\pi^3}\sum_{n=0}^{\infty}\frac{\sinh\frac{(2n+1)\pi(a-x)}{b}}{(2n+1)^3}\,\frac{\sin\frac{(2n+1)\pi y}{b}}{\sinh\frac{(2n+1)\pi a}{b}}$.

14. Find a solution to the equation $\frac{\partial^2 u}{\partial x^2}+\frac{\partial^2 u}{\partial y^2}=0$ inside an annulus bounded by the circles $x^2+y^2=R_1^2$, $x^2+y^2=R_2^2$ that satisfies the conditions

$$\left.\frac{\partial u}{\partial r}\right|_{r=R_1}=+\frac{Q}{\lambda 2\pi R_1}, \quad \left.u\right|_{r=R_2}=u_2$$

Give a hydrodynamic interpretation of the problem.

Hint. Solve the problem in polar coordinates.

Ans. $u=u_2-\frac{Q}{2\lambda\pi}\ln\frac{R_2}{r}$.

15. Prove that the function $u(x, y)=e^{-y}\sin x$ is a solution of the equation $\frac{\partial^2 u}{\partial x^2}+\frac{\partial^2 u}{\partial y^2}=0$ in the square $0\leqslant x\leqslant 1$, $0\leqslant y\leqslant 1$ that satisfies the conditions $u(0, y)=0$, $u(1, y)=e^{-y}\sin 1$, $u(x, 0)=\sin x$, $u(x, 1)=e^{-1}\sin x$.

16. In Problems 12-15 solve the Laplace equations for given boundary conditions by the finite-difference method for $h=0.25$. Compare the approximate solution with the exact solution.

CHAPTER 7

OPERATIONAL CALCULUS AND CERTAIN OF ITS APPLICATIONS

Operational calculus is an important branch of mathematical analysis. The methods of operational calculus are used in physics, mechanics, electrical engineering and elsewhere. Operational calculus finds especially broad applications in automation and telemechanics. In this chapter we give (on the basis of the foregoing material of this text) the fundamental concepts of operational calculus and operational methods of solving ordinary differential equations. *

7.1 THE ORIGINAL FUNCTION AND ITS TRANSFORM

Suppose we have a function of a real variable t defined for $t \geqslant 0$ [we shall sometimes consider that the function $f(t)$ is defined on an infinite interval $-\infty < t < \infty$, but $f(t) = 0$ when $t < 0$]. We shall assume that the function $f(t)$ is piecewise continuous, that is, such that in any finite interval it has a finite number of discontinuities of the first kind (see Sec. 2.9, Vol. I). To ensure the existence of certain integrals in the infinite interval $0 \leqslant t < \infty$ we impose an additional restriction on the function $f(t)$: namely, we suppose that there exist constant positive numbers M and s_0 such that

$$|f(t)| < Me^{s_0 t} \tag{1}$$

for any value t in the interval $0 \leqslant t < \infty$.

Let us consider the product of the function $f(t)$ by the complex function e^{-pt} of a real variable ** t, where $p = a + ib$ $(a > 0)$ is some complex number:

$$e^{-pt} f(t) \tag{2}$$

---

* The following books are recommended for further study of operational calculus: A. I. Lurie, *Operational Calculus and Its Applications to Problems of Mechanics*, Moscow-Leningrad, Gostekhizdat, 1950; V. A. Ditkin and P. I. Kuznetsov, *Handbook of Operational Calculus*, Moscow, Leningrad, Gostekhizdat, 1951; V. A. Ditkin and A. P. Prudnikov, *Integral Transformations and Operational Calculus*, Fizmatgiz, 1961; Jan G. Mikusiński, *Rachunek operatorów*, Warszawa, 1953, (Operational Calculus).

** See Sec. 7.4, Vol. I, concerning complex functions of a real variable.

Function (2) is also a complex function of a real variable t:

$$e^{-pt}f(t)=e^{-(a+ib)t}f(t)=e^{-at}f(t)\,e^{-ibt}$$
$$=e^{-at}f(t)\cos bt-ie^{-at}f(t)\sin bt$$

Let us further consider the improper integral

$$\int_0^\infty e^{-pt}f(t)\,dt=\int_0^\infty e^{-at}f(t)\cos bt\,dt-i\int_0^\infty e^{-at}f(t)\sin bt\,dt \tag{3}$$

We shall show that if the function $f(t)$ satisfied condition (1) and $a>s_0$, then the integrals on the right of (3) exist and the convergence of the integrals is absolute. We begin by evaluating the first of these integrals:

$$\left|\int_0^\infty e^{-at}f(t)\cos bt\,dt\right|\leqslant\int_0^\infty |e^{-at}f(t)\cos bt|\,dt$$
$$<M\int_0^\infty e^{-at}e^{s_0t}\,dt=M\int_0^\infty e^{-(a-s_0)t}\,dt=\frac{M}{a-s_0}$$

In similar fashion we evaluate the second integral. Thus, the integral $\int_0^\infty e^{-pt}f(t)\,dt$ exist. It defines a certain function of p, which we denote* by $F(p)$:

$$F(p)=\int_0^\infty e^{-pt}f(t)\,dt \tag{4}$$

The function $F(p)$ is called the *Laplace transform*, or the *L-transform*, or simply the *transform* of the function $f(t)$. The function $f(t)$ is known as the *original function*. If $F(p)$ is the transform of $f(t)$, then we write

$$F(p)\risingdotseq f(t) \tag{5}$$

or

$$f(t)\fallingdotseq F(p) \tag{6}$$

or

$$L\{f(t)\}=F(p) \tag{7}$$

As we shall presently see, with the help of transforms it is possible to simplify the solution of many problems, for instance, to reduce the solution of differential equations to simple algebraic

* The function $F(p)$, for $p\neq 0$, is a function of a complex variable (for example, see V. I. Smirnov's *Course of Higher Mathematics*, Vol. III, Part 2, in Russian). The transformation (4) is similar to the Fourier transformation considered in Sec. 5.14.

operations in finding a transform. Knowing the transform one can find the original function either from specially prepared "original function-transform" tables or by methods that will be given below. Certain natural questions arise.

Let there be given a certain function $F(p)$. Does there exist a function $f(t)$ for which $F(p)$ is a transform? If there does, then is this function the only one? The answer is yes to both questions, given certain definite assumptions with respect to $F(p)$ and $f(t)$. For example, the following theorem, which we give without proof, establishes that the transform is unique:

Uniqueness Theorem. *If two continuous functions* $\varphi(t)$ and $\psi(t)$ *have one and the same L-transform* $F(p)$, *then these functions are identically equal.*

This theorem will play an important role throughout the subsequent text. Indeed, if in the solution of some practical problem we have in some way determined the transform of the desired function, and from the transform the original function, then on the basis of the foregoing theorem we conclude that the function we have found is the solution of the given problem and that no other solutions exist.

7.2 TRANSFORMS OF THE FUNCTIONS $\sigma_0(t)$, $\sin t$, $\cos t$

1. The function $f(t)$, defined as

$$f(t)=1 \text{ for } t\geqslant 0$$
$$f(t)=0 \text{ for } t<0$$

is called the *Heaviside unit function* and is denoted by $\sigma_0(t)$. The graph of this function is given in Fig. 148. Let us find the L-transform of the Heaviside function:

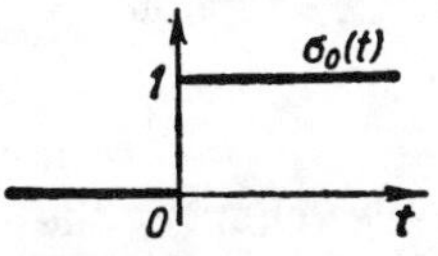

$$L\{\sigma_0(t)\}=\int_0^\infty e^{-pt}dt=-\frac{e^{-pt}}{p}\Big|_0^\infty=\frac{1}{p}$$

Fig. 148

Thus,*

$$1 \risingdotseq \frac{1}{p} \tag{1}$$

or, more precisely,

$$\sigma_0(t)\risingdotseq\frac{1}{p}$$

---

* In computing the integral $\int_0^\infty e^{-pt}\,dt$ it is possible to represent it as the sum of integrals of real functions; the same result would be obtained. This also holds for the next two integrals.

In some books on operational calculus the following expression is called the transform of the function $f(t)$:

$$F^*(p) = p\int_0^\infty e^{-pt} f(t)\,dt$$

With this definition we have $\sigma_0(t) \div 1$ and, consequently, $C \div C$, more exactly, $C\sigma_0(t) \div C$.

II. Let $f(t) = \sin t$; then

$$L\{\sin t\} = \int_0^\infty e^{-pt}\sin t\,dt = \left.\frac{e^{-pt}(-p\sin t - \cos t)}{p^2+1}\right|_0^\infty = \frac{1}{p^2+1}$$

And so

$$\sin t \div \frac{1}{p^2+1} \tag{2}$$

III. Let $f(t) = \cos t$; then

$$L\{\cos t\} = \int_0^\infty e^{-pt}\cos t\,dt = \left.\frac{e^{-pt}(\sin t - p\cos t)}{p^2+1}\right|_0^\infty = \frac{p}{p^2+1}$$

Thus,

$$\cos t \div \frac{p}{p^2+1} \tag{3}$$

7.3 THE TRANSFORM OF A FUNCTION WITH CHANGED SCALE OF THE INDEPENDENT VARIABLE. TRANSFORMS OF THE FUNCTIONS sin *at*, cos *at*

Let us consider the transform of the function $f(at)$, where $a > 0$:

$$L\{f(at)\} = \int_0^\infty e^{-pt} f(at)\,dt$$

We change the variable in the integral, putting $z = at$; hence, $dz = a\,dt$; then we get

$$L\{f(at)\} = \frac{1}{a}\int_0^\infty e^{-\frac{p}{a}z} f(z)\,dz$$

or

$$L\{f(at)\} = \frac{1}{a}F\left(\frac{p}{a}\right)$$

Thus, *if*

$$F(p) \div f(t)$$

then

$$\frac{1}{a}F\left(\frac{p}{a}\right) \div f(at) \tag{1}$$

Example 1. From (2) of Sec. 7.2, by (1), we straightway get

$$\sin at \doteqdot \frac{1}{a}\frac{1}{\left(\frac{p}{a}\right)^2+1}$$

or

$$\sin at \doteqdot \frac{a}{p^2+a^2} \tag{2}$$

Example 2. From (3), Sec. 7.2, by (1), we obtain

$$\cos at \doteqdot \frac{1}{a}\frac{\frac{p}{a}}{\left(\frac{p}{a}\right)^2+1}$$

or

$$\cos at \doteqdot \frac{p}{p^2+a^2} \tag{3}$$

7.4 THE LINEARITY PROPERTY OF A TRANSFORM

Theorem. *The transform of a sum of several functions multiplied by constants is equal to the sum of the transforms of these functions multiplied by the corresponding constants, that is, if*

$$f(t)=\sum_{i=1}^{n} C_i f_i(t) \tag{1}$$

(C_i are constants) and

$$F(p) \doteqdot f(t), \quad F_i(p) \doteqdot f_i(t)$$

then

$$F(p)=\sum_{i=1}^{n} C_i F_i(p) \tag{1'}$$

Proof. Multiplying all the terms of (1) by e^{-pt} and integrating with respect to t from 0 to ∞ (taking the factors C_i outside the integral sign), we get (1′).

Example 1. Find the transform of the function

$$f(t)=3\sin 4t-2\cos 5t$$

Solution. Applying formulas (2), (3), Sec. 7.3, and (1′), we have

$$L\{f(t)\}=3\frac{4}{p^2+16}-2\frac{p}{p^2+25}=\frac{12}{p^2+16}-\frac{2p}{p^2+25}$$

Example 2. Find the original function whose transform is expressed by the formula

$$F(p)=\frac{5}{p^2+4}+\frac{20p}{p^2+9}$$

Solution. We represent $F(p)$ as

$$F(p)=\frac{5}{2}\frac{2}{p^2+(2)^2}+20\frac{p}{p^2+(3)^2}$$

Hence, by (2), (3), Sec. 7.3, and (1'), we have

$$f(t)=\frac{5}{2}\sin 2t+20\cos 3t$$

From the uniqueness theorem, Sec. 7.1, it follows that this is the only original function that corresponds to the given $F(p)$.

7.5 THE SHIFT THEOREM

Theorem. *If* $F(p)$ *is the transform of the function* $f(t)$, *then* $F(p+\alpha)$ *is the transform of the function* $e^{-\alpha t}f(t)$, *that is,*

$$\left.\begin{array}{l}\text{if } F(p) \stackrel{\cdot}{\to} f(t)\\ \text{then } F(p+\alpha) \stackrel{\cdot}{\to} e^{-\alpha t}f(t)\end{array}\right\} \qquad (1)$$

[It is assumed here that $\operatorname{Re}(p+\alpha)>s_0$.]

Proof. Find the transform of the function $e^{-\alpha t}f(t)$:

$$L\{e^{-\alpha t}f(t)\}=\int_0^\infty e^{-pt-\alpha t}f(t)\,dt=\int_0^\infty e^{-(p+\alpha)t}f(t)\,dt$$

Thus,

$$L\{e^{-\alpha t}f(t)\}=F(p+\alpha)$$

This theorem makes it possible to expand considerably the class of transforms for which it is easy to find the original functions.

7.6 TRANSFORMS OF THE FUNCTIONS $e^{-\alpha t}$ sinh αt, cosh αt, $e^{-\alpha t}$ sin αt, $e^{-\alpha t}$ cos αt

From (1) (Sec. 7.2, on the basis of (1), Sec. 7.5, we straightway get

$$\frac{1}{p+\alpha}\stackrel{\cdot}{\to} e^{-\alpha t} \qquad (1)$$

Similarly,

$$\frac{1}{p-\alpha}\stackrel{\cdot}{\to} e^{\alpha t} \qquad (1')$$

Subtracting from the terms of (1') the corresponding terms of (1) and dividing the results by two, we get

$$\frac{1}{2}\left(\frac{1}{p-\alpha}-\frac{1}{p+\alpha}\right)\stackrel{\cdot}{\to}\frac{1}{2}(e^{\alpha t}-e^{-\alpha t})$$

or

$$\frac{\alpha}{p^2-\alpha^2}\stackrel{\cdot}{\to}\sinh\alpha t \qquad (2)$$

Similarly, by adding (1) and (1'), we obtain

$$\frac{p}{p^2-\alpha^2}\stackrel{\cdot}{\to}\cosh\alpha t \qquad (3)$$

From (2), Sec 7.3, by (1), Sec. 7.5, we have

$$\frac{a}{(p+\alpha)^2+a^2} \risingdotseq e^{-\alpha t} \sin at \tag{4}$$

From (3), Sec. 7.3, by (1), Sec. 7.5, we get

$$\frac{p+\alpha}{(p+\alpha)^2+a^2} \risingdotseq e^{-\alpha t} \cos at \tag{5}$$

Example 1. Find the original function whose transform is given by the formula

$$F(p) = \frac{7}{p^2+10p+41}$$

Solution. Transform $F(p)$ to the form of the expression on the left-hand side of (4):

$$\frac{7}{p^2+10p+41} = \frac{7}{(p+5)^2+16} = \frac{7}{4}\,\frac{4}{(p+5)^2+4^2}$$

Thus

$$F(p) = \frac{7}{4}\,\frac{4}{(p+5)^2+4^2}$$

Hence, by formula (4) we will have

$$F(p) \risingdotseq \frac{7}{4} e^{-5t} \sin 4t$$

Example 2. Find the original function whose transform is given by the formula

$$F(p) = \frac{p+3}{p^2+2p+10}$$

Solution. Transform the function $F(p)$:

$$\frac{p+3}{p^2+2p+10} = \frac{(p+1)+2}{(p+1)^2+9} = \frac{p+1}{(p+1)^2+3^2} + \frac{2}{(p+1)^2+3^2} = \frac{p+1}{(p+1)^2+3^2} + \frac{2}{3}\,\frac{3}{(p+1)^2+3^2}$$

Using formulas (4) and (5), we find the original function:

$$F(p) \risingdotseq e^{-t} \cos 3t + \frac{2}{3} e^{-t} \sin 3t$$

7.7 DIFFERENTIATION OF TRANSFORMS

Theorem. *If* $F(p) \risingdotseq f(t)$, *then*

$$(-1)^n \frac{d^n}{dp^n} F(p) \risingdotseq t^n f(t) \tag{1}$$

Proof. We first prove that if $f(t)$ satisfies condition (1), Sec. 7.1, then the integral

$$\int_0^\infty e^{-pt} (-t)^n f(t)\, dt \tag{2}$$

exists.

By hypothesis, $|f(t)| < Me^{s_0 t}$, $p = a + ib$, $a > s_0$; and $a > 0$, $s_0 > 0$. Obviously, there will be an $\varepsilon > 0$ such that the inequality $a > s_0 + \varepsilon$ will be fulfilled. As in Sec. 7.1, it is proved that the following integral exists:

$$\int_0^\infty e^{-(a-\varepsilon)t} |f(t)| \, dt$$

We then evaluate the integral (2)

$$\int_0^\infty |e^{-pt} t^n f(t)| \, dt = \int_0^\infty |e^{-(p-\varepsilon)t} e^{-\varepsilon t} t^n f(t)| \, dt$$

Since the function $e^{-\varepsilon t} t^n$ is bounded and, in absolute value, is less than some number N for any value $t > 0$, we can write

$$\int_0^\infty |e^{-pt} t^n f(t)| \, dt < N \int_0^\infty |e^{-(p-\varepsilon)t} f(t)| \, dt = N \int_0^\infty e^{-(p-\varepsilon)t} |f(t)| \, dt < \infty$$

It is thus proved that the integral (2) exists. But this integral may be regarded as an nth-order derivative with respect to the parameter * p of the integral

$$\int_0^\infty e^{-pt} f(t) \, dt$$

And so, from formula

$$F(p) = \int_0^\infty e^{-pt} f(t) \, dt$$

we get the formula

$$\int_0^\infty e^{-pt} (-t)^n f(t) \, dt = \frac{d^n}{dp^n} \int_0^\infty e^{-pt} f(t) \, dt$$

From these two equations we have

$$(-1)^n \frac{d^n}{dp^n} F(p) = \int_0^\infty e^{-pt} t^n f(t)$$

which is formula (1).

Let us use (2) to find the transform of a power function. We write the formula (1), Sec. 7.2:

$$\frac{1}{p} \doteq 1$$

* Earlier we found a formula for differentiating a definite integral with respect to a real parameter (see Sec. 11.10, Vol. I). Here, the parameter p is a complex number, but the differentiation formula holds true.

Using formula (1) of this section, from this formula we get

$$(-1)\frac{d}{dp}\left(\frac{1}{p}\right)\xrightarrow{\cdot}t$$

or

$$\frac{1}{p^2}\xrightarrow{\cdot}t$$

Similarly

$$\frac{2}{p^3}\xrightarrow{\cdot}t^2$$

For any n we have

$$\frac{n!}{p^{n+1}}\xrightarrow{\cdot}t^n \tag{3}$$

Example 1. From the formula [see (2), Sec. 7.3]

$$\frac{a}{p^2+a^2}=\int_0^\infty e^{-pt}\sin at\,dt$$

by differentiating the left and right sides with respect to the parameter p, we get

$$\frac{2pa}{(p^2+a^2)^2}\xrightarrow{\cdot}t\sin at \tag{4}$$

Example 2. From (3), Sec. 7.3, on the basis of (1), we have

$$-\frac{a^2-p^2}{(p^2+a^2)^2}\xrightarrow{\cdot}t\cos at \tag{5}$$

Example 3. From (1), Sec. 7.6, by (1), we have

$$\frac{1}{(p+\alpha)^2}\xrightarrow{\cdot}te^{-\alpha t} \tag{6}$$

7.8 THE TRANSFORMS OF DERIVATIVES

Theorem. *If* $F(p)\xrightarrow{\cdot}f(t)$, *then*

$$pF(p)-f(0)\xrightarrow{\cdot}f'(t) \tag{1}$$

Proof. From the definition of a transform we can write

$$L\{f'(t)\}=\int_0^\infty e^{-pt}f'(t)\,dt \tag{2}$$

We shall assume that all the derivatives $f'(t)$, $f''(t)$, ..., $f^{(n)}(t)$ which we encounter satisfy the condition (1), Sec. 7.1, and, consequently, the integral (2) and similar integrals for subsequent derivatives exist. Computing by parts the integral on the right of (2), we find

$$L\{f'(t)\}=\int_0^\infty e^{-pt}f'(t)\,dt=e^{-pt}f(t)\Big|_0^\infty+p\int_0^\infty e^{-pt}f(t)\,dt$$

But by condition (1), Sec. 7.1,

$$\lim_{t\to\infty} e^{-pt}f(t)=0$$

and

$$\int_0^\infty e^{-pt}f(t)\,dt=F(p)$$

Therefore

$$L\{f'(t)\}=-f(0)+pF(p)$$

The theorem is proved. Let us now consider the transforms of derivatives of any order. Substituting into (1) the expression $pF(p)-f(0)$ in place of $F(p)$ and the expression $f'(t)$ in place of $f(t)$, we get

$$p[pF(p)-f(0)]-f'(0)\mathrel{\dot{\to}}f''(t)$$

or, removing brackets,

$$p^2F(p)-pf(0)-f'(0)\mathrel{\dot{\to}}f''(t) \tag{3}$$

The transform for a derivative of order n will be

$$\begin{aligned}p^nF(p)-[p^{n-1}f(0)+p^{n-2}f'(0)+\ldots\\+pf^{(n-2)}(0)+f^{(n-1)}(0)]\mathrel{\dot{\to}}f^{(n)}(t)\end{aligned} \tag{4}$$

Note. Formulas (1), (3), and (4) are simplified if $f(0)=f'(0)=$ $=\ldots=f^{(n-1)}(0)=0$. In this case we get

$$F(p)\mathrel{\dot{\to}}f(t)$$

$$pF(p)\mathrel{\dot{\to}}f'(t)$$

$$\cdot\ \cdot\ \cdot\ \cdot\ \cdot\ \cdot\ \cdot\ \cdot$$

$$p^nF(p)\mathrel{\dot{\to}}f^{(n)}(t)$$

7.9 TABLE OF TRANSFORMS

For convenience, the transforms which we obtained are here given in the form of a table.

No.	$F(p)=\int\limits_0^\infty e^{-pt} f(t)\,dt$	$f(t)$
1	$\frac{1}{p}$	1
2	$\frac{a}{p^2+a^2}$	$\sin at$
3	$\frac{p}{p^2+a^2}$	$\cos at$
4	$\frac{1}{p+\alpha}$	$e^{-\alpha t}$
5	$\frac{\alpha}{p^2-\alpha^2}$	$\sinh \alpha t$
6	$\frac{p}{p^2-\alpha^2}$	$\cosh \alpha t$
7	$\frac{a}{(p+\alpha)^2+a^2}$	$e^{-\alpha t}\sin at$
8	$\frac{p+\alpha}{(p+\alpha)^2+a^2}$	$e^{-\alpha t}\cos at$
9	$\frac{n!}{p^{n+1}}$	t^n
10	$\frac{2pa}{(p^2+a^2)^2}$	$t\sin at$
11	$\frac{p^2-a^2}{(p^2+a^2)^2}$	$t\cos at$
12	$\frac{1}{(p+\alpha)^2}$	$te^{-\alpha t}$
13	$\frac{1}{(p^2+a^2)^2}$	$\frac{1}{2a^3}(\sin at-at\cos at)$
14	$(-1)^n\frac{d^n}{dp^n}F(p)$	$t^n f(t)$
15	$F_1(p)\,F_2(p)$	$\int\limits_0^t f_1(\tau)\,f_2(t-\tau)\,d\tau$

Note. Formulas 13 and 15 of this table will be derived later on. If for the transform of the function $f(t)$ we take

$$F^*(p)=p\int\limits_0^\infty e^{-pt} f(t)\,dt$$

then in the formulas 1-13 of the table the expressions in the first column must be multiplied by p (formulas 14 and 15 will change more fundamentally). Since $F^*(p)=pF(p)$, it follows that by substituting into the left side of formula 14 the expression $\frac{F^*(p)}{p}$ in place of $F(p)$ and multiplying by p, we get

14′ $$(-1)^n p \frac{d^n}{dp^n}\left(\frac{F^*(p)}{p}\right) \xrightarrow{\cdot} t^n f(t)$$

Substituting into the left side of formula 15

$$F_1(p)=\frac{F_1^*(p)}{p}, \quad F_2(p)=\frac{F_2^*(p)}{p}$$

and multiplying this product by p, we have

15′ $$\frac{1}{p} F_1^*(p) F_2^*(p) \xrightarrow{\cdot} \int_0^t f_1(\tau) f_2(t-\tau)\, d\tau$$

7.10 AN AUXILIARY EQUATION FOR A GIVEN DIFFERENTIAL EQUATION

Suppose we have a linear differential equation of order n with constant coefficients $a_0, a_1, \ldots, a_{n-1}, a_n$:

$$a_0 \frac{d^n x}{dt^n} + a_1 \frac{d^{n-1}x}{dt^{n-1}} + \ldots + a_{n-1}\frac{dx}{dt} + a_n x(t) = f(t) \tag{1}$$

It is required to find a solution of this equation $x=x(t)$ for $t \geqslant 0$ that satisfies the initial conditions

$$x(0)=x_0, \quad x'(0)=x_0', \quad \ldots, \quad x^{(n-1)}(0)=x_0^{(n-1)} \tag{2}$$

Before, we solved this problem as follows: we found the general solution of equation (1) containing n arbitrary constants; then we determined the constants so that they should satisfy the initial conditions (2).

Here we give a simpler method of solving this problem using operational calculus. We seek the L-transform of the solution $x(t)$ of (1) satisfying the conditions (2). We designate this L-transform by $\bar{x}(p)$; thus, $\bar{x}(p) \xrightarrow{\cdot} x(t)$.

Let us suppose that there exist transforms of the solution of (1) and of its derivatives up to order n inclusive (after finding the solution we can test the truth of this assumption). We multiply all terms of (1) by e^{-pt}, where $p=a+ib$, and integrate with respect to t from 0 to ∞:

$$a_0 \int_0^\infty e^{-pt}\frac{d^n x}{dt^n}\,dt + a_1 \int_0^\infty e^{-pt}\frac{d^{n-1}x}{dt^{n-1}}\,dt + \\ + \ldots + a_n \int_0^\infty e^{-pt} x(t)\,dt = \int_0^\infty e^{-pt} f(t)\,dt \tag{3}$$

On the left-hand side of the equation are the L-transforms of the function $x(t)$ and its derivatives, on the right, the L-transform of the function $f(t)$, which we denote by $F(p)$. Hence, equation (3) may be rewritten as

$$a_0 L\left\{\frac{d^n x}{dt^n}\right\} + a_1 L\left\{\frac{d^{n-1}x}{dt^{n-1}}\right\} + \ldots + a_n L\{x(t)\} = L\{f(t)\}$$

Substituting into this equation the expressions (1), (3), and (4), Sec. 7.8, in place of the transforms of the function and of its derivatives, we get

$$\begin{aligned} &a_0\{p^n\bar{x}(p) - [p^{n-1}x_0 + p^{n-2}x_0' + p^{n-3}x_0'' + \ldots + x_0^{(n-1)}]\} \\ &+ a_1\{p^{n-1}\bar{x}(p) - [p^{n-2}x_0 + p^{n-3}x_0' + \ldots + x_0^{(n-2)}]\} \\ &\cdot\ \cdot\ \cdot\ \cdot\ \cdot\ \cdot\ \cdot\ \cdot\ \cdot\ \cdot\ \cdot\ \cdot\ \cdot\ \cdot\ \cdot\ \cdot \\ &+ a_{n-1}\{p\bar{x}(p) - x_0\} + a_n\bar{x}(p) = F(p) \end{aligned} \tag{4}$$

Equation (4) is known as the *auxiliary equation*, or the *transform equation*. The unknown in this equation is the transform $\bar{x}(p)$, which is determined from it. Transform it leaving on the left the terms that contain $\bar{x}(p)$:

$$\begin{aligned} &\bar{x}(p)[a_0p^n + a_1p^{n-1} + \ldots + a_{n-1}p + a_n] \\ &\quad = a_0[p^{n-1}x_0 + p^{n-2}x_0' + \ldots + x_0^{(n-1)}] \\ &\quad + a_1[p^{n-2}x_0 + p^{n-3}x_0' + \ldots + x_0^{(n-2)}] \\ &\quad \cdot\ \cdot\ \cdot\ \cdot\ \cdot\ \cdot\ \cdot\ \cdot\ \cdot\ \cdot\ \cdot\ \cdot\ \cdot\ \cdot \\ &\quad + a_{n-2}[px_0 + x_0'] + a_{n-1}x_0 + F(p) \end{aligned} \tag{4'}$$

The coefficient of $\bar{x}(p)$ on the left of (4′) is an nth-degree polynomial in p, which results when in place of the derivatives we put the corresponding powers of p into the left-hand member of equation (1). We denote the polynomial by $\varphi_n(p)$:

$$\varphi_n(p) = a_0p^n + a_1p^{n-1} + \ldots + a_{n-1}p + a_n \tag{5}$$

The right-hand side of (4′) is formed as follows:

the coefficient a_{n-1} is multiplied by x_0,
the coefficient a_{n-2} is multiplied by $px_0 + x_0'$,
. . . .
the coefficient a_1 is multiplied by $p^{n-2}x_0 + p^{n-3}x_0' + \ldots + x_0^{(n-2)}$
the coefficient a_0 is multiplied by $p^{n-1}x_0 + p^{n-2}x_0' + \ldots + x_0^{(n-1)}$

All these products are combined. To this is also added the transform of the right side of the differential equation $F(p)$. All terms of the right side of (4′), with the exception of $F(p)$, form, after collecting like terms, a polynomial in p of degree $n-1$ with known coefficients. We denote it by $\psi_{n-1}(p)$. And so equation (4') can be written as follows:

$$\bar{x}(p)\varphi_n(p)=\psi_{n-1}(p)+F(p)$$

From this equation we determine $\bar{x}(p)$:

$$\bar{x}(p)=\frac{\psi_{n-1}(p)}{\varphi_n(p)}+\frac{F(p)}{\varphi_n(p)} \tag{6}$$

Determined in this way, $\bar{x}(p)$ is the transform of the solution $x(t)$ of the equation (1), which solution satisfies the initial conditions (2). If we now find the function $x^*(t)$ whose transform is the function $\bar{x}(p)$ determined by equation (6), then, by the uniqueness theorem formulated in Sec. 7.1, it will follow that $x^*(t)$ is the solution of equation (1) that satisfies the conditions (2), that is,

$$x^*(t)=x(t)$$

If we seek the solution of (1) for zero initial conditions: $x_0=x_0'=x_0''=\ldots=x_0^{(n-1)}=0$, then in (6) we will have $\psi_{n-1}(p)=0$ and the equation will take the form

$$\bar{x}(p)=\frac{F(p)}{\varphi_n(p)}$$

or

$$\bar{x}(p)=\frac{F(p)}{a_0p^n+a_1p^{n-1}+\ldots+a_n} \tag{6'}$$

Example 1. Find the solution of the equation

$$\frac{dx}{dt}+x=1$$

satisfying the initial conditions $x=0$ when $t=0$.

Solution. Form the auxiliary equation

$$\bar{x}(p)(p+1)=0+\frac{1}{p} \quad \text{or} \quad \bar{x}(p)=\frac{1}{(p+1)p}$$

Decomposing the fraction on the right into partial fractions, we get

$$\bar{x}(p)=\frac{1}{p}-\frac{1}{p+1}$$

Using formulas 1 and 4 of the table, we find the solution:

$$x(t)=1-e^{-t}$$

Example 2. Find the solution of the equation

$$\frac{d^2x}{dt^2}+9x=1$$

that satisfies the initial conditions: $x_0=x_0'=0$ for $t=0$.

Solution. Write the auxiliary equation (4')

$$\bar{x}(p)(p^2+9)=\frac{1}{p} \quad \text{or} \quad \bar{x}(p)=\frac{1}{p(p^2+9)}$$

Decomposing this fraction into partial fractions, we get

$$\bar{x}(p)=\frac{-\frac{1}{9}p}{p^2+9}+\frac{\frac{1}{9}}{p}$$

Using formulas 1 and 3 of the table we find the solution:

$$x(t)=-\frac{1}{9}\cos 3t+\frac{1}{9}$$

Example 3. Find the solution of the equation

$$\frac{d^2x}{dt^2}+3\frac{dx}{dt}+2x=t$$

that satisfies the initial conditions: $x_0=x_0'=0$ for $t=0$.

Solution. Write the auxiliary equation (4′)

$$\bar{x}(p)(p^2+3p+2)=\frac{1}{p^2}$$

or

$$\bar{x}(p)=\frac{1}{p^2}\,\frac{1}{(p^2+3p+2)}=\frac{1}{p^2(p+1)(p+2)}$$

Decomposing this fraction into partial fractions by the method of undetermined coefficients, we obtain

$$\bar{x}(p)=\frac{1}{2}\,\frac{1}{p^2}-\frac{3}{4}\,\frac{1}{p}+\frac{1}{p+1}-\frac{1}{4(p+2)}.$$

From formulas 9, 1 and 4 of the table we find the solution:

$$x(t)=\frac{1}{2}t-\frac{3}{4}+e^{-t}-\frac{1}{4}e^{-2t}$$

Example 4. Find the solution of the equation

$$\frac{d^2x}{dt^2}+2\frac{dx}{dt}+5x=\sin t$$

satisfying the initial conditions $x_0=1$, $x_0'=2$ for $t=0$.

Solution. Write the auxiliary equation (4′):

$$\bar{x}(p)(p^2+2p+5)=p\cdot 1+2+2\cdot 1+L\{\sin t\}$$

or

$$\bar{x}(p)(p^2+2p+5)=p+4+\frac{1}{p^2+1}$$

whence we find $\bar{x}(p)$:

$$\bar{x}(p)=\frac{p+4}{p^2+2p+5}+\frac{1}{(p^2+1)(p^2+2p+5)}$$

Decomposing the latter fraction on the right into partial fractions, we can write

$$\bar{x}(p)=\frac{\frac{11}{10}p+4}{p^2+2p+5}+\frac{-\frac{1}{10}p+\frac{1}{5}}{p^2+1}$$

or

$$\bar{x}(p)=\frac{11}{10}\cdot\frac{p+1}{(p+1)^2+2^2}+\frac{29}{10\cdot 2}\cdot\frac{2}{(p+1)^2+2^2}-\frac{1}{10}\cdot\frac{p}{p^2+1}+\frac{1}{5}\cdot\frac{1}{p^2+1}$$

Applying formulas 8, 7, 3, and 2 of the table, we get the solution

$$x(t)=\frac{11}{10}e^{-t}\cos 2t+\frac{29}{20}e^{-t}\sin 2t-\frac{1}{10}\cos t+\frac{1}{5}\sin t$$

or, finally,

$$x(t)=e^{-t}\left(\frac{11}{10}\cos 2t+\frac{29}{20}\sin 2t\right)-\frac{1}{10}\cos t+\frac{1}{5}\sin t$$

7.11 DECOMPOSITION THEOREM

From formula (6) of the previous section it follows that the transform of the solution of a linear differential equation consists of two terms: the first term is a proper rational fraction in p, the second term is a fraction whose numerator is the transform of the right side of the equation $F(p)$, while the denominator is the polynomial $\varphi_n(p)$. If $F(p)$ is a rational fraction, then the second term will also be a rational fraction. It is thus necessary to be able to find the original function whose transform is the proper rational fraction. We shall deal with this question in the present section. Let the L-transform of some function be a proper rational fraction in p:

$$\frac{\psi_{n-1}(p)}{\varphi_n(p)}$$

It is required to find the original function. In Sec. 10.7, Vol. I, it was shown that any proper rational fraction may be represented in the form of a sum of partial fractions of four types:

I. $\dfrac{A}{p-a}$;

II. $\dfrac{A}{(p-a)^k}$;

III. $\dfrac{Ap+B}{p^2+a_1p+a_2}$, where the roots in the denominator are complex, that is, $\frac{a_1^2}{4}-a_2<0$;

IV. $\dfrac{Ap+B}{(p^2+a_1p+a_2)^k}$, where $k\geqslant 2$, the roots of the denominator are complex.

Let us find the original functions for these partial fractions. For a type I fraction we get (on the basis of formula 4 of the table)

$$\frac{A}{p-a}\;\dot{\rightarrow}\;Ae^{at}$$

For a type II fraction, by formulas 9 and 4 of the table, we have

$$\frac{A}{(p-a)^k} \doteqdot A \frac{1}{(k-1)!} t^{k-1} e^{at} \tag{1}$$

Let us now consider a type III fraction. We perform identity transformations:

$$\frac{Ap+B}{p^2+a_1p+a_2} = \frac{Ap+B}{\left(p+\frac{a_1}{2}\right)^2+\left(\sqrt{a_2-\frac{a_1^2}{4}}\right)^2}$$

$$= \frac{A\left(p+\frac{a_1}{2}\right)+\left(B-\frac{Aa_1}{2}\right)}{\left(p+\frac{a_1}{2}\right)^2+\left(\sqrt{a_2-\frac{a_1^2}{4}}\right)^2} = A\frac{p+\frac{a_1}{2}}{\left(p+\frac{a_1}{2}\right)^2+\left(\sqrt{a_2-\frac{a_1^2}{4}}\right)^2}$$

$$+\left(B-\frac{Aa_1}{2}\right)\frac{1}{\left(p+\frac{a_1}{2}\right)^2+\left(\sqrt{a_2-\frac{a_1^2}{4}}\right)^2}$$

Denoting the first and second terms by M and N, respectively, we get (from formulas 8 and 7 of the table)

$$M \doteqdot Ae^{-\frac{a_1}{2}t} \cos t\sqrt{a_2-\frac{a_1^2}{4}}$$

$$N \doteqdot \left(B-\frac{Aa_1}{2}\right)\frac{1}{\sqrt{a_2-\frac{a_1^2}{4}}}e^{-\frac{a_1}{2}t} \sin t\sqrt{a_2-\frac{a_1^2}{4}}$$

And, finally,

$$\frac{Ap+B}{p^2+a_1p+a_2}$$

$$\doteqdot e^{-\frac{a_1}{2}t}\left[A\cos t\sqrt{a_2-\frac{a_1^2}{4}}+\frac{B-\frac{Aa_1}{2}}{\sqrt{a_2-\frac{a_1^2}{4}}}\sin t\sqrt{a_2-\frac{a_1^2}{4}}\right] \tag{2}$$

We shall not consider a type IV partial fraction, since it would involve considerable calculations. We shall consider certain special cases below. For further information see the list of books at the beginning of this chapter.

7.12 EXAMPLES OF SOLUTIONS OF DIFFERENTIAL EQUATIONS AND SYSTEMS OF DIFFERENTIAL EQUATIONS BY THE OPERATIONAL METHOD

Example 1. Find the solution of the equation

$$\frac{d^2x}{dt^2}+4x=\sin 3x$$

that satisfies the initial conditions: $x_0=0$, $x_0'=0$ when $t=0$.

Solution. Form the auxiliary equation

$$\bar{x}(p)(p^2+4)=\frac{3}{p^2+9}, \qquad \bar{x}(p)=\frac{3}{(p^2+9)(p^2+4)}$$

or

$$\bar{x}(p)=\frac{-\frac{3}{5}}{p^2+9}+\frac{\frac{3}{5}}{p^2+4}=-\frac{1}{5}\cdot\frac{3}{p^2+9}+\frac{3}{10}\cdot\frac{2}{p^2+4}$$

whence we get the solution

$$x(t)=\frac{3}{10}\sin 2t-\frac{1}{5}\sin 3t$$

Example 2. Find the solution of the equation

$$\frac{d^3x}{dt^3}+x=0$$

that satisfies the initial conditions: $x_0=1$, $x_0'=3$, $x_0''=8$ when $t=0$.

Solution. Forming the auxiliary equation

$$\bar{x}(p)(p^3+1)=p^2\cdot 1+p\cdot 3+8$$

we find

$$\bar{x}(p)=\frac{p^2+3p+8}{p^3+1}=\frac{p^2+3p+8}{(p+1)(p^2-p+1)}$$

Decomposing the rational fraction obtained into partial fractions, we get

$$\frac{p^2+3p+8}{(p+1)(p^2-p+1)}=\frac{2}{p+1}+\frac{-p+6}{p^2-p+1}$$

$$=2\cdot\frac{1}{p+1}-\frac{p-\frac{1}{2}}{\left(p-\frac{1}{2}\right)^2+\left(\sqrt{\frac{3}{4}}\right)^2}+\frac{11}{\sqrt{3}}\cdot\frac{\frac{\sqrt{3}}{2}}{\left(p-\frac{1}{2}\right)^2+\left(\frac{\sqrt{3}}{2}\right)^2}$$

Using Table of Transforms (Sec. 7.9), we write the solution:

$$x(t)=2e^{-t}+e^{\frac{1}{2}t}\left(-\cos\frac{\sqrt{3}}{2}t+\frac{11}{\sqrt{3}}\sin\frac{\sqrt{3}}{2}t\right)$$

Example 3. Find the solution of the equation

$$\frac{d^2x}{dt^2}+x=t\cos 2t$$

that satisfies the initial conditions: $x=0$, $x_0'=0$ when $t=0$.

Solution. Write the auxiliary equation

$$\bar{x}(p)(p^2+1)=\frac{1}{p^2+4}-\frac{8}{(p^2+4)^2}$$

whence, after some manipulating, we get

$$\bar{x}(p)=-\frac{5}{9}\,\frac{1}{p^2+1}+\frac{5}{9}\,\frac{1}{p^2+4}+\frac{8}{3}\,\frac{1}{(p^2+4)^2}$$

Consequently,

$$x(t)=-\frac{5}{9}\sin t+\frac{5}{18}\sin 2t+\frac{1}{3}\left(\frac{1}{2}\sin 2t-t\cos 2t\right)$$

Obviously, the operational method may also be used to solve systems of linear differential equations. The following is an illustration.

Example 4. Find the solution of the set of equations

$$3\frac{dx}{dt}+2x+\frac{dy}{dt}=1$$

$$\frac{dx}{dt}+4\frac{dy}{dt}+3y=0$$

that satisfies the initial conditions: $x=0$, $y=0$ when $t=0$.

Solution. We denote $x(t)\leftarrow\!\!\div\,\bar{x}(p)$, $y(t)\leftarrow\!\!\div\,\bar{y}(p)$ and write the system of auxiliary equations:

$$(3p+2)\,\bar{x}(p)+p\bar{y}(p)=\frac{1}{p}$$

$$p\bar{x}(p)+(4p+3)\,\bar{y}(p)=0$$

Solving this system, we find

$$\bar{x}(p)=\frac{4p+3}{p(p+1)(11p+6)}=\frac{1}{2p}-\frac{1}{5(p+1)}-\frac{33}{10(11p+6)}$$

$$\bar{y}(p)=-\frac{1}{(11p+6)(p+1)}=\frac{1}{5}\left(\frac{1}{p+1}-\frac{11}{11p+6}\right)$$

From the transforms we find the original functions, i.e., the sought-for solutions of the system:

$$x(t)=\frac{1}{2}-\frac{1}{5}e^{-t}-\frac{3}{10}e^{-\frac{6}{11}t}$$

$$y(t)=\frac{1}{5}\left(e^{-t}-e^{-\frac{6}{11}t}\right)$$

Linear systems of higher orders are solved in similar fashion.

7.13 THE CONVOLUTION THEOREM

The following convolution theorem is frequently useful when solving differential equations by the operational method.

Convolution Theorem. *If $F_1(p)$ and $F_2(p)$ are the transforms of the functions $f_1(t)$ and $f_2(t)$, that is,*

$$F_1(p)\div\!\!\rightarrow f_1(t) \text{ and } F_2(p)\div\!\!\rightarrow f_2(t)$$

then $F_1(p)\cdot F_2(p)$ is the transform of the function

$$\int_0^t f_1(\tau) f_2(t-\tau)\,d\tau$$

that is,

$$F_1(p)\,F_2(p) \risingdotseq \int_0^t f_1(\tau) f_2(t-\tau)\,d\tau \tag{1}$$

Proof. We find the transform of the function

$$\int_0^t f_1(\tau) f_2(t-\tau)\,d\tau$$

from the definition of a transform:

$$L\left\{\int_0^t f_1(\tau) f_2(t-\tau)\,d\tau\right\} = \int_0^\infty e^{-pt}\left[\int_0^t f_1(\tau) f_2(t-\tau)\,d\tau\right] dt$$

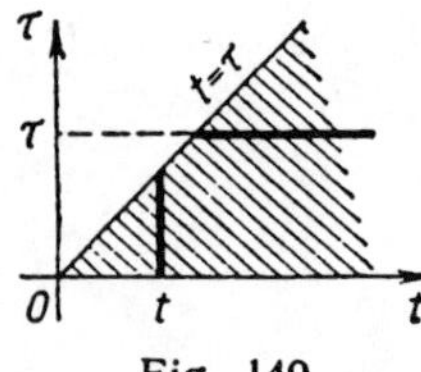

Fig. 149

The integral on the right is a twofold iterated integral of the form $\iint\limits_D \Phi(\tau, t)\,dt\,d\tau$, which is taken over a region bounded by the straight lines $\tau=0$, $\tau=t$ (Fig. 149). Changing the order of integration in this integral, we get

$$L\left\{\int_0^t f_1(\tau) f_2(t-\tau)\,d\tau\right\} = \int_0^\infty \left[f_1(\tau)\int_\tau^\infty e^{-pt} f_2(t-\tau)\,dt\right] d\tau$$

Changing the variable $t-\tau=z$ in the inner integral, we obtain

$$\int_\tau^\infty e^{-pt} f_2(t-\tau)dt = \int_0^\infty e^{-p(z+\tau)} f_2(z)\,dz = e^{-p\tau}\int_0^\infty e^{-pz} f_2(z)dz = e^{-p\tau}F_2(p)$$

Hence,

$$L\left\{\int_0^t f_1(\tau) f_2(t-\tau)\,d\tau\right\} = \int_0^\infty f_1(\tau)\,e^{-p\tau}F_2(p)\,d\tau$$

$$= F_2(p)\int_0^\infty e^{-p\tau} f_1(\tau)\,d\tau = F_2(p)\,F_1(p)$$

And so

$$\int_0^t f_1(\tau) f_2(t-\tau)\,d\tau \fallingdotseq F_1(p)\,F_2(p)$$

This is formula 15 of Table of Transforms (see Sec. 7.9).

Note 1. The expression $\int_0^t f_1(\tau) f_2(t-\tau)\,d\tau$ is called the *convolution* (*Faltung, resultant*) of two functions $f_1(t)$ and $f_2(t)$. The operation of obtaining it is also known as the *convolution* of two functions; here

$$\int_0^t f_1(\tau) f_2(t-\tau)\,d\tau = \int_0^t f_1(t-\tau) f_2(\tau)\,d\tau$$

That this equation is true is evident if we make the change of variable $t-\tau=z$ in the right-hand integral.

Example. Find the solution to the equation

$$\frac{d^2x}{dt^2}+x=f(t)$$

that satisfies the initial conditions: $x_0=x_0'=0$ for $t=0$.

Solution. Write the auxiliary equation

$$\bar{x}(p)(p^2+1)=F(p)$$

where $F(p)$ is the transform of the function $f(t)$. Hence,

$$\bar{x}(p)=\frac{1}{p^2+1}F(p), \quad \text{but} \quad \frac{1}{p^2+1} \mathrel{\dot{\to}} \sin t \quad \text{and} \quad F(p) \mathrel{\dot{\to}} f(t).$$

Applying the convolution formula (1) and denoting $\frac{1}{p^2+1}=F_2(p)$, $F(p)=F_1(p)$, we get

$$x(t)=\int_0^t f(\tau)\sin(t-\tau)\,dt \tag{2}$$

Note 2. On the basis of the convolution theorem it is easy to find the transform of the integral of the given function if we know the transform of this function; namely, if $F(p) \mathrel{\dot{\to}} f(t)$, then

$$\frac{1}{p}F(p) \mathrel{\dot{\to}} \int_0^t f(\tau)\,d\tau \tag{3}$$

Indeed, if we denote

$$f_1(t)=f(t), \; f_2(t)=1, \quad \text{then} \quad F_1(p)=F(p), \quad F_2(p)=\frac{1}{p}$$

Putting these functions into (1), we get formula (3).

7.14 THE DIFFERENTIAL EQUATIONS OF MECHANICAL VIBRATIONS. THE DIFFERENTIAL EQUATIONS OF ELECTRIC-CIRCUIT THEORY

From mechanics we know that the vibrations of a material point of mass m are described by the equation*

$$\frac{d^2x}{dt^2}+\frac{\lambda}{m}\frac{dx}{dt}+\frac{k}{m}x=\frac{1}{m}f_1(t) \tag{1}$$

where x is the deflection of the point from a certain position and k is the rigidity of the elastic system, for instance, a spring (a car spring), the force of resistance to motion is proportional (the proportionality constant is λ) to the first power of the velocity, and $f_1(t)$ is the outer (or disturbing) force.

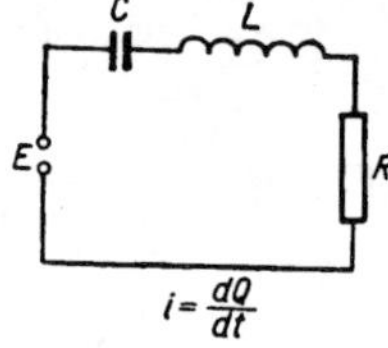

Fig. 150

Equations of type (1) describe small vibrations of other mechanical systems with one degree of freedom, for example, the torsional vibrations of a flywheel on an elastic shaft, if x is the angle of rotation of the flywheel, m is the moment of inertia of the flywheel, k is the torsional rigidity of the shaft, and $mf_1(t)$ is the moment of the outer forces relative to the axis of rotation. Equations of type (1) describe not only mechanical vibrations but also phenomena that occur in electric circuits.

Suppose we have an electric circuit consisting of an inductance L, a resistance R and a capacitance C, to which is applied an emf E (Fig. 150). We denote by i the current in the circuit, by Q the charge of the capacitor; then, as we know from electrical engineering, i and Q satisfy the following equations:

$$L\frac{di}{dt}+Ri+\frac{Q}{C}=E \tag{2}$$

$$\frac{dQ}{dt}=i \tag{3}$$

From (3) we get

$$\frac{d^2Q}{dt^2}=\frac{di}{dt} \tag{3'}$$

Substituting (3) and (3′) into (2), we get for Q an equation of type (1):

$$L\frac{d^2Q}{dt^2}+R\frac{dQ}{dt}+\frac{1}{C}Q=E \tag{4}$$

---

* See, for example, Sec. 1.26, where such an equation is derived in considering the vibration of a weight on a car spring.

Differentiating both sides of (2) and utilizing (3), we obtain an equation for determining the current i:

$$L\frac{d^2i}{dt^2}+R\frac{di}{dt}+\frac{1}{C}i=\frac{dE}{dt} \tag{5}$$

Equations (4) and (5) are type (1) equations.

7.15 SOLUTION OF THE DIFFERENTIAL EQUATION OF OSCILLATIONS

Let us write the oscillation equation in the form

$$\frac{d^2x}{dt^2}+a_1\frac{dx}{dt}+a_2x=f(t) \tag{1}$$

where the mechanical and physical meaning of the desired function x, of the coefficients a_1, a_2 and of the function $f(t)$ is readily established by comparing this equation with equations (1), (4), (5) of the preceding section. Let us find the solution to equation (1) that satisfies the initial conditions $x=x_0$, $x'=x_0'$ when $t=0$.

We form the auxiliary equation for equation (1):

$$\bar{x}(p)(p^2+a_1p+a_2)=x_0p+x_0'+a_1x_0+F(p) \tag{2}$$

where $F(p)$ is the transform of the function $f(t)$. From (2) we find

$$\bar{x}(p)=\frac{x_0p+x_0'+a_1x_0}{p^2+a_1p+a_2}+\frac{F(p)}{p^2+a_1p+a_2} \tag{3}$$

Thus, for a solution $Q(t)$ of equation (4), Sec. 7.14, that satisfies the initial conditions $Q=Q_0$, $Q'=Q_0'$ when $t=0$, the transform will have the form

$$\bar{Q}(p)=\frac{L(Q_0p+Q_0')+RQ_0}{Lp^2+Rp+\frac{1}{C}}+\frac{\bar{E}(p)}{Lp^2+Rp+\frac{1}{C}}$$

The type of solution is significantly dependent on whether the roots of the trinomial $p^2+a_1p+a_2$ are complex, or real and distinct, or real and equal. Let us examine in detail the case where the roots of the trinomial are complex; that is, when $\left(\frac{a_1}{2}\right)^2-a_2<0$. The other cases are considered in similar fashion.

Since the transform of a sum of two functions is equal to the sum of their transforms, it follows from formula (2), Sec. 7.1:, that the original function for the first fraction on the right of (3) will have the form

$$\frac{x_0 p + x_0' + a_1 x_0}{p^2 + a_1 p + a_2} \doteq e^{-\frac{a_1}{2} t} \left[x_0 \cos t \sqrt{a_2 - \frac{a_1^2}{4}} + \frac{x_0' + \frac{x_0 a_1}{2}}{\sqrt{a_2 - \frac{a_1^2}{4}}} \sin t \sqrt{a_2 - \frac{a_1^2}{4}} \right] \tag{4}$$

Let us then find the original function corresponding to the fraction

$$\frac{F(p)}{p^2 + a_1 p + a_2}$$

Here, we take advantage of the convolution theorem, first noting that

$$\frac{1}{p^2 + a_1 p + a_2} \doteq \frac{e^{-\frac{a_1}{2} t}}{\sqrt{a_2 - \frac{a_1^2}{4}}} \sin t \sqrt{a_2 - \frac{a_1^2}{4}}, \qquad F(p) \doteq f(t)$$

Hence, from (1), Sec. 7.13, we get

$$\frac{F(p)}{p^2 + a_1 p + a_2} \doteq \frac{1}{\sqrt{a_2 - \frac{a_1^2}{4}}} \int_0^t f(\tau) e^{-\frac{a_1}{2}(t-\tau)} \sin (t-\tau) \sqrt{a_2 - \frac{a_1^2}{4}}\, d\tau \tag{5}$$

And so, from (3), taking into account (4) and (5), we get

$$x(t) = e^{-\frac{a_1}{2} t} \left[x_0 \cos t \sqrt{a_2 - \frac{a_1^2}{4}} + \frac{x_0' + \frac{x_0 a_1}{2}}{\sqrt{a_2 - \frac{a_1^2}{4}}} \sin t \sqrt{a_2 - \frac{a_1^2}{4}} \right]$$
$$+ \frac{1}{\sqrt{a_2 - \frac{a_1^2}{4}}} \int_0^t f(\tau) e^{-\frac{a_1}{2}(t-\tau)} \sin (t-\tau) \sqrt{a_2 - \frac{a_1^2}{4}}\, d\tau \tag{6}$$

If the external force $f(t) \equiv 0$, which means that if we have free mechanical or electrical oscillations, then the solution is given by the first term on the right-hand side of expression (6). If the initial data are equal to zero, i.e., if $x_0 = x_0' = 0$, then the solution is given by the second term on the right side of (6). Let us consider these cases in more detail.

7.16 INVESTIGATING FREE OSCILLATIONS

Let equation (1) of the preceding section describe *free oscillations*, that is, $f(t) \equiv 0$. For convenience in writing we introduce the notation $a_1 = 2n$, $a_2 = k^2$, $k_1^2 = k^2 - n^2$. Then it will have the form

$$\frac{d^2x}{dt^2} + 2n\frac{dx}{dt} + k^2x = 0 \tag{1}$$

The solution of this equation x_{fr} that satisfies the initial conditions $x = x_0$, $x' = x_0'$ for $t = 0$ is given by the formula (4), Sec. 7.15, or by the first term of (6):

$$x_{fr}(t) = e^{-nt}\left[x_0 \cos k_1 t + \frac{x_0' + x_0 n}{k_1}\sin k_1 t\right] \tag{2}$$

We denote $x_0 = a$, $\frac{x_0' + x_0 n}{k_1} = b$. It is obvious that for any a and b we can select M and δ such that the following equalities will be fulfilled:

$$a = M \sin \delta, \quad b = M \cos \delta$$

here,

$$M^2 = a^2 + b^2, \quad \tan \delta = \frac{a}{b}$$

We rewrite formula (2) as

$$x_{fr} = e^{-nt}\left[M \cos k_1 t \sin \delta + M \sin k_1 t \cos \delta\right]$$

or, in final form, the solution may be written thus:

$$x_{fr} = \sqrt{a^2 + b^2}\, e^{-nt} \sin(k_1 t + \delta) \tag{3}$$

Solution (3) corresponds to *damped oscillations*.

If $2n = a_1 = 0$, that is, if there is no internal friction, then the solution will be of the form

$$x_{fr} = \sqrt{a^2 + b^2} \sin(k_1 t + \delta)$$

In this case *harmonic oscillations* occur. (In Sec. 1.27, Fig. 27 and 29 give graphs of harmonic and damped oscillations.)

7.17 INVESTIGATING MECHANICAL AND ELECTRICAL OSCILLATIONS IN THE CASE OF A PERIODIC EXTERNAL FORCE

When studying elastic vibrations of mechanical systems and, in particular, when studying electrical oscillations, one has to consider different types of external force $f(t)$. Let us consider in detail the case of a periodic external force. Let equation (1), Sec. 7.15, have the form

$$\frac{d^2x}{dt^2}+2n\frac{dx}{dt}+k^2x=A\sin\omega t \tag{1}$$

To determine the nature of the motion it is sufficient to consider the case when $x_0=x_0'=0$. One could obtain the solution of the equation by formula (6), Sec. 7.15, but we obtain the solution by carrying out all the intermediate calculations.

Let us write the transform equation

$$\bar{x}(p)(p^2+2np+k^2)=A\frac{\omega}{p^2+\omega^2}$$

from which we get

$$\bar{x}(p)=\frac{A\omega}{(p^2+2np+k^2)(p^2+\omega^2)} \tag{2}$$

We consider the case when $2n\neq 0$ $(n^2<k^2)$. Decompose the fraction on the right into partial fractions:

$$\frac{A\omega}{(p^2+2np+k^2)(p^2+\omega^2)}=\frac{Np+B}{p^2+2np+k^2}+\frac{Cp+D}{p^2+\omega^2} \tag{3}$$

We determine the constants B, C, D, N by the method of undetermined coefficients. Using formula (2). Sec. 7.11, and (2) of this section, we find the original function:

$$x(t)=\frac{A}{(k^2-\omega^2)^2+4n^2\omega^2}\left\{(k^2-\omega^2)\sin\omega t-2n\omega\cos\omega t\right.$$
$$\left.+e^{-nt}\left[(2n^2-k^2+\omega^2)\frac{\omega}{k_1}\sin k_1t+2n\omega\cos k_1t\right]\right\} \tag{4}$$

here again, $k_1=\sqrt{k^2-n^2}$. This is the solution of equation (1) that satisfies the initial conditions $x_0=x_0'=0$ when $t=0$.

Let us consider a special case when $2n=0$. This corresponds to a mechanical system with no internal resistance (no damper), or to an electric circuit where $R=0$ (no internal resistance in the circuit). Equation (1) then takes the form

$$\frac{d^2x}{dt^2}+k^2x=A\sin\omega t \tag{5}$$

and we get the solution of this equation satisfying the conditions $x_0=x_0'=0$ for $t=0$ if in (4) we put $n=0$:

$$x(t)=\frac{A}{(k^2-\omega^2)k}[-\omega\sin kt+k\sin\omega t] \tag{6}$$

Here we have the sum of two harmonic oscillations: natural oscillations with frequency k:

$$x_nt=-\frac{A}{k^2-\omega^2}\frac{\omega}{k}\sin kt$$

and forced oscillations with frequency ω:

$$x_f(t)=\frac{A}{k^2-\omega^2}\sin\omega t$$

The type of oscillations for the case $k \gg \omega$ is shown in Fig. 151.

Let us again return to formula (4). If $2n>0$ (which occurs in the mechanical and electrical systems under consideration), then the term containing the factor e^{-nt}, which represents damped natural oscillations, rapidly decreases for increasing t. For t sufficiently large, the character of the oscillations will be determined by the term that does not contain the factor e^{-nt}; that is, by the term

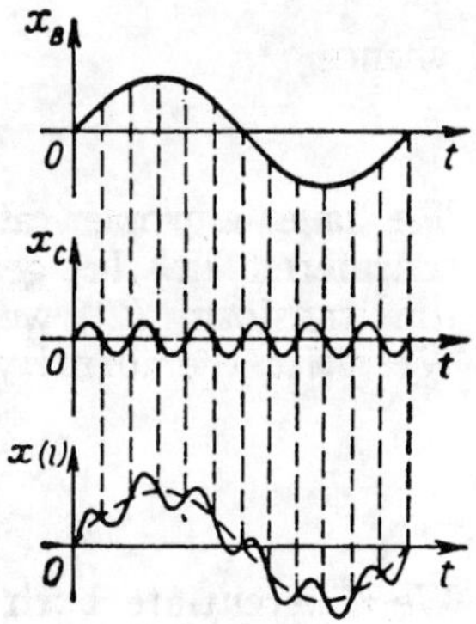

Fig. 151

$$x(t)=\frac{A}{(k^2-\omega^2)^2+4n^2\omega^2}\{(k^2-\omega^2)\sin\omega t \\ -2n\omega\cos\omega t\} \tag{7}$$

We introduce the notation

$$\frac{A(k^2-\omega^2)}{(k^2-\omega^2)^2+4n^2\omega^2}=M\cos\delta,\quad -\frac{A\cdot 2n\omega}{(k^2-\omega^2)^2+4n^2\omega^2}=M\sin\delta \tag{8}$$

where

$$M=\frac{A}{\sqrt{(k^2-\omega^2)^2+4n^2\omega^2}}$$

The solution (7) may be rewritten as follows:

$$x(t)=\frac{A}{k^2\sqrt{\left(1-\frac{\omega^2}{k^2}\right)^2+4n^2\frac{\omega^2}{k^4}}}\sin(\omega t+\delta) \tag{9}$$

From formula (9) it follows that the frequency k of forced oscillations does not coincide with the frequency ω of the external force. If the internal resistance, characterized by the number n, is small and the frequency ω is not very different from k, then the amplitude of oscillations may be made as great as one pleases, since the denominator may be arbitrarily small. For $n=0$, $\omega^2=k^2$, the solution is not expressed by formula (9).

7.18 SOLVING THE OSCILLATION EQUATION IN THE CASE OF RESONANCE

Let us consider the special case when $a_1=2n=0$, that is, when there is no resistance and the frequency of the external force coincides with that of the natural oscillations, $k=\omega$. The equation then takes the form

$$\frac{d^2x}{dt^2}+k^2x=A\sin kt \tag{1}$$

We shall seek the solution that satisfies the initial conditions $x_0=0$, $x'_0=0$ for $t=0$. The auxiliary equation will be

$$\bar{x}(p)(p^2+k^2)=A\frac{k}{p^2+k^2}$$

whence

$$\bar{x}(p)=\frac{Ak}{(p^2+k^2)^2} \tag{2}$$

We have a proper rational fraction of type IV, which we have not considered in the general form. To find the original function for the transform (2), we take advantage of the following procedure. We write the identity (formula 2 of Table of Transforms, Sec. 7.9)

$$\frac{k}{p^2+k^2}=\int_0^\infty e^{-pt}\sin kt\,dt \tag{3}$$

We differentiate both sides of this equation with respect to k (the integral on the right may be represented in the form of a sum of two integrals of a real variable, each of which depends on the parameter k):

$$\frac{1}{p^2+k^2}-\frac{2k^2}{(p^2+k^2)^2}=\int_0^\infty e^{-pt}\,t\cos kt\,dt$$

Utilizing (3) we can rewrite this equation as

$$-\frac{2k^2}{(p^2+k^2)^2}=\int_0^\infty e^{-pt}\left[t\cos kt-\frac{1}{k}\sin kt\right]dt$$

Whence it follows directly that

$$\frac{Ak}{(p^2+k^2)^2}\overset{\cdot}{\underset{\cdot}{\rightarrow}}\frac{A}{2k}\left(\frac{1}{k}\sin kt-t\cos kt\right)$$

(from this formula we obtain formula 13 of the table). Thus, the solution of equation (1) will be

$$x(t)=\frac{A}{2k}\left(\frac{1}{k}\sin kt-t\cos kt\right) \tag{4}$$

Let us study the second term of this equation:

$$x_2(t)=-\frac{A}{2k}t\cos kt \tag{4'}$$

This quantity is not bounded as t increases. The amplitude of oscillations that correspond to formula (4′) increases without bound as t increases without bound. Hence, the amplitude of oscillations corresponding to formula (4) also increases without bound. This is *resonance*; it occurs when the frequency of the natural oscillations coincides with that of the external force (see also Sec. 1.28, Figs. 30, 31).

7.19 THE DELAY THEOREM

Let a function $f(t)$, for $t<0$, be identically equal to zero (Fig. 152, *a*). Then the function $f(t-t_0)$ will be identically zero for $t<t_0$ (Fig. 152, *b*). We shall prove a theorem which is known as the delay theorem.

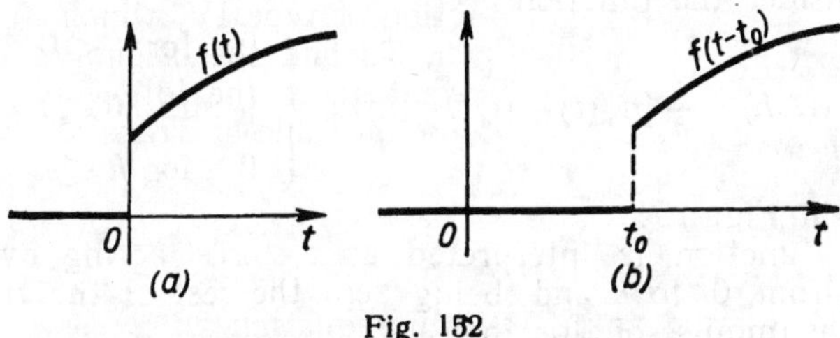

Fig. 152

Theorem. *If $F(p)$ is the transform of the function $f(t)$, then $e^{-pt_0}F(p)$ is the transform of the function $f(t-t_0)$; that is, if $f(t) \leftarrow\!\!\!\!\div F(p)$, then*

$$f(t-t_0) \leftarrow\!\!\!\!\div e^{-pt_0}F(p) \tag{1}$$

Proof. By the definition of a transform we have

$$L\{f(t-t_0)\}=\int_0^{\infty} e^{-pt} f(t-t_0)\,dt$$

$$=\int_0^{t_0} e^{-pt} f(t-t_0)\,dt+\int_{t_0}^{\infty} e^{-pt} f(t-t_0)\,dt$$

The first integral on the right of the equation is zero since $f(t-t_0)=0$ for $t<t_0$. In the last integral we change the variable, putting $t-t_0=z$:

$$L\{f(t-t_0)\}=\int_0^{\infty} e^{-p(z+t_0)} f(z)\,dz=e^{-pt_0}\int_0^{\infty} e^{-pz} f(z)\,dz=e^{-pt_0}F(p)$$

Thus, $f(t-t_0) \leftarrow\!\!\!\!\div e^{-pt_0}F(p)$.

Example. In Sec. 7.2 it was established for the Heaviside unit function that

$$\sigma_0(t) \leftarrow\!\!\!\!\div \frac{1}{p}$$

It follows, from the theorem that has just been proved, that for the function $\sigma_0(t-h)$ depicted in Fig. 153, the L-transform is

$$\frac{1}{p}e^{-ph}$$

that is,

$$\sigma_0(t-h) \leftarrow\!\!\!\div \frac{1}{p}e^{-ph} \tag{2}$$

7.20 THE DELTA FUNCTION AND ITS TRANSFORM

We consider the function

$$\sigma_1(t, h) = \frac{1}{h}[\sigma_0(t) - \sigma_0(t-h)] = \begin{cases} 0 & \text{for } t < 0 \\ \frac{1}{h} & \text{for } 0 \leqslant t < h \\ 0 & \text{for } h \leqslant t \end{cases} \tag{1}$$

depicted in Fig. 154.

If this function is interpreted as a force acting over a time interval from 0 to h and being zero the rest of the time, then, clearly, the impulse of this force will be equal to unity.

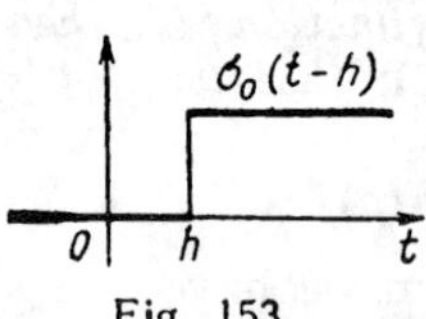

Fig. 153

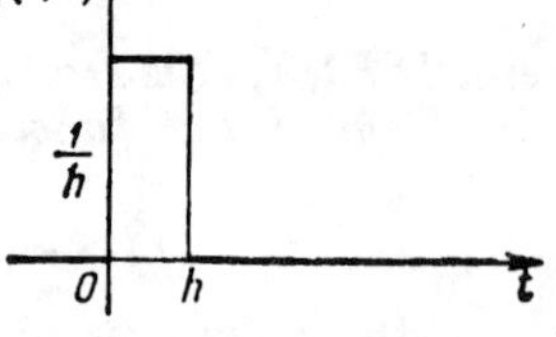

Fig. 154

By formulas (1), Sec. 7.2, and (2) of the preceding section, the transform of this function is

$$\frac{1}{h}\left(\frac{1}{p} - \frac{1}{p}e^{-ph}\right)$$

that is,

$$\sigma_1(t, h) \leftarrow\!\!\!\div \frac{1}{p}\left(\frac{1-e^{-ph}}{h}\right) \tag{2}$$

In mechanics it is convenient to regard forces acting over very brief time intervals as forces acting instantaneously but having a finite impulse. For this reason, the function $\delta(t)$ is introduced as the limit of the function $\sigma_1(t, h)$ when $h \to 0$:

$$\delta(t) = \lim_{h \to 0} \sigma_1(t, h) \tag{3}$$

This function is called the *unit impulse function*, or the *delta function*. (It should be noted that $\delta(t)$ is not a function in the ordinary meaning of the word; physicists often call $\delta(t)$ the Dirac function or the Dirac delta function.)

It is natural to put

$$\int_{-\infty}^{\infty} \delta(t)\, dt = 1 \tag{4}$$

One also writes

$$\int_{0}^{0} \delta(t)\, dt = 1 \tag{5}$$

The function $\delta(x)$ finds application not only in mechanics but in many divisions of mathematics, in particular in the solution of numerous problems involving the equations of mathematical physics.

Let us consider the action of $\delta(t)$ if it is represented as a force. We find the solution to the equation

$$\frac{d^2s}{dt^2} = \delta(t) \tag{6}$$

that satisfies the conditions: $s=0$, $\frac{ds}{dt}=0$, when $t=0$. From (6) we find, taking account of (5),

$$v = \frac{ds}{dt} = \int_{0}^{t} \delta(\tau)\, d\tau = 1 \tag{7}$$

for any t, in particular for $t=0$. Hence, by defining $\delta(x)$ by the equation (3), we can interpret this function as a force imparting to a unit mass at time $t=0$ a velocity equal to unity.

We define the L-transform of the function $\delta(t)$ as the limit of the transform of the function $\sigma_1(t, h)$ as $h \to 0$:

$$L\{\delta(x)\} = \lim_{h \to 0} \frac{1}{p} \cdot \frac{1-e^{-ph}}{h} = \frac{1}{p} \cdot p = 1$$

(here we made use of l'Hospital's rule for finding a limit). Thus,

$$\delta(t) \leftarrow\!\!\div 1 \tag{8}$$

We then define the function $\delta(t-t_0)$, which is interpreted as a force that instantaneously, at time $t=t_0$, imparts a unit velocity to a unit mass. Obviously, on the basis of the delay theorem we will have

$$\delta(t-t_0) \leftarrow\!\!\div e^{-pt} \tag{9}$$

As in the case of (5), we can write

$$\int_{t_0}^{t_0} \delta(t-t_0)\, dt = 1 \tag{10}$$

From the mechanical interpretation of the delta function it follows that the presence of the delta function in the right member of the equation can be replaced by an appropriate change in the initial conditions. We will illustrate this by a simple example.

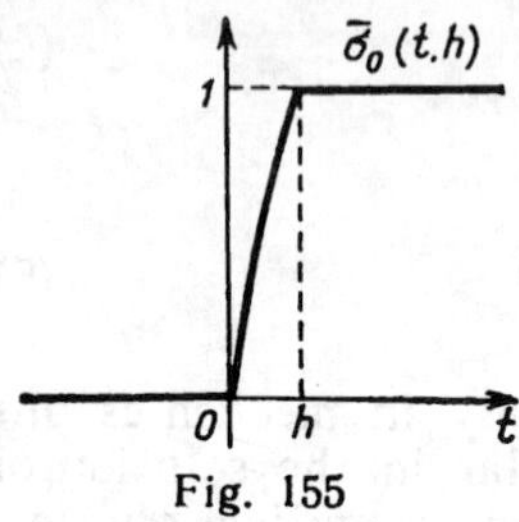

Fig. 155

Suppose we have a differential equation

$$\frac{d^2x}{dt^2} = f(t) + \delta(t) \tag{11}$$

with initial conditions $x_0 = 0$, $x'_0 = 0$ when $t = 0$. The auxiliary equation is

$$p^2\bar{x}(p) = F(p) + 1 \tag{12}$$

whence

$$\bar{x}(p) = \frac{F(p)}{p^2} + \frac{1}{p^2}$$

Using formulas 9 and 15 of the table, we get

$$x(t) = \int_0^t f(\tau)(t - \tau)\,d\tau + t \tag{13}$$

We would arrive at the same result if we found the solution to the equation

$$\frac{d^2x}{dt^2} = f(t)$$

with initial conditions $x_0 = 0$, $x'_0 = 1$, when $t = 0$. In this case the auxiliary equation would have the form

$$p^2\bar{x}(p) - 1 = F(p) \tag{14}$$

It is equivalent to the auxiliary equation (12) and, hence, the solution will coincide with the solution (13).

In conclusion, we note the following important property of the delta function. On the basis of (4) and (5), we can write

$$\int_{-\infty}^{t} \delta(\tau)\,d\tau = \begin{cases} 0 & \text{for } -\infty < t < 0 \\ 1 & \text{for } \quad 0 \leqslant t < \infty \end{cases} \tag{15}$$

In other words, this integral is equal to the Heaviside unit function $\sigma_0(t)$. Thus,

$$\sigma_0(t) = \int_{-\infty}^{t} \delta(\tau)\,d\tau \tag{16}$$

Differentiating the right and left sides of the equation with respect to t, we get the conditional equation

$$\sigma'_0(t) = \delta(t) \tag{17}$$

To get at the meaning of (17), let us examine the function $\bar{\sigma}_0(t, h)$ shown in Fig. 155. It is clear that

$$\bar{\sigma}'_0(t, h) = \sigma_1(t, h) \tag{18}$$

(with the exception of the points $t=0$ and $t=h$). Passing to the limit as $h \to 0$ in equation (18), we see that $\overline{\sigma}_0(t, h) \to \sigma_0(t)$ and we write

$$\overline{\sigma}_0'(t, h) \to \sigma_0'(t) \quad \text{as } h \to 0 \cdot$$

The right member of (18) $\sigma_1(t, h) \to \delta(t)$ as $h \to 0$. Thus, equation (18) passes into the conditional equation (17).

Exercises on Chapter 7

Find solutions to the following equations for the indicated initial conditions:

1. $\frac{d^2x}{dt^2}+3\frac{dx}{dt}+2x=0$, $x=1$, $x'=2$ for $t=0$. *Ans.* $x=4e^{-t}-3e^{-2t}$.

2. $\frac{d^3x}{dt^3}-\frac{d^2x}{dt^2}=0$, $x=2$, $x'=0$, $x''=1$ for $t=0$. *Ans.* $x=1-t+e^t$.

3. $\frac{d^2x}{dt^2}-2a\frac{dx}{dt}+(a^2+b^2)x=0$, $x=x_0$, $x'=x_0'$ for $t=0$.

Ans. $x=\frac{e^{at}}{b}\left[x_0 b\cos bt+(x_0'-x_0 a)\sin bt\right]$.

4. $\frac{d^2x}{dt^2}-3\frac{dx}{dt}+2x=e^{5t}$, $x=1$, $x'=2$ for $t=0$. *Ans.* $x=\frac{1}{12}e^{5t}+\frac{1}{4}e^t+\frac{2}{3}e^{2t}$.

5. $\frac{d^2x}{dt^2}+m^2x=a\cos nt$, $x=x_0$, $x'=x_0'$ for $t=0$.

Ans. $x=\frac{a}{m^2-n^2}(\cos nt-\cos mt)+x_0\cos mt+\frac{x_0'}{m}\sin mt$.

6. $\frac{d^2x}{dt^2}-\frac{dx}{dt}=t^2$, $x=0$, $x'=0$ for $t=0$. *Ans.* $x=2e^t-\frac{1}{3}t^3-t^2-2t-2$.

7. $\frac{d^3x}{dt^3}+x=\frac{1}{2}t^2e^t$, $x=x'=x''=0$ for $t=0$.

Ans. $x=\frac{1}{4}\left(t^2-3t+\frac{3}{2}\right)e^t-\frac{1}{24}e^{-t}-\frac{1}{3}\left\{\cos\frac{t\sqrt{3}}{2}-\sqrt{3}\sin\frac{t\sqrt{3}}{2}\right\}e^{\frac{1}{2}t}$.

8. $\frac{d^3x}{dt^3}+x=1$, $x_0=x_0'=x_0''=0$ for $t=0$. *Ans.* $x=1-\frac{1}{3}e^{-t}-\frac{2}{3}e^{\frac{t}{2}}\cos\frac{t\sqrt{3}}{2}$.

9. $\frac{d^4x}{dt^4}-2\frac{d^2x}{dt^2}+x=\sin t$, $x_0=x_0'=x_0''=x_0'''=0$ for $t=0$.

Ans. $x=\frac{1}{8}[e^t(t-2)+e^{-t}(t+2)+2\sin t]$.

10. Find solutions to the system of differential equations

$$\frac{d^2x}{dt^2}+y=1, \qquad \frac{d^2y}{dt^2}+x=0$$

that satisfy the initial conditions $x_0=y_0=x_0'=y_0'=0$ for $t=0$.

Ans. $x(t)=-\frac{1}{2}\cos t+\frac{1}{4}e^t+\frac{1}{4}e^{-t}$,

$y(t)=-\frac{1}{2}\cos t-\frac{1}{4}e^t-\frac{1}{4}e^{-t}+1$.

CHAPTER 8

ELEMENTS OF THE THEORY OF PROBABILITY AND MATHEMATICAL STATISTICS

Our daily experience in ordinary life, in practical situations, and also in scientific investigations constantly supplies us with examples of the breakdown of the familiar regularities of strict determinism that we are accustomed to. For instance, suppose we wish to know how many telephone calls a first-aid station will receive within twenty four hours.

Long-term observations indicate that there is no way of predicting the number of such calls. This number is subject to appreciable (and, what is more, random) fluctuations. Likewise, the time the doctor will have to spend with the patient in each instance is quite random.

Suppose we want to test N machine parts, all items having been manufactured under identical conditions and out of the same materials. The time from start of testing to their breakdown (failure) turns out to be a random quantity subject to an extremely broad spread of values.

In shooting at a target we have what is called shell dispersion. The departure of the point of impact of a shell from the centre of the target cannot be predicted because it is a random quantity.

It is not sufficient merely to indicate the fact of randomness in order to make use of a particular phenomenon of nature or to control a technological process. We have to learn to evaluate random events numerically and predict the course they will take. Such, at the present time, are the insistent demands of theoretical and practical problems. Two divisions of mathematics are engaged in the solution of such problems and in constructing the requisite general mathematical theory: they are the theory of probability and mathematical statistics.

In this chapter we deal with the basic essentials of probability theory and mathematical statistics.

8.1 RANDOM EVENT. RELATIVE FREQUENCY OF A RANDOM EVENT. THE PROBABILITY OF AN EVENT. THE SUBJECT OF PROBABILITY THEORY

The basic concept of probability theory is that of a *random (chance) event*. A random event is an event which may occur or fail to occur under the realization of a certain set of conditions.

Example 1. In coin tossing, the occurrence of heads is a random event.

Example 2. In firing at a target from a particular gun, hitting the target or a given area on it is a random event.

Example 3. In manufacturing a cylinder with a specified diameter of 20 cm, errors less than 0.2 mm represent a random event for a given set of production conditions.

Definition 1. *The relative frequency* p^* of a random event A is the ratio of the number m^* of occurrences of the given event to the total number n^* of identical trials, in each of which the given event could occur or fail to occur. We will write

$$\mathbf{P}^*(A) = p^* = \frac{m^*}{n^*} \tag{1}$$

Example 4. Suppose, under identical conditions, we fire 6 sequences of shots at a given target:

in the first sequence there were 5 shots and 2 hits,
in the second sequence there were 10 shots and 6 hits,
in the third sequence there were 12 shots and 7 hits,
in the fourth sequence there were 50 shots and 27 hits,
in the fifth sequence there were 100 shots and 49 hits,
in the sixth sequence there were 200 shots and 102 hits.

Event A is a hit. The relative frequency of hits in the sequences will be

$$\text{first, } \frac{2}{5} = 0.40,$$

$$\text{second, } \frac{6}{10} = 0.60,$$

$$\text{third, } \frac{7}{12} = 0.58,$$

$$\text{fourth, } \frac{27}{50} = 0.54,$$

$$\text{fifth, } \frac{49}{100} = 0.49,$$

$$\text{sixth, } \frac{102}{200} = 0.51.$$

From observations of a variety of phenomena, we can conclude that if the number of trials in each sequence is small, then the relative frequencies of the occurrence of event A in the different sequences can differ substantially from one another. However, if the number of trials in the sequences is great, then, as a rule, the relative frequencies of the occurrence of event A in different sequences will differ but slightly, and the difference is the smaller, the greater the number of trials in the sequences. We say that the relative frequency in a large number of trials ceases more and more to be accidental (of a random nature). However, it must be noted that there are events the relative frequency of which is not stable and may experience considerable fluctuations even in sequences with a large number of trials.

Experiments show that in most cases there is a constant p such that the relative frequencies of occurrence of an event A, given a large number of trials, differ but slightly from p, except in rare cases. This experimental fact is symbolized as follows:

$$\frac{m^*}{n^*} \xrightarrow[n^* \to \infty]{} p \tag{2}$$

The number p is called the *probability* of occurrence of a random event A. This statement is symbolized as

$$\mathbf{P}(A) = p \tag{3}$$

The probability p is an objective characteristic of the possibility of occurrence of event A under given trials. It is determined by the nature of A.

Given a large number of trials, the relative frequency differs very slightly from the probability, except in rare cases, which may be ignored.

Relation (2) can be formulated briefly as follows.

If the number n^* of trials is increased without bound, the relative frequency of event A converges to the probability p of occurrence of the event.

Note. In the foregoing, we postulated relation (2) on the basis of experiments. But other natural conditions that follow from experiment have been postulated. From them relation (2) is derived, which then becomes a theorem. This is the familiar theorem of probability theory that bears the name of Jakob Bernoulli (1654—1705).

Since probability is an objective characteristic of the possibility of occurrence of a certain event, to predict the course of numerous processes that one has to consider in military affairs, in the organization of production, in economic situations, etc., it is necessary to be able to determine the probability of occurrence of certain compound events. Determining the probability of occurrence of an event on the basis of the probabilities of the elementary events governing the given compound event, and the study of probabilistic regularities of various random events constitute the subject of the *theory of probability*.

8.2 THE CLASSICAL DEFINITION OF PROBABILITY AND THE CALCULATION OF PROBABILITIES

In many cases it is possible to calculate the probability of a random event by proceeding from an analysis of the trial. An example will help to illustrate this idea.

Example 1. A homogeneous cube with faces labelled 1 to 6 is called a die. We will consider the random event of the occurrence of a number l $(1 \leqslant l \leqslant 6)$ on the upper face for each throw of the die. By virtue of the symmetry of the die, the events (the appearance of any number from 1 to 6) are equally probable. Hence they are called *equally probable events*. Given a large number of throws, n, it can be expected that the number l (and any other number from 1 to 6) will turn up in roughly $n/6$ cases. Experiments corroborate this fact.

The relative frequency will be close to the number $p^* = 1/6$. It is therefore considered that the probability of the number l $(1 \leqslant l \leqslant 6)$ turning up is equal to 1/6.

Below, we will analyze random events whose probability can be calculated directly.

Definition 1. Random events in a given trial are called *disjoint* (*mutually exclusive*) if no two can occur at the same time.

Definition 2. We will say that random events form a *complete group* if in each trial any one of them can occur but no disjoint event can occur.

We consider a **complete group of equally probable disjoint random events.** We give the name *cases* to such events.

An event (case) of such a group is termed *favourable* to the occurrence of event A if the occurrence of the case implies the occurrence of A.

Example 2. We have 8 balls in an urn. Each ball is numbered from 1 to 8. Balls labelled 1, 2, 3 are red, the others are black. The occurrence of a ball labelled 1 (or 2 or 3) is an event favourable to the occurrence of a red ball.

For this case, we can give a definition of probability that differs from that given in Sec. 8.1.

Definition 3. The *probability* p of event A is the ratio of the number m of favourable cases to the number n of all possible cases forming a complete group of equally probable disjoint events, or, symbolically,

$$\mathbf{P}(A) = p = \frac{m}{n} \tag{1}$$

Definition 4. If relative to some event there are n favourable cases forming a complete group of equally probable disjoint events, then such an event is called a *certain event*. A certain event has probability $p = 1$.

If not a single one of n cases forming a complete group of equally probable disjoint events is favourable to an event, then it is termed an *impossible event* and has probability $p = 0$.

Note 1. The converse assertions also hold true here. In other cases, however, for instance in the case of a continuous random variable (Sec. 8.12), the converse assertions may not hold true; in other words, from the fact that the probability of some event is equal to 1 or 0, it does not yet follow that this event is certain or impossible.

From the definition of probability it follows that the relation

$$0 \leqslant p \leqslant 1$$

holds true.

Example 3. One card is drawn from a deck of 36 cards. What is the probability of drawing a spade?

Solution. This is a scheme of cases. Event A is the occurrence of a spade. We have a total of $n=36$ cases. The number of cases favouring A is $m=9$. Consequently, $p=\frac{9}{36}=\frac{1}{4}$.

Example 4. Two coins are tossed at the same time. What is the probability of getting 2 heads?

Solution. Let us compile a scheme of all possible cases.

	First coin	Second coin
1st case	head	head
2nd case	head	tail
3rd case	tail	head
4th case	tail	tail

There are 4 cases in all, of which one is favourable. Hence, the probability of obtaining two heads is

$$p=\frac{1}{4}$$

Example 5. The probability of hitting a target from one gun is $\frac{8}{10}$, from another gun, $\frac{7}{10}$. Find the probability of destroying the target in a simultaneous firing from both guns. The target will be destroyed if at least one of the guns makes a hit.

Solution. This problem can be modelled as follows. Two urns contain 10 balls each, numbered from 1 to 10, with 8 red and 2 black in the first urn, and 7 red and 3 black in the second. One ball is drawn from each urn. What is the probability that at least one of 2 drawn balls will be red?

Since each ball of the first urn can be drawn with any ball of the second urn, there will be a total of 100 cases: $n=100$.

Let us calculate the favourable cases.

When each of the 8 red balls of the first urn is drawn together with any ball of the second urn, we will have at least one red ball, drawn. There will be $10\times 8=80$ such cases. Drawing each of the two black balls of the first urn together with any one of the 7 red balls of the second urn gives us one red ball. There will be $2\times 7=14$ such cases. This makes a total of $m=80+14=94$ favourable cases.

The probability that there will be at least one red ball from among those drawn is

$$p=\frac{m}{n}=\frac{94}{100}$$

Such also is the probability of destroying the target.

Note 2. In this example, we reduced the problem of probability in gunfire to a problem of the probability of a given ball being drawn from an urn. Many problems in probability theory can be reduced to the "urn model". This permits us to regard urn-model problems (drawing balls from urns) as *generalized problems.*

Example 6. Ten items out of a set of 100 are defective. What is the probability that 3 out of any 4 chosen items will not be defective?

Solution. Four items out of 100 can be chosen in the following number of ways: $n=C_{100}^4$. The number of cases where 3 out of 4 items are nondefective is equal to $m=C_{90}^3\cdot C_{10}^1$.

The desired probability is

$$p=\frac{m}{n}=\frac{C_{90}^3\cdot C_{10}^1}{C_{100}^4}=\frac{1424}{4753}\approx 0.3$$

8.3 THE ADDITION OF PROBABILITIES. COMPLEMENTARY RANDOM EVENTS

Definition 1. *The logical sum (union)* of two events A_1 and A_2 is an event C consisting in the occurrence of at least one of the events.

Below we consider the probability of the union of two disjoint events A_1 and A_2. The union of these events is denoted by

$$A_1+A_2$$

or

$$A_1 \text{ or } A_2\text{*}$$

The following theorem, which is called the *theorem on the addition of probabilities*, holds true.

Theorem 1. *Suppose, in a given trial (phenomenon, experiment), a random event A_1 can occur with probability $\mathbf{P}(A_1)$ and an event A_2 with probability $\mathbf{P}(A_2)$. The events A_1 and A_2 are exclusive. Then the probability of the union of the events, that is, the probability that either event A_1 or event A_2 will take place, is computed from the formula*

$$\mathbf{P}(A_1 \text{ or } A_2)=\mathbf{P}(A_1)+\mathbf{P}(A_2) \tag{1}$$

* Note here that the "or" is inclusive and signifies the occurrence of at least one of the events, in accord with Definition 1.

Proof. Suppose

$$\mathbf{P}(A_1)=\frac{m_1}{n}, \quad \mathbf{P}(A_2)=\frac{m_2}{n}$$

Since the events A_1 and A_2 are exclusive, it follows that for a total number n of cases, the number of cases favouring the occurrence of A_1 and A_2 together is equal to 0, and the number of cases favouring the occurrence of A_1 or A_2 is equal to m_1+m_2. Consequently,

$$\mathbf{P}(A_1 \text{ or } A_2)=\frac{m_1+m_2}{n}=\frac{m_1}{n}+\frac{m_2}{n}=\mathbf{P}(A_1)+\mathbf{P}(A_2)$$

and the proof is complete.

The proof of this theorem is the same for any number of terms:

$$\mathbf{P}(A_1 \text{ or } A_2 \text{ or } \ldots \text{or } A_n)=\mathbf{P}(A_1)+\mathbf{P}(A_2)+\ldots+\mathbf{P}(A_n) \qquad (1')$$

This equation may be written thus:

$$\mathbf{P}\left(\sum_{i=1}^{n} A_i\right)=\sum_{i=1}^{n}\mathbf{P}(A_i) \qquad (1'')$$

Note. We proved the addition theorem for a scheme of cases where the probability is determined by direct computation. In the sequel we will assume that the addition theorem holds true also for the case where direct computation of probabilities is impossible. This assertion is based on the following reasoning. For large numbers of trials, the probabilities of events are (with rare exceptions) close to the relative frequencies, and for the latter the proof is the same as that given above. This remark will hold true also in the proof of subsequent theorems that we will prove by means of the urn scheme.

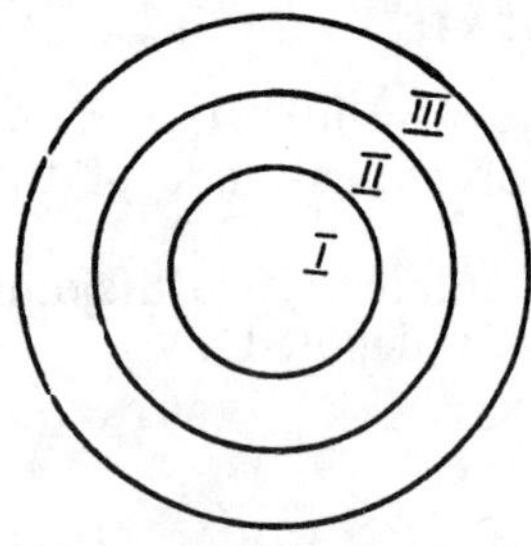

Fig. 156

Example 1. Shots are fired at a certain domain D consisting of three non-overlapping zones (Fig. 156). The probability of hitting Zone I is $\mathbf{P}(A_1)=\frac{5}{100}$, Zone II, $\mathbf{P}(A_2)=\frac{10}{100}$, and Zone III, $\mathbf{P}(A_3)=\frac{17}{100}$. What is the probability of hitting D? Event A is a hit scored in domain D. By formula (1′) we have

$$\mathbf{P}(A_1)+\mathbf{P}(A_2)+\mathbf{P}(A_3)=\frac{5}{100}+\frac{10}{100}+\frac{17}{100}=\frac{32}{100}$$

Definition 2. Two events are called *complementary events* if they are exclusive and form a complete group.

If one event is denoted by A, the complement (complementary event) is denoted by $\overline{A}$.

Let the probability of the occurrence of event A be p, the probability of the nonoccurrence of event A, that is, the probability of the occurrence of event $\overline{A}$, be $\mathbf{P}(\overline{A})=q$.

On trial, either A or $\overline{A}$ will occur, therefore Theorem 1 gives

$$\mathbf{P}(A)+\mathbf{P}(\overline{A})=1$$

That is, the union of the probabilities of complementary events is equal to unity:

$$p+q=1 \tag{2}$$

Example 2. A shot is fired at a target. Event A represents a hit. The probability p of a hit is $\mathbf{P}(A)=p$. Determine the probability of a miss. A miss is given by event $\overline{A}$, the complement of A. The probability of a miss is thus $q=1-p$.

Example 3. A measurement is made. Event A will denote an error less than λ. Let $\mathbf{P}(A)=p$. The complementary event is an error exceeding λ or equal to λ, and it is denoted by event $\overline{A}$. The probability of this event is $\mathbf{P}(\overline{A})=q=1-p$.

Corollary 1. *If random events $A_1, A_2, \ldots, A_n$ form a complete group of exclusive events, then the following equation holds true:*

$$\mathbf{P}(A_1)+\mathbf{P}(A_2)+\ldots+\mathbf{P}(A_n)=1 \tag{3}$$

Proof. Since the events $A_1, A_2, \ldots, A_n$ form a complete group of events, the occurrence of one of them is a certain event. Consequently,

$$\mathbf{P}(A_1 \text{ or } A_2 \text{ or } \ldots \text{ or } A_n)=1$$

Transforming the left member by formula (1′), we get (3).

Definition 3. Random events A and B are called *compatible* if in a given trial both events can occur, which is to say we have a *logical product (intersection)* of events A and B.

The event which consists in the intersection of A and B is denoted by (A and B) or (AB). The probability of the intersection of events A and B will be denoted by $\mathbf{P}(A \text{ and } B)$.

Theorem 2. *The probability of the union of compatible events is computed from the formula*

$$\mathbf{P}(A \text{ or } B)=\mathbf{P}(A)+\mathbf{P}(B)-\mathbf{P}(A \text{ and } B) \tag{4}$$

The truth of formula (4) can be illustrated geometrically. We first give the following definition.

Definition 4. Given a certain domain D with area S. Consider a subdomain d of D. Let $\overline{S}$ be the area of d. Then the probability of a point falling in d (the falling of a point in D is taken to be a certain event) is, by definition, $\bar{S}/S$, or $p=\bar{S}/S$. This is called *geometric probability*.

Then, assuming as certain the falling of a point in a square with unit side, we have (Fig. 157):

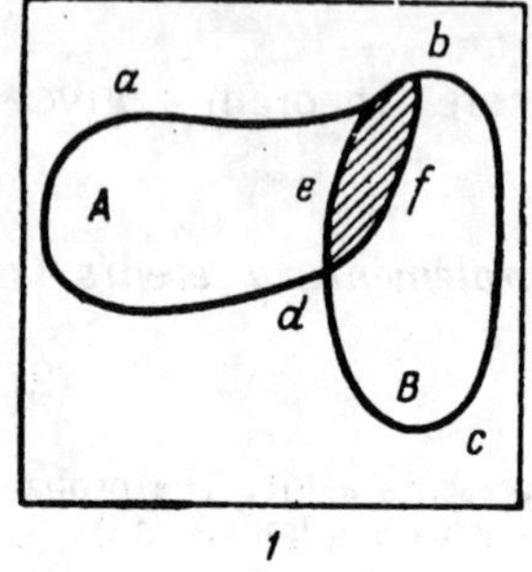

Fig. 157

$$\begin{aligned} \mathbf{P}(A \text{ or } B) &= \text{area } abcda, \\ \mathbf{P}(A) &= \text{area } abfda, \\ \mathbf{P}(B) &= \text{area } bcdeb, \\ \mathbf{P}(A \text{ and } B) &= \text{area } debfd \end{aligned} \tag{5}$$

We clearly have the equation

$$\text{area } abcda = \text{area } abfda + \text{area } bcdeb - \text{area } debfd$$

Putting into this equation the left members of (5), we get (4). In similar fashion we can compute the probability of the union of any number of compatible random events.

Note that Theorem 2 can be proved by proceeding from the foregoing definitions and rules of the operations.

8.4 MULTIPLICATION OF PROBABILITIES OF INDEPENDENT EVENTS

Definition 1. An event A is said to be *independent* of B if the probability of occurrence of A does not depend on whether event B took place or not.

Theorem 1. *If random events A and B are independent, then the probability of the intersection of events A and B is equal to the product of the probabilities of occurrence of A and B:*

$$\mathbf{P}(A \text{ and } B) = \mathbf{P}(A) \cdot \mathbf{P}(B) \tag{1}$$

Proof. Let us carry out the proof by the urn scheme. Each of two urns has, respectively, n_1 and n_2 balls. There are m_1 red balls in the first urn, and the remaining balls are black. There are m_2 red balls in the second urn, all others being black. One ball is drawn from each urn. What is the probability that both balls will be red?

Let event A be a drawing of a red ball from the first urn. Event B will be a drawing of a red ball from the second urn. These events are independent. Clearly,

$$\mathbf{P}(A) = \frac{m_1}{n_1}, \quad \mathbf{P}(B) = \frac{m_2}{n_2} \tag{2}$$

Altogether, there will be $n_1 n_2$ possible cases of simultaneous drawing of one ball from each urn. The number of cases favouring red balls being drawn from both urns will be m_1m_2. The probability of the intersection of events A and B will be

$$\mathbf{P}(A \text{ and } B) = \frac{m_1m_2}{n_1n_2} = \frac{m_1}{n_1} \cdot \frac{m_2}{n_2}$$

If in this formula we replace $\frac{m_1}{n_1}$ and $\frac{m_2}{n_2}$ by their expressions in (2), we get (1). The theorem is illustrated in Fig. 158.

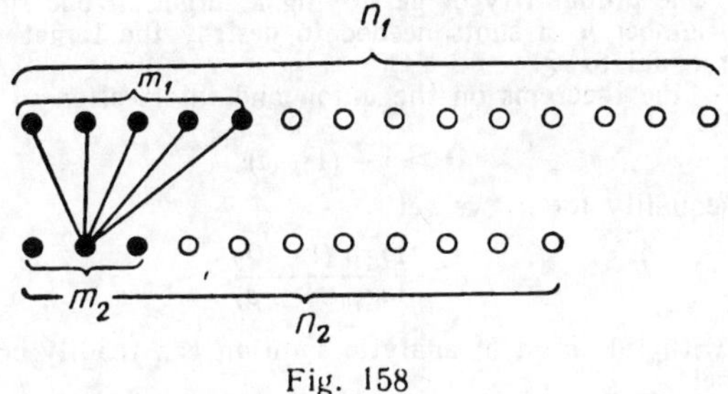

Fig. 158

If we have n independent events $A_1, A_2, \ldots, A_n$, then in similar fashion we can prove the validity of the equation

$$\mathbf{P}(A_1 \text{ and } A_2 \text{ and } \ldots \text{ and } A_n) = \mathbf{P}(A_1) \cdot \mathbf{P}(A_2) \cdot \ldots \cdot \mathbf{P}(A_n) \quad (3)$$

Example 1. Two tanks are firing at one and the same target. Tank One has a probability of $\frac{9}{10}$ of hitting the target, Tank Two, a probability of $\frac{5}{6}$. One shot is fired from each tank at the same time. Determine the probability that two hits will be scored.

Solution. Here, $\mathbf{P}(A) = \frac{9}{10}$, $\mathbf{P}(B) = \frac{5}{6}$; $\mathbf{P}(A \text{ and } B)$ is the probability of two hits and is found from formula (1):

$$\mathbf{P}(A \text{ and } B) = \frac{9}{10} \cdot \frac{5}{6} = \frac{3}{4}$$

Example 2. No-failure operation of a device is determined by trouble-free operation of each of three component units. The probabilities of no-failure operation of the units during a certain cycle are to $p_1 = 0.6$, $p_2 = 0.7$, $p_3 = 0.9$. Find the probability that the device will not break down during the indicated operation cycle.

Solution. By Theorem (3) on the multiplication of probabilities, we have

$$p = p_1 \cdot p_2 \cdot p_3 = 0.6 \times 0.7 \times 0.9 = 0.378$$

Note. Theorem 2, Sec. 8.3 [formula (4)], on the probability of the union of compatible events, with regard for formula (1) of this section, is then written as

$$\mathbf{P}(A \text{ or } B) = \mathbf{P}(A) + \mathbf{P}(B) - \mathbf{P}(A) \cdot \mathbf{P}(B) \quad (4)$$

Example 3. Solve the problem given in Example 5, Sec. 8.2, using formula (4).

Solution. Event A is a hit scored by the first gun. Event B is a hit scored by the second gun. It is obvious that

$$\mathbf{P}(A)=\frac{8}{10}, \quad \mathbf{P}(B)=\frac{7}{10},$$

$$\mathbf{P}(A \text{ or } B)=\frac{8}{10}+\frac{7}{10}-\frac{8}{10}\cdot\frac{7}{10}=\frac{94}{100}$$

We naturally obtained the same result as before.

Example 4. The probability of destroying a target in one shot is equal to p. Determine the number n of shots needed to destroy the target with probability greater than or equal to Q.

Solution. By the theorems on the union and intersection of probabilities we can write

$$Q \geqslant 1-(1-p)^n$$

Solving this inequality for n, we get

$$n \geqslant \frac{\log_{10}(1-Q)}{\log_{10}(1-p)}$$

A problem with this kind of analytic solution can readily be stated in terms of the urn model.

8.5 DEPENDENT EVENTS. CONDITIONAL PROBABILITY. TOTAL PROBABILITY

Definition 1. Event A is said to be *dependent* on event B if the probability of occurrence of A depends on whether B took place or not.

The probability that event A occurred, provided that B took place, will be denoted by $\mathbf{P}(A/B)$ and will be called the *conditional probability of event A* provided that B has occurred.

Example 1. An urn contains 3 white balls and 2 black balls. One ball is drawn (first drawing) and then a second is drawn (second drawing). Event B is the occurrence of a white ball in the first drawing, event A, the occurrence of a white ball in the second drawing.

It is clear that the probability of A, provided B has occurred, is

$$\mathbf{P}(A/B)=\frac{2}{4}=\frac{1}{2}$$

The probability of event A, provided B has not taken place (a black ball appears in the first drawing), will be

$$\mathbf{P}(A/\overline{B})=\frac{3}{4}$$

We see that

$$\mathbf{P}(A/B) \neq \mathbf{P}(A/\overline{B})$$

Theorem 1. *The probability of the intersection of two events is equal to the product (logical intersection) of the probability of one by the conditional probability of the other computed on the condition that the first event has taken place, that is,*

$$\mathbf{P}(A \text{ and } B)=\mathbf{P}(B)\cdot\mathbf{P}(A/B) \tag{1}$$

Proof. We carry out the proof for events that reduce to the urn model (that is, for the case where the classical definition of probability is applicable).

An urn contains n balls, of which n_1 are white and n_2 are black. Suppose the n_1 white balls include l balls marked with a star (Fig. 159).

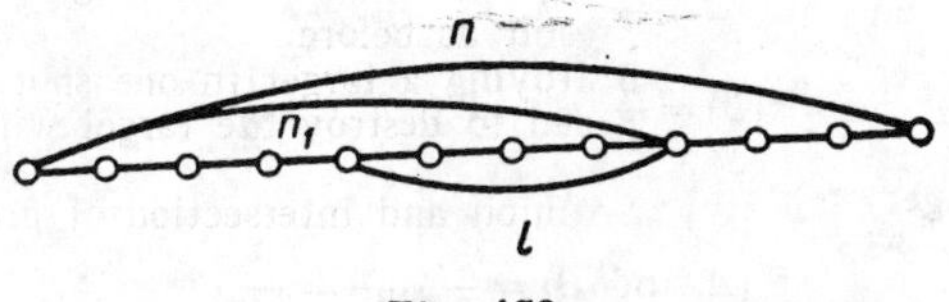

Fig. 159

One ball is drawn from the urn. What is the probability of drawing a marked white ball?

Let B be the event of a white ball being drawn, A the event of a marked white ball being drawn. It is then clear that

$$\mathbf{P}(B)=\frac{n_1}{n} \tag{2}$$

The probability of a marked white ball being drawn, provided that a white ball has already been drawn, is

$$\mathbf{P}(A/B)=\frac{l}{n_1} \tag{3}$$

The probability of a marked white ball being drawn is $\mathbf{P}$ (A and B). Obviously,

$$\mathbf{P}(A \text{ and } B)=\frac{l}{n} \tag{4}$$

But

$$\frac{l}{n}=\frac{n_1}{n}\cdot\frac{l}{n_1} \tag{5}$$

Substituting into (5) the left-hand members of (2), (3) and (4), we get

$$\mathbf{P}(A \text{ and } B)=\mathbf{P}(B)\cdot\mathbf{P}(A/B)$$

Equation (1) is proved.

If the events under consideration do not fit the classical scheme, then formula (1) serves to define conditional probability. Namely, *the conditional probability of A, provided event B has occurred, is determined by means of the formula*

$$\mathbf{P}(A/B)=\frac{\mathbf{P}(A \text{ and } B)}{\mathbf{P}(B)} \quad [\text{for } \mathbf{P}(B)\neq 0]$$

Note 1. Let us apply this formula to the expression $\mathbf{P}$ (B and A):

$$\mathbf{P}(B \text{ and } A)=\mathbf{P}(A)\cdot\mathbf{P}(B/A) \tag{6}$$

In equations (1) and (6), the left-hand members are equal, since this is one and the same probability; hence, the right-hand members are also equal. We can therefore write the equation

$$\mathbf{P}(A \text{ and } B) = \mathbf{P}(B)\cdot\mathbf{P}(A/B) = \mathbf{P}(A)\cdot\mathbf{P}(B/A) \tag{7}$$

Example 2. For the case of Example 1, given at the beginning of this section we have

$$\mathbf{P}(B) = \frac{3}{5}, \qquad \mathbf{P}(A/B) = \frac{1}{2}$$

By formula (1) we get

$$\mathbf{P}(A \text{ and } B) = \frac{3}{5}\cdot\frac{1}{2} = \frac{3}{10}$$

The probability $\mathbf{P}(A \text{ and } B)$ can also be readily calculated directly.

Example 3. The probability of manufacturing a nondefective (acceptable) item by a given machine is equal to 0.9. The probability of the occurrence of quality articles of grade one among the nondefective items is 0.8. Determine the probability of turning out grade-one articles by this machine.

Solution. Event B stands for manufacturing a nondefective article on the given machine. Event A stands for the occurrence of a grade-one item.

Here, $\mathbf{P}(B) = 0.9$, $\mathbf{P}(A/B) = 0.8$.

Substituting into formula (1), we get the desired probability:

$$\mathbf{P}(A \text{ and } B) = 0.9\cdot 0.8 = 0.72$$

Theorem 2. *If event A can be realized only when one of the events $B_1, B_2, \ldots, B_n$, which form a complete group of exclusive events, occurs, the probability of event A is computed from the formula*

$$\mathbf{P}(A) = \mathbf{P}(B_1)\cdot\mathbf{P}(A/B_1) + \mathbf{P}(B_2)\,\mathbf{P}(A/B_2) + \ldots + \mathbf{P}(B_n)\,\mathbf{P}(A/B_n) \tag{8}$$

Formula (8) is called the *formula of total probability.*

Proof. Event A can occur if one of the compatible events

$$(B_1 \text{ and } A),\quad (B_2 \text{ and } A),\quad \ldots,\quad (B_n \text{ and } A)$$

is realized. Consequently, by the theorem of addition of probabilities, we get

$$\mathbf{P}(A) = \mathbf{P}(B_1 \text{ and } A) + \mathbf{P}(B_2 \text{ and } A) + \ldots + \mathbf{P}(B_n \text{ and } A)$$

Replacing the terms of the right side in accordance with formula (1), we get equation (8).

Example 4. Three shots are fired at a target in succession. The probability of a hit in the first shot is $p_1 = 0.3$, in the second, $p_2 = 0.6$, in the third, $p_3 = 0.8$. In the case of one hit, the probability of destroying the target is $\lambda_1 = 0.4$, in the case of two hits, $\lambda_2 = 0.7$, in the case of three hits, $\lambda_3 = 1.0$. Determine the probability of destroying the target in three shots (event A).

Solution. Let us consider the complete group of exclusive events:

B_1 means one hit,
B_2 means two hits,
B_3 means three hits,
B_4 means no hits.

We determine the probability of each event. One hit will occur if: either the first shot scores a hit and the second and third miss; or the first shot is a miss, the second is a hit, and the third is a miss; or the first shot is a miss, the second is a miss, and the third is a hit. Therefore, by the theorem of multiplication and addition of probabilities, we get the following expression for the probability of one hit:

$$\mathbf{P}(B_1)=p_1(1-p_2)(1-p_3)+(1-p_1)p_2(1-p_3)+(1-p_1)(1-p_2)p_3=0.332$$

Reasoning in like fashion, we have

$$\mathbf{P}(B_2)=p_1p_2(1-p_3)+p_1(1-p_2)p_3+(1-p_1)p_2p_3=0.468$$
$$\mathbf{P}(B_3)=p_1p_2p_3=0.144$$
$$\mathbf{P}(B_4)=(1-p_1)(1-p_2)(1-p_3)=0.056$$

Let us write down the conditional probabilities of destroying the target upon the realization of each of these events:

$$\mathbf{P}(A/B_1)=0.4,\quad \mathbf{P}(A/B_2)=0.7,\quad \mathbf{P}(A/B_3)=1.0,\quad \mathbf{P}(A/B_4)=0$$

Substituting the resulting expressions into formula (8), we get the probability of target destruction:

$$\mathbf{P}(A)=\mathbf{P}(B_1)\cdot\mathbf{P}(A/B_1)+\mathbf{P}(B_2)\cdot\mathbf{P}(A/B_2)+\mathbf{P}(B_3)\cdot\mathbf{P}(A/B_3)+\mathbf{P}(B_4)\cdot\mathbf{P}(A/B_4)=0.332\cdot0.4+0.468\cdot0.7+0.144\cdot1.0+0.056\cdot0=0.6044$$

Note 2. If event A does not depend on event B, then

$$\mathbf{P}(A/B)=\mathbf{P}(A)$$

and formula (1) takes the form

$$\mathbf{P}(A \text{ and } B)=\mathbf{P}(B)\cdot\mathbf{P}(A)$$

That is, we obtain formula (1), Sec. 8.4.

8.6 PROBABILITY OF CAUSES. BAYES'S FORMULA

Statement of the problem. As in Theorem 2, Sec. 8.5, we will consider a complete group of exclusive events $B_1, B_2, \ldots, B_n$, the probabilities of occurrence of which are $\mathbf{P}(B_1), \mathbf{P}(B_2), \ldots, \mathbf{P}(B_n)$. Event A can occur only together with some one of the events $B_1, B_2, \ldots, B_n$, which we will call *causes*.

The probability of the occurrence of event A is, in accord with formula (8) of Sec. 8.5,

$$\mathbf{P}(A)=\mathbf{P}(B_1)\mathbf{P}(A/B_1)+\mathbf{P}(B_2)\mathbf{P}(A/B_2)+\ldots+\mathbf{P}(B_n)\mathbf{P}(A/B_n) \qquad (1)$$

Suppose that event A has taken place. This fact will alter the probability of the causes, $\mathbf{P}(B_1), \ldots, \mathbf{P}(B_n)$. It is required to determine the conditional probabilities of the realization of these causes on the assumption that event A has occurred, that is, to determine

$$\mathbf{P}(B_1/A),\quad \mathbf{P}(B_2/A),\ \ldots\quad \mathbf{P}(B_n/A)$$

Solution of the problem. By formula (7), Sec. 8.5, we will find the probability $\mathbf{P}(A \text{ and } B_1)$:

$$\mathbf{P}(A \text{ and } B_1) = \mathbf{P}(B_1)\cdot\mathbf{P}(A/B_1) = \mathbf{P}(A)\cdot\mathbf{P}(B_1/A)$$

whence

$$\mathbf{P}(B_1/A) = \frac{\mathbf{P}(B_1)\cdot\mathbf{P}(A/B_1)}{\mathbf{P}(A)}$$

Substituting for $\mathbf{P}(A)$ its expression (1), we get

$$\mathbf{P}(B_1/A) = \frac{\mathbf{P}(B_1)\cdot\mathbf{P}(A/B_1)}{\sum_{i=1}^{n}\mathbf{P}(B_i)\cdot P(A/B_i)} \tag{2}$$

The probabilities $\mathbf{P}(B_2/A)$, $\mathbf{P}(B_3/A)$, ..., $\mathbf{P}(B_n/A)$ are determined in similar fashion:

$$\mathbf{P}(B_k/A) = \frac{\mathbf{P}(B_k)\cdot\mathbf{P}(A/B_k)}{\sum_{i=1}^{n}\mathbf{P}(B_i)\cdot\mathbf{P}(A/B_i)} \tag{3}$$

Formula (2) is called *Bayes's formula* or the *theorem of causes. (Bayes's rule for the probability of causes.)*

Note. From formula (3) it follows that in the expression of the probability $\mathbf{P}(B_k/A)$ — the probability of the realization of cause B_k provided that event A has occurred — the denominator is independent of the number k.

Example 1. Suppose that prior to an experiment there were four equally probable causes:

$$B_1,\ B_2,\ B_3,\ B_4:\ \mathbf{P}(B_1) = \mathbf{P}(B_2) = \mathbf{P}(B_3) = \mathbf{P}(B_4) = 0.25$$

The conditional probabilities of the occurrence of event A are, respectively, equal to

$$\mathbf{P}(A/B_1) = 0.7, \qquad \mathbf{P}(A/B_2) = 0.1$$
$$\mathbf{P}(A/B_3) = 0.1, \qquad \mathbf{P}(A/B_4) = 0.02$$

Suppose the outcome of a trial yields event A. Then, by formula (3), we get

$$\mathbf{P}(B_1/A) = \frac{0.25\cdot 0.7}{0.25\cdot 0.7 + 0.25\cdot 0.1 + 0.25\cdot 0.1 + 0.25\cdot 0.02} = \frac{0.175}{0.23} \approx 0.76$$

$$\mathbf{P}(B_2/A) = \frac{0.25\cdot 0.1}{0.23} = 0.11$$

$$\mathbf{P}(B_3/A) = \frac{0.25\cdot 0.1}{0.23} = 0.11$$

$$\mathbf{P}(B_4/A) = \frac{0.25\cdot 0.02}{0.23} = 0.02$$

Here $\mathbf{P}(B_1) = 0.25$, $\mathbf{P}(B_1/A) = 0.76$ became greater because event A occurred. In this case, the probability $\mathbf{P}(A/B_1) = 0.7$ is greater than the other conditional probabilities.

Example 2. Each of two tanks fired independently at a target. The probability of the first tank destroying the target is $p_1 = 0.8$, that of the second,

$p_2=0{,}4$. The target is destroyed by a single hit. Determine the probability that it was destroyed by the first tank.

Solution. Event A is the destruction of the target by a single hit. Prior to firing, we have the following possible causes:

B_1, both tanks missed,

B_2, both tanks scored a hit,

B_3, the first tank scored a hit, the second missed,

B_4, the first tank missed, the second scored a hit.

We find the probabilities of these causes by the theorem of multiplication of probabilities:

$$\mathbf{P}(B_1)=(1-p_1)(1-p_2)=0.2\cdot 0.6=0.12$$
$$\mathbf{P}(B_2)=p_1p_2=0.8\cdot 0.4=0.32$$
$$\mathbf{P}(B_3)=p_1(1-p_2)=0.8\cdot 0.6=0.48$$
$$\mathbf{P}(B_4)=(1-p_1)p_2=0.2\cdot 0.4=0.08$$

We determine the conditional probabilities of occurrence of the event:

$$\mathbf{P}(A/B_1)=0,\quad \mathbf{P}(A/B_2)=0,\quad \mathbf{P}(A/B_3)=1,\quad \mathbf{P}(A/B_4)=1$$

By formula (2) we find the conditional probability of the causes:

$$\mathbf{P}(B_1/A)=\frac{0.12\cdot 0}{0.12\cdot 0+0.32\cdot 0+0.48\cdot 1+0.08\cdot 1}=\frac{0}{0.56}=0$$

$$\mathbf{P}(B_2/A)=\frac{0.32\cdot 0}{0.56}=0$$

$$\mathbf{P}(B_3/A)=\frac{0.48\cdot 1}{0.56}=\frac{6}{7}$$

$$\mathbf{P}(B_4/A)=\frac{0.08\cdot 1}{0.56}=\frac{1}{7}$$

Example 3. At a factory, 30% of the instruments are assembled by specialists of high qualification, 70% by those of medium qualification. The reliability of an instrument assembled by the former is 0.90, that assembled by the latter, 0.80. An instrument picked off the shelf turns out to be reliable. Determine the probability that it was assembled by the specialist of higher qualification.

Solution. Event A stands for reliable operation of the instrument. Prior to checking the instrument, we have the following possible causes:

B_1, the instrument is assembled by the highly qualified specialist,

B_2, the instrument is assembled by the specialist of medium qualification.

We write down the probabilities of these causes:

$$\mathbf{P}(B_1)=0.3,\qquad \mathbf{P}(B_2)=0.7$$

The conditional probabilities of the event are

$$\mathbf{P}(A/B_1)=0.9,\qquad \mathbf{P}(A/B_2)=0.8$$

We now determine the probabilities of the causes B_1 and B_2, provided that event A has occurred.

By formula (2) we have

$$\mathbf{P}(B_1/A)=\frac{0.3\cdot 0.9}{0.3\cdot 0.9+0.7\cdot 0.8}=\frac{0.27}{0.83}=0.325$$

$$\mathbf{P}(B_2/A)=\frac{0.7\cdot 0.8}{0.3\cdot 0.9+0.7\cdot 0.8}=\frac{0.56}{0.83}=0.675$$

8.7 A DISCRETE RANDOM VARIABLE. THE DISTRIBUTION LAW OF A DISCRETE RANDOM VARIABLE

Definition. A variable quantity x which, in a trial, assumes one value out of a finite or infinite sequence $x_1, x_2, \ldots, x_k, \ldots$ is called a *discrete random quantity* (or variable), if to each value x_k there corresponds a definite probability p_k that the variable x will assume the value x_k.

It follows from the definition that to every value x_k there corresponds a probability p_k.

The functional relationship between p_k and x_k is called the *distribution law of probabilities of a discrete random variable* x: *

Possible values of the random variable	x_1	x_2	...	...	...	x_k	...
Probabilities of these values	1	p_2	...	...	...	p_k	...

The distribution law can also be represented graphically in the form of a *polygon of probability distribution* (also called a *frequency polygon*): in a rectangular coordinate system, points are constructed with coordinates (x_k, p_k) and are joined by a polygonal line (Fig. 160).

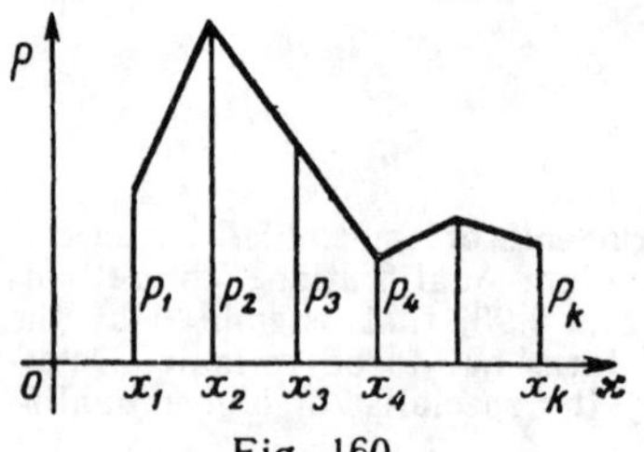

Fig. 160

The distribution law can also be specified analytically:

$$p_k = f(x_k)$$

The fact that a random variable x assumes one of the values of a sequence $x_1, x_2, \ldots, x_k, \ldots$ is a certain event, and so the condition

$$\sum_{i=1}^{N} p_i = 1 \tag{1}$$

must hold true for the case of a finite sequence of N values, or

$$\sum_{i=1}^{\infty} p_i = 1 \tag{1'}$$

for the case of an infinite sequence. The value x_i of the random variable having the greatest probability is called the *mode*. The random variable x depicted in Fig. 160 has mode x_2.

---

* Or, briefly, the distribution law of a random variable.

Example 1. A variable x denotes the number that turns up in a single throw of a die. The variable x can assume one of the following values: 1, 2, 3, 4, 5, 6. The probability of any one value turning up is 1/6. We thus get the following table for the distribution of this random variable:

x	1	2	3	4	5	6
p	1/6	1/6	1/6	1/6	1/6	1/6

Example 2. The probability of the occurrence of event A in each of an infinite sequence of trials is equal to p. The random variable x is the number of the trial in which A occurred for the first time. Find the law of distribution of x.

Solution. The random variable x can assume any integral positive value 1, 2, 3, ... The probability p_1 that event A will occur in the first trial is

$$p_1 = \mathbf{P}(A) = p$$

The probability p_2 that the event will not occur in the first trial but will occur in the second is

$$p_2 = \mathbf{P}(\bar{A} \text{ and } A) = (1-p)\,p$$

The probability p_3 that A will not occur in the first or the second trial but will occur in the third is

$$p_3 = \mathbf{P}(\bar{A} \text{ and } \bar{A} \text{ and } A) = (1-p)(1-p)\,p = (1-p)^2\,p$$

and so on,

$$p_k = (1-p)^{k-1}\,p \tag{2}$$

Here is the table of distribution of probabilities:

x	1	2	3	...	...	k	...
p_k	p	$(1-p)\,p$	$(1-p)^2\,p$	...	...	$(1-p)^{k-1}\,p$	...

We also have

$$\sum_{k=1}^{\infty} p_k = \sum_{k=1}^{\infty} (1-p)^{k-1}\,p = \frac{p}{1-(1-p)} = 1$$

Problem on firing to a first hit. The foregoing problem finds applications, in particular in the theory of gunfire.

Suppose we have a case of gunfire to a first hit. The probability of a hit scored on each shot is p.

The random variable x is the number of the shot in which the first hit was scored. The distribution table of the probabilities of this random variable will be the same as in Example 2.

Example 3. The probability of a hit on each shot is $p=0.8$. We have three shells. Determine the probability that one, two and three shells will be used if the gunfire is continued to a first hit or to a miss by all three shells; compile a table of the distribution of the random variable x (the number of shells used).

Solution. Let x be a random variable indicating the number of shells fired; $\mathbf{P}(x=x_1)$ is the probability that x_1 shells will be used. Then $\mathbf{P}(x=1)=p=0.8$ is the probability of a hit scored on the first shot.

$$\mathbf{P}(x=2)=(1-p)\,p=(1-0.8)\cdot 0.8=0.16$$

is the probability that the first shot was a miss, the second a hit.

$$\mathbf{P}(x=3)=(1-p)^2=(1-0.8)\cdot(1-0.8)=0.2\cdot 0.2=0.04$$

because there are only three shells and gunfire ceases irrespective of whether the third shot is a hit or a miss. This last probability could also have been computed as a difference:

$$1-\mathbf{P}(x=1)-\mathbf{P}(x=2)=1-0.8-0.16=0.04$$

The distribution table will look like this:

x	1	2	3
$\mathbf{P}(x=x_k)$	0.8	0.16	0.04

Note. This problem can be stated in terms of the urn model, which means it can also be applied to other problems. This remark applies to certain other problems as well.

8.8 RELATIVE FREQUENCY AND THE PROBABILITY OF RELATIVE FREQUENCY IN REPEATED TRIALS

Suppose we have a sequence of n trials, in each of which event A can occur with probability p. Let x be a random variable denoting the relative frequency of occurrence of event A in the sequence consisting of n trials. It is required to determine the law of distribution of the random variable x in the n-trial sequence.

It is quite obvious that in n trials x will assume one of the following values:

$$\frac{0}{n},\quad \frac{1}{n},\quad \frac{2}{n},\ \ldots,\ \frac{n}{n}$$

Theorem 1. *The probability* $\mathbf{P}\left(x=\frac{m}{n}\right)$ *that the random variable* x *will assume the value* $\frac{m}{n}$, *that is, that in* n *trials event* A *will occur* m *times and the event* $\bar{A}$ *(nonoccurrence of* A*) will occur* $n-m$ *times is equal to* $C_n^m p^m q^{n-m}$, *where* C_n^m *is the number of combinations of* n *elements taken* m *at a time;* p *is the probability of the occurrence of event* A, $p=\mathbf{P}(A)$; q *is the probability of the nonoccurrence of event* A, *that is,* $q=1-p=\mathbf{P}(\bar{A})$.

Proof. Event A will occur m times in n trials, for instance, if the alternation of events A and $\bar{A}$ is as given below:

$$\underbrace{AA\ldots A}_{m}\underbrace{\bar{A}\bar{A}\ldots\bar{A}}_{n-m}$$

which is to say, A occurs in the first m trials, and does not occur in the next $n-m$ trials (event $\bar{A}$ occurs). Since

$$\mathbf{P}(A)=p,\quad \mathbf{P}(\bar{A})=1-p=q$$

then by the multiplication theorem the probability of such an alternation of events A and $\bar{A}$ will be

$$p^m\cdot q^{n-m}$$

But event A can appear m times in n trials if we have a different sequence of events $\bar{A}$ and A. For example, given the alternation $\underbrace{AA\ldots A}_{m-1}\underbrace{\bar{A}\bar{A}\ldots\bar{A}}_{n-m}\underbrace{A}_{1}$. But it is necessary that A occur m times, and $\bar{A}$, $n-m$ times. The probability of the occurrence of this kind of alternation of events A and $\bar{A}$ is

$$p^{m-1}q^{n-m}p=p^mq^{n-m}$$

How many distinct alternations of events A and $\bar{A}$ can there be in n trials in which A occurs m times? It is clearly equal to the number of combinations of n elements taken m at a time:

$$C_n^m=\frac{n(n-1)(n-2)\ldots[n-(m-1)]}{1\cdot2\cdot3\ldots m}$$

Thus, by the addition theorem, we have

$$\mathbf{P}\left(x=\frac{m}{n}\right)=\underbrace{p^mq^{n-m}+p^mq^{n-m}+\ldots+p^mq^{n-m}}_{C_n^m}$$

or

$$\mathbf{P}\left(x=\frac{m}{n}\right)=C_n^mp^mq^{n-m} \tag{1}$$

The proof of the theorem is complete.

Having proved the theorem, we have thus determined the law of distribution of the random variable x; we give the law in the form of a table:

x	$\frac{0}{n}$	$\frac{1}{n}$	$\frac{2}{n}$	$\ldots$	$\frac{m}{n}$	$\ldots$	$\ldots$	$\frac{n}{n}$
$\mathbf{P}\left(x=\frac{m}{n}\right)$	$1\cdot q^n$	$C_n^1pq^{n-1}$	$C_n^2p^2q^{n-2}$	$\ldots$	$C_n^mp^mq^{n-m}$	$\ldots$	$\ldots$	$1\cdot p^n$

This distribution law is known as the *binomial distribution* because the probabilities $\mathbf{P}\left(x=\frac{m}{n}\right)$ are equal to the corresponding terms in the binomial expansion of the expression $(q+p)^n$:

$$(q+p)^m=\sum_{m=0}^{n} C_n^m p^m q^{n-m} \tag{2}$$

As was to be expected, the union (logical sum) of the probabilities of all possible values of the random variable is equal to 1 since $(p+q)^n=1^n=1$.

Note. In the study of many problems it is often necessary to determine the probability that event A will occur "at least once", that is, the relative frequency of the event, $x\geqslant\frac{1}{n}$. It is quite obvious that this probability $\mathbf{P}\left(x\geqslant\frac{1}{n}\right)$ can be determined from the equation

$$\mathbf{P}\left(x\geqslant\frac{1}{n}\right)=1-\mathbf{P}\left(x=\frac{0}{n}\right)=1-q^n \tag{3}$$

From the distribution table it also follows that the probability $\mathbf{P}\left(x\geqslant\frac{k}{n}\right)$ that the event will occur no less than k times is found from the formula

$$\mathbf{P}\left(x\geqslant\frac{k}{n}\right)=\sum_{m=k}^{n} C_n^m p^m q^{n-m} \tag{4}$$

or

$$\mathbf{P}\left(x\geqslant\frac{k}{n}\right)=1-\sum_{m=0}^{k-1} C_n^m p^m q^{n-m}$$

Example 1. Give a graphical representation of the binomial distribution of the random variable x for $n=8$, $p=1/2$, $q=1/2$.

Solution. We determine all the values of the probabilities that appear in the table:

$$\mathbf{P}(x=0)=C_8^0 q^8=1\cdot\left(\frac{1}{2}\right)^8=\frac{1}{256}$$

$$\mathbf{P}\left(x=\frac{1}{8}\right)=C_8^1\frac{1}{2}\cdot\left(\frac{1}{2}\right)^7=\frac{8}{1}\cdot\frac{1}{256}=\frac{1}{32}$$

$$\mathbf{P}\left(x=\frac{2}{8}\right)=C_8^2\left(\frac{1}{2}\right)^8=\frac{8\cdot 7}{1\cdot 2}\cdot\frac{1}{2^8}=\frac{7}{64}$$

$$\mathbf{P}\left(x=\frac{3}{8}\right)=C_8^3\left(\frac{1}{2}\right)^8=\frac{8\cdot 7\cdot 6}{1\cdot 2\cdot 3}\frac{1}{2^8}=\frac{7}{32}$$

$$\mathbf{P}\left(x=\frac{4}{8}\right)=C_8^4\left(\frac{1}{2}\right)^8=\frac{8\cdot 7\cdot 6\cdot 5}{1\cdot 2\cdot 3\cdot 4}\cdot\frac{1}{2^8}=\frac{35}{128}$$

$$\mathbf{P}\left(x=\frac{5}{8}\right)=C_8^5\frac{1}{2^8}=\frac{7}{32}$$

$$\mathbf{P}\left(x=\frac{6}{8}\right)=C_8^6\frac{1}{2^8}=\frac{7}{64}$$

$$\mathbf{P}\left(x=\frac{7}{8}\right)=C_8^7\frac{1}{2^8}=\frac{1}{32}$$

$$\mathbf{P}\left(x=\frac{8}{8}\right)=C_8^8\frac{1}{2^8}=\frac{1}{256}$$

Let us construct the distribution polygon, or frequency polygon (Fig. 161).

Example 2. What is the probability that event A will occur twice (a) in two trials, (b) in three trials, (c) in 10 trials, if the probability of the occurrence of the event in each trial is equal to 0.4?

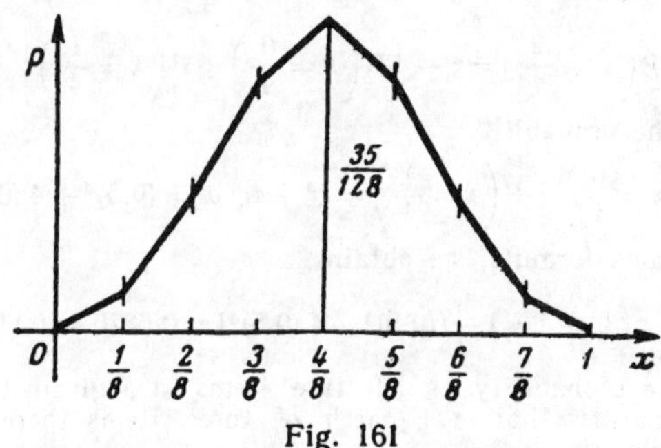

Fig. 161

Solution. (a) Here, $n=2$, $p=0.4$, $q=0.6$:

$$\mathbf{P}\left(x=\frac{2}{2}\right)=C_2^2p^2q^0=\frac{2\cdot 1}{1\cdot 2}(0.4)^2=0.16$$

(b) here, $n=3$, $p=0.4$, $q=0.6$:

$$\mathbf{P}\left(x=\frac{2}{3}\right)=C_3^2p^2q^1=\frac{3\cdot 2}{1\cdot 2}(0.4)^2\cdot 0.6=0.288$$

(c) here, $n=10$, $p=0.4$, $q=0.6$:

$$\mathbf{P}\left(x=\frac{2}{10}\right)=C_{10}^2p^2q^8=\frac{10\cdot 9}{1\cdot 2}(0.4)^2\cdot(0.6)^8=0.121$$

Example 3. Five independent shots are fired at a target. The probability of a hit in each shot is 0.2. Three hits suffice to destroy the target. Determine the probability of target destruction.

Solution. Here, $n=5$, $p=0.2$, $q=0.8$. It is clear that the probability of target destruction should be computed from the formula

$$p_{des}=\mathbf{P}\left(x=\frac{3}{5}\right)+\mathbf{P}\left(x=\frac{4}{5}\right)+\mathbf{P}\left(x=\frac{5}{5}\right)$$

or from the formula

$$p_{des}=1-\left[\mathbf{P}\left(x=\frac{0}{5}\right)+\mathbf{P}\left(x=\frac{1}{5}\right)+\mathbf{P}\left(x=\frac{2}{5}\right)\right]$$

By the first formula we have

$$p_{des}=C_5^3p^3q^2+C_5^4p^4q^1+C_5^5p^5=\frac{5\cdot4\cdot3}{1\cdot2\cdot3}\cdot(0.2)^3\cdot(0.8)^2$$
$$+\frac{5\cdot4\cdot3\cdot2}{1\cdot2\cdot3\cdot4}(0.2)^4\cdot0.8+1\cdot(0.2)^5=0.05792\approx0.06$$

Example 4. Four independent trials are carried out. The probability of the occurrence of event A in each trial is 0.5. Determine the probability that A will occur at least twice.

Solution. Here, $n=4$, $p=0.5$, $q=0.5$:

$$\mathbf{P}\left(x\geqslant\frac{2}{4}\right)=\mathbf{P}\left(x=\frac{2}{4}\right)+\mathbf{P}\left(x=\frac{3}{4}\right)+\mathbf{P}\left(x=\frac{4}{4}\right)$$

or

$$\mathbf{P}\left(x\geqslant\frac{2}{4}\right)=1-\left[\mathbf{P}\left(x=\frac{0}{4}\right)+\mathbf{P}\left(x=\frac{1}{4}\right)\right]$$

We compute the probability

$$\mathbf{P}\left(x<\frac{2}{4}\right)=\mathbf{P}\left(x=\frac{0}{4}\right)+\mathbf{P}\left(x=\frac{1}{4}\right)=q^4+4q^3p^1=(0.5)^4+4\,(0.5)^4=0.3125$$

Hence, by the second formula, we obtain

$$\mathbf{P}\left(x\geqslant\frac{2}{4}\right)=1-[(0.5)^4+4\,(0.5)^4]=0.6875\approx0.69$$

Example 5. The probability of defective items in a given batch is $p=0.1$. What is the probability that in a batch of three items there will be $m=0$, 1, 2, 3 defective items?

Solution.

$$\mathbf{P}\left(x=\frac{0}{3}\right)=C_3^0q^3=1\cdot0.9^3=0.729$$
$$\mathbf{P}\left(x=\frac{1}{3}\right)=C_3^1pq^2=\frac{3}{1}\cdot0.1\cdot0.9^2=0.243$$
$$\mathbf{P}\left(x=\frac{2}{3}\right)=C_3^2p^2q=\frac{3\cdot2}{1\cdot2}\cdot0.1^2\cdot0.9=0.027$$
$$\mathbf{P}\left(x=\frac{3}{3}\right)=C_3^3p^3=1\cdot0.1^3=0.001$$

8.9 THE MATHEMATICAL EXPECTATION OF A DISCRETE RANDOM VARIABLE

Suppose we have a discrete random variable x with an appropriate distribution law:

x	x_1	x_2	...	x_k	...	x_n
$\mathbf{P}(x=x_k)$	p_1	p_2	...	p_k	...	p_n

Definition 1. *The mathematical expectation* (or, simply, *expectation*) of a random variable x (we symbolize expectation by $\mathbf{M}\,[x]$

or m_x) is the sum of the products of all possible values of the random variable by the probabilities of these values:

$$M[x] = x_1p_1 + x_2p_2 + \ldots + x_np_n$$

or, briefly,

$$M[x] = \sum_{k=1}^{n} x_kp_k \tag{1}$$

Here, as has already been pointed out, $\sum_{k=1}^{n} p_k = 1$.

If the values of the random variable form an infinite sequence, then

$$m_x = \sum_{k=1}^{\infty} x_kp_k \tag{1'}$$

We will consider only those random variables for which the series converges.

We will now see what connection there is between the mathematical expectation of a random variable and the arithmetic mean of that variable for a large number of trials; namely, we will demonstrate that *in a large number of trials, the arithmetic mean of the observed values is close to the expectation*; or, in the terms of Sec. 8.1, we can say that *the arithmetic mean of the observed values of a random variable tends to the expectation when the number of trials increases without bound.*

Suppose that N independent experiments are performed and that

the value x_1 occurred n_1 times,
the value x_2 occurred n_2 times,
. . . .
the value x_ν occurred n_ν times.

The random variable x assumes the values $x_1, x_2, \ldots, x_\nu$.

We compute the arithmetic mean of the variables x (we denote it by $\overline{M}[x]$ or $\overline{m}_x$):

$$\overline{m}_x = \frac{x_1n_1 + x_2n_2 + \ldots + x_\nu n_\nu}{N} = x_1\frac{n_1}{N} + x_2\frac{n_2}{N} + \ldots + x_\nu\frac{n_\nu}{N} \tag{2}$$

However, since in the case of a large number of trials N the relative frequency $\frac{n_k}{N}$ tends to the probability of the occurrence of the value x_k, it follows that

$$\sum_{k=1}^{\nu} x_k\frac{n_k}{N} \approx \sum_{k=1}^{\nu} x_kp_k$$

Under rather natural assumptions we have

$$\overline{\mathbf{M}}[x] \underset{n\to\infty}{\longrightarrow} \mathbf{M}[x] \qquad (3)$$

Note 1. If we considered an urn scheme with N balls where there are n_1 balls labelled x_1, n_2 balls labelled x_2 and so on, the expected value on drawing one ball will be expressed by formula (2), that is, it will be equal to $\overline{m}_x$.

Example 1. Determine the expectation of the random variable x of the number of hits in three shots if the probability of a hit in each shot is $p=0.4$.

Solution. The random variable x can take on the values

$$x_1=0, \quad x_2=1, \quad x_3=2, \quad x_4=3$$

We set up a distribution table of the given random variable.

The probabilities of these values are found by the theorem on repeated trials ($n=3$, $p=0.4$, $q=0.6$):

$$\mathbf{P}(x=0)=C_3^0\,(0.6)^3=0.216$$
$$\mathbf{P}(x=1)=C_3^1\,(0.4)\,(0.6)^2=0.432$$
$$\mathbf{P}(x=2)=C_3^2\,(0.4)^2\cdot(0.6)=0.288$$
$$\mathbf{P}(x=3)=C_3^3\,(0.4)^3=0.064$$

The distribution table of the random variable is

x	0	1	2	3
$\mathbf{P}(x=x_k)$	0.216	0.432	0.228	0.064

We calculate the expectation from formula (1):

$$m_x=0\cdot 0.216+1\cdot 0.432+2\cdot 0.288+3\cdot 0.064=1.2 \text{ hits}$$

Example 2. One shot is fired at a target. The probability of a hit is p. Determine the expectation of the random variable x of the number of hits.

Form the distribution table of the random variable

x	0	1
p_k	$1-p$	p

Consequently, $m_x=0\cdot(1-p)+1\cdot p=p$.

Note 2. In the sequel it will be established that the expectation $\mathbf{M}[x]$ of the number of occurrences of event A in n independent trials is equal to the product of the number of trials by the probability p of occurrence of A in each trial:

$$\mathbf{M}[x]=np \qquad (4)$$

If in (4), n is the number of shots and p is the probability of a hit, then the solution of the problem of Example 1 will be:

$$\mathbf{M}[x] = np = 3 \cdot 0.4 = 1.2 \text{ hits}$$

If in formula (4) we know $\mathbf{M}[x]$ and p, then we can find n—the number of trials that yields the given expectation of the number of the occurrences of the event:

$$n = \frac{\mathbf{M}[x]}{p}$$

Example 3. The probability of scoring a hit in one shot is $p = 0.2$. Determine the shell consumption that will ensure expectation 5 of the number of hits:

$$n = \frac{5}{0.2} = 25 \text{ shells}$$

(Note once again that similar problems occur in a variety of investigations. The word "hit" is replaced by "occurrence of an event" and "shot" is replaced by "trial".)

Example 4. Determine the expectation of the random variable x with the following distribution table:

x	1	2	3	...	k	...
p_k	p	$(1-p)\,p$	$(1-p)^2\,p$	...	$(1-p)^{k-1}\,p$	...

See Example 2, Sec. 8.7

Solution. From formula (1) we have (denoting $1-p=q$)

$$\begin{aligned} m_x &= 1 \cdot p + 2qp + 3q^2p + \ldots + kq^{k-1}p + \ldots \\ &= p\,(1 + 2q + 3q^2 + \ldots + kq^{k-1} + \ldots) \\ &= p\,(q + q^2 + q^3 + \ldots + q^k + \ldots)' = p\left(\frac{q}{1-q}\right)' \\ &= p \cdot \frac{1-q+q}{(1-q)^2} = \frac{p}{(1-q)^2} = \frac{p}{p^2} = \frac{1}{p} \end{aligned}$$

Thus

$$m_x = \frac{1}{p}$$

Note that

$$m_x \to 1 \quad \text{as} \quad p \to 1$$
$$m_x \to \infty \quad \text{as} \quad p \to 0$$

These relations can be explained by the essence of the problem. Indeed, if the probability of occurrence of event A in each trial is close to 1 ($p \approx 1$), then it may be expected that A will occur in the first trial ($m_x \approx 1$). But if the probability p is small ($p \approx 0$), then one can expect that in order for event A to occur it will be necessary to perform a large number of trials ($m_x \approx \infty$).

The mathematical expectation of a random variable x is called the *centre of probability distribution of the random variable* x.

Note 3. The term "centre of probability distribution" is introduced by analogy with the term "centre of gravity". If on the x-axis we place masses $p_1, p_2, \ldots, p_n$ at points with abscissas $x_1, x_2, \ldots, x_n$, then from analytic geometry it is clear that the abscissa of the centre of gravity of these masses is determined from the formula

$$x_C = \frac{\sum_{k=1}^{n} x_k p_k}{\sum_{k=1}^{n} p_k}$$

If $\sum_{k=1}^{n} p_k = 1$, then

$$x_C = \sum_{k=1}^{n} x_k p_k \tag{5}$$

Formula (5) coincides (in appearance) with formula (1) for mathematical expectation.

We have thus established that the centre of gravity of a series of masses and the expectation are computed from analogous formulas, whence the term "centre of probability distribution".

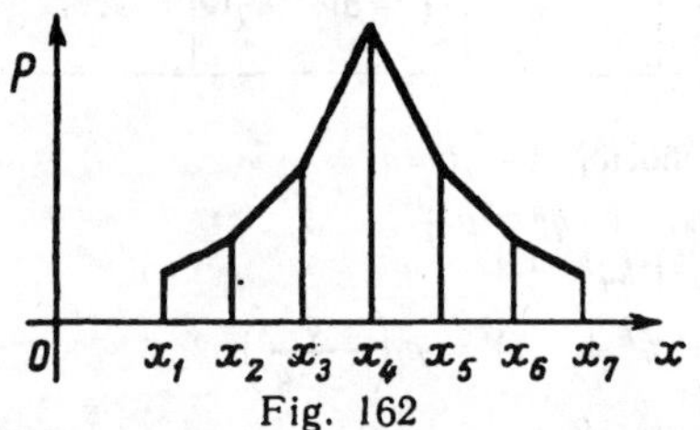

Fig. 162

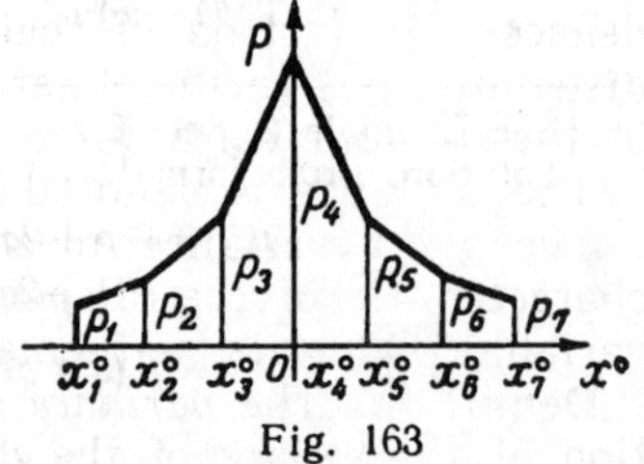

Fig. 163

Suppose we have a random variable x with an associated distribution (Fig. 162); let its expectation be m_x. Now consider the difference between the random variable x and its expectation: $x - m_x$.

We will call this quantity the *centred random variable*, or the *deviation*, and we will denote it by x^0.

It is quite clear that the distribution law of this random variable x^0 will be

x^0	$x_1^0 = x_1 - m_x$	$x_2^0 = x_2 - m_x$	...	$x_k^0 = x_k - m_x$
p_k	p_1	p_2	...	p_k

(see Fig. 163).

Let us find the expectation of a centred random variable:

$$\mathbf{M}[x-m_x]=\sum_{k=1}^{n}(x_k-m_x)\,p_k=\sum_{k=1}^{n}x_kp_k-\sum_{k=1}^{n}m_xp_k$$

$$=m_x-m_x\sum_{k=1}^{n}p_k=m_x-m_x\cdot 1=0$$

Thus, *the expectation of a centred random variable is equal to zero.*

Note 4. It is sometimes advisable to regard a nonrandom constant (a certain quantity) c as a random variable which assumes the value c with probability 1 and assumes all other values with probability 0.

Then it is meaningful to speak of the expectation of a constant:

$$\mathbf{M}[c]=c\cdot 1=c \tag{6}$$

The expectation of a constant is equal to that constant.

8.10 VARIANCE. ROOT-MEAN-SQUARE (STANDARD) DEVIATION. MOMENTS

In addition to the expectation of a random variable x, which defines the position of the centre of a probability distribution, a distribution is further characterized quantitatively by the *variance* of the random variable x.

The variance is denoted by $\mathbf{D}[x]$ or σ_x^2.

The word variance means dispersion. Variance is a numerical characteristic of the dispersion, or spread of values, of a random variable about its mathematical expectation.

Definition. The *variance* of a random variable x is the expectation of the square of the difference between x and its expectation (that is, the expectation of the square of the appropriate centred random variable):

$$\mathbf{D}[x]=\mathbf{M}[(x-m_x)^2] \tag{1}$$

or

$$\mathbf{D}[x]=\sum_{k=1}^{n}(x_k-m_x)^2\,p_k \tag{2}$$

Variance has the dimensions of the square of the random variable. It is sometimes more convenient in describing dispersion to make use of a quantity whose dimensions coincide with those of the random variable. This quantity is termed the *root-mean-square deviation (standard deviation)*.

$$\sigma[x]=\sqrt{\mathbf{D}[x]}$$

or, in expanded form,

$$\sigma[x]=\sqrt{\sum_{k=1}^{n}(x_k-m_x)^2 p_k} \tag{3}$$

The standard deviation is also denoted by σ_x.

Note 1. In computing variance, it is often convenient to transform formula (1) as follows:

$$\mathbf{D}[x]=\sum_{k=1}^{n}(x_k-m_x)^2 p_k=\sum_{k=1}^{n}x_k^2 p_k-2\sum_{k=1}^{n}x_k m_x p_k+\sum_{k=1}^{n}m_x^2 p_k$$

$$=\sum_{k=1}^{n}x_k^2 p_k-2m_x\sum_{k=1}^{n}x_k p_k+m_x^2\sum_{k=1}^{n}p_k$$

$$=\mathbf{M}[x^2]-2m_x\cdot m_x+m_x^2\cdot 1=\mathbf{M}[x^2]-m_x^2$$

Thus

$$\mathbf{D}[x]=\mathbf{M}[x^2]-m_x^2 \tag{4}$$

That is, the variance is equal to the difference between the expectation of the square of the random variable and the square of the expectation of the random variable.

Example 1. One shot is fired at a target. The probability of a hit is p. Determine the expectation, the variance and the standard deviation.

Solution. We construct a table of the values of the number of hits:

x	1	0
p_k	p	q

$q=1-p$. Hence,

$$\left.\begin{aligned}&\mathbf{M}[x]=1\cdot p+0\cdot q=p\\&\mathbf{D}(x)=(1-p)^2p+(0-p)^2q=q^2p+p^2q=pq\\&\sigma(x)=\sqrt{pq}\end{aligned}\right\} \tag{5}$$

Let us consider some more examples in order to get a good idea of what the concepts of variance and standard deviation mean as measures of dispersion of a random variable.

Example 2. A random variable x is specified by the following distribution law (see table and Fig. 164).

x	2	3	4
p_k	0.3	0.4	0.3

Determine the (1) expectation, (2) variance, (3) standard deviation.

Solution.

1. $\mathbf{M}[x]=2\cdot 0.3+3\cdot 0.4+4\cdot 0.3=3$,
2. $\mathbf{D}[x]=(2-3)^2\cdot 0.3+(3-3)^2\cdot 0.4+(4-3)^2\cdot 0.3=0.6$,
3. $\sigma[x]=\sqrt{\mathbf{D}[x]}=\sqrt{0.6}=0.77$.

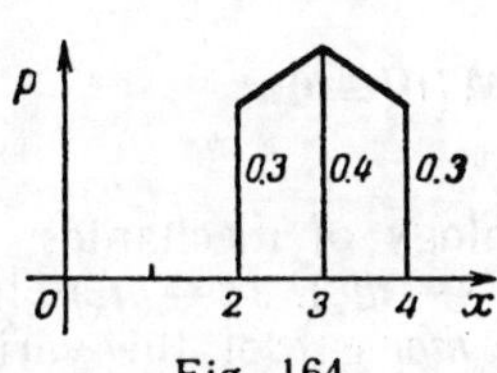

Fig. 164

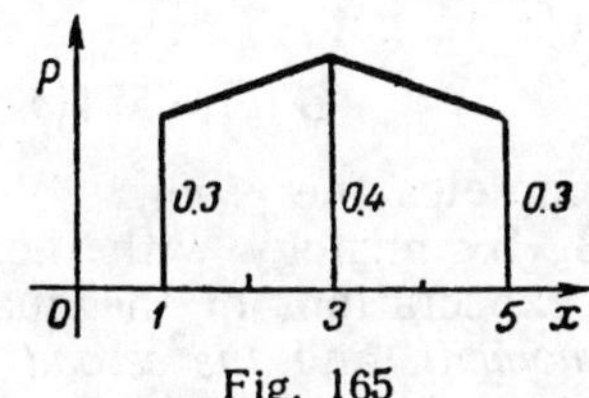

Fig. 165

Example 3. A random variable x has the distribution law (see table and Fig. 165):

x	1	3	5
p_k	0.3	0.4	0.3

Determine the (1) expectation, (2) variance, (3) standard deviation.

Solution.

1. $\mathbf{M}[x]=1\cdot 0.3+3\cdot 0.4+5\cdot 0.3=3$,
2. $\mathbf{D}[x]=(1-3)^2\cdot 0.3+(3-3)^2\cdot 0.4+(5-3)^2\cdot 0.3=2.4$,
3. $\sigma[x]=\sqrt{2.4}=1.55$.

The dispersion of the random variable in the first example is less than that in the second example (see Figs. 164, 165). The variances of these quantities are equal to 0.6 and 2.4, respectively.

Example 4. A random variable x has the following distribution law (see table and Fig. 166):

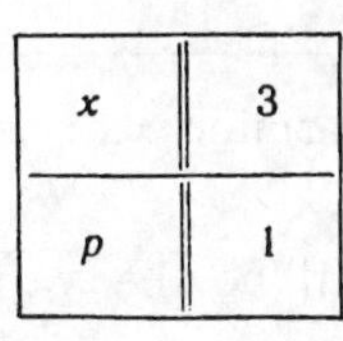

x	3
p	1

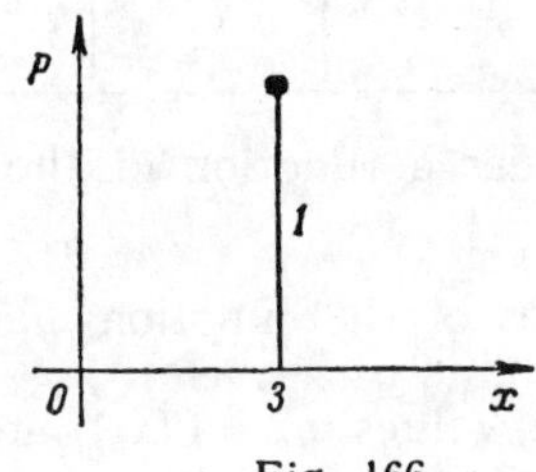

Fig. 166

Determine the (1) expectation, (2) variance, (3) standard deviation.

Solution.

1. $\mathbf{M}[x]=3\cdot 1=3$,
2. $\mathbf{D}[x]=(3-3)^2\cdot 1=0$,
3. $\sigma[x]=0$.

There is no dispersion of the random variable in this case.

Note 2. If we regard a constant as a random variable which assumes a value c with probability 1, then it is easy to demonstrate that $\mathbf{D}[c]=0$.

Proof. It has been shown [see (6), Sec. 8.9] that $\mathbf{M}[c]=c$. From formula (1) we get

$$\mathbf{D}[c]=\mathbf{M}[(c-c)^2]=\mathbf{M}[0]=0$$

which completes the proof.

Note 3. By analogy with the terminology of mechanics, we can call the expectations of the quantities $(x-m_x)$, $(x-m_x)^2$ the *first central moment* and *the second central moment* of the variable x. A third central moment is also considered:

$$\sum_{k=1}^{n}(x_k-m_x)^3 p_k$$

If a random variable is distributed symmetrically about the centre of the probability distribution (Fig. 162), it is then obvious that its third central moment will be equal to zero. If the third moment is different from zero, the random variable cannot be distributed symmetrically.

8.11 FUNCTIONS OF RANDOM VARIABLES

Let a random variable x be given by the following distribution law:

x	x_1	x_2	...	x_k	...	x_n
p_k	p_1	p_2	...	p_k	...	p_n

We consider a function of the random variable x:

$$y=f(x).$$

The values of the function $y_k=f(x_k)$ will be the values of the random variable y.

If all the values $y_k=f(x_k)$ are distinct, then the distribution law of the random variable y is given by the table

$y=f(x)$	$y_1=f(x_1)$	$y_2=f(x_2)$	...	$y_k=f(x_k)$	...	$y_n=f(x_n)$
p_k	p_1	p_2	...	p_k	...	p_n

If some of the values $y_k = f(x_k)$ are equal, the appropriate columns may be combined into one by combining the corresponding probabilities.

The expectation of the function $y = f(x)$ of the random variable x will be found from a formula similar to formula (1) of Sec. 8.9:

$$\mathbf{M}[f(x)] = \sum_{k=1}^{n} f(x_k)\, p_k \tag{1}$$

In similar fashion we determine the variance of the function:

$$\mathbf{D}[f(x)] = \mathbf{M}[(f(x) - \mathbf{M}[f(x)])^2] = \sum_{k=1}^{n} (f(x_k) - m_{f(x)})^2\, p_k$$

Example. A random variable φ is given by the following distribution law:

φ	$-\frac{\pi}{2}$	$-\frac{\pi}{4}$	0	$\frac{\pi}{4}$	$\frac{\pi}{2}$
p_k	0.1	0.1	0.2	0.3	0.3

We consider a function of this random variable:

$$y = A \sin \varphi$$

We form the distribution table for the random variable y:

y	$-A$	$-\frac{A\sqrt{2}}{2}$	0	$\frac{A\sqrt{2}}{2}$	A
p_k	0.1	0.1	0.2	0.3	0.3

Let us find the expectation of the function:

$$\mathbf{M}[A \sin \varphi] = -A \cdot 0.1 - \frac{A\sqrt{2}}{2} 0.1 + 0 \cdot 0.2 + \frac{A\sqrt{2}}{2} \cdot 0.3 + A \cdot 0.3$$

$$= A\left(0.2 + \frac{\sqrt{2}}{2} 0.2\right) = A\,(0.2 + 0.14) = 0.34A$$

Problems of this kind occur in the consideration of oscillatory processes.

8.12 CONTINUOUS RANDOM VARIABLE. PROBABILITY DENSITY FUNCTION OF A CONTINUOUS RANDOM VARIABLE. THE PROBABILITY OF THE RANDOM VARIABLE FALLING IN A SPECIFIED INTERVAL.

An example will help to cast some light on the problem at hand.

Example. The amount of wear of a cylinder is measured after a certain period of operation. This quantity is determined by the value of the increase in diameter of the cylinder. We denote it by $\bar{x}$. From the essence of the problem, it follows that $\bar{x}$ can assume any value in a certain interval (a, b) of possible values.

This quantity is termed a *continuous random variable.*

We consider the continuous random variable $\bar{x}$ specified on a certain interval (a, b) which can also be an infinite interval, $(-\infty, +\infty)$. We divide this interval into subintervals of length $\Delta x_{i-1} = x_i - x_{i-1}$ by the arbitrary points $x_0, x_1, x_2, \ldots, x_n$.

Suppose we know the probability that the random variable $\bar{x}$ will fall in the interval (x_{i-1}, x_i). We denote this probability by $\mathbf{P}(x_{i-1} < \bar{x} < x_i)$ and represent it as the area of a rectangle with base Δx_i (Fig. 167).

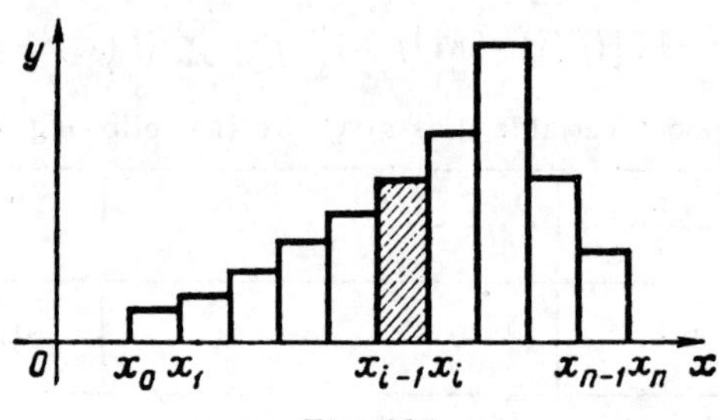

Fig. 167

For each interval (x_{i-1}, x_i) we determine the probability of $\bar{x}$ falling in this interval and, hence, we can construct an appropriate rectangle. The result is a step-like broken line.

Definition 1. If there exists a function $y = f(x)$ such that

$$\lim_{\Delta x \to 0} \frac{\mathbf{P}(x < \bar{x} < x + \Delta x)}{\Delta x} = f(x) \tag{1}$$

then this function $f(x)$ is termed the *probability density function* (or, simply, *density function*) *of the random variable* x, or the distribution. (It is also called the *frequency function*, *distribution density*, or the *probability density*.) We will use $\bar{x}$ to denote the continuous random variable, x or x_k to denote the values of this random variable. If it is clear from the context, we will occasionally drop the bar on x. The curve $y = f(x)$ is called the *probability curve* or the *distribution curve* (Fig. 168). Using the definition of limit, from equation (1) follows, to within infinitesimals of higher order than Δx, the approximate equation

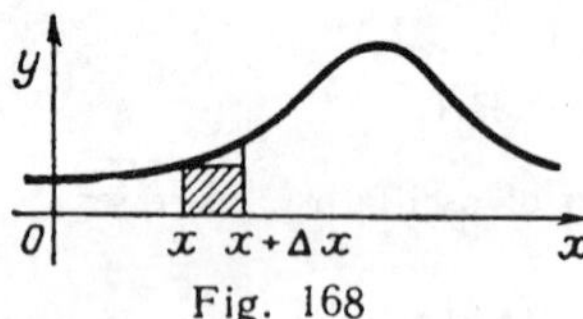

Fig. 168

$$\mathbf{P}(x < \bar{x} < x + \Delta x) \approx f(x)\,\Delta x \tag{2}$$

We will now prove the following theorem.

Theorem 1. *Let $f(x)$ be the density function of the random variable $\bar{x}$. Then the probability that a value of the random variable $\bar{x}$ will fall in some interval (α, β) is equal to the definite integral of the function $f(x)$ from α to β; that is, we have the following equation:*

$$\mathbf{P}(\alpha < \bar{x} < \beta) = \int_{\alpha}^{\beta} f(x)\,dx \tag{3}$$

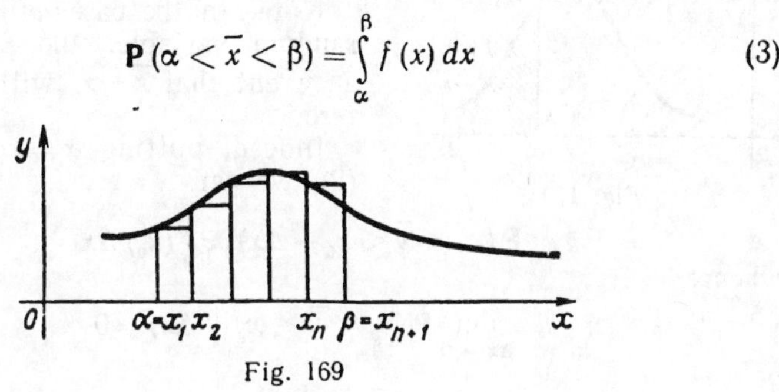

Fig. 169

Proof. Partition the interval (α, β) into n subintervals by the points $\alpha = x_1, x_2, \ldots, x_{n+1} = \beta$ (Fig. 169). Apply formula (2) to each subinterval:

$$\mathbf{P}(x_1 < \bar{x} < x_2) \approx f(x_1)\,\Delta x_1$$
$$\mathbf{P}(x_2 < \bar{x} < x_3) \approx f(x_2)\,\Delta x_2$$
$$\cdots\cdots\cdots\cdots\cdots$$
$$\mathbf{P}(x_n < \bar{x} < x_{n+1}) \approx f(x_n)\,\Delta x_n$$

Adding the left and right members of the equations, we will clearly get $\mathbf{P}(\alpha < \bar{x} < \beta)$ on the left. We thus have the approximate equation

$$\mathbf{P}(\alpha < \bar{x} < \beta) \approx \sum_{i=1}^{n} f(x_i)\,\Delta x_i$$

Passing to the limit in the right member as $\max \Delta x_i \to 0$, we get, on the basis of the properties of integral sums, the exact equation

$$\mathbf{P}(\alpha < \bar{x} < \beta) = \lim_{\max \Delta x_i \to 0} \sum_{i=1}^{n} f(x_i)\,\Delta x_i$$

(We assume that $f(x)$ is such that the limit on the right exists.) The limit on the right is the definite integral of the function $f(x)$ from α to β. Thus

$$\mathbf{P}(\alpha < \bar{x} < \beta) = \int_{\alpha}^{\beta} f(x)\,dx$$

which completes the proof.

Thus, knowing the probability density function of a random variable, we can determine the probability that a value of the random variable will lie in a given interval. Geometrically, this probability is equal to the area of the resulting curvilinear trapezoid (Fig. 170).

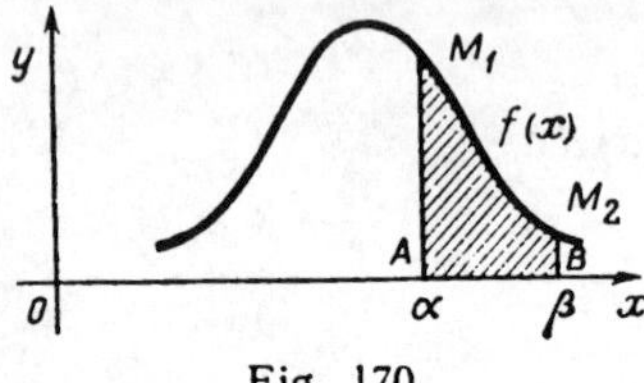

Fig. 170

Note. In the case of a continuous random variable, the probability of an event that $\bar{x}=x_0$ will be equal to zero.

Indeed, putting $x=x_0$ in equation (2), we get

$$\mathbf{P}(x_0<\bar{x}<x_0+\Delta x)\approx f(x_0)\,\Delta x$$

whence

$$\lim_{\Delta x\to 0}\mathbf{P}(x_0<\bar{x}<x_0+\Delta x)=0$$

or

$$\mathbf{P}(\bar{x}=x_0)=0$$

(Also see Note 1 on page 448.) For this reason, in (3) and the preceding equations we can write not only $\mathbf{P}(\alpha<\bar{x}<\beta)$ but also $\mathbf{P}(\alpha\leqslant\bar{x}\leqslant\beta)$, since

$$\mathbf{P}(\alpha\leqslant\bar{x}\leqslant\beta)=\mathbf{P}(\bar{x}=\alpha)+\mathbf{P}(\alpha<\bar{x}<\beta)+\mathbf{P}(\bar{x}=\beta)=\mathbf{P}(\alpha<\bar{x}<\beta)$$

If all possible values of the random variable $\bar{x}$ lie in the interval (a, b), then

$$\int_a^b f(x)\,dx=1 \tag{4}$$

since we are certain that a value of the random variable will lie in the interval (a, b).

If the interval of possible values is $(-\infty, +\infty)$, then

$$\int_{-\infty}^{\infty} f(x)\,dx=1 \tag{5}$$

Note that if it follows from the essence of the problem that the function $f(x)$ is defined on the finite interval (a, b), then we can regard the function to be defined on the whole infinite interval $(-\infty, +\infty)$, but

$$f(x)=0$$

outside the interval (a, b). In this case, both (4) and (5) hold true. The density function of the random variable fully determines the random variable.

8.13 THE DISTRIBUTION FUNCTION. LAW OF UNIFORM DISTRIBUTION

Definition 1. Let $f(x)$ be the density function of some random variable $\bar{x}$ $(-\infty < x < +\infty)$; then the function

$$F(x) = \int_{-\infty}^{x} f(x)\,dx \tag{1}$$

is called the *distribution function*.

For a discrete random variable, the distribution function is equal to the sum of the probabilities of those values x_k which are less than x:

$$F(x) = \sum_{x_k < x} p_k \, .$$

From equation (3), Sec. 8.12, it follows that the distribution function $F(x)$ is the probability that the random variable $\bar{x}$ will assume a value less than x (Fig. 172):

$$F(x) = \mathbf{P}(-\infty < \bar{x} < x) \tag{2}$$

It follows from Fig. 171 that for a given value of x, the value of the distribution function is numerically equal to the area bounded by the distribution curve lying to the left of the ordinate drawn through the point x.

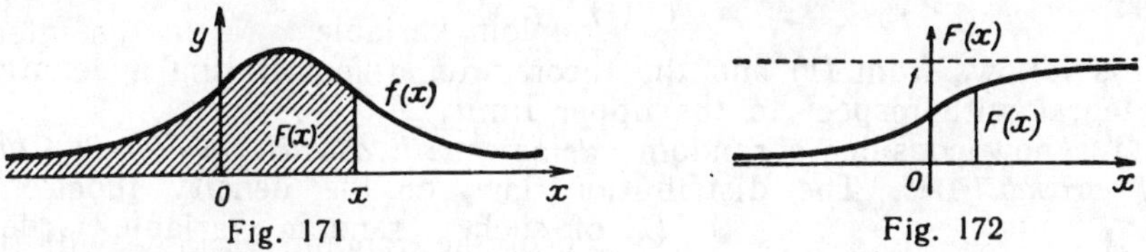

Fig. 171 Fig. 172

The graph of the function $F(x)$ is termed the *probability distribution curve* (Fig. 172).

Passing to the limit as $x \to +\infty$ in (1), with regard for (5), Sec. 8.12, we get

$$\lim_{x \to +\infty} F(x) = \lim_{x \to +\infty} \int_{-\infty}^{x} f(x)\,dx = \int_{-\infty}^{\infty} f(x)\,dx = 1$$

Now let us prove the following theorem.

Theorem 1. *The probability of a random variable $\bar{x}$ lying in a given interval* (α, β) *is equal to the increment in the distribution function over that interval:*

$$\mathbf{P}(\alpha < \bar{x} < \beta) = F(\beta) - F(\alpha)$$

Proof. Let us express the probability of $\overline{x}$ to assume a value lying in the given interval (α, β). We rewrite formula (3), Sec. 8.12, as

$$\mathbf{P}(\alpha < \overline{x} < \beta) = \int_{\alpha}^{\beta} f(x)\,dx = \int_{-\infty}^{\beta} f(x)\,dx - \int_{-\infty}^{\alpha} f(x)\,dx$$

(see Fig. 173). Or, using (1), we can write

$$\mathbf{P}(\alpha < \overline{x} < \beta) = F(\beta) - F(\alpha)$$

which completes the proof (see Fig. 174).

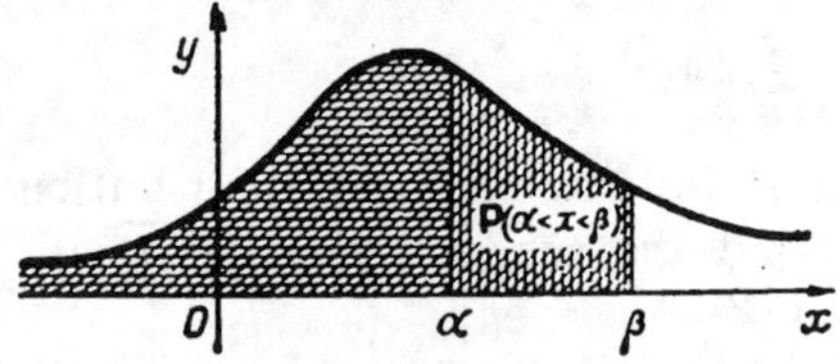

Fig. 173

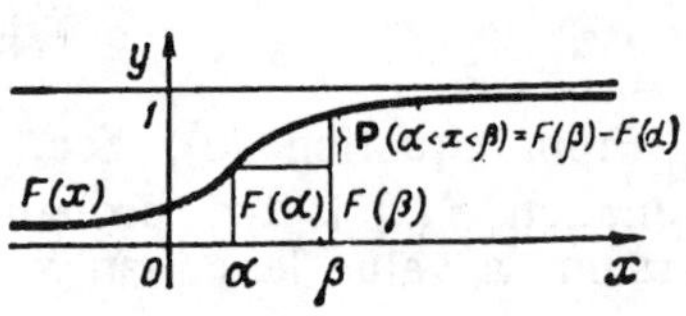

Fig. 174

Note that the density function $f(x)$ and the corresponding distribution function $F(x)$ are connected by the relation

$$F'(x) = f(x) \tag{3}$$

This follows from (1) and the theorem on differentiating a definite integral with respect to the upper limit.

We now consider a random variable *with a uniform distribution of probabilities*. The distribution law, or the density function, $f(x)$ of such a random variable is defined as follows:

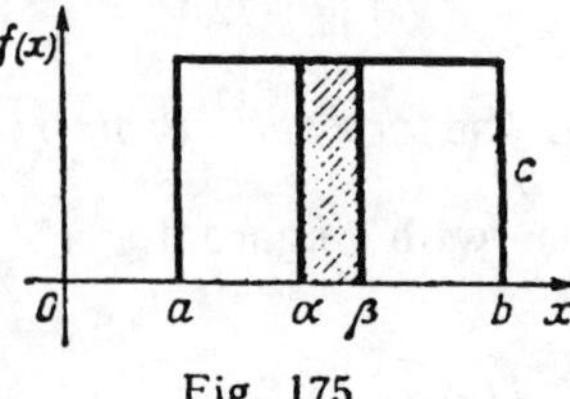

Fig. 175

$$f(x) = 0 \quad \text{for } x < a$$
$$f(x) = c \quad \text{for } a < x < b$$
$$f(x) = 0 \quad \text{for } b < x$$

On the interval (a, b) the density function $f(x)$ has a constant value c (Fig. 175); outside this interval it is zero. Such a distribution is also called the *law of uniform distribution*.

From the condition $\int_{-\infty}^{\infty} f(x)\,dx = 1$ we find the value of c:

$$\int_{-\infty}^{\infty} f(x)\,dx = \int_{a}^{b} c\,dx = c(b-a) = 1$$

hence

$$c=\frac{1}{b-a}, \quad b-a=\frac{1}{c}$$

From the last equation it follows that the interval (a, b) of uniform distribution is necessarily finite. Let us determine the probability that the random variable $\bar{x}$ will assume a value lying in the interval (α, β):

$$\mathbf{P}(\alpha<\bar{x}<\beta)=\int_{\alpha}^{\beta} f(x)\,dx=\int_{\alpha}^{\beta}\frac{1}{b-a}\cdot dx=\frac{\beta-\alpha}{b-a}$$

Thus, the desired probability is

$$\mathbf{P}(\alpha<\bar{x}<\beta)=\frac{\beta-\alpha}{b-a}$$

(this relation is similar to the definition of geometrical probability for the two-dimensional case given on page 452).

We now determine the distribution function

$$F(x)=\int_{-\infty}^{x} f(x)\,dx$$

If $x<a$, then $f(x)=0$, and, hence,

$$F(x)=0$$

If $a<x<b$, then $f(x)=\frac{1}{b-a}$ and, hence,

$$F(x)=\int_{a}^{x}\frac{1}{b-a}\,dx=\frac{x-a}{b-a}$$

If $b<x$, then

$$f(x)=0, \quad \int_{b}^{\infty} f(x)\,dx=0$$

hence

$$F(x)=\int_{-\infty}^{x} f(x)\,dx=\int_{a}^{b}\frac{1}{b-a}\,dx=\frac{b-a}{b-a}=1$$

(see Fig. 176).

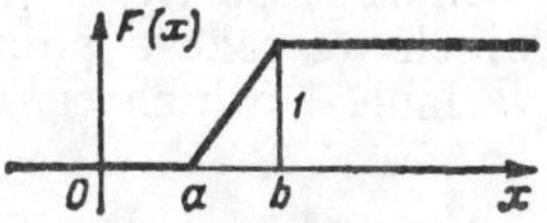

Fig. 176

The following are examples of random variables that obey the law of uniform distribution.

Example 1. The measurements of a certain quantity are rounded to the nearest scale division. The rounding errors are a random variable with uniform distribution of probabilities. If $2l$ is the number of certain units in one scale division, then the density function of this random variable is

$$f(x)=\begin{cases} 0 & \text{for} \quad x<-l \\ \dfrac{1}{2l} & \text{for} \quad -l<x<l \\ 0 & \text{for} \quad l<x \end{cases}$$

Here, $a=-l$, $b=l$, $c=\frac{1}{2l}$.

Example 2. A symmetric wheel in rotation comes to a stop due to friction. When the wheel comes to rest, the angle θ formed by a certain definite moving radius of the wheel with a fixed radius is a random variable with a density function

$$f(\theta)=\begin{cases} 0 & \text{for} \quad \theta<0 \\ \dfrac{1}{2\pi} & \text{for} \quad 0<\theta<2\pi \\ 0 & \text{for} \quad 2\pi<\theta \end{cases}$$

8.14 NUMERICAL CHARACTERISTICS OF A CONTINUOUS RANDOM VARIABLE

In the same manner that we considered a discrete (discontinuous) random variable, let us now examine the numerical characteristics of a continuous random variable $\bar{x}$ with density function $f(x)$.

Definition 1. The *mathematical expectation* of a continuous random variable x with density function $f(x)$ is the expression

$$M[\bar{x}]=\int_{-\infty}^{\infty} xf(x)\,dx \tag{1}$$

If a random variable x can assume values only on a finite interval $[a, b]$, then the expectation $M[\bar{x}]$ is expressed by the formula

$$M[\bar{x}]=\int_{a}^{b} xf(x)\,dx \tag{1'}$$

Formula (1′) may be regarded as a generalization of formula (1), Sec. 8.9.

Indeed, partition $[a, b]$ into subintervals (x_{k-1}, x_k), in each of which we take a point ξ_k. Consider the auxiliary *discrete* random variable ξ which can take on the values

$$\xi_1, \xi_2, \ldots, \xi_k, \ldots, \xi_n$$

Let the probabilities of the corresponding values of the discrete random variable be $p_1, p_2, \ldots, p_k, \ldots, p_n$:

$$p_1=f(\xi_1)\,\Delta x_1, \ p_2=f(\xi_2)\,\Delta x_2, \ \ldots,$$
$$p_k=f(\xi_k)\,\Delta x_k, \ \ldots, \ p_n=f(\xi_n)\,\Delta x_n \ *$$

* At the same time, $f(\xi_k)\,\Delta x_k$ is the probability that the continuous random variable x will assume a value in the interval (x_{k-1}, x_k).

The expectation of the given discrete variable ξ will be

$$\mathbf{M}[\xi] = \sum_{k=1}^{n} \xi_k p_k$$

or

$$\mathbf{M}[\xi] = \xi_1 f(\xi_1)\,\Delta x_1 + \xi_2 f(\xi_2)\,\Delta x_2 + \ldots + \xi_k f(\xi_k)\,\Delta x_k + \ldots$$
$$+ \xi_n f(\xi_n)\,\Delta x_n = \sum_{k=1}^{n} \xi_k f(\xi_k)\,\Delta x_k$$

Passing to the limit as $\max \Delta x_k \to 0$, we get

$$\lim_{\max \Delta x_k \to 0} \sum_{k=1}^{n} \xi_k f(\xi_k)\,\Delta x_k = \int_a^b x f(x)\,dx$$

The expression on the right is the expectation of the *continuous* random variable x which can take on any value of x in the interval $[a, b]$. The reasoning is similar for an infinite interval, that

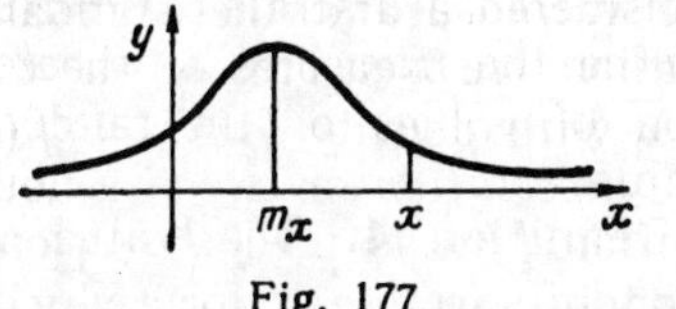

Fig. 177

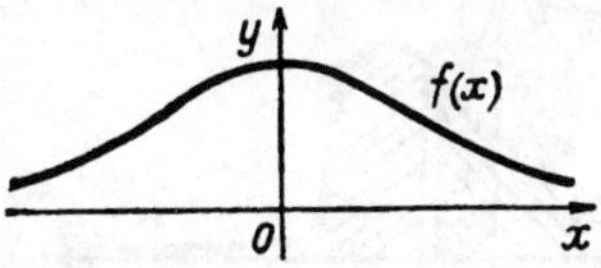

Fig. 178

is, for the expression (1). Formulas (1) and (1′) are similar to formula (1) of Sec. 8.9 for a discrete random variable. We use the symbol m_x for the expectation.

The expectation is called the *centre of probability distribution* of the random variable $\bar{x}$ (Fig. 177). If the distribution curve is symmetric about the y-axis, that is, if $f(x)$ is an even function, then clearly

$$\mathbf{M}[\bar{x}] = \int_{-\infty}^{\infty} x f(x)\,dx = 0$$

In this case the centre of the distribution will coincide with the coordinate origin (Fig. 178). Let us consider a centred random variable $\bar{x} - m_x$. We will find its expectation:

$$\mathbf{M}[\bar{x} - m_x] = \int_{-\infty}^{\infty} (x - m_x) f(x)\,dx = \int_{-\infty}^{\infty} x f(x)\,dx - m_x \int_{-\infty}^{\infty} f(x)\,dx$$
$$= m_x - m_x \cdot 1 = 0$$

The expectation of a centred random variable is zero.

Definition 2. The *variance* of a random variable $\bar{x}$ is the expectation of the square of the corresponding centred random variable:

$$\mathbf{D}[\bar{x}] = \int_{-\infty}^{\infty} (x-m_x)^2 f(x)\,dx \tag{2}$$

Formula (2) is similar to formula (2) of Sec. 8.10.

Definition 3. The *standard deviation* of a random variable $\bar{x}$ is the square root of the variance:

$$\sigma[\bar{x}] = \sqrt{\mathbf{D}[\bar{x}]} = \sqrt{\int_{-\infty}^{\infty} (x-m_x)^2 f(x)\,dx} \tag{3}$$

This formula is similar to formula (3) of Sec. 8.10. When we consider some illustrative examples, we will see that as in the case of a discrete random variable, the variance and the standard deviation are measures of the dispersion of values of the random variable.

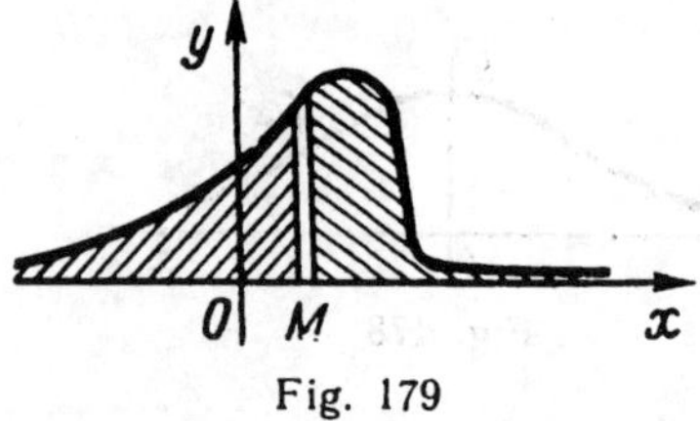

Fig. 179

Definition 4. The value of the random variable $\bar{x}$ for which the density function is a maximum is termed the *mode* (symbolized by M_0). For the random variable $\bar{x}$, the distribution curve of which is shown in Figs. 177 and 178, the mode coincides with the expectation.

Definition 5. A number (symbolized by M_e) is called the *median* if it satisfies the equation

$$\int_{-\infty}^{M_e} f(x)\,dx = \int_{M_e}^{\infty} f(x)\,dx = \frac{1}{2} \tag{4}$$

(Fig. 179). This equation may be rewritten as

$$\mathbf{P}(\bar{x} < M_e) = \mathbf{P}(M_e < \bar{x}) = \frac{1}{2}$$

This means that there is an equal probability that the random variable $\bar{x}$ will assume a value less than M_e and greater than M_e.

Note that the random variable $\bar{x}$ may not take on the value M_e at all.

8.15 NORMAL DISTRIBUTION. THE EXPECTATION OF A NORMAL DISTRIBUTION

Studies of various phenomena show that many random variables, such, for example, as measurement errors, the lateral deviation and range deviation of the point of impact from a certain centre in gunfire, and the amount of wear in machine parts, have a density function given by the formula

$$f(x) = \frac{1}{\sigma\sqrt{2\pi}} \exp\left[-\frac{(x-a)^2}{2\sigma^2}\right] \quad (1)$$

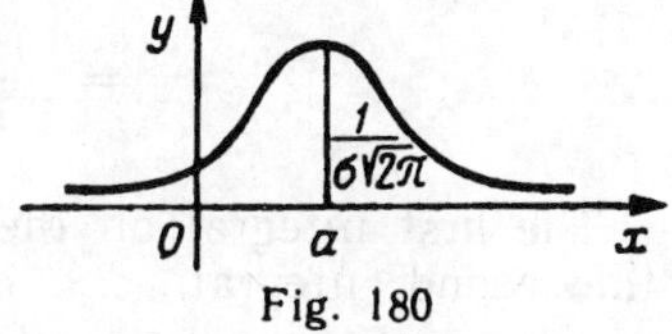

Fig. 180

We say the random variable has *normal distribution* or is normally distributed (the term Gaussian distribution is also used). The so-called normal curve (normal distribution curve) in shown Fig. 180. A table of the values of the function (1) for $a=0$, $\sigma=1$ is given in the Appendix, Table 2. A similar curve was investigated in detail in Sec. 5.9 of Vol. I.

First of all, we will show that the density function (1) satisfies the basic relation (5) of Sec. 8.12:

$$\int_{-\infty}^{\infty} f(x)\,dx = 1$$

Indeed, introducing the notation

$$\frac{x-a}{\sigma\sqrt{2}} = t, \qquad dx = \sigma\sqrt{2}\,dt$$

we can write

$$\int_{-\infty}^{\infty} \frac{1}{\sigma\sqrt{2\pi}} \exp\left[-\frac{(x-a)^2}{2\sigma^2}\right] dx = \frac{1}{\sqrt{\pi}} \int_{-\infty}^{\infty} e^{-t^2}\,dt = \frac{1}{\sqrt{\pi}}\sqrt{\pi} = 1$$

since (see Sec. 2.5, Example 2)

$$\int_{-\infty}^{\infty} e^{-t^2}\,dt = \sqrt{\pi}$$

We will determine the expectation of a random variable with normal distribution [see (1)]. By formula (1) of Sec. 8.14 we have

$$m_x = \int_{-\infty}^{\infty} x \frac{1}{\sigma\sqrt{2\pi}} \exp\left[-\frac{(x-a)^2}{2\sigma^2}\right] dx \quad (2)$$

Making the change of variable

$$\frac{x-a}{\sigma\sqrt{2}} = t$$

we get

$$x=a+\sqrt{2}\sigma t, \quad dx=\sigma\sqrt{2}\,dt$$

Hence

$$m_x=\frac{1}{\sqrt{\pi}}\int_{-\infty}^{\infty}(a+\sqrt{2}\sigma t)\,e^{-t^2}\,dt$$

$$=\frac{1}{\sqrt{\pi}}\,a\int_{-\infty}^{\infty}e^{-t^2}\,dt+\frac{\sigma\sqrt{2}}{\sqrt{\pi}}\int_{-\infty}^{\infty}te^{-t^2}\,dt$$

The first integral on the right is equal to $\sqrt{\pi}$. Let us compute the second integral:

$$\int_{-\infty}^{\infty}t\cdot e^{-t^2}dt=-\frac{1}{2}e^{-t^2}\Big|_{-\infty}^{\infty}=0$$

Thus

$$m_x=a \tag{3}$$

The value of the parameter a in formula (1) is equal to the expectation of the random variable under consideration. The point $x=a$ is the centre of the distribution or the centre of dispersion. When $x=a$ the function $f(x)$ has a maximum and so the value $x=a$ is the *mode* of the random variable. Since the curve of (1) is symmetric about the straight line $x=a$, then

$$\int_{-\infty}^{a}f(x)\,dx=\int_{a}^{\infty}f(x)\,dx$$

That is, the value $x=a$ is the *median* of the normal distribution.

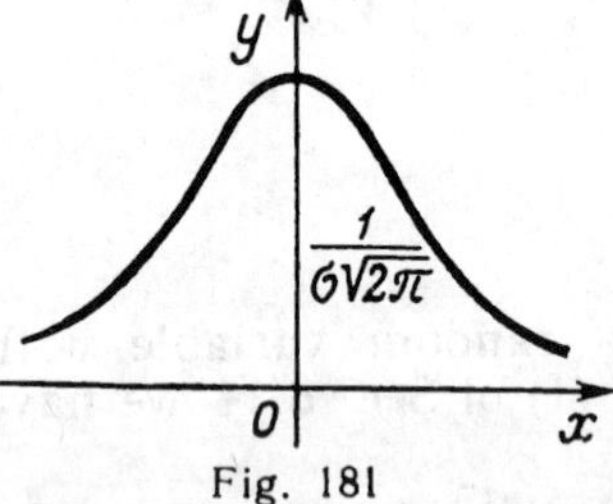

Fig. 181

If in formula (1) we put $a=0$, then we get

$$f(x)=\frac{1}{\sigma\sqrt{2\pi}}\exp\left(-\frac{x^2}{2\sigma^2}\right) \tag{4}$$

The corresponding curve is symmetric about the y-axis. The function $f(x)$ is the density function of normal distribution of a random variable with centre of probability distribution coincident with the origin (Fig. 181). The numerical characteristics of random variables with distribution laws (1) and (4) defining the type of dispersion of values of the random variable about the centre of dispersion are determined by the shape of the curve, which is independent of a, and therefore coincide. The quantity a determines the amount of shift of the curve (1) to the right (for $a>0$)

or to the left (for $a < 0$). To save space, in most of our discussions we will henceforth use the probability density function defined by formula (4).

8.16 VARIANCE AND STANDARD DEVIATION OF A NORMALLY DISTRIBUTED RANDOM VARIABLE

Suppose the probability density function (the frequency function) of the random variable $\bar{x}$ is given by

$$f(x)=\frac{1}{\sigma\sqrt{2\pi}}\exp\left(-\frac{x^2}{2\sigma^2}\right) \tag{1}$$

The variance of a continuous random variable is found from formula (2), Sec. 8.14.

In our case

$$m_x=a=0$$

We have

$$\mathbf{D}[\bar{x}]=\int_{-\infty}^{\infty}x^2\frac{1}{\sigma\sqrt{2\pi}}\exp\left(-\frac{x^2}{2\sigma^2}\right)dx$$

Making the substitution $\frac{x}{\sigma\sqrt{2}}=t$, we get

$$\mathbf{D}[\bar{x}]=\frac{2\sigma^2}{\sqrt{\pi}}\int_{-\infty}^{\infty}t^2\cdot e^{-t^2}\,dt=\frac{\sigma^2}{\sqrt{\pi}}\int_{-\infty}^{\infty}t\cdot 2t\cdot e^{-t^2}\,dt$$

Integrating by parts, we obtain

$$\mathbf{D}[\bar{x}]=\frac{\sigma^2}{\sqrt{\pi}}\left[-t\cdot e^{-t^2}\Big|_{-\infty}^{\infty}+\int_{-\infty}^{\infty}e^{-t^2}\,dt\right]$$

Since

$$\lim_{t\to\infty}t\cdot e^{-t^2}=0,\quad \int_{-\infty}^{\infty}e^{-t^2}\,dt=\sqrt{\pi}$$

we finally have

$$\mathbf{D}[\bar{x}]=\sigma^2 \tag{2}$$

The standard deviation, in accordance with formula (3), Sec. 8.14, is

$$\sigma[\bar{x}]=\sqrt{\mathbf{D}[\bar{x}]}=\sigma \tag{3}$$

Thus, the variance is equal to the parameter σ^2 in the density-function formula (1). We have already pointed out that the variance characterizes the dispersion of values of the random variable about the centre of dispersion. Let us now see how the value of the parameter σ^2 affects the shape of the distribution curve (frequency curve). Fig. 182 gives the distribution curves for the values $\sigma = \frac{1}{2}$, $\sigma = 1$, $\sigma = 2$. We see here that the smaller σ, the greater the maximum of the function $f(x)$, also the greater the probability of values close to the centre of dispersion ($x = 0$) and the smaller the probability of values distant from the origin. We can thus say that the smaller the variance σ^2, the smaller the dispersion of values of the random variable.

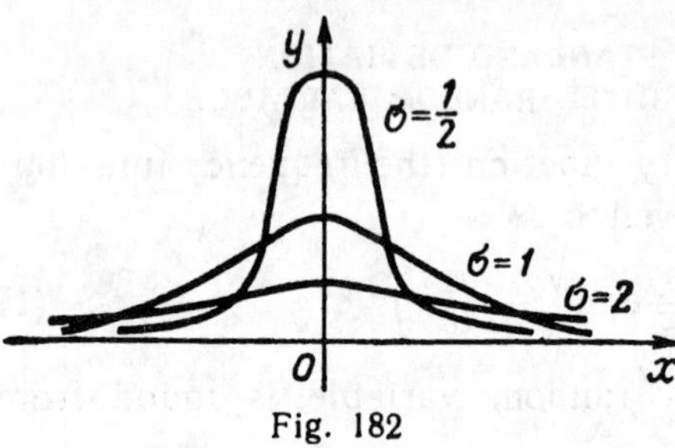

Fig. 182

8.17 THE PROBABILITY OF A VALUE OF THE RANDOM VARIABLE FALLING IN A GIVEN INTERVAL. THE LAPLACE FUNCTION. NORMAL DISTRIBUTION FUNCTION

In accordance with formula (3), Sec. 8.12, we determine the probability that a value of the random variable $\bar{x}$ having the density function

$$f(x) = \frac{1}{\sigma\sqrt{2\pi}} \exp\left[-\frac{(x-a)^2}{2\sigma^2}\right]$$

will fall in the interval (α, β)

$$\mathbf{P}(\alpha < \bar{x} < \beta) = \int_{\alpha}^{\beta} f(x)\,dx$$

or

$$\mathbf{P}(\alpha < \bar{x} < \beta) = \frac{1}{\sigma\sqrt{2\pi}} \int_{\alpha}^{\beta} \exp\left[-\frac{(x-a)^2}{2\sigma^2}\right] dx \tag{1}$$

(Fig. 183). Making the change of variable

$$\frac{x-a}{\sigma\sqrt{2}} = t$$

we get

$$\mathbf{P}(\alpha < \bar{x} < \beta) = \frac{1}{\sqrt{\pi}} \int_{\frac{\alpha-a}{\sigma\sqrt{2}}}^{\frac{\beta-a}{\sigma\sqrt{2}}} e^{-t^2}\,dt \tag{1'}$$

The integral on the right is not expressible in terms of elementary functions. The values of this integral can be expressed in terms of the values of the *probability integral*

$$\Phi(x)=\frac{2}{\sqrt{\pi}}\int_0^x e^{-t^2}\,dt \tag{2}$$

Here are some of the properties of the function $\Phi(x)$ which we will use later on.

1. $\Phi(x)$ is defined for all values of x.
2. $\Phi(0)=0$.
3. $\Phi(+\infty)=\frac{2}{\sqrt{\pi}}\int_0^\infty e^{-t^2}\,dt=\frac{2}{\sqrt{\pi}}\cdot\frac{\sqrt{\pi}}{2}=1.$
4. $\Phi(x)$ is monotonic increasing on the interval $(0, \infty)$.
5. $\Phi(x)$ is an odd function since

$$\Phi(-x)=-\Phi(x) \tag{3}$$

6. The graph of the function $\Phi(x)$ is shown in Fig. 184.

Detailed tables of the values of this function have been compiled (a short table is given in the Appendix, Table 1).

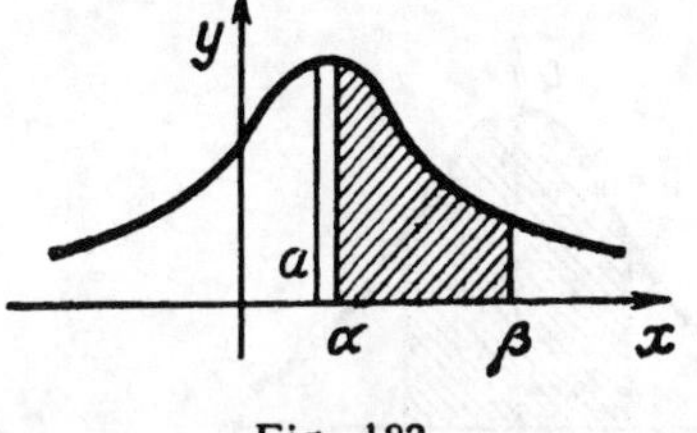

Fig. 183

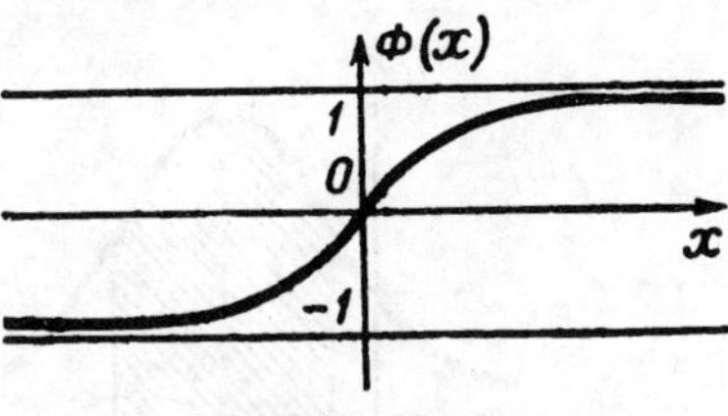

Fig. 184

Rewrite equation (1′) using the theorem on the partition of the interval of integration:

$$\mathbf{P}(\alpha<\bar{x}<\beta)=\frac{1}{\sqrt{\pi}}\left[\int_{\frac{\alpha-a}{\sigma\sqrt{2}}}^{0} e^{-t^2}\,dt+\int_0^{\frac{\beta-a}{\sigma\sqrt{2}}} e^{-t^2}\,dt\right]$$

$$=\frac{1}{\sqrt{\pi}}\left[-\int_0^{\frac{\alpha-a}{\sigma\sqrt{2}}} e^{-t^2}\,dt+\int_0^{\frac{\beta-a}{\sigma\sqrt{2}}} e^{-t^2}\,dt\right]$$

This equation can be rewritten as follows:

$$\mathbf{P}(\alpha<\bar{x}<\beta)=\frac{1}{2}\left[\frac{2}{\sqrt{\pi}}\int_{0}^{\frac{\beta-a}{\sigma\sqrt{2}}}e^{-t^2}\,dt-\frac{2}{\sqrt{\pi}}\int_{0}^{\frac{\alpha-a}{\sigma\sqrt{2}}}e^{-t^2}\,dt\right]$$

With the aid of the function $\Phi(x)$ [see (2)], we give a final expression for the probability of the normally distributed random variable $\bar{x}$ falling in the interval (α, β):

$$\mathbf{P}(\alpha<\bar{x}<\beta)=\frac{1}{2}\left[\Phi\left(\frac{\beta-a}{\sigma\sqrt{2}}\right)-\Phi\left(\frac{\alpha-a}{\sigma\sqrt{2}}\right)\right] \tag{4}$$

When $a=0$ we have

$$\mathbf{P}(\alpha<\bar{x}<\beta)=\frac{1}{2}\left[\Phi\left(\frac{\beta}{\sigma\sqrt{2}}\right)-\Phi\left(\frac{\alpha}{\sigma\sqrt{2}}\right)\right] \tag{5}$$

Equating the right members of (1) for the case $a=0$ and of (5), we get

$$\int_{\alpha}^{\beta}\frac{1}{\sigma\sqrt{2\pi}}\exp\left(-\frac{x^2}{2\sigma^2}\right)dx=\frac{1}{2}\left[\Phi\left(\frac{\beta}{\sigma\sqrt{2}}\right)-\Phi\left(\frac{\alpha}{\sigma\sqrt{2}}\right)\right] \tag{5'}$$

We frequently have to compute the probability that a value of the random variable will fall in the interval $(a-l,\ a+l)$ symmetric

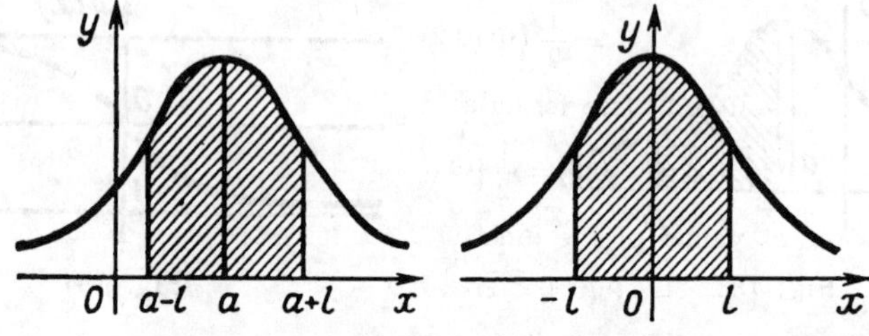

Fig. 185

about the point $x=a$ (Fig. 185). Formula (4) then takes the form

$$\mathbf{P}(a-l<\bar{x}<a+l)=\frac{1}{2}\left[\Phi\left(\frac{l}{\sigma\sqrt{2}}\right)-\Phi\left(-\frac{l}{\sigma\sqrt{2}}\right)\right]$$

Noting that $\Phi\left(-\frac{l}{\sigma\sqrt{2}}\right)=-\Phi\left(\frac{l}{\sigma\sqrt{2}}\right)$ [see formula (3)], we finally get

$$\mathbf{P}(a-l<\bar{x}<a+l)=\Phi\left(\frac{l}{\sigma\sqrt{2}}\right) \tag{6}$$

The right side does not depend on the position of the centre of dispersion, and so for $a=0$ as well we get

$$\mathbf{P}(-l<\bar{x}<l)=\Phi\left(\frac{l}{\sigma\sqrt{2}}\right) \tag{7}$$

Example 1. A random variable $\bar{x}$ obeys the normal distribution law with centre of dispersion $a=0.5$ and variance $\sigma^2=\frac{1}{8}$. Determine the probability that a value of the random variable $\bar{x}$ will fall in the interval (0.4, 0.6) (Fig. 186).

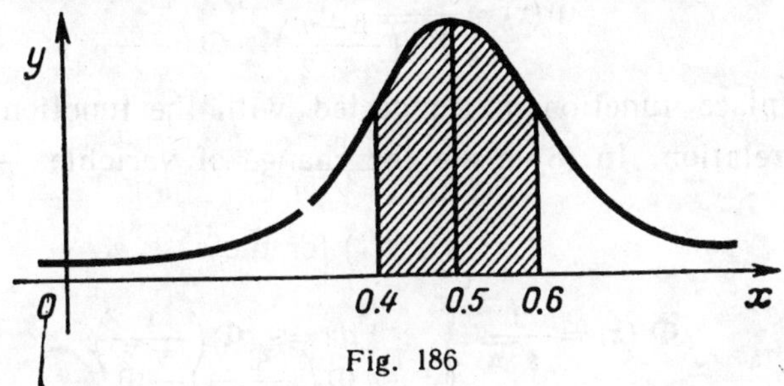

Fig. 186

Solution. Here $\frac{1}{\sigma\sqrt{2}}=2$. By formula (4) we have

$$\mathbf{P}(0.4<\bar{x}<0.6)=\frac{1}{2}\{\Phi[2(0.6-0.5)]-\Phi[2(0.4-0.5)]\}$$

$$=\frac{1}{2}\{\Phi(0.2)-\Phi(-0.2)\}$$

But $\Phi(-0.2)=-\Phi(0.2)$ [see formula (3)] and so we can write

$$\mathbf{P}(0.4<\bar{x}<0.6)=\frac{1}{2}[\Phi(0.2)+\Phi(0.2)]=\Phi(0.2)$$

Using the table of values of the function $\Phi(x)$ in the Appendix, we find

$$\mathbf{P}(0.4<\bar{x}<0.6)=0.223$$

Example 2. The length of an item manufactured on an automatic machine tool is a normally distributed random variable with parameters $\mathbf{M}[\bar{x}]=10$, $\sigma^2=\frac{1}{200}$. Find the probability of defective production if the tolerance is 10 ± 0.05.

Solution. Here, $a=10$, $\frac{1}{\sigma\sqrt{2}}=10$, $\sigma=\frac{1}{10\sqrt{2}}$. The probability of defective production, p_{def} is, in accordance with (4),

$$p_{def}=1-\mathbf{P}(9.95<\bar{x}<10.05)$$

$$=1-\frac{1}{2}\{\Phi[10(10.05-10)]-\Phi[10(9.95-10)]\}.$$

$$=1-\frac{1}{2}\{\Phi(0.5)-\Phi(-0.5)\}=1-\Phi(0.5)=1-0.52=0.48$$

Example 3. Find the probability of hitting a strip of width $2l = 3.5$ metres if the errors of gunfire obey the normal distribution law with parameters $a = 0$, $\sigma = 1.9$.

Solution. Here, $\alpha = -1.75$, $\beta = 1.75$, $\frac{1}{\sigma\sqrt{2}} = 0.372$. From formula (7) we get

$$\mathbf{P}(-1.75 < \bar{x} < 1.75) = \Phi(1.75 \cdot 0.372) = \Phi(0.651) = 0.643$$

Note. In place of the function $\Phi(x)$, (2), frequent use is made of the *Laplace function*

$$\overline{\Phi}(x) = \frac{1}{\sqrt{2\pi}} \int_0^x e^{-\frac{t^2}{2}} dt \tag{8}$$

The Laplace function is connected with the function $\Phi(x)$ by a simple relation. In (8) make the change of variable $\frac{t}{\sqrt{2}} = z$ to get

$$\overline{\Phi}(x) = \frac{1}{\sqrt{\pi}} \int_0^{\frac{x}{\sqrt{2}}} e^{-z^2} dz = \frac{1}{2} \Phi\left(\frac{x}{\sqrt{2}}\right)$$

Thus

$$\overline{\Phi}(x) = \frac{1}{2} \Phi\left(\frac{x}{\sqrt{2}}\right) \tag{9}$$

and, obviously,

$$\overline{\Phi}(x\sqrt{2}) = \frac{1}{2} \Phi(x) \tag{10}$$

Using the function $\overline{\Phi}(x)$ and relation (9), we write (5) as follows:

$$\mathbf{P}(\alpha < \bar{x} < \beta) = \overline{\Phi}\left(\frac{\beta}{\sigma}\right) - \overline{\Phi}\left(\frac{\alpha}{\sigma}\right) \tag{11}$$

and when $\sigma = 1$

$$\mathbf{P}(\alpha < \bar{x} < \beta) = \overline{\Phi}(\beta) - \overline{\Phi}(\alpha)$$

A table of the values of the Laplace function $\overline{\Phi}(x)$ is given in the Appendix, Table 3.

Now let us define the *normal distribution function*. From formula (1), Sec. 8.13, we have

$$F(x) = \int_{-\infty}^{x} f(x)\, dx = \frac{1}{\sigma\sqrt{2\pi}} \int_{-\infty}^{x} \exp\left[-\frac{(x-a)^2}{2\sigma^2}\right] dx = \mathbf{P}(-\infty < \bar{x} < x)$$

Using formula (4) for the case $\alpha = -\infty$, $\beta = x$, we get

$$F(x) = \frac{1}{2}\left[\Phi\left(\frac{x-a}{\sigma\sqrt{2}}\right) - \Phi(-\infty)\right]$$

but $\Phi(-\infty)=-1$ [see formula (3)]. Hence,

$$F(x)=\frac{1}{2}\left[\Phi\left(\frac{x-a}{\sigma\sqrt{2}}\right)+1\right] \quad (12)$$

The graph of the function $F(x)$ for $a=0$ is shown in Fig. 187.

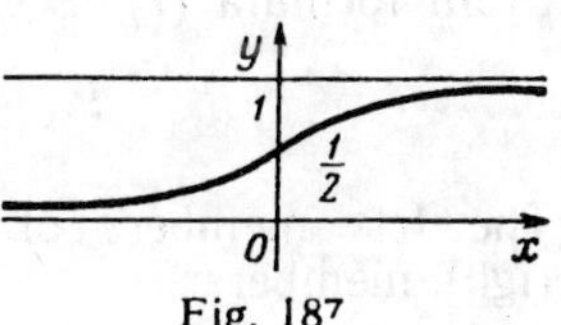

Fig. 187

8.18 PROBABLE ERROR

The *probable error* (or *mean error*) is a measure of dispersion that has found numerous applications in the probability, gunfire and errors theories.

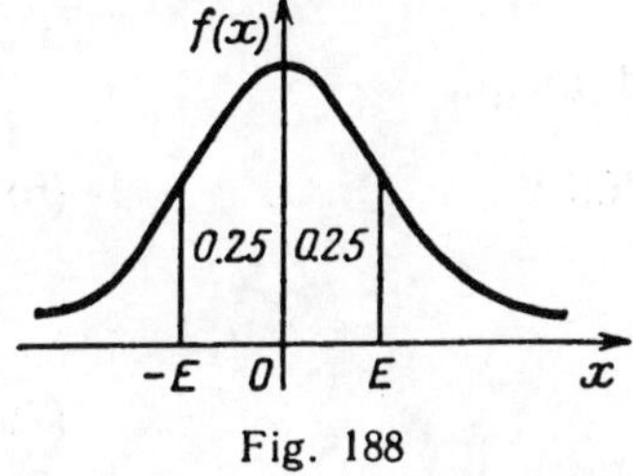

Fig. 188

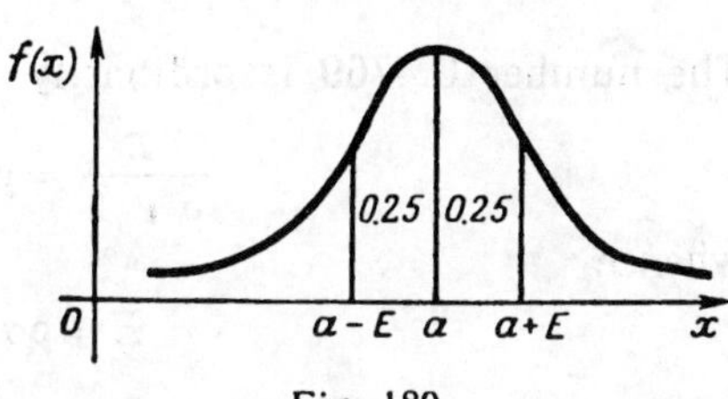

Fig. 189

Definition 1. The *probable error* is a number E such that the probability that a random variable obeying the normal distribution law

$$f(x)=\frac{1}{\sigma\sqrt{2\pi}}\exp\left(-\frac{x^2}{2\sigma^2}\right)$$

will fall in the interval $(-E, E)$ is $\frac{1}{2}$ (Fig. 188); that is,

$$\mathbf{P}(-E<\bar{x}<E)=\frac{1}{2} \quad (1)$$

For any random variable $\bar{x}$ subject to the normal distribution law with centre of dispersion at $x=a$, the probable error E (Fig. 189) satisfies the relation

$$\mathbf{P}(a-E<\bar{x}<a+E)=\frac{1}{2} \quad (2)$$

Let us express the standard deviation σ in terms of the probable error E.

We express the left member of (1) in terms of the function $\Phi(x)$:

$$\mathbf{P}(-E<\bar{x}<E)=\int_{-E}^{E}\frac{1}{\sigma\sqrt{2\pi}}\exp\left(-\frac{x^2}{2\sigma^2}\right)dx \quad (3)$$

From formula (7), Sec. 8.17, we get

$$\mathbf{P}(-E<\bar{x}<E)=\Phi\left(\frac{E}{\sigma\sqrt{2}}\right) \tag{4}$$

The left members of (1) and (4) are equal, and so also are the right members:

$$\Phi\left(\frac{E}{\sigma\sqrt{2}}\right)=\frac{1}{2} \tag{5}$$

Using the table of values of the function $\Phi(x)$, we find the value of the argument $x=0.4769$ for which $\Phi(x)=\frac{1}{2}$. Hence

$$\frac{E}{\sigma\sqrt{2}}=0.4769$$

The number 0.4769 is ordinarily denoted by ρ:

$$\frac{E}{\sigma\sqrt{2}}=\rho=0.4769 \tag{6}$$

whence

$$\left.\begin{aligned} E &= \rho\sigma\sqrt{2} \\ \sigma &= \frac{E}{\rho\sqrt{2}} \end{aligned}\right\} \tag{7}$$

8.19 EXPRESSING THE NORMAL DISTRIBUTION IN TERMS OF THE PROBABLE ERROR. THE REDUCED LAPLACE FUNCTION

Expressing the parameter σ in terms of the parameter E by formula (7), Sec. 8.18, and substituting into (4), Sec. 8.15, we get an expression of the distribution in terms of the probable error:

$$f(x)=\frac{\rho}{E\sqrt{\pi}}\exp\left(-\rho^2\frac{x^2}{E^2}\right) \tag{1}$$

The probability of the random variable (an error, for example) falling in the interval (α, β) will, in accordance with (5), Sec. 8.17, be

$$\mathbf{P}(\alpha<\bar{x}<\beta)=\frac{1}{2}\left[\Phi\left(\rho\frac{\beta}{E}\right)-\Phi\left(\rho\frac{\alpha}{E}\right)\right] \tag{2}$$

and, in accordance with formula (7) of Sec. 8.17,

$$\mathbf{P}(-l<\bar{x}<l)=\Phi\left(\rho\frac{l}{E}\right) \tag{3}$$

The numbers $\frac{\beta}{E}$ and $\frac{\alpha}{E}$ in the right member of (2) are determined by the nature of the problem; $\rho=0.4769$.

In order to avoid multiplying by ρ in each computation, tables have been compiled for the function $\Phi(\rho x)$. This function is denoted by $\hat{\Phi}(x)$:

$$\hat{\Phi}(x) = \Phi(\rho x) \tag{4}$$

and is called the *reduced Laplace function*. See Table 1 in the Appendix for values of this function.

On the basis of (2), Sec. 8.17, the function $\hat{\Phi}(x)$ is defined by the integral

$$\hat{\Phi}(x) = \frac{2}{\sqrt{\pi}} \int_0^{\rho x} e^{-t^2}\, dt$$

Making the change of variable $t = \rho z$, we get

$$\hat{\Phi}(x) = \frac{2\rho}{\sqrt{\pi}} \int_0^{x} e^{-\rho^2 z^2}\, dz \tag{5}$$

Let us express the right member of (2) in terms of the reduced Laplace function·

$$\mathbf{P}(\alpha < \bar{x} < \beta) = \frac{1}{2}\left[\hat{\Phi}\left(\frac{\beta}{E}\right) - \hat{\Phi}\left(\frac{\alpha}{E}\right)\right] \tag{6}$$

In particular, the probability of a value of a random variable falling in a symmetric interval about the centre of dispersion $(-l, l)$ is, by (3),

$$\mathbf{P}(-l < \bar{x} < l) = \hat{\Phi}\left(\frac{l}{E}\right) \tag{7}$$

and

$$\mathbf{P}(0 < \bar{x} < l) = \frac{1}{2}\hat{\Phi}\left(\frac{l}{E}\right) \tag{8}$$

Note that if the expectation $a \neq 0$, the probability of the random variable $\bar{x}$ falling in the interval (α, β), will be expressed as follows in terms of the probable error E [see formula (4), Sec. 8.17]:

$$\mathbf{P}(\alpha < \bar{x} < \beta) = \frac{1}{2}\left[\Phi\left(\rho\frac{\beta - a}{E}\right) - \Phi\left(\rho\frac{\alpha - a}{E}\right)\right] \tag{9}$$

This equation is expressed in terms of the reduced Laplace function as follows:

$$\mathbf{P}(\alpha < \bar{x} < \beta) = \frac{1}{2}\left[\hat{\Phi}\left(\frac{\beta - a}{E}\right) - \hat{\Phi}\left(\frac{\alpha - a}{E}\right)\right] \tag{10}$$

8.20 THE THREE-SIGMA RULE. ERROR DISTRIBUTION

In practical computations, the unit of measurement of the deviation of a normally distributed random variable from the centre of dispersion (the mathematical expectation) is taken to be the root-mean-square (standard) deviation σ. Then, by formula (7), Sec. 8.17, we get some useful equations:

$$\mathbf{P}(-\sigma < \bar{x} < \sigma) = \Phi\left(\frac{1}{\sqrt{2}}\right) = 0.683$$

$$\mathbf{P}(-2\sigma < \bar{x} < 2\sigma) = \Phi(\sqrt{2}) = 0.954$$

$$\mathbf{P}(-3\sigma < \bar{x} < 3\sigma) = \Phi\left(\frac{3}{\sqrt{2}}\right) = 0.997$$

These results are depicted geometrically in Fig. 190.

We can be almost certain that the random variable (error) will not depart from the absolute value of the expectation by more than 3σ. This proposition is called the *three-sigma rule.*

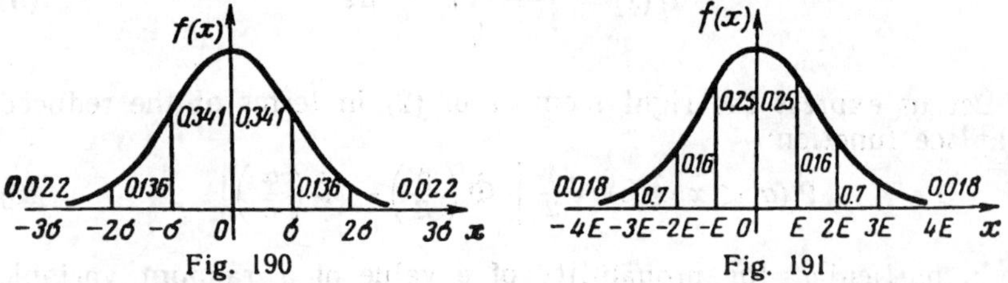

Fig. 190 Fig. 191

In gunfire theory and in the processing of various statistical data it is often useful to know the probability of the random variable $\bar{x}$ falling in the intervals $(0, E)$, $(E, 2E)$, $(2E, 3E)$, $(3E, 4E)$, $(4E, 5E)$ for the density function defined by (1), Sec. 8.19. In many cases, a knowledge of these probabilities reduces computations and helps in the analysis of the phenomena.

In computing these probabilities, we will use formula (8), Sec. 8.19, and the table of the function $\hat{\Phi}(x)$:

$$\mathbf{P}(0 < \bar{x} < E) = \frac{1}{2}\hat{\Phi}(1) = 0.2500$$

$$\mathbf{P}(E < \bar{x} < 2E) = \frac{1}{2}[\hat{\Phi}(2) - \hat{\Phi}(1)] = 0.1613$$

$$\mathbf{P}(2E < \bar{x} < 3E) = \frac{1}{2}[\hat{\Phi}(3) - \hat{\Phi}(2)] = 0.0672$$

$$\mathbf{P}(3E < \bar{x} < 4E) = \frac{1}{2}[\hat{\Phi}(4) - \hat{\Phi}(3)] = 0.0180$$

$$\mathbf{P}(4E < \bar{x} < \infty) = \frac{1}{2}[\hat{\Phi}(\infty) - \hat{\Phi}(4)] = \frac{1}{2}(1 - 0.9930) = 0.0035$$

The results of the computations are geometrically represented in Fig. 191, which is called the *error distribution*. These calculations imply that it is practically a certainty that a value of the random variable will fall in the interval $(-4E, 4E)$. The probability that this value of the random variable will fall outside that interval is less than 0.01.

Example 1. One shot is fired at a strip 100 metres wide aimed at the middle line of the strip perpendicular to the plane of the path of the projectile. Dispersion obeys the normal law with longitudinal probable error $E = 20$ metres. Determine the probability of hitting the strip (Fig. 192). In gunfire theory, the longitudinal probable error is denoted by B_{lon}, the lateral probable error by B_{lat}.

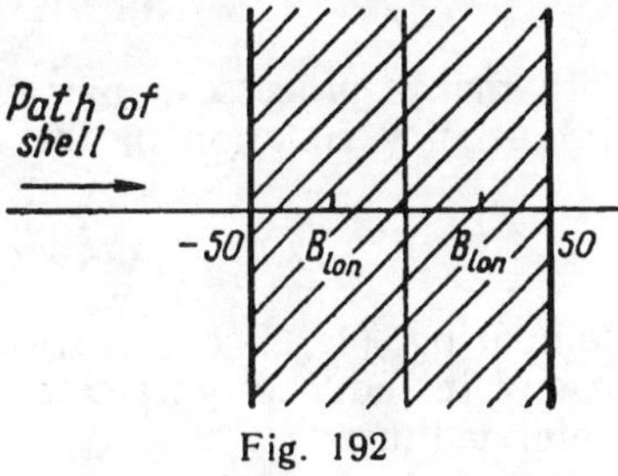

Fig. 192

Solution. Take advantage of formula (7), Sec. 8.19. In our case $l = 50$ metres, $E = B_{lon} = 20$ metres. Hence

$$\mathbf{P}(-50 < \bar{x} < 50) = \hat{\Phi}\left(\frac{50}{20}\right) = \hat{\Phi}(2.5) = 0.9082 \approx 0.91$$

Note. An approximate solution may be obtained without resorting to tables of the function $\hat{\Phi}(x)$, simply by using the error distribution (Fig. 191). In our case, $l = 2.5E$, hence

$$\mathbf{P}(-50 < \bar{x} < 50) = 2(0.25 + 0.16 + 0.01) = 0.90$$

Example 2. Experiment shows that the error of an instrument used for measuring distances obeys the normal distribution law with probable error $E = 10$ metres. Find the probability that the distance determined by this instrument will depart from the true distance by no more than 15 metres.

Solution. Here, $l = 15$ m, $E = 10$ m. By formula (7), Sec. 8.19, we get

$$\mathbf{P}(-15 < \bar{x} < 15) = \hat{\Phi}\left(\frac{15}{10}\right) = \hat{\Phi}(1.5) = 0.6883 \approx 0.69$$

8.21 MEAN ARITHMETIC ERROR

To characterize errors we introduce the concept of the *mean arithmetic error*, which is equal to the expectation of the absolute value of the errors. We denote the mean arithmetic error by d. Let us determine the mean arithmetic error if the errors x obey the normal law (4), Sec. 8.15. By a formula similar to (2) of Sec. 8.15, we get $(a = 0)$

$$d = \int_{-\infty}^{\infty} |x| \frac{1}{\sigma\sqrt{2\pi}} \exp\left(-\frac{x^2}{2\sigma^2}\right) dx = \frac{2}{\sigma\sqrt{2\pi}} \int_0^{\infty} x \exp\left(-\frac{x^2}{2\sigma^2}\right) dx$$

$$= \frac{2}{\sigma\sqrt{2\pi}} \left[-\sigma^2 \exp\left(-\frac{x^2}{2\sigma^2}\right)\right]_0^{\infty} = \frac{2\sigma}{\sqrt{2\pi}}$$

Thus, the mean arithmetic error is expressed in terms of the root-mean-square (standard) deviation σ thus:

$$d=\frac{2\sigma}{\sqrt{2\pi}}=\sigma\sqrt{\frac{2}{\pi}} \tag{1}$$

8.22 MODULUS OF PRECISION. RELATIONSHIPS BETWEEN THE CHARACTERISTICS OF THE DISTRIBUTION OF ERRORS

In many processes, particularly in gunfire theory, the probability density function of the normal law is given as

$$f(x)=\frac{h}{\sqrt{\pi}}e^{-h^2x^2} \tag{1}$$

Comparing (4), Sec. 8.15, and (1) of this section we see that the newly introduced parameter h is expressed in terms of the parameter σ thus:

$$h=\frac{1}{\sigma\sqrt{2}} \tag{2}$$

The quantity h is inversely proportional to σ, that is, it is inversely proportional to the mean-square error or the standard deviation. The smaller the variance σ^2, that is, the smaller the dispersion, the greater the value of h. For this reason, h is termed the *modulus of precision*.

From (2) of this section and (1) of Sec. 8.21 we get

$$\sigma=\frac{1}{h\sqrt{2}} \tag{3}$$

$$d=\frac{1}{h\sqrt{\pi}} \tag{4}$$

The probable error E is expressed in terms of the modulus of precision h via formula (7), Sec. 8.18, and (3):

$$E=\frac{\rho}{h} \tag{5}$$

It is sometimes necessary to express one characteristic of error distribution in terms of another. The following equations are found to be useful:

$$\left.\begin{aligned}\frac{E}{\sigma}=\rho\sqrt{2}=0.6745,\quad \frac{E}{d}=\rho\sqrt{\pi}=0.8453,\quad \frac{\sigma}{d}=\sqrt{\frac{\pi}{2}}=1.2533\\ \frac{\sigma}{E}=\frac{1}{\rho\sqrt{2}}=1.4826,\quad \frac{d}{E}=\frac{1}{\rho\sqrt{\pi}}=1.1829\end{aligned}\right\} \tag{6}$$

8.23 TWO-DIMENSIONAL RANDOM VARIABLES

Two-dimensional random variables are involved for example when considering the process of target destruction on a plane (xOy).

The value of a two-dimensional random variable is determined by two numbers: x and y. We will denote the two-dimensional random variable itself by $(\bar{x}, \bar{y})$. Let $\bar{x}$ and $\bar{y}$ assume discrete values x_i and y_i. Let each pair of values (x_i, y_i) of a certain set be associated with a definite probability p_{ij}. We can form a table of probability distribution of the discrete two-dimensional random variable:

$\bar{y}$ \ $\bar{x}$	x_1	x_2	...	...	...	x_n
y_1	p_{11}	p_{21}				p_{n1}
y_2	p_{12}	p_{22}				p_{n2}
⋮						
⋮						
y_m	p_{1m}	p_{2m}				p_{nm}

Clearly, the following equation must hold:

$$\sum_{j=1}^{m}\sum_{i=1}^{n} p_{ij}=1 \tag{1}$$

Now let us determine a continuous two-dimensional random variable. The probability that a value of the two-dimensional random variable $(\bar{x}, \bar{y})$ satisfies the inequalities $x<\bar{x}<x+\Delta x$, $y<\bar{y}<y+\Delta y$ will be denoted thus: $\mathbf{P}(x<\bar{x}<x+\Delta x, y<\bar{y}<y+\Delta y)$.

Definition 1. The function $f(x, y)$ is called the *probability density function* of the two-dimensional random variable $(\bar{x}, \bar{y})$ if, to within infinitesimals of higher order than $\Delta\rho=\sqrt{\Delta x^2+\Delta y^2}$, the following equation holds true:

$$\mathbf{P}(x<\bar{x}<x+\Delta x, \quad y<\bar{y}<y+\Delta y)\approx f(x, y)\,\Delta x\,\Delta y \tag{2}$$

Formula (2) is similar to formula (2) of Sec. 8.12.

We consider a rectangular coordinate system (xOy). If the values of the random variable $(\bar{x}, \bar{y})$ are denoted by points of the plane having the appropriate coordinates x and y, then the expression

$\mathbf{P}(x < \bar{x} < x + \Delta x,\ y < \bar{y} < y + \Delta y)$ denotes the probability that the two-dimensional random variable $(\bar{x}, \bar{y})$ will assume a value denoted by a point lying in the hatched rectangle Δs (Fig. 193). We will say that the "value of the random variable lies in the domain Δs".*

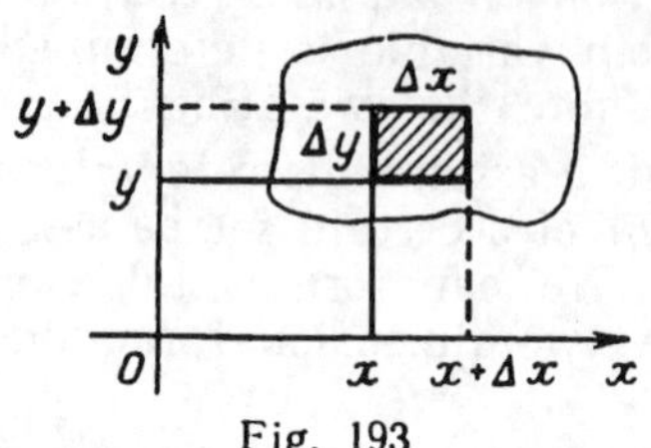

Fig. 193

The probability $\mathbf{P}(x < \bar{x} < x + \Delta x,\ y < \bar{y} < y + \Delta y)$ will also be denoted by $\mathbf{P}[(\bar{x}, \bar{y}) \subset \Delta s]$. Using this notation, we can rewrite (2) as

$$\mathbf{P}[(\bar{x}, \bar{y}) \subset \Delta s] \approx f(x, y)\,\Delta s \tag{3}$$

We will now prove the following theorem, which is analogous to Theorem 1, Sec. 8.12.

Theorem 1. *The probability* $\mathbf{P}[(\bar{x}, \bar{y}) \subset D]$ *that a two-dimensional random variable* $(\bar{x}, \bar{y})$ *with density function* $f(x, y)$ *will be in domain* D *is expressed by the double integral of the function* $f(x, y)$ *over* D, *that is,*

$$\mathbf{P}[(\bar{x}, \bar{y}) \subset D] = \iint\limits_D f(x, y)\,dx\,dy \tag{4}$$

Proof. Partition D into subdomains Δs, as is done in the theory of double integrals. For each subdomain we write (3) and add the left-hand and right-hand members of the resulting equations. Since

$$\sum \Delta s = D \text{ and } \sum \mathbf{P}[(\bar{x}, \bar{y}) \subset \Delta s] = \mathbf{P}[(\bar{x}, \bar{y}) \subset D]$$

we get an approximate equation to within infinitesimals of higher order than Δs:

$$\mathbf{P}[(\bar{x}, \bar{y}) \subset D] \approx \sum f(x, y)\,\Delta s$$

Passing to the limit in the right-hand member of this equation as $\Delta s \to 0$, we get on the right a double integral and, on the basis of the properties of an integral sum, the exact equation

$$\mathbf{P}[(\bar{x}, \bar{y}) \subset D] = \iint\limits_D f(x, y)\,dx\,dy$$

The proof is complete.

Note 1. If the domain D is a rectangle bounded by the straight lines $x = \alpha$, $x = \beta$, $y = \gamma$, $y = \delta$ (Fig. 194), then

$$\mathbf{P}[\alpha < \bar{x} < \beta,\ \gamma < \bar{y} < \delta] = \int_\alpha^\beta \int_\gamma^\delta f(x, y)\,dx\,dy \tag{5}$$

* The shape of the subdomain in (3) may be arbitrary.

Note 2. Like (1), the following equation

$$\int_{-\infty}^{\infty}\int_{-\infty}^{\infty} f(x, y)\,dx\,dy = 1 \tag{6}$$

holds true since it is a certainty that the two-dimensional variable will assume some value. We put $f(x, y) = 0$ wherever the function $f(x, y)$ is not defined by the meaning of the problem.

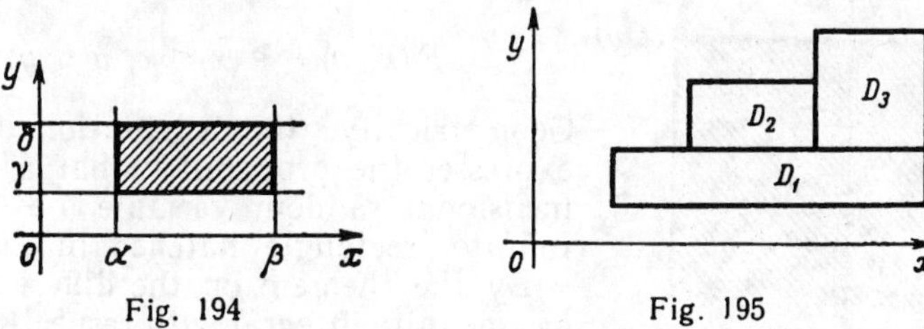

Fig. 194 Fig. 195

If the domain D is the sum of rectangles of the kind shown in Fig. 195, then the probability of the random variable falling in such a domain is defined as the sum of the probabilities for the separate rectangles, that is, as the sum of definite integrals over each rectangle:

$$\mathbf{P}[(\bar{x}, \bar{y}) \subset D] = \mathbf{P}[(\bar{x}, \bar{y}) \subset D_1] + \mathbf{P}[(\bar{x}, \bar{y}) \subset D_2] + \mathbf{P}[(\bar{x}, \bar{y}) \subset D_3]$$

Example. The density function of a two-dimensional random variable is given by the formula

$$f(x, y) = \frac{1}{\pi^2 (1+x^2)(1+y^2)}$$

Find the probability that a value of the random variable will fall in the rectangle bounded by the straight lines $x = 0$, $x = 1$, $y = \frac{1}{\sqrt{3}}$, $y = \sqrt{3}$.

Solution. By formula (5) we obtain

$$\mathbf{P}\left[0 < \bar{x} < 1, \frac{1}{\sqrt{3}} < \bar{y} < \sqrt{3}\right] = \frac{1}{\pi^2}\int_0^1 \int_{\frac{1}{\sqrt{3}}}^{\sqrt{3}} \frac{dx\,dy}{(1+x^2)(1+y^2)}$$

$$= \frac{1}{\pi^2}\int_0^1 \frac{dx}{1+x^2}\int_{\frac{1}{\sqrt{3}}}^{\sqrt{3}} \frac{dy}{1+y^2} = \frac{1}{\pi^2}\arctan x\Big|_0^1 \arctan y\Big|_{\frac{1}{\sqrt{3}}}^{\sqrt{3}}$$

$$= \frac{1}{\pi^2}\left(\frac{\pi}{4} - 0\right)\left(\frac{\pi}{3} - \frac{\pi}{6}\right) = \frac{1}{24}$$

Definition 2. The function

$$F(x, y) = \int_{-\infty}^{y} \int_{-\infty}^{x} f(u, v)\, du\, dv \tag{7}$$

is termed the *distribution function* of probabilities of the two-dimensional random variable $(\bar{x}, \bar{y})$.

It is clear that the distribution function expresses the probability that $\bar{x} < x$, $\bar{y} < y$, that is,

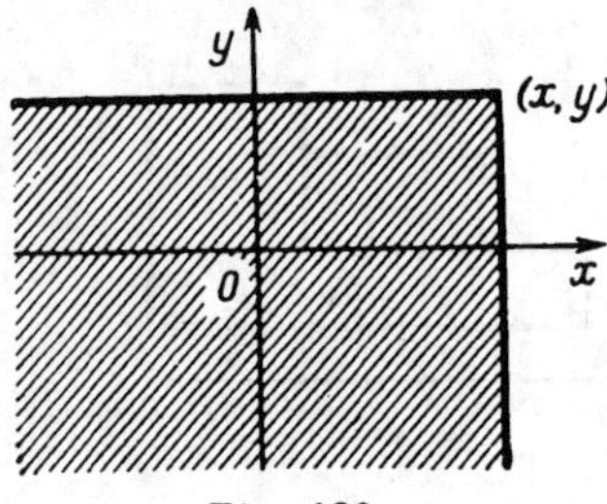

Fig. 196

$$F(x, y) = \mathbf{P}(\bar{x} < x, \bar{y} < y)$$

Geometrically, the distribution function expresses the probability that a two-dimensional random variable lies in the infinite rectangle hatched in Fig. 196.

By the theorem on the differentiation of a definite integral with respect to a parameter, a relation is established between the density function and the distribution function:

$$\left.\begin{aligned} \frac{\partial F}{\partial x} &= \int_{-\infty}^{y} f(x, v)\, dv \\ \frac{\partial^2 F}{\partial x\, \partial y} &= f(x, y) \end{aligned}\right\} \tag{8}$$

The probability density function of a two-dimensional random variable is the mixed second derivative of the distribution function.

8.24 NORMAL DISTRIBUTION IN THE PLANE

Definition 1. The distribution of a two-dimensional random variable is termed *normal* if the density function of this variable is expressed by the formula

$$f(x, y) = \frac{1}{2\pi\sigma_x\sigma_y} \exp\left(-\frac{x^2}{2\sigma_x^2} - \frac{y^2}{2\sigma_y^2}\right) \tag{1}$$

The graph of this function is a surface (see Fig. 197).

The *centre of dispersion* of a random variable with distribution law (1) is the point (0, 0).* σ_x and σ_y are called the *principal standard deviations*.

---

* If the centre of dispersion lies at the point (a, b), then the distribution is given by the formula

$$f(x, y) = \frac{1}{2\pi\sigma_x\sigma_y} \exp\left[-\frac{(x-a)^2}{2\sigma_x^2} - \frac{(y-b)^2}{2\sigma_y^2}\right] \tag{1'}$$

Rewrite formula (1) as follows:

$$f(x, y)=\frac{1}{\sigma_x \sqrt{2\pi}} \exp\left(-\frac{x^2}{2\sigma_x^2}\right) \frac{1}{\sigma_y \sqrt{2\pi}} \exp\left(-\frac{y^2}{2\sigma_y^2}\right) \tag{2}$$

Thus, $f(x, y)$ may be regarded as a product of two density functions of normally distributed random variables $\bar{x}$ and $\bar{y}$. As in the

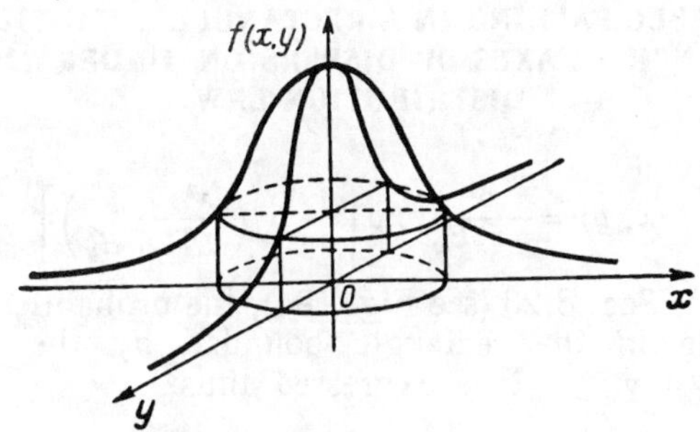

Fig. 197

case of a one-dimensional random variable, we define the *principal probable* errors, E_x and E_y [see formula (7), Sec. 8.18] as follows:

$$E_x = \rho\sigma_x \sqrt{2}, \quad E_y = \rho\sigma_y \sqrt{2} \tag{3}$$

Substituting σ_x and σ_y, expressed in terms of E_x and E_y, into formula (1), we get

$$f(x, y)=\frac{\rho^2}{\pi E_x E_y} \exp\left[-\rho^2\left(\frac{x^2}{E_x^2}+\frac{y^2}{E_y^2}\right)\right] \tag{4}$$

Let us consider the level lines of the surface (4):

$$\frac{x^2}{E_x^2}+\frac{y^2}{E_y^2}=k^2=\text{constant} \tag{5}$$

[here, $f(x, y)=\text{const}$]. The level lines are ellipses with semiaxes equal to kE_x and kE_y. The centres of the ellipses coincide with the centre of dispersion. These ellipses are called *ellipses of dispersion*. Their axes are termed *axes of dispersion*. The *unit ellipse of dispersion* is an ellipse whose semiaxes are equal to the probable errors E_x and E_y. The equation of the unit ellipse is obtained by putting $k=1$ in equation (5):

$$\frac{x^2}{E_x^2}+\frac{y^2}{E_y^2}=1 \tag{6}$$

The *total ellipse of dispersion* is an ellipse whose semiaxes are equal to $4E_x$ and $4E_y$. The equation of this ellipse is

$$\frac{x^2}{(4E_x)^2}+\frac{y^2}{(4E_y)^2}=1 \tag{7}$$

In the next section we will establish the fact that the probability of a two-dimensional random variable falling in the total ellipse of dispersion is equal to 0.97, which makes a hit a practical certainty.

8.25 THE PROBABILITY OF A TWO-DIMENSIONAL RANDOM VARIABLE FALLING IN A RECTANGLE WITH SIDES PARALLEL TO THE PRINCIPAL AXES OF DISPERSION UNDER THE NORMAL DISTRIBUTION LAW

Let

$$f(x, y)=\frac{\rho^2}{\pi E_x E_y}\exp\left[-\rho^2\left(\frac{x^2}{E_x^2}+\frac{y^2}{E_y^2}\right)\right]$$

By formula (5), Sec. 8.23 (see Fig. 194), the probability of a random variable falling in the rectangle bounded by the straight lines $x=\alpha$, $x=\beta$, $y=\gamma$, $y=\delta$ is expressed thus:

$$\mathbf{P}(\alpha<\bar{x}<\beta, \gamma<\bar{y}<\delta)=\int_\alpha^\beta\int_\gamma^\delta \frac{\rho^2}{\pi E_x E_y}\exp\left[-\rho^2\left(\frac{x^2}{E_x^2}+\frac{y^2}{E_y^2}\right)\right]dx\,dy \quad (1)$$

Representing the integrand as a product of two functions, we can write

$$\mathbf{P}(\alpha<\bar{x}<\beta, \gamma<\bar{y}<\delta)$$
$$=\int_\alpha^\beta \frac{\rho}{\sqrt{\pi}E_x}\exp\left(-\rho^2\frac{x^2}{E_x^2}\right)dx\int_\gamma^\delta \frac{\rho}{\sqrt{\pi}E_y}\exp\left(-\rho^2\frac{y^2}{E_y^2}\right)dy \quad (2)$$

and, on the basis of formula (6), Sec. 8.19, we finally get

$$\mathbf{P}(\alpha<\bar{x}<\beta, \gamma<\bar{y}<\delta)$$
$$=\frac{1}{4}\left[\hat{\Phi}\left(\frac{\beta}{E_x}\right)-\hat{\Phi}\left(\frac{\alpha}{E_x}\right)\right]\left[\hat{\Phi}\left(\frac{\delta}{E_y}\right)-\hat{\Phi}\left(\frac{\gamma}{E_y}\right)\right] \quad (3)$$

If we put $\alpha=-l_1$, $\beta=l_1$, $\gamma=-l_2$, $\delta=l_2$ here, that is, if we regard a rectangle with centre at the origin, then by (7), Sec. 8.19, formula (3) will take the form

$$\mathbf{P}(-l_1<\bar{x}<l_1, -l_2<\bar{y}<l_2)=\hat{\Phi}\left(\frac{l_1}{E_x}\right)\hat{\Phi}\left(\frac{l_2}{E_y}\right) \quad (4)$$

Note. The problem on the probability of a random variable falling in a rectangle with sides parallel to the coordinate axes could also be solved in the following manner. Falling in the rectangle is a compound event consisting in the coincidence of two independent events: falling in the strip $-l_1<\bar{x}<l_1$ and falling in the strip $-l_2<\bar{y}<l_2$. (To save space, we consider a rectangle with centre at the origin.) Let the density function of

the random variable $\bar{x}$ be

$$f_1(x) = \frac{\rho}{\sqrt{\pi} E_x} \exp\left(-\rho^2 \frac{x^2}{E_x^2}\right)$$

The density function of the random variable $\bar{y}$ is

$$f_2(y) = \frac{1}{\sqrt{\pi} E_y} \exp\left(-\rho^2 \frac{y^2}{E_y^2}\right)$$

We compute the probability of the random variable falling in the strip $-l_1 < \bar{x} < l_1$ and in the strip $-l_2 < \bar{y} < l_2$. By formula (7), Sec. 8.19, we get

$$\mathbf{P}(-l_1 < \bar{x} < l_1) = \hat{\Phi}\left(\frac{l_1}{E_x}\right)$$

$$\mathbf{P}(-l_2 < \bar{y} < l_2) = \hat{\Phi}\left(\frac{l_2}{E_y}\right)$$

The probability of a compound event—falling in the rectangle—is equal to the product of the probabilities:

$$\mathbf{P}(\alpha < \bar{x} < \beta,\ \gamma < \bar{y} < \delta) = \mathbf{P}(-l_1 < \bar{x} < l_1)\, \mathbf{P}(-l_2 < \bar{y} < l_2)$$
$$= \hat{\Phi}\left(\frac{l_1}{E_x}\right) \hat{\Phi}\left(\frac{l_2}{E_y}\right)$$

We have obtained formula (4).

Example. Shots are fired at a rectangle with sides 200 m and 100 m bounded by the lines

$$x = -100,\ x = 100,\ y = -50,\ y = 50$$

The principal mean errors are equal to $E_x = B_{lon} = 50$ metres and $E_y = B_{lat} = 10$ metres, respectively. Find the probability of hitting the rectangle in a single shot (Fig. 198).

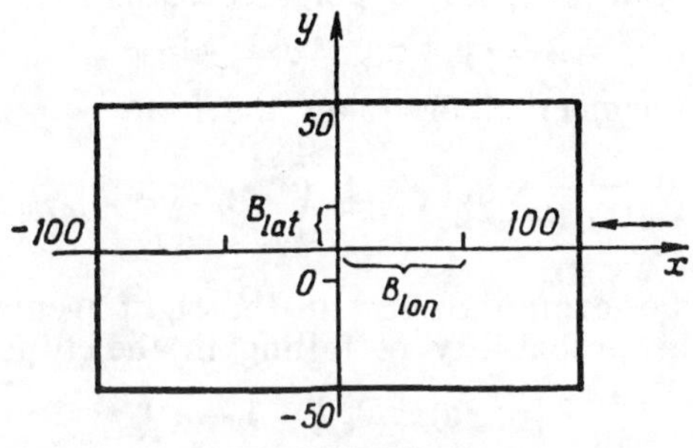

Fig. 198

Solution. In our case

$$l_1 = 100,\ l_2 = 50,\ E_x = 50,\ E_y = 10$$

Putting these values into (4) and using the table of values of the function $\hat{\Phi}(x)$ (see Table 1 of the Appendix), we find

$$\mathbf{P} = \hat{\Phi}\left(\frac{100}{50}\right) \cdot \hat{\Phi}\left(\frac{50}{10}\right) = \hat{\Phi}(2) \cdot \hat{\Phi}(5) = 0.8227 \cdot 0.9993 = 0.8221$$

8.26 THE PROBABILITY OF A TWO-DIMENSIONAL RANDOM VARIABLE FALLING IN THE ELLIPSE OF DISPERSION

In the theory of errors one comes up against the following problem. It is required to compute the probability that a random variable, say, an error in the plane, will fall in the ellipse of dispersion

$$\frac{x^2}{E_x^2}+\frac{y^2}{E_y^2}=k^2 \tag{1}$$

if the density function is given by formula (4), Sec. 8.24. By (4) of Sec. 8.23, we get

$$\mathbf{P}[(\bar{x},\ \bar{y})\subset D]=\iint\limits_{D_{el}}\frac{\rho^2}{\pi E_x E_y}\exp\left[-\rho^2\left(\frac{x^2}{E_x^2}+\frac{y^2}{E_y^2}\right)\right]dx\,dy \tag{2}$$

where the domain D_{el} is bounded by the ellipse (1). We make the change of variables

$$x=E_x u,\qquad y=E_y v$$

Then the ellipse D_{el} will be carried into the circle

$$u^2+v^2=k^2 \tag{3}$$

Since the Jacobian of the transformation is equal to $I=E_x\cdot E_y$, equation (2) becomes

$$\mathbf{P}[(\bar{x},\ \bar{y})\subset D_{el}]=\frac{1}{\pi}\iint\limits_{D_k}\rho^2 e^{-\rho^2(u^2+v^2)}\,du\,dv \tag{4}$$

In the last integral we pass to polar coordinates

$$u=r\cos\varphi,\qquad v=r\sin\varphi$$

Then the right member of (4) takes the form

$$\mathbf{P}[(\bar{x},\ \bar{y})\subset D_{el}]=\frac{1}{\pi}\int\limits_0^{2\pi}\int\limits_0^{k}\rho^2 e^{-\rho^2 r^2}\,r\,dr\,d\varphi$$

Carrying out the computations in the right member, we get the expression of the probability of falling in the ellipse of dispersion:

$$\mathbf{P}[(\bar{x},\ \bar{y})\subset D_{el}]=1-e^{-\rho^2k^2} \tag{5}$$

Let us consider some particular cases. The probability of falling in the unit ellipse of dispersion occurs if we put $k=1$ in formula (5)

$$\mathbf{P}[(\bar{x},\ \bar{y})\subset D_{el}]_{k=1}=1-e^{-\rho^2}=0.203 \tag{6}$$

The probability of falling in the total ellipse of dispersion (7), Sec. 8.24, occurs if in (5) we put $k=4$:

$$\mathbf{P}[(\bar{x},\ \bar{y})\subset D_{el}]_{k=4}=1-e^{-16\rho^2}=0.974 \tag{7}$$

Let us consider the particular case where in (4), Sec. 8.24, $E_x = E_y = E$. The ellipse of dispersion (5), Sec. 8.24, turns into circle

$$x^2 + y^2 = k^2 E^2 \tag{8}$$

with radius $R = kE$. The probability of a two-dimensional random variable falling in a circle of radius R is, in accord with (5),

$$\mathbf{P}\left[(\bar{x}, \bar{y}) \subset D_R\right] = 1 - \exp\left(-\rho^2 \frac{R^2}{E^2}\right) \tag{9}$$

Definition 1. The *mean radial error* is a number E_R such that the probability of a two-dimensional variable falling in a circle of radius $R = E_R$ is equal to $\frac{1}{2}$.

The definition implies that the quantity $R = E_R$ is determined from the relation

$$1 - \exp\left(-\rho^2 \frac{E_R^2}{E^2}\right) = \frac{1}{2}$$

From a table of values of the exponential function we find

$$E_R = 1.75E$$

8.27 PROBLEMS OF MATHEMATICAL STATISTICS. STATISTICAL DATA

Statistical data are obtained as a result of observation and recording of mass random (chance) phenomena. An instance of such data are the errors of measurement.

If an observed quantity is a random variable, then it is studied by the methods of probability theory. To grasp the nature of this random variable, one has to know its distribution law. Determining the laws of distribution of such quantities and estimating the values of the parameters of distribution on the basis of observed values constitutes a problem in mathematical statistics.

Another problem of mathematical statistics is the establishment of methods of processing and analyzing statistical data to obtain certain conclusions necessary in organizing an optimal process involving the quantities under consideration.

The following are some illustrative examples of observations of phenomena that yield statistical data.

Example 1. The distance to an object is measured a large number of times with the aid of a measuring instrument yielding a series of values of the observed quantity. They are called *observed values* (we will use this term for any value obtained in studying any phenomenon).

The values thus obtained require systematization and processing before any conclusions can be drawn from them.

As has already been pointed out, the difference δ between an observed value x and the true value of the observed quantity a $(x-a=\delta)$ is called the *error of measurement*. The errors of measurement require mathematical processing in order to obtain specific conclusions.

Example 2. In mass production one has to consider the amount of deviation of a certain dimension of a finished article (say, length) from a given dimension (fabrication error).

Example 3. The difference between the coordinate of the point of impact in gunfire and the coordinate of the target is the error of gunfire (dispersion). These errors require mathematical investigation.

Example 4. The results of measurements of the amount of departure of the dimensions of an item after a period of operation from its dimensions prior to operation (design dimensions) require mathematical analysis. Such deviations can also be regarded as "errors".

It is clear from the foregoing that these quantities are random variables and each observed value should be regarded as a particular value of the random variable.

For instance, the range error (dispersion) in gunfire is determined by the error obtained when weighing the charge, the error in weight in the manufacturing process, the aiming error, the range determining error, variations in weather conditions, and so forth. All these are random variables, and dispersion, as the result of their joint action, is also a random variable.

8.28 STATISTICAL SERIES. HISTOGRAM

The statistical findings obtained in observations (measurements) are tabulated in two rows. The first row contains the number i of the measurement, the second, the value x_i of the quantity x being measured:

i	1	2	3	...	i	...	n
x_i	x_1	x_2	x_3	...	x_i	...	x_n

This table is called a *simple statistical series*. If there are a large number of measurements, the statistical data is not readily surveyable and an analysis is complicated. Therefore, the data obtained from a simple statistical series are given in a *frequency table*. This is done in the following manner.

The whole interval of values of the quantity x is partitioned into subintervals of equal length (a_0, a_1), (a_1, a_2), ..., $(a_{\lambda-1}, a_\lambda)$ and then we count the number of values m_k of x falling in the interval (a_{k-1}, a_k). The values at the end of the interval are referred either to the left or right intervals (sometimes half is placed in the left interval and half in the right). The number

$$\frac{m_k}{n} = p_k^* \tag{1}$$

is the relative frequency corresponding to the interval (a_{k-1}, a_k). Clearly,

$$\sum_{k=1}^{\lambda} p_k^* = 1 \tag{2}$$

Having processed the data in this manner, we then construct a table consisting of three rows. The first row indicates the intervals in order of increasing a_k, the second row, the corresponding numbers m_k, and the third row, the frequencies $p_k = \frac{m_k}{n}$:

Intervals	(a_0, a_1)	(a_1, a_2)	...	(a_{k-1}, a_k)	...	$(a_{\lambda-1}, a_\lambda)$
m_k	m_1	m_2	...	m_k	...	m_λ
p_k^*	p_1^*	p_2^*	...	p_k^*	...	p_λ^*

This is what is called the *frequency table*. It may be converted into a diagram, as follows. On the x-axis we mark off points a_0, a_1, ..., a_k, ..., a_λ. On each interval $[a_{k-1}, a_k]$ as a base, we construct a rectangle whose area is equal to p_k^*. The resulting figure is called a *histogram* (Fig. 199).

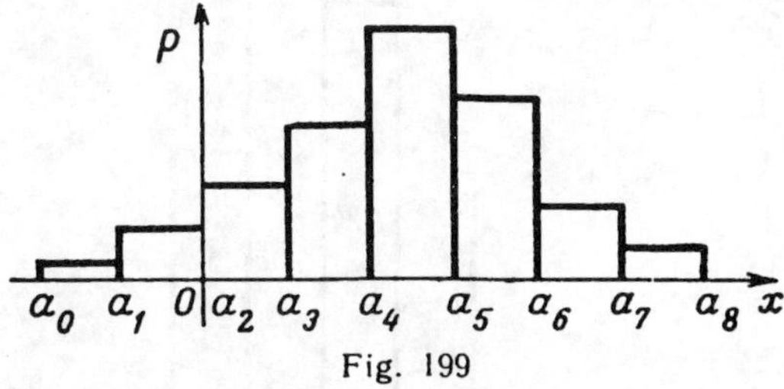

Fig. 199

On the basis of the frequency table and the histogram we can give an approximate construction of a statistical distribution function.

The processing is continued as follows. The midvalues of the intervals (a_{k-1}, a_k) are denoted by $\tilde{x}_k$ and this value is taken to be the value of the measurement, which is repeated m_k times. Then in place of the frequency table we construct the following table:

$\tilde{x}_k$	$\tilde{x}_1$	$\tilde{x}_2$	...	$\tilde{x}_k$	...	$\tilde{x}_\lambda$
m_k	m_1	m_2	...	m_k	...	m_λ
p_k^*	p_1^*	p_2^*	...	p_k^*	...	p_λ^*

This procedure is based on the fact that all values in the interval (a_{k-1}, a_k) are close-lying values and are therefore considered to be equal to the abscissa of the midvalue of the interval x_k^*.

Example. One hundred range measurements were made yielding the following results (frequency table):

Intervals	80–110	110–140	140–170	170–200	200–230	230–260	260–290	290–320
m_k	2	5	16	24	28	18	6	1
p_k^*	0.02	0.05	0.16	0.24	0.28	0.18	0.06	0.01

From this frequency table we compile a histogram (Fig. 200).

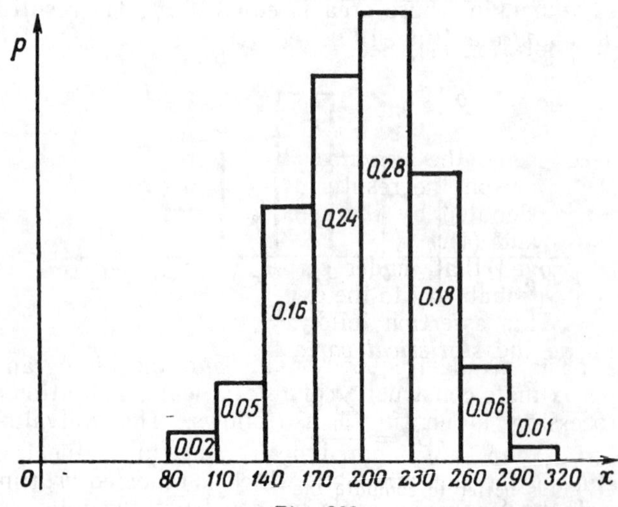

Fig. 200

We then construct the following table:

$\tilde{x}_k$	95	125	155	185	215	245	275	305
m_k	2	5	16	24	28	18	6	1
p_k^*	0.02	0.05	0.16	0.24	0.28	0.18	0.06	0.01

8.29 DETERMINING A SUITABLE VALUE OF A MEASURED QUANTITY

Suppose that in measuring a certain quantity we obtained the measurements $x_1, x_2, \ldots, x_n$. These values may be regarded as particular values of the random variable x. For the suitable value of the quantity being measured we take the arithmetic mean of the values:

$$m_x^* = \frac{\sum_{i=1}^{n} x_i}{n} \tag{1}$$

The quantity m_x^* is called the *statistical mean.*

If the number of measurements n is great, then use is made of the tabulated data considered in Sec. 8.28, and m_x^* is computed as follows:

$$m_x^* = \frac{\tilde{x}_1 m_1 + \tilde{x}_2 m_2 + \ldots + \tilde{x}_k m_k + \ldots + \tilde{x}_\lambda m_\lambda}{n}$$

or, using the notation (1), Sec. 8.28,

$$m_x^* = \sum_{k=1}^{\lambda} \tilde{x}_k p_k^* \tag{2}$$

This value is termed the *weighted mean.*

Note. From now on the results of computations via formulas (1) and (2) will be denoted by the same letter. This will also be true of formulas (3) and (4).

It can be proved that, under certain restrictions, the statistical mean tends in probability to the expectation of the random variable x as $n \to \infty$. This assertion follows from the Chebyshev theorem.

Let us define the *statistical variance:* *

$$D^* = \frac{\sum_{i=1}^{n} (x_i - m_x^*)^2}{n} \tag{3}$$

* Actually, it is better to compute the statistical variance by means of formula (5) of Sec. 8.30.

This quantity characterizes the dispersion of values of an observed variable.

If we make use of the material of the tables in Sec. 8.28, then the statistical variance is determined from the formula

$$D^* = \sum_{k=1}^{\lambda} (\tilde{x}_k - m_k^*)^2 p_k^* \tag{4}$$

This formula is similar to formula (2) of Sec. 8.10.

Example. Determine the statistical mean and the statistical variance on the basis of the statistical data of the Example in Sec. 8.28.

Solution. By formula (2) we get

$$m_x^* = \frac{\sum_{i=1}^{n} x_i}{n} = \sum_{k=1}^{\lambda} x_k p_k^* = 95 \cdot 0.02 + 125 \cdot 0.05 + 155 \cdot 0.16$$
$$+ 185 \cdot 0.24 + 215 \cdot 0.28 + 245 \cdot 0.18 + 275 \cdot 0.06 + 305 \cdot 0.01 = 201.20$$

By formula (4) we obtain

$$D^* [\bar{x}] = \frac{\sum_{k=1}^{n} (x_k - m_x^*)^2}{n} = \sum_{k=1}^{\lambda} (\tilde{x}_k - m_x^*)^2 p_k^* = \sum_{k=1}^{\lambda} \tilde{x}_k^2 p_k^* - m_x^{*2}$$
$$= 95^2 \cdot 0.02 + 125^2 \cdot 0.05 + 155^2 \cdot 0.16 + 185^2 \cdot 0.24 + 215^2 \cdot 0.28$$
$$+ 245^2 \cdot 0.18 + 275^2 \cdot 0.06 + 305^2 \cdot 0.01 - (201.20)^2 = 1753.56$$

8.30 DETERMINING THE PARAMETERS OF A DISTRIBUTION LAW. LYAPUNOV'S THEOREM. LAPLACE'S THEOREM

Let $\bar{x}$ be a random variable, say, the result of a measurement, let a be the quantity being measured, and δ, the error of measurement. Then these quantities are connected by the relation

$$\delta = \bar{x} - a, \quad \bar{x} = a + \delta \tag{1}$$

Numerous experiments and observations have shown that the errors of measurements (after excluding crude errors and after excluding any systematic error, that is, an error that is constant for all measurements—say the instrument error—or that varies by a known law from measurement to measurement) obey the normal distribution law with centre of distribution at the origin. This is confirmed by theoretical arguments as well.

If a random variable is the sum of a large number of random variables, then for certain restrictions the sum obeys the normal distribution law. This assertion is stated in the so-called central limit theorem of A. M. Lyapunov (1857-1918). The theorem is given here in simplified form.

Theorem 1. *If independent random variables $\bar{x}_1, \bar{x}_2, \ldots, \bar{x}_n$ obey one and the same distribution law with expectation a (without loss of generality, we can assume $a=0$) and variance σ^2, then, as n increases without bound, the sum $\bar{y}_n = \dfrac{\sum_{i=1}^{n} \bar{x}_i}{\sigma\sqrt{n}}$ is asymptotically normally distributed ($\bar{y}_n$ is normalized so that $\mathbf{M}[\bar{y}_n]=0$, $\mathbf{D}[\bar{y}_n]=1$).*

The practical significance of the Lyapunov theorem consists in the following. We consider a random variable, say, the deviation of a certain quantity from the given quantity. This deviation is due to many factors, each of which makes a certain contribution to the deviation, say in the case of gunfire, the departure of the point of impact from the aiming point is due to the aiming error, the range error, the shell-fabrication error, etc. We do not even know all of the contributing factors, neither do we know the distribution laws of the component random variables. But from the Lyapunov theorem it follows that the random variable of the total deviation obeys the normal distribution law.

Lyapunov's theorem implies that if $\bar{x}_1, \bar{x}_2, \ldots, \bar{x}_n$ are the results of measurements of a certain quantity (each of the $\bar{x}_i$ is a random variable), then the random variable, which is the arithmetic mean,

$$\bar{x} = \frac{\bar{x}_1 + \bar{x}_2 + \ldots + \bar{x}_n}{n}$$

is asymptotically normally distributed for sufficiently large n if the random variables $\bar{x}_i$ obey one and the same distribution law.

The theorem holds true also for a sum of random variables having unlike distribution laws, subject to certain additional conditions, which as a rule are fulfilled for random variables ordinarily encountered in practical situations. Experience shows that for the number of terms of the order of 10, the sum may be regarded as normally distributed.

Let us denote by $\bar{a}$ and $\bar{\sigma}^2$ the approximate values of the expectation and variance. We can then write down approximately the laws of distribution of the random variables $\bar{\delta}$ and $\bar{x}$:

$$\bar{f}(\delta) = \frac{1}{\bar{\sigma}\sqrt{2\pi}} \exp\left(-\frac{\delta^2}{2\bar{\sigma}^2}\right) \tag{2}$$

$$f(x) = \frac{1}{\bar{\sigma}\sqrt{2\pi}} \exp\left[-\frac{(x-\bar{a})^2}{2\bar{\sigma}^2}\right] \tag{3}$$

On the basis of experimental findings, the parameter $\bar{a}$ is determined from formula (1), Sec. 8.29:

$$\bar{a} = \frac{\sum_{i=1}^{n} x_i}{n} \tag{4}$$

This follows from the so-called theorem of Chebyshev (1821-1894). Without dwelling on the proof, we note that the parameter σ is more naturally determined from the following formula instead of from (3) of Sec. 8.29:

$$\bar{\sigma}^2 = \frac{\sum_{i=1}^{n} (x_i - \bar{a})^2}{n-1} \tag{5}$$

Note that the right member in (5) and the right member of (3), Sec. 8.29, differ by the factor $\frac{n}{n-1}$, which is close to 1 in practical problems.

Example 1. Write down the expression for the distribution law of the random variable on the basis of the measurements given in the Example of Sec. 8.28, and the results of computations carried out in the Example of Sec. 8.29.

Solution. On the basis of the computations carried out in the Example of Sec. 8.29, we obtain

$$\bar{a} = m_x^* = 201$$

$$\bar{\sigma}^2 = \frac{n}{n-1} D^* = \frac{100}{99} \cdot 1754 = 1771$$

$$\bar{\sigma} = \sqrt{1771} \approx 41$$

Substituting into (3), we get

$$f(x) = \frac{1}{41\sqrt{2\pi}} \exp\left[-\frac{(x-201)^2}{2 \cdot 1771}\right]$$

Note. If a statistical distribution function has been obtained for some random variable x, then the question of whether the given random variable obeys the normal law or not is sometimes settled in the following manner.

Suppose we have the following values of a random variable:

$$x_1, x_2, \ldots, x_n$$

Find the arithmetic mean of $\bar{a}$ by formula (4). Find the values of the centred random variable

$$y_1, y_2, \ldots, y_n$$

The absolute values of y_i are then arranged in a series in increasing order. If n is odd, then for the probable (mean) error E_m in the constructed series of absolute values, we take the absolute value $|y_m|$ that occupies the $\left(\frac{n-1}{2}\right)$ th + 1 site; if n is even, then for E_m we take the arithmetic mean of the absolute values occupying sites with numbers $\frac{n}{2}$ and $\frac{n}{2}+1$.

We then form the mean arithmetic error using the formula

$$d=\frac{\sum_{i=1}^{n}|y_i|}{n} \tag{6}$$

From (5) we determine the standard deviation

$$\bar{\sigma}=\sqrt{\frac{\sum_{i=1}^{n} y_i^2}{n-1}} \tag{7}$$

And then we find the ratios $\frac{E_m}{d}$ and $\frac{E_m}{\bar{\sigma}}$.

For a random variable that obeys the normal law, the ratios $\frac{E}{d}$ and $\frac{E}{\bar{\sigma}}$ are equal, respectively, to 0.8453 and 0.6745 [see formula (6), Sec. 8.22]. If the ratios $\frac{E_m}{d}$ and $\frac{E_m}{\bar{\sigma}}$ differ from 0.8453 and 0.6745 by a quantity of the order of 10%, then it is agreed, conditionally, that the random variable y obeys the normal law.

A corollary to the central limit theorem is an important theorem of Laplace concerning the probability that an event will occur not less than α times and no more than β times. We give this theorem without proof.

Theorem 2 (Laplace). *If n independent trials are made in each of which the probability of occurrence of event A is p, then the following relation holds true:*

$$\mathbf{P}(\alpha < m < \beta)=\frac{1}{2}\left[\Phi\left(\frac{\beta-np}{\sqrt{2}\sqrt{npq}}\right)-\Phi\left(\frac{\alpha-np}{\sqrt{2}\sqrt{npq}}\right)\right] \tag{8}$$

where m is the number of occurrences of A, $q=1-p$, and $\mathbf{P}(\alpha < m < \beta)$ is the probability that the number of occurrences of A lies between α and β.

The function $\Phi(x)$ is defined in Sec. 8.17.

We will demonstrate the use of the Laplace theorem in the solution of problems.

Example 2. The probability of defective items in the production of certain articles is $p=0.01$. Determine the probability that in 1000 articles there will be no more than 20 defective items.

Solution. Here,

$$n=1000, \quad p=0.01, \quad q=0.99, \quad \alpha=0, \quad \beta=20$$

We then find

$$\frac{\alpha-np}{\sqrt{2}\sqrt{npq}}=\frac{0-10}{\sqrt{2}\sqrt{9.9}}=-2.25$$

$$\frac{\beta-np}{\sqrt{2}\sqrt{npq}}=\frac{20-10}{\sqrt{2}\sqrt{9.9}}=2.25$$

By formula (8) we get

$$\mathbf{P}\,(0\leqslant m\leqslant 20)=\frac{1}{2}\left[\Phi\,(2.25)-\Phi\,(-2.25)\right]=\Phi\,(2.25)$$

Using the table of the function $\Phi(x)$, we find

$$\mathbf{P}\,(0\leqslant m\leqslant 20)=0.9985$$

Note that the theorems of Bernoulli, Lyapunov, Chebyshev, and Laplace discussed above constitute what is known as the *law of large numbers in probability theory*.

Exercises on Chapter 8

1. Two dice are thrown at one time. Find the probability that a sum of 5 will turn up. *Ans.* 1/9.

2. There are 10 tickets in a lottery: 5 wins and 5 losses. Take two tickets. What is the probability of a win? *Ans.* 7/9.

3. A die is thrown 5 times. What is the probability that a number of four will turn up at least once? *Ans.* 0.99987.

4. The probability of hitting an aircraft with a rifle is 0.004. How many men must shoot so that the probability of a hit becomes $> 70\%$? *Ans.* $n > 300$.

5. Two guns fire at the same target. The probability of a hit from the first gun is 0.7, from the second one, 0.6. Determine the probability of at least one hit. *Ans.* 0.88.

6. One hundred cards are numbered from 1 to 100. Find the probability that a randomly chosen card has the digit 5. *Ans.* 0.19.

7. There are 4 machines. The probability that a machine is in operation at an arbitrary time t is equal to 0.9. Find the probability that at time t at least one machine is working *Ans.* 0.9999.

8. The probability of hitting a target is $p=0.9$. Find the probability that in three shots there will be three hits. *Ans.* ≈ 0.73.

9. Box One contains 30% first-grade articles, Box Two contains 40% first-grade articles. One article is drawn from each box. Find the probability that both drawn articles are first-grade. *Ans.* 0.12.

10. A mechanism consists of three parts. The probability of defective items in the manufacture of the first part is $p_1=0.008$, the probability of defectives in the second part is $p_2=0.012$, the probability of defectives in the third part is $p_3=0.01$. Find the probability that the whole mechanism is defective. *Ans.* 0.03.

11. The probability of a hit in a single shot is $p=0.6$. Determine the probability that three shots will yield at least one hit. *Ans.* 0.936.

12. Out of a total of 350 machines, there are 160 of grade One, 110 of grade Two, and 80 of grade Three. The probability of defectives in the grade-one category is 0.01, in the grade-two category, 0.02, in the grade-three category, 0.04. Take one machine. Determine the probability that it is acceptable. *Ans.* 0.98.

13. Suppose it is known that due to errors committed in preparation for gunfire, the centre of dispersion of the shells (CDS) in the first shot can lie at (in range) one of five points. The probabilities that the CDS will lie at these points are equal to $p_1 = 0.1$, $p_2 = 0.2$, $p_3 = 0.4$, $p_4 = 0.2$, $p_5 = 0.1$. It is also known that if the CDS lies in the first point, the probability (in range) of hitting the target will be $\bar{p}_1 = 0.15$. For the other points we will have for this case: $\bar{p}_2 = 0.25$, $\bar{p}_3 = 0.60$, $\bar{p}_4 = 0.25$, $\bar{p}_5 = 0.15$.

In the initial adjustment of sight, a shot was fired with a miss in range. Determine the probability that the shot is made at adjustment of sight corresponding to each of the five indicated points of the CDS, that is, determine the probabilities of causes (hypotheses) concerning distinct errors in the position of the CDS after a trial (shot). *Ans.* 0.85, 0.75, 0.40, 0.75, 0.85.

14. A die is thrown 5 times. What is the probability that a six will turn up twice and a non-six three times? *Ans.* 625 3888.

15. Six shots are fired. Find the probability that not all are plus rounds if the probability of a plus round is $p = 1/2$, the probability of a minus round is $q = 1/2$ (gunfire at a "narrow" target). *Ans.* 31/32.

16. Referring to the conditions of the preceding problem, determine the probability that there will be three plus rounds and three minus rounds. *Ans.* 5/16.

17. Find the mathematical expectation of the number that turns up in a single toss of a die. *Ans.* 7/2.

18. Find the variance of a random variable x given by the following distribution table·

x	2	3	5
p	0.1	0.6	0.3

Ans. 1.05.

19. The probability of occurrence of event A in a single trial is equal to 0.4. Five independent trials are carried out. Find the variance of the number of occurrences of A. *Ans.* 1.2.

20. In a case of gunfire at a target, the probability of a hit is 0.8. The fire is kept up till a first hit. There are 4 shells. Find the expectation of the number of shells used. *Ans.* 1.242.

21. In firing at a "narrow" target, the probability of a plus round is $p = 1/4$, the probability of a minus round is $q = 3/4$. Find the probability of a combination of two plus rounds and 4 minus rounds in six shots. *Ans.* 0.297.

22. The probability that an article will be defective is $p = 0.01$. What is the probability that a batch of 10 articles will yield 0, 1, 2, 3 defectives? *Ans.* 0.9045, 0,0904, 0.0041, 0.0011.

23. Find the probabilities of obtaining at least one hit in the case of 10 shots if the probability of hitting the target in a single shot is $p = 0.15$. *Ans.* $1 - (0.85)^{10} \approx 0.803$.

24. A random variable x is given by the distribution function

$$F(x) = \begin{cases} 0 & \text{for } x < 0 \\ x & \text{for } 0 \leqslant x \leqslant 1 \\ 1 & \text{for } 1 < x \end{cases}$$

Find the density function $f(x)$, $\mathbf{M}[x]$, $\mathbf{D}[x]$.

Ans. $$f(x) = \begin{cases} 0 & \text{for } x < 0 \\ 1 & \text{for } 0 \leqslant x \leqslant 1, \\ 0 & \text{for } 1 < x \end{cases} \qquad \mathbf{M}[x] = \frac{1}{2}, \qquad \mathbf{D}[x] = \frac{1}{12}.$$

25. A random variable x obeys the normal law with expectation 30 and variance 100. Find the probability that the value of the random variable lies in the interval (10, 50). *Ans.* 0.954.

26. A random variable obeys the normal distribution law with variance $\sigma^2 = 0.16$. Find the probability that a value of the random variable will differ in absolute value from the mathematical expectation by less than 0.3. *Ans.* 0.5468.

27. A random variable x is normally distributed with centre of dispersion $a = 0.3$ and modulus of precision $h = 2$. Find the probability of a hit in the interval (0.5, 2.0). *Ans.* 0.262.

28. Gunfire is conducted at a strip of width 4 metres. There is a systematic error in aiming of 1 metre (undershooting). The probable error is 5 metres. Find the probability of hitting the strip in the case of normal dispersion. *Ans.* 0.211.

29. Fire is conducted at a rectangle bounded by the straight lines $x_1 = 10$ m, $x_2 = 20$ m, $y_1 = 15$ m, $y_2 = 35$ m, in the direction of the straight line bisecting the short side. The probable errors of normal distribution on the plane are $E_x = 5$ m, $E_y = 10$ m. Find the probability of a hit in a single shot. *Ans.* 0.25.

30. An error in the fabrication of an article with a given length 20 cm is a normally distributed random variable; $\sigma = 0.2$ cm. Determine the probability that the length of a manufactured article will differ from the given value by less than 0.3 cm. *Ans.* 0.866.

31. Under the hypotheses of Example 30, determine the fabrication error that has probability 0.95 of not being exceeded. *Ans.* 0.392.

32. A random variable x is normally distributed with parameters $\mathbf{M}[x] = 5$ and $\sigma = 2$. Determine the probability that the random variable will lie in the interval (1, 10). Make a drawing. *Ans.* 0.971.

33. The length of an article made on an automatic machine is a normally distributed random variable with parameters $\mathbf{M}[x] = 15$, $\sigma = 0.2$. Find the probability of defective output if the tolerance is 15 ± 0.3. What accuracy of length of the article can be guaranteed with probability 0.97? Make a drawing.

34. In measuring a certain quantity we get the following statistical series:

x	1	2	3	4
Relative frequency	20	15	10	5

Determine the statistical mean and the statistical variance. *Ans.* 2 and 1.

35. The results of measurements are tabulated as follows:

x	0.18	0.20	0.22	0.24	0.26	0.28
Relative frequency	4	18	33	35	9	1

Find the statistical mean $\bar{a}$ and the statistical variance $\bar{\sigma}^2$. *Ans.* 0.226, 0.004.

36. The probability of defective production is $p = 0.02$. Find the probability that in a batch of 400 parts there will be from 7 to 10 defectives. *Ans.* 0.414.

37. The probability of a hit is $p = \frac{1}{2}$. What is the probability that in 250 shots the number of hits will lie between 100 and 150? *Ans.* 0.998.

38. The probability of defective production is $p = 0.02$. Find the probability that among 1000 items there will be not more than 25 defectives. *Ans.* 0.87.

CHAPTER 9

MATRICES

9.1 LINEAR TRANSFORMATIONS. MATRIX NOTATION

Consider two planes P and Q. Given in the P plane is a rectangular system of coordinates x_1Ox_2, in the Q plane, a system y_1Oy_2. The P and Q planes can be brought to coincidence and so can the coordinate systems. Let us consider the following system of equations:

$$\left.\begin{aligned} y_1 &= a_{11}x_1 + a_{12}x_2 \\ y_2 &= a_{21}x_1 + a_{22}x_2 \end{aligned}\right\} \qquad (1)$$

By (1), to each point $M(x_1, x_2)$ of the x_1x_2-plane there corresponds a point $\overline{M}(y_1, y_2)$ of the y_1y_2-plane.

We say that equations (1) constitute *linear transformations* of the coordinates. These equations map the x_1x_2-plane onto the y_1y_2-plane (not necessarily onto the whole plane). Since the equations (1) are linear, the mapping is termed a *linear mapping*.

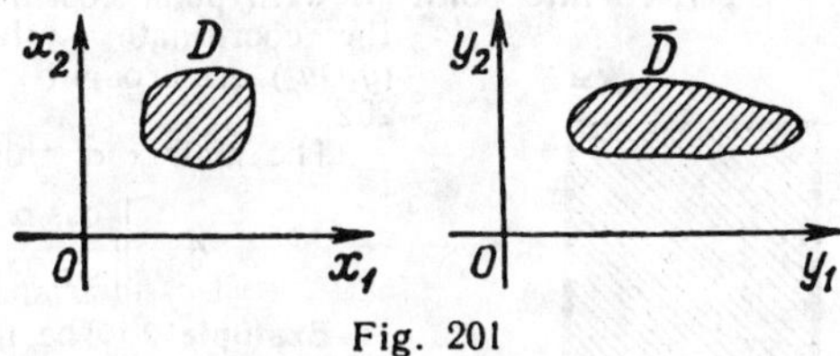

Fig. 201

If in the x_1x_2-plane we consider some domain D, then using (1) we can define some set of points $\overline{D}$ of the y_1y_2-plane (Fig. 201).

Note that one can also consider nonlinear mappings

$$y_1 = \varphi(x_1, x_2), \quad y_2 = \psi(x_1, x_2)$$

In this text we confine ourselves to linear mappings.

The mapping (1) is fully defined by the set of coefficients a_{11} a_{12}, a_{21}, a_{22}.

A rectangular array of these coefficients can be written hus:

$$\begin{Vmatrix} a_{11} & a_{12} \\ a_{21} & a_{22} \end{Vmatrix} \quad \text{or} \quad \begin{pmatrix} a_{11} & a_{12} \\ a_{21} & a_{22} \end{pmatrix}$$

It is called a *matrix* of the mapping (1). The symbol $\| \; \|$ or ( ) is the matrix symbol.

Matrices are also designated by a single letter, for instance $\boldsymbol{A}$ or $\| A \|$:

$$\boldsymbol{A} = \begin{Vmatrix} a_{11} & a_{12} \\ a_{21} & a_{22} \end{Vmatrix} \tag{2}$$

A determinant made up of the elements of a matrix, in the order they are given in the matrix, is called the *determinant of the matrix* [we denote the determinant by $\Delta(A)$]:

$$\Delta = \begin{Vmatrix} a_{11} & a_{12} \\ a_{21} & a_{22} \end{Vmatrix} \tag{3}$$

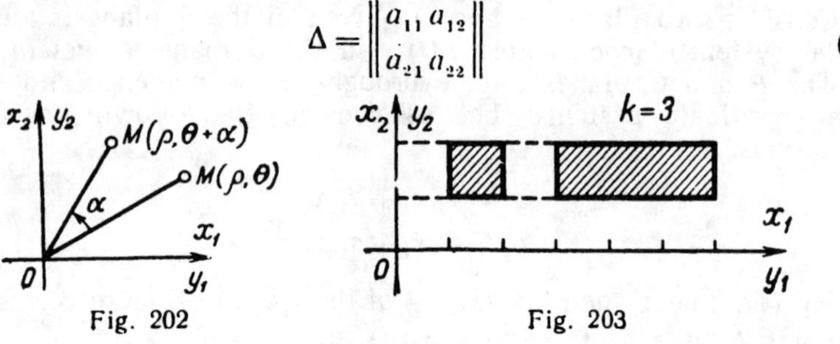

Fig. 202

Fig. 203

Example 1. The mapping

$$y_1 = x_1 \cos \alpha - x_2 \sin \alpha$$
$$y_2 = x_1 \sin \alpha + x_2 \cos \alpha$$

is a *rotation through the angle* α. In this mapping, every point M with polar coordinates (ρ, θ) is carried into point $\overline{M}$ with polar coordinates $(\rho, \theta + \alpha)$ if the coordinate systems (x_1Ox_2) and (y_1Oy_2) are brought to coincidence (Fig. 202).

The matrix of this mapping is

$$\boldsymbol{A} = \begin{Vmatrix} \cos \alpha & -\sin \alpha \\ \sin \alpha & \cos \alpha \end{Vmatrix}$$

Example 2. The mapping

$$y_1 = kx_1$$
$$y_2 = x_2$$

represents a *stretching along the* x_1*-axis with stretching factor* k (Fig. 203).

The matrix of this mapping is

$$\boldsymbol{A} = \begin{Vmatrix} k & 0 \\ 0 & 1 \end{Vmatrix}$$

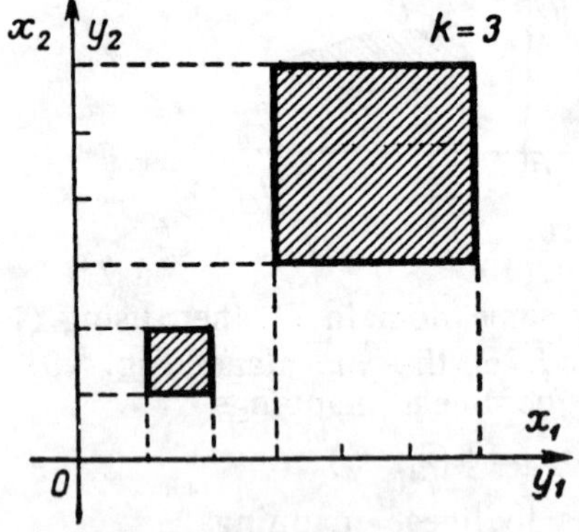

Fig. 204

Example 3. The mapping

$$y_1 = kx_1$$
$$y_2 = kx_2$$

is a *k-fold stretching both in the direction of the* x_1*-axis and in the direction of the* x_2*-axis* (Fig. 204).

The matrix of this mapping is

$$A=\begin{Vmatrix} k & 0 \\ 0 & k \end{Vmatrix}$$

Example 4. The transformation

$$y_1 = -x_1$$
$$y_2 = x_2$$

is called a *mirror reflection in the x_2-axis* (Fig. 205).

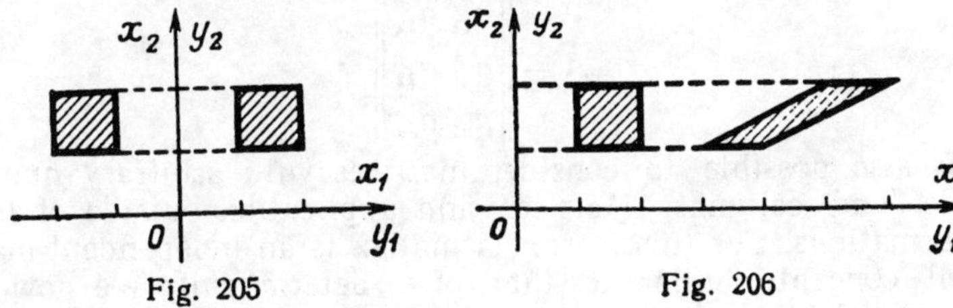

Fig. 205 Fig. 206

The matrix of this transformation is

$$A=\begin{Vmatrix} -1 & 0 \\ 0 & 1 \end{Vmatrix}$$

Example 5. The transformation

$$y_1 = x_1 + \lambda x_2$$
$$y_2 = x_2$$

is termed a *simple shear transformation along the x_2-axis* (Fig. 206). The matrix of this transformation is

$$A=\begin{Vmatrix} 1 & \lambda \\ 0 & 1 \end{Vmatrix}$$

We can consider a linear transformation with an arbitrary number of variables.

Thus, the transformation

$$\left.\begin{aligned} y_1 &= a_{11}x_1 + a_{12}x_2 + a_{13}x_3 \\ y_2 &= a_{21}x_1 + a_{22}x_2 + a_{23}x_3 \\ y_3 &= a_{31}x_1 + a_{32}x_2 + a_{33}x_3 \end{aligned}\right\} \quad (4)$$

is a mapping of three-dimensional space (x_1, x_2, x_3) into the three-dimensional space (y_1, y_2, y_3). The matrix of this transformation is

$$A=\begin{Vmatrix} a_{11} & a_{12} & a_{13} \\ a_{21} & a_{22} & a_{23} \\ a_{31} & a_{32} & a_{33} \end{Vmatrix} \quad (5)$$

It is also possible to consider linear transformations with a non-square matrix, which is a matrix where the number of rows differs from the number of columns. The transformation

$$\left.\begin{aligned} y_1 &= a_{11}x_1 + a_{12}x_2 \\ y_2 &= a_{21}x_1 + a_{22}x_2 \\ y_3 &= a_{31}x_1 + a_{32}x_2 \end{aligned}\right\} \tag{6}$$

is a mapping of the x_1x_2-plane into a certain set of points in the space (y_1, y_2, y_3).

Here, the matrix of the transformation is

$$A = \begin{Vmatrix} a_{11} & a_{12} \\ a_{21} & a_{22} \\ a_{31} & a_{32} \end{Vmatrix} \tag{7}$$

It is also possible to consider matrices with arbitrary numbers of rows and columns. Matrices find application outside of linear transformations. For this reason, a matrix is an independent mathematical concept similar to that of a determinant. We now give some definitions involving the matrix concept.

9.2 GENERAL DEFINITIONS INVOLVING MATRICES

Definition 1. A rectangular array of mn numbers consisting of m rows and n columns,

$$A = \begin{Vmatrix} a_{11} & a_{12} & \dots & a_{1n} \\ \cdot & \cdot & \cdot & \cdot \\ a_{m1} & a_{m2} & \dots & a_{mn} \end{Vmatrix} \tag{1}$$

is termed a *matrix*. A matrix is also denoted briefly as

$$A = \| a_{ij} \| \quad (i = 1, 2, \dots, m; \; j = 1, 2, \dots, n) \tag{2}$$

where a_{ij} are the entries (elements) of the matrix.

If the number of rows in a matrix is equal to the number of columns, $m = n$, then we have a *square matrix*:

$$A = \begin{Vmatrix} a_{11} & a_{12} & \dots & a_{1n} \\ a_{21} & a_{22} & \dots & a_{2n} \\ \cdot & \cdot & \cdot & \cdot \\ a_{n1} & a_{n2} & \dots & a_{nn} \end{Vmatrix} \tag{3}$$

Definition 2. A determinant consisting of the elements of a square matrix (in the order given in the matrix) is called the *determinant of the matrix*. We denote it by $\Delta(A)$:

$$\Delta(A) = \begin{vmatrix} a_{11} & a_{12} & \dots & a_{1n} \\ a_{21} & a_{22} & \dots & a_{2n} \\ a_{n1} & a_{n2} & \dots & a_{nn} \end{vmatrix} \tag{4}$$

Note that a nonsquare matrix does not have a determinant.

Definition 3. A matrix A^* is called the *transpose* of A if the columns of A are the rows of A^*.

Example. Let

$$A = \begin{Vmatrix} a_{11} & a_{12} \\ a_{21} & a_{22} \\ a_{31} & a_{32} \end{Vmatrix}$$

The transposed matrix A^* is then

$$A^* = \begin{Vmatrix} a_{11} & a_{21} & a_{31} \\ a_{12} & a_{22} & a_{32} \end{Vmatrix}$$

Definition 4. A square matrix A is said to be *symmetric about the principal diagonal* if $a_{ij} = a_{ji}$. It is clear from this that a symmetric matrix coincides with its transpose.

Definition 5. A square matrix whose elements outside the principal diagonal are all zero is called a *diagonal matrix*. If the elements of a diagonal matrix on the principal diagonal are all unity, then the matrix is termed a *unit matrix* (*identity matrix*) and is denoted by E:

$$E = \begin{Vmatrix} 1 & 0 & \ldots & 0 \\ 0 & 1 & \ldots & 0 \\ . & . & . & . \\ 0 & 0 & \ldots & 1 \end{Vmatrix} \tag{5}$$

Definition 6. We can also consider matrices consisting of a single column or a single row:

$$X = \begin{Vmatrix} x_1 \\ x_2 \\ \vdots \\ x_m \end{Vmatrix}, \quad Y = \| y_1 \; y_2 \; \ldots \; y_m \| \tag{6}$$

The former is called a *column matrix*, the latter, a *row matrix*.

Definition 7. Two matrices A and B are considered *equal* if they have the same number of rows and columns and if the corresponding elements are equal:

$$A = B \tag{7}$$

or

$$\| a_{ij} \| = \| b_{ij} \| \quad (i = 1, 2, \ldots, m; \quad j = 1, 2, \ldots, n) \tag{8}$$

if

$$a_{ij} = b_{ij} \tag{9}$$

It is sometimes convenient to identify a column matrix and a vector in a space with the corresponding number of dimensions, where the elements of the matrix are the projections of the vector on the appropriate coordinate axes. Thus, we can write

$$\begin{Vmatrix} x_1 \\ x_2 \\ x_3 \end{Vmatrix} = x_1\boldsymbol{i} + x_2\boldsymbol{j} + x_3\boldsymbol{k} \tag{10}$$

A row matrix is also occasionally conveniently identified with a vector.

9.3 INVERSE TRANSFORMATION

From equations (1) of Sec. 9.1,

$$\left.\begin{aligned} y_1 &= a_{11}x_1 + a_{12}x_2 \\ y_2 &= a_{21}x_1 + a_{22}x_2 \end{aligned}\right\} \tag{1}$$

it follows that the mapping of x_1x_2-plane into y_1y_2-plane is *one-to-one*, since each point of the x_1x_2-plane is associated with a unique point of the y_1y_2-plane.

If the determinant of the transformation matrix is nonzero,

$$\Delta(\boldsymbol{A}) = \begin{vmatrix} a_{11} & a_{12} \\ a_{21} & a_{22} \end{vmatrix} \neq 0 \quad \text{or} \quad a_{11}a_{22} - a_{21}a_{12} \neq 0 \tag{2}$$

then, as we know, the system of equations (1) has a unique solution for x_1 and x_2:

$$x_1 = \frac{\begin{vmatrix} y_1 & a_{12} \\ y_2 & a_{22} \end{vmatrix}}{\begin{vmatrix} a_{11} & a_{12} \\ a_{21} & a_{22} \end{vmatrix}}, \qquad x_2 = \frac{\begin{vmatrix} a_{11} & y_1 \\ a_{21} & y_2 \end{vmatrix}}{\begin{vmatrix} a_{11} & a_{12} \\ a_{21} & a_{22} \end{vmatrix}}$$

or, expanded,

$$\left.\begin{aligned} x_1 &= \frac{a_{22}}{\Delta} y_1 + \frac{-a_{12}}{\Delta} y_2 \\ x_2 &= \frac{-a_{21}}{\Delta} y_1 + \frac{a_{11}}{\Delta} y_1 \end{aligned}\right\} \tag{3}$$

To each point $M(y_1, y_2)$ of the y_1y_2-plane there corresponds a definite point $M(x_1, x_2)$ of the x_1x_2-plane. In this case, the mapping (1) is termed a one-to-one (*nonsingular*) mapping. The transformation of coordinates (y_1, y_2) into the coordinates (x_1, x_2) (3) is called an *inverse transformation*. Here, the inverse mapping is linear. Note that a linear nonsingular mapping is called an *affine mapping*. The matrix of an inverse transformation is a matrix which we denote by $\boldsymbol{A}^{-1}$:

$$\boldsymbol{A}^{-1} = \begin{Vmatrix} \frac{a_{22}}{\Delta} & \frac{-a_{12}}{\Delta} \\ \frac{-a_{21}}{\Delta} & \frac{a_{11}}{\Delta} \end{Vmatrix} \tag{4}$$

If the determinant of the matrix A is equal to zero,

$$a_{11}a_{22}-a_{21}a_{12}=0 \tag{5}$$

then the transformation (1) is said to be *singular*. In that case, it is not a one-to-one transformation.

We will prove this. Let us consider two possible cases:

(1) If $a_{11}=a_{12}=a_{21}=a_{22}=0$, then for arbitrary x_1 and x_2 it will be true that $y_1=0$, $y_2=0$. In this case, any point (x_1, x_2) of the x_1x_2-plane will pass into the origin of the y_1y_2-plane.

(2) Let at least one of the coefficients of the transformation be different from zero, say $a_{11}\neq 0$.

Multiplying the first equation of (1) by a_{21}, the second by a_{11} and then subtracting, we get [with regard for equation (5)]

$$\begin{array}{r|l} a_{21} & y_1=a_{11}x_1+a_{12}x_2 \\ a_{11} & y_2=a_{21}x_1+a_{22}x_2 \\ \hline \multicolumn{2}{c}{a_{21}y_1-a_{11}y_2=0} \end{array} \tag{6}$$

Thus, given arbitrary x_1, x_2, we get (6) for the values y_1 and y_2, that is, the corresponding point of the x_1x_2-plane falls on the straight line (6) of the y_1y_2-plane. It is clear that this mapping is not a one-to-one mapping since each point of the straight line (6) of the y_1y_2-plane is associated with a set of points of the x_1x_2-plane lying on the straight line $y_1=a_{11}x_1+a_{12}x_2$.

In neither case is the mapping one-to-one.

Example 1. The transformation

$$y_1=2x_1+x_2$$
$$y_2=x_1-x_2$$

is one-to-one since the determinant $\Delta(A)$ of the transformation matrix A is nonzero:

$$\Delta(A)=\begin{vmatrix} 2 & 1 \\ 1 & -1 \end{vmatrix}=-3$$

The inverse transformation is

$$x_1=\frac{1}{3}y_1+\frac{1}{3}y_2$$
$$x_2=\frac{1}{3}y_1-\frac{2}{3}y_2$$

The matrix of the inverse transformation is, by formula (4),

$$A^{-1}=\begin{Vmatrix} \frac{1}{3} & \frac{1}{3} \\ \frac{1}{3} & -\frac{2}{3} \end{Vmatrix}$$

Example 2. The linear transformation

$$y_1=x_1+2x_2$$
$$y_2=2x_1+4x_2$$

in singular, since the determinant of the transformation matrix

$$\Delta(\boldsymbol{A}) = \begin{vmatrix} 1 & 2 \\ 2 & 4 \end{vmatrix} = 0$$

This transformation carries all points of the x_1x_2-plane into the straight line $y_2 - 2y_1 = 0$ of the y_1y_2-plane.

9.4 OPERATIONS ON MATRICES. ADDITION OF MATRICES

Definition 1. The *sum* of two matrices $\|a_{ij}\|$ and $\|b_{ij}\|$ having the same number of rows and the same number of columns is a matrix $\|c_{ij}\|$ whose element c_{ij} is the sum $a_{ij} + b_{ij}$ of the corresponding elements of the matrices $\|a_{ij}\|$ and $\|b_{ij}\|$; that is,

$$\|a_{ij}\| + \|b_{ij}\| = \|c_{ij}\| \tag{1}$$

if

$$a_{ij} + b_{ij} = c_{ij} \quad (i = 1, 2, \ldots, m; \quad j = 1, 2, \ldots, n) \tag{2}$$

Example 1.

$$\begin{Vmatrix} a_{11} & a_{12} \\ a_{21} & a_{22} \end{Vmatrix} + \begin{Vmatrix} b_{11} & b_{12} \\ b_{21} & b_{22} \end{Vmatrix} = \begin{Vmatrix} a_{11}+b_{11} & a_{12}+b_{12} \\ a_{21}+b_{21} & a_{22}+b_{22} \end{Vmatrix}$$

The *difference* of two matrices is defined in similar fashion.

The expedience of such a definition of the sum of two matrices follows, in particular, from the conception of a vector as a column matrix.

Multiplication of a matrix by a scalar (number). To multiply a matrix by a scalar λ, multiply each element of the matrix by that scalar:

$$\lambda \|a_{ij}\| = \|\lambda a_{ij}\| \tag{3}$$

If λ is integral, then formula (3) is obtained as a consequence of the rule for adding matrices.

Example 2.

$$\lambda \begin{Vmatrix} a_{11} & a_{12} \\ a_{21} & a_{22} \end{Vmatrix} = \begin{Vmatrix} \lambda a_{11} & \lambda a_{12} \\ \lambda a_{21} & \lambda a_{22} \end{Vmatrix}$$

Product of two matrices. Suppose we have a linear transformation of the x_1x_2-plane into the y_1y_2-plane:

$$\left.\begin{aligned} y_1 &= a_{11}x_1 + a_{12}x_2 \\ y_2 &= a_{21}x_1 + a_{22}x_2 \end{aligned}\right\} \tag{4}$$

with the transformation matrix

$$\boldsymbol{A} = \begin{Vmatrix} a_{11} & a_{12} \\ a_{21} & a_{22} \end{Vmatrix} \tag{5}$$

Furthermore, suppose a linear transformation has been performed of the y_1y_2-plane into the z_1z_2-plane:

$$\left.\begin{aligned} z_1 &= b_{11}y_1 + b_{12}y_2 \\ z_2 &= b_{21}y_1 + b_{22}y_2 \end{aligned}\right\} \tag{6}$$

with transformation matrix

$$B = \begin{Vmatrix} b_{11} & b_{12} \\ b_{21} & b_{22} \end{Vmatrix} \tag{7}$$

It is required to determine the matrix of the transformation of the x_1x_2-plane into the z_1z_2-plane. Substituting the expressions (4) into (6) we get

$$\begin{aligned} z_1 &= b_{11}(a_{11}x_1 + a_{12}x_2) + b_{12}(a_{21}x_1 + a_{22}x_2) \\ z_2 &= b_{21}(a_{11}x_1 + a_{12}x_2) + b_{22}(a_{21}x_1 + a_{22}x_2) \end{aligned}$$

or

$$\left.\begin{aligned} z_1 &= (b_{11}a_{11} + b_{12}a_{21})\,x_1 + (b_{11}a_{12} + b_{12}a_{22})\,x_2 \\ z_2 &= (b_{21}a_{11} + b_{22}a_{21})\,x_1 + (b_{21}a_{12} + b_{22}a_{22})\,x_2 \end{aligned}\right\} \tag{8}$$

The matrix of the resulting transformation is

$$C = \begin{Vmatrix} b_{11}a_{11} + b_{12}a_{21} & b_{11}a_{12} + b_{12}a_{22} \\ b_{21}a_{11} + b_{22}a_{21} & b_{21}a_{12} + b_{22}a_{22} \end{Vmatrix} \tag{9}$$

or, briefly,

$$C = \begin{Vmatrix} c_{11} & c_{12} \\ c_{21} & c_{22} \end{Vmatrix} \tag{10}$$

The matrix (9) is called the *product of the matrices* (7) and (5) and we write

$$\begin{Vmatrix} b_{11} & b_{12} \\ b_{12} & b_{22} \end{Vmatrix} \cdot \begin{Vmatrix} a_{11} & a_{12} \\ a_{21} & a_{22} \end{Vmatrix} = \begin{Vmatrix} b_{11}a_{11} + b_{12}a_{21} & b_{11}a_{12} + b_{12}a_{22} \\ b_{21}a_{11} + b_{22}a_{21} & b_{21}a_{12} + b_{22}a_{22} \end{Vmatrix} \tag{11}$$

or, briefly,

$$B \cdot A = C \tag{12}$$

Let us now state the rule for multiplying two matrices B and A if the first contains m rows and k columns, and the second, k rows and n columns.

Schematically we have the equation

$$\begin{Vmatrix} b_{11}\, b_{12} \ldots b_{1k} \\ \cdot\;\cdot\;\cdot\;\cdot\;\cdot\;\cdot\;\cdot \\ b_{i1}\, b_{i2} \ldots b_{ik} \\ \cdot\;\cdot\;\cdot\;\cdot\;\cdot\;\cdot\;\cdot \\ b_{m1}\, b_{m2} \ldots b_{mk} \end{Vmatrix} \cdot \begin{Vmatrix} a_{11}\, a_{12} \ldots a_{1j} \ldots a_{1n} \\ a_{21}\, a_{22} \ldots a_{2j} \ldots a_{2n} \\ \\ \cdot\;\cdot\;\cdot\;\cdot\;\cdot\;\cdot\;\cdot\;\cdot\;\cdot\;\cdot \\ a_{k1}\, a_{k2} \ldots a_{kj} \ldots a_{kn} \end{Vmatrix} = \begin{Vmatrix} c_{11} \ldots\ldots c_{1n} \\ \vdots \qquad\quad \vdots \\ \ldots c_{ij} \ldots \\ \vdots \qquad\quad \vdots \\ c_{m1} \ldots\ldots c_{mn} \end{Vmatrix} \tag{13}$$

Element c_{ij} of matrix $\boldsymbol{C}$, which is the product of matrix $\boldsymbol{B}$ by matrix $\boldsymbol{A}$, is equal to the sum of the products of the elements of row i of matrix $\boldsymbol{B}$ by the corresponding elements of column j of matrix A; that is,

$$c_{ij}=\sum_{\lambda=1}^{k} b_{i\lambda}a_{\lambda j} \qquad (i=1, 2, \ldots, m;\ \ j=1, 2, \ldots, n)$$

Example 3. Suppose

$$\boldsymbol{B}=\begin{Vmatrix} 1 & 0 \\ 0 & 0 \end{Vmatrix}, \qquad \boldsymbol{A}=\begin{Vmatrix} 0 & 0 \\ 1 & 0 \end{Vmatrix}$$

then

$$(1)\ \ \boldsymbol{BA}=\begin{Vmatrix} 1 & 0 \\ 0 & 0 \end{Vmatrix}\cdot\begin{Vmatrix} 0 & 0 \\ 1 & 0 \end{Vmatrix}=\begin{Vmatrix} 0 & 0 \\ 0 & 0 \end{Vmatrix}$$

$$(2)\ \ \boldsymbol{AB}=\begin{Vmatrix} 0 & 0 \\ 1 & 0 \end{Vmatrix}\cdot\begin{Vmatrix} 1 & 0 \\ 0 & 0 \end{Vmatrix}=\begin{Vmatrix} 0 & 0 \\ 1 & 0 \end{Vmatrix}$$

In this example,

$$\boldsymbol{BA}\neq\boldsymbol{AB}$$

We therefore conclude that *in matrix multiplication the commutative law does not hold true.*

Example 4. Given the matrices

$$\boldsymbol{A}=\begin{Vmatrix} 1 & 0 & 0 \\ 0 & 2 & 1 \\ 3 & 0 & 0 \end{Vmatrix}, \qquad \boldsymbol{B}=\begin{Vmatrix} 0 & 1 & 0 \\ 2 & 0 & 1 \\ 1 & 0 & 1 \end{Vmatrix}$$

Find $\boldsymbol{AB}$ and $\boldsymbol{BA}$.

Solution. By formula (13) we have

$$\boldsymbol{AB}=\begin{Vmatrix} 1\cdot0+0\cdot2+0\cdot1 & 1\cdot1+0\cdot0+0\cdot0 & 1\cdot0+0\cdot1+0\cdot1 \\ 0\cdot0+2\cdot2+1\cdot1 & 0\cdot1+2\cdot0+1\cdot0 & 0\cdot0+2\cdot1+1\cdot1 \\ 3\cdot0+0\cdot2+0\cdot1 & 3\cdot1+0\cdot0+0\cdot0 & 3\cdot0+0\cdot1+0\cdot1 \end{Vmatrix}=\begin{Vmatrix} 0 & 1 & 0 \\ 5 & 0 & 3 \\ 0 & 3 & 0 \end{Vmatrix}$$

$$\boldsymbol{BA}=\begin{Vmatrix} 0\cdot1+1\cdot0+0\cdot3 & 0\cdot0+1\cdot2+0\cdot0 & 0\cdot0+1\cdot1+0\cdot0 \\ 2\cdot1+0\cdot0+1\cdot3 & 2\cdot0+0\cdot2+1\cdot0 & 2\cdot0+0\cdot1+1\cdot0 \\ 1\cdot1+0\cdot0+1\cdot3 & 1\cdot0+0\cdot2+1\cdot0 & 1\cdot0+0\cdot1+1\cdot0 \end{Vmatrix}=\begin{Vmatrix} 0 & 2 & 1 \\ 5 & 0 & 0 \\ 4 & 0 & 0 \end{Vmatrix}$$

Example 5. Find the product of the matrices

$$\begin{Vmatrix} a_{11} & a_{12} & a_{13} \\ a_{21} & a_{22} & a_{23} \end{Vmatrix}\cdot\begin{Vmatrix} b_{11} & b_{12} & b_{13} \\ b_{21} & b_{22} & b_{23} \\ b_{31} & b_{32} & b_{33} \end{Vmatrix}$$

$$=\begin{Vmatrix} a_{11}b_{11}+a_{12}b_{21}+a_{13}b_{31} & a_{11}b_{12}+a_{12}b_{22}+a_{13}b_{32} & a_{11}b_{13}+a_{12}b_{23}+a_{13}b_{33} \\ a_{21}b_{11}+a_{22}b_{21}+a_{23}b_{31} & a_{21}b_{12}+a_{22}b_{22}+a_{23}b_{32} & a_{21}b_{13}+a_{22}b_{23}+a_{23}b_{33} \end{Vmatrix}$$

By direct verification we see that the following relations for matrices are valid (k is a scalar, $\boldsymbol{A}$, $\boldsymbol{B}$, $\boldsymbol{C}$ are matrices):

$$(k\boldsymbol{A})\cdot\boldsymbol{B}=\boldsymbol{A}\cdot(k\boldsymbol{B}) \qquad (14)$$

$$(\boldsymbol{A}+\boldsymbol{B})\cdot\boldsymbol{C}=\boldsymbol{A}\cdot\boldsymbol{C}+\boldsymbol{B}\cdot\boldsymbol{C} \qquad (15)$$

$$\boldsymbol{C}\cdot(\boldsymbol{A}+\boldsymbol{B})=\boldsymbol{CA}+\boldsymbol{CB} \qquad (16)$$

$$\boldsymbol{A}(\boldsymbol{BC})=(\boldsymbol{AB})\boldsymbol{C} \qquad (17)$$

From the rules for multiplying a square matrix A by a scalar k and from the rule for taking out a common factor of the elements of the columns of the determinant of a matrix of order n follows

$$\Delta(kA) = k^n \Delta(A) \tag{18}$$

Since multiplication of two square matrices A and B yields a square matrix whose elements are formed by the rule for multiplying determinants, the following equation is clearly valid:

$$\Delta(AB) = \Delta(A) \cdot \Delta(B) \tag{19}$$

Multiplication by the unit matrix. A square matrix whose elements on the principal diagonal are all equal to unity and elsewhere are zero (as mentioned above) is a *unit matrix.*

For instance,

$$E = \begin{Vmatrix} 1 & 0 \\ 0 & 1 \end{Vmatrix} \tag{20}$$

is a unit matrix of order two.

By the rule for multiplication of matrices we get

$$AE = \begin{Vmatrix} a_{11} & a_{12} \\ a_{21} & a_{22} \end{Vmatrix} \cdot \begin{Vmatrix} 1 & 0 \\ 0 & 1 \end{Vmatrix} = \begin{Vmatrix} a_{11} & a_{12} \\ a_{21} & a_{22} \end{Vmatrix}$$

That is,

$$AE = A \tag{21}$$

and also

$$EA = A \tag{22}$$

It is easy to see that the product of a square matrix of any order by the corresponding unit matrix is equal to the original matrix, which means (21) and (22) hold true. In matrix multiplication, the unit matrix plays the role of unity, whence the name "unit matrix" (or identity matrix).

The unit matrix (2) is associated with the transformation

$$y_1 = x_1$$
$$y_2 = x_2$$

This is the *identity transformation.* Conversely, to an identity transformation there corresponds a unit (identity) matrix. In similar fashion we define the identity transformation of any number of variables.

9.5 TRANSFORMING A VECTOR INTO ANOTHER VECTOR BY MEANS OF A MATRIX

Suppose we have a vector

$$X = x_1 i + x_2 j + x_3 k$$

which we will write in the form of a column matrix:

$$\boldsymbol{X}=\begin{Vmatrix} x_1 \\ x_2 \\ x_3 \end{Vmatrix} \tag{1}$$

We transform the projections of this vector with the aid of the matrix

$$\boldsymbol{A}=\begin{Vmatrix} a_{11} & a_{12} & a_{13} \\ a_{21} & a_{22} & a_{23} \\ a_{31} & a_{32} & a_{33} \end{Vmatrix} \tag{2}$$

to get

$$\left.\begin{aligned} y_1 &= a_{11}x_1+a_{12}x_2+a_{13}x_3 \\ y_2 &= a_{21}x_1+a_{22}x_2+a_{23}x_3 \\ y_3 &= a_{31}x_1+a_{32}x_2+a_{33}x_3 \end{aligned}\right\} \tag{3}$$

This yields a new vector

$$\boldsymbol{Y}=y_1\boldsymbol{i}+y_2\boldsymbol{j}+y_3\boldsymbol{k}$$

which can be written as a column vector thus:

$$\boldsymbol{Y}=\begin{Vmatrix} y_1 \\ y_2 \\ y_3 \end{Vmatrix}=\begin{Vmatrix} a_{11}x_1+a_{12}x_2+a_{13}x_3 \\ a_{21}x_1+a_{22}x_2+a_{23}x_3 \\ a_{31}x_1+a_{32}x_2+a_{33}x_3 \end{Vmatrix} \tag{4}$$

Using the matrix-multiplication rule, we can write down the transformation operation as follows:

$$\begin{Vmatrix} a_{11} & a_{12} & a_{13} \\ a_{21} & a_{22} & a_{23} \\ a_{31} & a_{32} & a_{33} \end{Vmatrix}\cdot\begin{Vmatrix} x_1 \\ x_2 \\ x_3 \end{Vmatrix}=\begin{Vmatrix} a_{11}x_1+a_{12}x_2+a_{13}x_3 \\ a_{21}x_1+a_{22}x_2+a_{23}x_3 \\ a_{31}x_1+a_{32}x_2+a_{33}x_3 \end{Vmatrix} \tag{5}$$

That is,

$$\boldsymbol{Y}=\boldsymbol{A}\boldsymbol{X} \tag{6}$$

Multiplication of a square matrix by a column matrix yields a column matrix with the same number of rows.

Clearly, the transformation of a three-dimensional vector $\boldsymbol{X}$ into the vector $\boldsymbol{Y}$ is another way of saying that three-dimensional space is transformed into three-dimensional space.

Note that the system of equations (3) is obtained from the matrix equation (4) by equating the elements of the left-hand matrix and the right-hand matrix.

Equation (4) yields a transformation of the vector $\boldsymbol{X}$ into the vector $\boldsymbol{Y}$ by the matrix $\boldsymbol{A}$.

All the reasoning pertaining to a vector in three-dimensional space can be carried over to a transformation of vectors in a space of any number of dimensions.

9.6 INVERSE MATRIX

Suppose we have a vector X. Let us transform it by means of a square matrix A to get a vector Y:

$$Y = AX \tag{1}$$

Let the determinant of A be nonzero, $\Delta(A) \neq 0$. Then the inverse transformation of the vector Y into the vector X exists. This transformation is found by solving the system of equations (3), Sec. 9.5, for x_1, x_2, x_3. The matrix of an inverse transformation is an *inverse matrix* of A. The inverse of A is denoted by A^{-1}. We can thus write

$$X = A^{-1}Y \tag{2}$$

Here, X is a column matrix, Y is a column matrix, AX is a column matrix, A^{-1} is a square matrix. Substituting into the right member of (2) the right member of (1) in place of Y, we get

$$X = A^{-1}AX \tag{3}$$

Thus, on the vector X we performed a succession of transformations with the matrices A and A^{-1}. That is, we carried out a transformation by means of a matrix that is equal to a product of matrices $(A^{-1}A)$. The result is an identity transformation. Hence, the matrix $A^{-1}A$ is an identity (unit) matrix:

$$A^{-1}A = E \tag{4}$$

Equation (3) takes the form

$$X = EX \tag{5}$$

Theorem 1. *If the matrix A^{-1} is the inverse of the matrix A, then A is the inverse of A^{-1}, and the following equation holds true:*

$$A^{-1}A = AA^{-1} = E \tag{6}$$

Proof. Transform both members of (3) by means of A:

$$AX = A(A^{-1}A)X$$

Using the associative property for matrix multiplication, we can write the last equation as

$$AX = (AA^{-1})AX$$

From this we get

$$AA^{-1} = E \tag{7}$$

which completes the proof.

From equations (4) and (7) it follows that the matrices A and A^{-1} are inverses of each other. It also follows therefrom that

$$(A^{-1})^{-1} = A \tag{8}$$

Indeed, equation (7) implies that

$$A^{-1}(A^{-1})^{-1}=E$$

Comparing this equation with (4), we get equation (8).

9.7 MATRIX INVERSION

Suppose we have a nonsingular matrix

$$A=\begin{Vmatrix} a_{11} & a_{12} & a_{13} \\ a_{21} & a_{22} & a_{23} \\ a_{31} & a_{32} & a_{33} \end{Vmatrix} \tag{1}$$

$$\Delta=\Delta(A)=\begin{vmatrix} a_{11} & a_{12} & a_{13} \\ a_{21} & a_{22} & a_{23} \\ a_{31} & a_{32} & a_{33} \end{vmatrix} \neq 0 \tag{2}$$

We will prove that the inverse matrix A^{-1} will be

$$A^{-1}=\begin{Vmatrix} \frac{A_{11}}{\Delta} & \frac{A_{21}}{\Delta} & \frac{A_{31}}{\Delta} \\ \frac{A_{12}}{\Delta} & \frac{A_{22}}{\Delta} & \frac{A_{32}}{\Delta} \\ \frac{A_{13}}{\Delta} & \frac{A_{23}}{\Delta} & \frac{A_{33}}{\Delta} \end{Vmatrix} \tag{3}$$

where A_{ij} is the cofactor of the element a_{ij} of the determinant $\Delta=\Delta(A)$.

We find the matrix C as the product of the matrices AA^{-1}:

$$C=AA^{-1}=\begin{Vmatrix} a_{11} & a_{12} & a_{13} \\ a_{21} & a_{22} & a_{23} \\ a_{31} & a_{32} & a_{33} \end{Vmatrix} \cdot \begin{Vmatrix} \frac{A_{11}}{\Delta} & \frac{A_{21}}{\Delta} & \frac{A_{31}}{\Delta} \\ \frac{A_{12}}{\Delta} & \frac{A_{22}}{\Delta} & \frac{A_{32}}{\Delta} \\ \frac{A_{13}}{\Delta} & \frac{A_{23}}{\Delta} & \frac{A_{33}}{\Delta} \end{Vmatrix} = \begin{Vmatrix} 1 & 0 & 0 \\ 0 & 1 & 0 \\ 0 & 0 & 1 \end{Vmatrix}$$

Indeed, by the matrix-multiplication rule, the diagonal elements of matrix C are each equal to the sum of the products of the elements of a row of the determinant Δ by the corresponding cofactors divided by the determinant Δ, that is, equal to unity. For example, the element c_{11} is defined as follows:

$$c_{11}=a_{11}\frac{A_{11}}{\Delta}+a_{12}\frac{A_{12}}{\Delta}+a_{13}\frac{A_{13}}{\Delta}=\frac{a_{11}A_{11}+a_{12}A_{12}+a_{13}A_{13}}{\Delta}=1$$

Each off-diagonal element is the sum of the products of the elements of a row by the cofactors of another row divided by the determinant Δ; thus, for example, element c_{23} is

$$c_{23}=a_{21}\frac{A_{31}}{\Delta}+a_{22}\frac{A_{32}}{\Delta}+a_{23}\frac{A_{33}}{\Delta}=\frac{a_{21}A_{31}+a_{22}A_{32}+a_{23}A_{33}}{\Delta}=\frac{0}{\Delta}=0$$

This completes the proof.

Note. The matrix

$$\tilde{A}=\begin{Vmatrix} A_{11} & A_{21} & A_{31} \\ A_{12} & A_{22} & A_{32} \\ A_{13} & A_{23} & A_{33} \end{Vmatrix} \tag{4}$$

is the *adjoint* of A. The inverse matrix A^{-1} is expressed in terms of the adjoint $\tilde{A}$ as follows:

$$A^{-1}=\frac{1}{\Delta(A)}\tilde{A} \tag{5}$$

The truth of this equation follows from (3).

Example. Given a matrix

$$A=\begin{Vmatrix} 1 & 2 & 0 \\ 0 & 3 & 1 \\ 0 & 1 & 2 \end{Vmatrix}$$

Find the inverse A^{-1} and the adjoint matrix $\tilde{A}$.

Solution. We find the determinant of A:

$$\Delta(A)=5$$

Then we find the cofactors

$$\begin{array}{lll} A_{11}=5, & A_{12}=0, & A_{13}=0 \\ A_{21}=-4, & A_{22}=2, & A_{23}=-1 \\ A_{31}=2, & A_{32}=-1, & A_{33}=3 \end{array}$$

Hence, by (3),

$$A^{-1}=\begin{Vmatrix} \frac{5}{5} & -\frac{4}{5} & \frac{2}{5} \\ 0 & \frac{2}{5} & -\frac{1}{5} \\ 0 & -\frac{1}{5} & \frac{3}{5} \end{Vmatrix}$$

Using formula (4) we find the adjoint matrix

$$\tilde{A}=\begin{Vmatrix} 5 & -4 & 2 \\ 0 & 2 & -1 \\ 0 & -1 & 3 \end{Vmatrix}$$

9.8 MATRIX NOTATION FOR SYSTEMS OF LINEAR EQUATIONS AND SOLUTIONS OF SYSTEMS OF LINEAR EQUATIONS

The discussion will be conducted for the case of three-dimensional space. Suppose we have the following system of linear equations:

$$\left.\begin{array}{l} a_{11}x_1+a_{12}x_2+a_{13}x_3=d_1 \\ a_{21}x_1+a_{22}x_2+a_{23}x_3=d_2 \\ a_{31}x_1+a_{32}x_2+a_{33}x_3=d_3 \end{array}\right\} \qquad (1)$$

We consider three matrices:

$$A=\left\|\begin{array}{ccc} a_{11} & a_{12} & a_{13} \\ a_{21} & a_{22} & a_{23} \\ a_{31} & a_{32} & a_{33} \end{array}\right\| \qquad (2)$$

$$X=\left\|\begin{array}{c} x_1 \\ x_2 \\ x_3 \end{array}\right\| \qquad (3)$$

$$D=\left\|\begin{array}{c} d_1 \\ d_2 \\ d_3 \end{array}\right\| \qquad (4)$$

Then, using the matrix-multiplication rule, we can write system (1) in matrix form as follows:

$$\left\|\begin{array}{ccc} a_{11} & a_{12} & a_{13} \\ a_{21} & a_{22} & a_{23} \\ a_{31} & a_{32} & a_{33} \end{array}\right\| \cdot \left\|\begin{array}{c} x_1 \\ x_2 \\ x_3 \end{array}\right\| = \left\|\begin{array}{c} d_1 \\ d_2 \\ d_3 \end{array}\right\| \qquad (5)$$

Indeed, in (5), on the left we have a product of two matrices which is equal to a column matrix whose elements are defined by (5). On the right we again have a column matrix. The two matrices are equal if their elements are equal. Equating the corresponding elements, we get the system of equations (1). The matrix equation (5) is compactly written as

$$AX=D \qquad (6)$$

Example. Write the following system of equations in matrix form:

$$\begin{array}{l} x_1+2x_2 \qquad\quad =5 \\ \qquad 3x_2+\ x_3=9 \\ \qquad\ \ x_2+2x_3=8 \end{array}$$

Solution. Write matrix A of the system, the solution matrix X and the matrix D of constant terms:

$$A=\left\|\begin{array}{ccc} 1 & 2 & 0 \\ 0 & 3 & 1 \\ 0 & 1 & 2 \end{array}\right\|, \quad X=\left\|\begin{array}{c} x_1 \\ x_2 \\ x_3 \end{array}\right\|, \quad D=\left\|\begin{array}{c} 5 \\ 9 \\ 8 \end{array}\right\|$$

In matrix form, this system of linear equations can be written thus:

$$\begin{Vmatrix} 1 & 2 & 0 \\ 0 & 3 & 1 \\ 0 & 1 & 2 \end{Vmatrix} \cdot \begin{Vmatrix} x_1 \\ x_2 \\ x_3 \end{Vmatrix} = \begin{Vmatrix} 5 \\ 9 \\ 8 \end{Vmatrix}$$

9.9 SOLVING SYSTEMS OF LINEAR EQUATIONS BY THE MATRIX METHOD

Let the determinant of a matrix A be $\Delta(A) \neq 0$. Premultiply the left and right members of (6), Sec. 9.8, by A^{-1}, the inverse of A, to get

$$A^{-1}AX = A^{-1}D \tag{1}$$

But

$$A^{-1}A = E, \quad EX = X$$

and so from (1) follows

$$X = A^{-1}D \tag{2}$$

This equation can, with regard for (5) of Sec. 9.7, be written as

$$X = \frac{1}{\Delta(A)} \tilde{A} D \tag{3}$$

or, expanded,

$$\begin{Vmatrix} x_1 \\ x_2 \\ x_3 \end{Vmatrix} = \frac{1}{\Delta(A)} \begin{Vmatrix} A_{11} & A_{21} & A_{31} \\ A_{12} & A_{22} & A_{32} \\ A_{13} & A_{23} & A_{33} \end{Vmatrix} \cdot \begin{Vmatrix} d_1 \\ d_2 \\ d_3 \end{Vmatrix} \tag{4}$$

Multiplying together the matrices on the right, we get

$$\begin{Vmatrix} x_1 \\ x_2 \\ x_3 \end{Vmatrix} = \frac{1}{\Delta(A)} \begin{Vmatrix} d_1A_{11} + d_2A_{21} + d_3A_{31} \\ d_1A_{12} + d_2A_{22} + d_3A_{32} \\ d_1A_{13} + d_2A_{23} + d_3A_{33} \end{Vmatrix} \tag{5}$$

Equating the elements of the matrices on the left and on the right, we get

$$\left.\begin{aligned} x_1 &= \frac{d_1A_{11} + d_2A_{21} + d_3A_{31}}{\Delta} \\ x_2 &= \frac{d_1A_{12} + d_2A_{22} + d_3A_{32}}{\Delta} \\ x_3 &= \frac{d_1A_{13} + d_2A_{23} + d_3A_{33}}{\Delta} \end{aligned}\right\} \tag{6}$$

The solution (6) can be written in terms of determinants

$$\left.\begin{aligned} x_1 = \frac{\begin{vmatrix} d_1 & a_{12} & a_{13} \\ d_2 & a_{22} & a_{23} \\ d_3 & a_{32} & a_{33} \end{vmatrix}}{\begin{vmatrix} a_{11} & a_{12} & a_{13} \\ a_{21} & a_{22} & a_{23} \\ a_{31} & a_{32} & a_{33} \end{vmatrix}}, \quad x_2 = \frac{\begin{vmatrix} a_{11} & d_1 & a_{13} \\ a_{21} & d_2 & a_{23} \\ a_{31} & d_3 & a_{33} \end{vmatrix}}{\begin{vmatrix} a_{11} & a_{12} & a_{13} \\ a_{21} & a_{22} & a_{23} \\ a_{31} & a_{32} & a_{33} \end{vmatrix}} \\ x_3 = \frac{\begin{vmatrix} a_{11} & a_{12} & d_1 \\ a_{21} & a_{22} & d_2 \\ a_{31} & a_{32} & d_3 \end{vmatrix}}{\begin{vmatrix} a_{11} & a_{12} & a_{13} \\ a_{21} & a_{22} & a_{23} \\ a_{31} & a_{32} & a_{33} \end{vmatrix}} \end{aligned}\right\} \tag{7}$$

Example 1. Solve the system of equations

$$\begin{aligned} x_1 + 2x_2 \qquad &= 5 \\ 3x_2 + x_3 &= 9 \\ x_2 + 2x_3 &= 8 \end{aligned}$$

by the matrix method.

Solution. Let us find the determinant of the matrix of the system:

$$\Delta(A) = \begin{Vmatrix} 1 & 2 & 0 \\ 0 & 3 & 1 \\ 0 & 1 & 2 \end{Vmatrix} = 5$$

Let us determine the inverse matrix by formula (3), Sec. 8.7:

$$A^{-1} = \begin{Vmatrix} 1 & -\frac{4}{5} & \frac{2}{5} \\ 0 & \frac{2}{5} & -\frac{1}{5} \\ 0 & -\frac{1}{5} & \frac{3}{5} \end{Vmatrix}$$

The matrix D will be

$$D = \begin{Vmatrix} 5 \\ 9 \\ 8 \end{Vmatrix}$$

We write down the solution in matrix form, via formula (2), as follows:

$$\begin{Vmatrix} x_1 \\ x_2 \\ x_3 \end{Vmatrix} = \begin{Vmatrix} 1 & -\frac{4}{5} & \frac{2}{5} \\ 0 & \frac{2}{5} & -\frac{1}{5} \\ 0 & -\frac{1}{5} & \frac{3}{5} \end{Vmatrix} \cdot \begin{Vmatrix} 5 \\ 9 \\ 8 \end{Vmatrix} = \begin{Vmatrix} 1\cdot 5 - \frac{4}{5}\cdot 9 + \frac{2}{5}\cdot 8 \\ 0\cdot 5 + \frac{2}{5}\cdot 9 - \frac{1}{5}\cdot 8 \\ 0\cdot 5 - \frac{1}{5}\cdot 9 + \frac{3}{5}\cdot 8 \end{Vmatrix}$$

Equating the rows of the matrices on the left and on the right, we get

$$x_1 = 1\cdot 5 - \frac{4}{5}\cdot 9 + \frac{2}{5}\cdot 8 = 1$$

$$x_2 = 0\cdot 5 + \frac{2}{5}\cdot 9 - \frac{1}{5}\cdot 8 = 2$$

$$x_3 = 0\cdot 5 - \frac{1}{5}\cdot 9 + \frac{3}{5}\cdot 8 = 3$$

Example 2. Solve the system of equations

$$x_1 + 2x_2 + x_3 = 0$$
$$2x_1 + x_2 + x_3 = 1$$
$$x_1 + 3x_2 + x_3 = 2$$

by the matrix method.

Solution. Find the determinant of the matrix of the system

$$\Delta(A) = \left\| \begin{matrix} 1 & 2 & 1 \\ 2 & 1 & 1 \\ 1 & 3 & 1 \end{matrix} \right\| = 1 \neq 0$$

We find the inverse matrix

$$A^{-1} = \left\| \begin{matrix} -2 & 1 & 1 \\ -1 & 0 & 1 \\ 5 & -1 & -3 \end{matrix} \right\|$$

Then write the solution of the system in matrix form:

$$\left\| \begin{matrix} x_1 \\ x_2 \\ x_3 \end{matrix} \right\| = \left\| \begin{matrix} -2 & 1 & 1 \\ -1 & 0 & 1 \\ 5 & -1 & -3 \end{matrix} \right\| \cdot \left\| \begin{matrix} 0 \\ 1 \\ 2 \end{matrix} \right\| = \left\| \begin{matrix} 3 \\ 2 \\ -7 \end{matrix} \right\|$$

Equating the rows of the matrices on the right and on the left, we get

$$x_1 = 3, \quad x_2 = 2, \quad x_3 = -7$$

9.10 ORTHOGONAL MAPPINGS. ORTHOGONAL MATRICES

Suppose in three-dimensional space we have two rectangular systems of coordinates (x_1, x_2, x_3) and (x_1', x_2', x_3') having a common origin O. Let a point M have coordinates (x_1, x_2, x_3) and (x_1', x_2', x_3') in the first and second coordinate systems (the origins need not coincide).

Denote by $\boldsymbol{e}_1, \boldsymbol{e}_2, \boldsymbol{e}_3$ the unit vectors on the coordinate axes in the first coordinate system, by $\boldsymbol{e}_1', \boldsymbol{e}_2', \boldsymbol{e}_3'$ the unit vectors in the second coordinate system. The vectors $\boldsymbol{e}_1, \boldsymbol{e}_2, \boldsymbol{e}_3$ are base vectors in the system (x_1, x_2, x_3), the vectors $\boldsymbol{e}_1', \boldsymbol{e}_2', \boldsymbol{e}_3'$ are base vectors in the system (x_1', x_2', x_3').

Then, in the first system the vector $\overline{OM}$ will be written thus:

$$\overline{OM} = x_1\boldsymbol{e}_1 + x_2\boldsymbol{e}_2 + x_3\boldsymbol{e}_3 \qquad (1)$$

in the second system thus:

$$\overline{OM} = x_1' e_1' + x_2' e_2' + x_3' e_3' \tag{2}$$

We consider a transformation of the coordinates x_1, x_2, x_3 of an arbitrary point M into the coordinates x_1', x_2', x_3' of that point. We can say that we are considering a transformation of the space (x_1, x_2, x_3) into the space (x_1', x_2', x_3').

This transformation has the property that a line-segment of length l is carried into a line-segment of the same length l. A triangle is carried into an equal triangle and hence two vectors emanating from a single point with an angle ψ between them go into two vectors of the same length and with the same angle between them.

A transformation having this property is called an *orthogonal transformation*.

We can say that under an orthogonal transformation, the whole space is displaced like a rigid solid, or we have a displacement and a mirror reflection. We now determine the matrix of this transformation.

We express the unit vectors e_1', e_2', e_3' in terms of the unit vectors e_1, e_2, e_3:

$$\left.\begin{aligned} e_1' &= \alpha_{11}e_1 + \alpha_{21}e_2 + \alpha_{31}e_3 \\ e_2' &= \alpha_{12}e_1 + \alpha_{22}e_2 + \alpha_{32}e_3 \\ e_3' &= \alpha_{13}e_1 + \alpha_{23}e_2 + \alpha_{33}e_3 \end{aligned}\right\} \tag{3}$$

Here,

$$\left.\begin{aligned} \alpha_{11} &= \cos(e_1, e_1'),\ \alpha_{12} = \cos(e_1, e_2'),\ \alpha_{13} = \cos(e_1, e_3') \\ \alpha_{21} &= \cos(e_2, e_1'),\ \alpha_{22} = \cos(e_2, e_2'),\ \alpha_{23} = \cos(e_2, e_3') \\ \alpha_{31} &= \cos(e_3, e_1'),\ \alpha_{32} = \cos(e_3, e_2'),\ \alpha_{33} = \cos(e_3, e_3') \end{aligned}\right\} \tag{4}$$

We write the nine direction cosines in the form of a matrix:

$$S = \left\| \begin{matrix} \alpha_{11} & \alpha_{21} & \alpha_{31} \\ \alpha_{12} & \alpha_{22} & \alpha_{32} \\ \alpha_{13} & \alpha_{23} & \alpha_{33} \end{matrix} \right\| \tag{5}$$

Using the relations (4), we can also write

$$\left.\begin{aligned} e_1 &= \alpha_{11}e_1' + \alpha_{12}e_2' + \alpha_{13}e_3' \\ e_2 &= \alpha_{21}e_1' + \alpha_{22}e_2' + \alpha_{23}e_3' \\ e_3 &= \alpha_{31}e_1' + \alpha_{32}e_2' + \alpha_{33}e_3' \end{aligned}\right\} \tag{6}$$

It is clear that the matrix

$$S^* = \left\| \begin{matrix} \alpha_{11} & \alpha_{12} & \alpha_{13} \\ \alpha_{21} & \alpha_{22} & \alpha_{23} \\ \alpha_{31} & \alpha_{32} & \alpha_{33} \end{matrix} \right\| \tag{7}$$

is the transpose of $\boldsymbol{S}$. Since $\boldsymbol{e}_1'$, $\boldsymbol{e}_2'$, $\boldsymbol{e}_3'$ are mutually perpendicular unit vectors, their triple scalar product is equal to ± 1. Thus,

$$(\boldsymbol{e}_1'\boldsymbol{e}_2'\boldsymbol{e}_3') = \begin{Vmatrix} \alpha_{11} & \alpha_{21} & \alpha_{31} \\ \alpha_{12} & \alpha_{22} & \alpha_{32} \\ \alpha_{13} & \alpha_{23} & \alpha_{33} \end{Vmatrix} = \pm 1 \tag{8}$$

Similarly

$$(\boldsymbol{e}_1\boldsymbol{e}_2\boldsymbol{e}_3) = \Delta(\boldsymbol{S}^*) = \begin{Vmatrix} \alpha_{11} & \alpha_{12} & \alpha_{13} \\ \alpha_{21} & \alpha_{22} & \alpha_{23} \\ \alpha_{31} & \alpha_{32} & \alpha_{33} \end{Vmatrix} = \pm 1 \tag{9}$$

We compute the product of the matrices

$$\boldsymbol{S}\boldsymbol{S}^* = \begin{Vmatrix} \alpha_{11} & \alpha_{21} & \alpha_{31} \\ \alpha_{12} & \alpha_{22} & \alpha_{32} \\ \alpha_{13} & \alpha_{23} & \alpha_{33} \end{Vmatrix} \cdot \begin{Vmatrix} \alpha_{11} & \alpha_{12} & \alpha_{13} \\ \alpha_{21} & \alpha_{22} & \alpha_{23} \\ \alpha_{31} & \alpha_{32} & \alpha_{33} \end{Vmatrix} = \begin{Vmatrix} 1 & 0 & 0 \\ 0 & 1 & 0 \\ 0 & 0 & 1 \end{Vmatrix} = \boldsymbol{E} \tag{10}$$

Indeed, if we denote by c_{ij} the elements of the matrix of the product, then

$$\left.\begin{aligned} c_{11} &= \alpha_{11}^2 + \alpha_{21}^2 + \alpha_{31}^2 = 1 \\ c_{22} &= \alpha_{12}^2 + \alpha_{22}^2 + \alpha_{32}^2 = 1 \\ c_{33} &= \alpha_{13}^2 + \alpha_{23}^2 + \alpha_{33}^2 = 1 \end{aligned}\right\} \tag{11}$$

$$c_{12} = \alpha_{11}\alpha_{12} + \alpha_{21}\alpha_{22} + \alpha_{31}\alpha_{32} = (\boldsymbol{e}_1'\boldsymbol{e}_2') = 0$$

Similarly

$$c_{ij} = \boldsymbol{e}_i'\boldsymbol{e}_j' = 0 \quad \text{for } i \neq j \quad (i = 1, 2, 3; \quad j = 1, 2, 3) \tag{12}$$

Thus

$$\boldsymbol{S}\boldsymbol{S}^* = \boldsymbol{E} \tag{13}$$

The transpose $\boldsymbol{S}^*$ coincides with the inverse $\boldsymbol{S}^{-1}$:

$$\boldsymbol{S}^* = \boldsymbol{S}^{-1} \tag{14}$$

A matrix satisfying the conditions (13) or (14)—that is, the inverse of its transpose—is called an *orthogonal matrix*. Let us now find the formulas for transforming the coordinates (x_1, x_2, x_3) into the coordinates (x_1', x_2', x_3'), and vice versa. By virtue of formulas (3) and (6), the right members of (1) and (2) can be expressed either in terms of the basis $(\boldsymbol{e}_1, \boldsymbol{e}_2, \boldsymbol{e}_3)$ or in terms of the basis $(\boldsymbol{e}_1', \boldsymbol{e}_2', \boldsymbol{e}_3')$. Hence, we can write the equation

$$x_1\boldsymbol{e}_1 + x_2\boldsymbol{e}_2 + x_3\boldsymbol{e}_3 = x_1'\boldsymbol{e}_1' + x_2'\boldsymbol{e}_2' + x_3'\boldsymbol{e}_3' \tag{15}$$

Multiplying in succession all terms of (15) by the vector $\boldsymbol{e}_1'$, by the vector $\boldsymbol{e}_2'$, by the vector $\boldsymbol{e}_3'$ and noting that

$$\left.\begin{aligned} \boldsymbol{e}_i'\boldsymbol{e}_j' &= 0 \quad \text{for} \quad i \neq j \\ \boldsymbol{e}_i'\boldsymbol{e}_j' &= 1 \quad \text{for} \quad i = j \\ \boldsymbol{e}_i\boldsymbol{e}_j' &= \alpha_{ij} \end{aligned}\right\} \tag{16}$$

we get

$$\left.\begin{aligned} x_1' &= \alpha_{11}x_1 + \alpha_{21}x_2 + \alpha_{31}x_3 \\ x_2' &= \alpha_{12}x_1 + \alpha_{22}x_2 + \alpha_{32}x_3 \\ x_3' &= \alpha_{13}x_1 + \alpha_{23}x_2 + \alpha_{33}x_3 \end{aligned}\right\} \tag{17}$$

Multiplying the terms of (15) in succession by $\boldsymbol{e}_1$, $\boldsymbol{e}_2$, $\boldsymbol{e}_3$, we get

$$\left.\begin{aligned} x_1 &= \alpha_{11}x_1' + \alpha_{12}x_2' + \alpha_{13}x_3' \\ x_2 &= \alpha_{21}x_1' + \alpha_{22}x_2' + \alpha_{23}x_3' \\ x_3 &= \alpha_{31}x_1' + \alpha_{32}x_2' + \alpha_{33}x_3' \end{aligned}\right\} \tag{18}$$

Thus, $\boldsymbol{S}$ is the matrix of orthogonal transformations (17) and $\boldsymbol{S}^*$ is the matrix of the inverse transformation (18).

It is thus proved that in a Cartesian coordinate system, an *orthogonal transformation* is associated with an *orthogonal matrix*. It can be proved that if the matrices of direct and inverse transformations (17) and (18) satisfy the relation (13) or (14), that is, are orthogonal, then the transformation is orthogonal as well.

If we introduce the column matrices

$$\boldsymbol{X}' = \begin{Vmatrix} x_1' \\ x_2' \\ x_3' \end{Vmatrix}, \quad \boldsymbol{X} = \begin{Vmatrix} x_1 \\ x_2 \\ x_3 \end{Vmatrix} \tag{19}$$

then the systems (17) and (18) can be written thus:

$$\boldsymbol{X}' = \boldsymbol{S}\boldsymbol{X} \tag{20}$$

$$\boldsymbol{X} = \boldsymbol{S}^{-1}\boldsymbol{X}' \tag{21}$$

If we introduce the transposed matrices of (19),

$$\boldsymbol{X}'^* = \| x_1' \ x_2' \ x_3' \|, \quad \boldsymbol{X}^* = \| x_1 \ x_2 \ x_3 \| \tag{22}$$

then we can write

$$\boldsymbol{X}'^* = \boldsymbol{X}^*\boldsymbol{S}^{-1}, \quad \boldsymbol{X}^* = \boldsymbol{X}'^*\boldsymbol{S} \tag{23}$$

9.11 THE EIGENVECTOR OF A LINEAR TRANSFORMATION

Definition 1. Suppose we have a vector $\boldsymbol{X}$:

$$\boldsymbol{X} = \begin{Vmatrix} x_1 \\ x_2 \\ x_3 \end{Vmatrix} \tag{1}$$

where

$$x_1^2 + x_2^2 + x_3^2 \neq 0$$

If after transforming the vector $\boldsymbol{X}$ by means of matrix $\boldsymbol{A}$ [see (2), Sec. 9.5] we get a vector $\boldsymbol{Y}$

$$\boldsymbol{Y} = \boldsymbol{A}\boldsymbol{X} \tag{2}$$

parallel to **X**,

$$\boldsymbol{Y}=\lambda \boldsymbol{X} \tag{3}$$

where λ is a scalar, then the vector **X** is termed the *eigenvector of matrix* ***A*** or the *eigenvector of the given linear transformation*; the scalar λ is termed the *eigenvalue*.

Let us find the eigenvector

$$\boldsymbol{X}=\left\|\begin{matrix} x_1 \\ x_2 \\ x_3 \end{matrix}\right\|$$

of the given linear transformation or of the given matrix **A**. For **X** to be an eigenvector of matrix **A**, it is necessary that (2) and (3) hold true. Equating the right members of these equations, we get

$$\boldsymbol{AX}=\lambda \boldsymbol{X} \tag{4}$$

or

$$\boldsymbol{AX}=\lambda \boldsymbol{EX}$$

that is,

$$(\boldsymbol{A}-\lambda \boldsymbol{E})\boldsymbol{X}=0 \tag{5}$$

From this equation it follows that the vector **X** is defined to within a constant.

Equation (4) can clearly be expanded to

$$\left.\begin{aligned} a_{11}x_1+a_{12}x_2+a_{13}x_3&=\lambda x_1 \\ a_{21}x_1+a_{22}x_2+a_{23}x_3&=\lambda x_2 \\ a_{31}x_1+a_{32}x_2+a_{33}x_3&=\lambda x_3 \end{aligned}\right\} \tag{6}$$

and (5) to

$$\left.\begin{aligned} (a_{11}-\lambda)x_1+a_{12}x_2+a_{13}x_3&=0 \\ a_{21}x_1+(a_{22}-\lambda)x_2+a_{23}x_3&=0 \\ a_{31}x_1+a_{32}x_2+(a_{33}-\lambda)x_3&=0 \end{aligned}\right\} \tag{7}$$

We obtain a system of homogeneous linear equations for determining the coordinates x_1, x_2, x_3 of the vector **X**. For system (7) to have nonzero solutions, it is necessary and sufficient for the determinant of the system to be equal to zero:

$$\begin{vmatrix} a_{11}-\lambda & a_{12} & a_{13} \\ a_{21} & a_{22}-\lambda & a_{23} \\ a_{31} & a_{32} & a_{33}-\lambda \end{vmatrix}=0 \tag{8}$$

or

$$\Delta(\boldsymbol{A}-\lambda \boldsymbol{E})=0 \tag{9}$$

This is a cubic equation in λ. It is called the *characteristic equa-*

tion of the matrix A. From this equation we can find the eigenvalues λ.

Let us consider a case where all roots of the characteristic equation are real and distinct. We denote them by λ_1, λ_2, λ_3.

To each eigenvalue λ there corresponds an eigenvector whose coordinates are determined from the system (7) for an appropriate value of λ. Let us denote the eigenvectors by τ_1, τ_2, τ_3. It can be shown that these vectors are linearly independent, that is to say, none can be expressed in terms of the others. Hence, any vector can be expressed in terms of the vectors τ_1, τ_2, τ_3, that is, they can be taken for base vectors.

We note without proof that all roots of the characteristic equation of a symmetric matrix are real.

Example 1. Find the eigenvectors and the corresponding eigenvalues of the matrix

$$\begin{Vmatrix} 1 & 1 \\ 8 & 3 \end{Vmatrix}$$

Solution. Form the characteristic equation and find the eigenvalues:

$$\begin{vmatrix} 1-\lambda & 1 \\ 8 & 3-\lambda \end{vmatrix} = 0, \text{ i.e., } \lambda^2-4\lambda-5=0, \ \lambda_1=-1, \ \lambda_2=5$$

We find the eigenvector corresponding to the eigenvalue $\lambda_1=-1$ from the appropriate system of equations (7):

$$\begin{aligned} (1-\lambda_1)x_1 + \quad x_2 &= 0 \quad \text{or} \quad 2x_1 + x_2 = 0 \\ 8x_1 + (3-\lambda_1)x_2 &= 0 \qquad\quad 8x_1 + 4x_2 = 0 \end{aligned}$$

Solving this system, we get $x_1=m$, $x_2=-2m$, where m is an arbitrary number. The eigenvector is

$$\tau_1 = mi - 2mj$$

For the eigenvalue $\lambda_2=5$, we write the system of equations

$$\begin{aligned} -4x_1 + x_2 &= 0 \\ 8x_1 - 2x_2 &= 0 \end{aligned}$$

The eigenvector is

$$\tau_2 = mi + 4mj$$

Example 2. Find the eigenvalues and the eigenvectors of the matrix

$$\begin{Vmatrix} 7 & -2 & 0 \\ -2 & 6 & -2 \\ 0 & -2 & 5 \end{Vmatrix}$$

Solution. We write the characteristic equation

$$\begin{vmatrix} 7-\lambda & -2 & 0 \\ -2 & 6-\lambda & -2 \\ 0 & -2 & 5-\lambda \end{vmatrix} = 0, \text{ i.e., } -\lambda^3+18\lambda^2-99\lambda+162=0$$

The roots of this equation are: $\lambda_1=3$, $\lambda_2=6$, $\lambda_3=9$.

For $\lambda_1=3$ the eigenvector is determined from the system of equations

$$\begin{aligned} 4x_1-2x_2 \qquad &=0 \\ -2x_1+3x_2-2x_3&=0 \\ -2x_2+2x_3&=0 \end{aligned}$$

Setting $x_1=m$, we get $x_2=2m$, $x_3=2m$. The eigenvector

$$\tau_1=mi+2mj+2mk$$

Similarly we find

$$\tau_2=mi+\frac{1}{2}mj-mk$$

$$\tau_3=-mi+mj-\frac{1}{2}mk$$

9.12 THE MATRIX OF A LINEAR TRANSFORMATION UNDER WHICH THE BASE VECTORS ARE EIGENVECTORS

We now define the matrix of a linear transformation when the basis consists of eigenvectors τ_1, τ_2, τ_3. The following relations must hold under this transformation:

$$\left.\begin{aligned} \tau_1^*&=\lambda_1\tau_1 \\ \tau_2^*&=\lambda_2\tau_2 \\ \tau_3^*&=\lambda_3\tau_3 \end{aligned}\right\} \qquad (1)$$

where τ_1^*, τ_2^*, τ_3^* are the images of the vectors τ_1, τ_2, τ_3.

Let the transformation matrix be

$$A'=\left\|\begin{matrix} a'_{11} & a'_{12} & a'_{13} \\ a'_{21} & a'_{22} & a'_{23} \\ a'_{31} & a'_{32} & a'_{33} \end{matrix}\right\| \qquad (2)$$

Let us determine the elements of this matrix. Relative to the basis τ_1, τ_2, τ_3, we can write

$$\tau_1=1\cdot\tau_1+0\cdot\tau_2+0\cdot\tau_3=\left\|\begin{matrix}1\\0\\0\end{matrix}\right\|$$

Since the vector τ_1 goes into the vector $\tau_1^*=\lambda_1\tau_1$ after a transformation by the matrix A':

$$\tau_1^*=\lambda_1\tau_1+0\cdot\tau_2+0\cdot\tau_3$$

we can write

$$\tau_1^*=\lambda_1\tau_1=A'\tau_1$$

Hence

$$\left\|\begin{matrix}\lambda_1\\0\\0\end{matrix}\right\|=\left\|\begin{matrix} a'_{11} & a'_{12} & a'_{13} \\ a'_{21} & a'_{22} & a'_{23} \\ a'_{31} & a'_{32} & a'_{33} \end{matrix}\right\|\cdot\left\|\begin{matrix}1\\0\\0\end{matrix}\right\| \qquad (3)$$

or, as a system of equations

$$\left.\begin{aligned} \lambda_1 &= a'_{11}\cdot 1 + a'_{12}\cdot 0 + a'_{13}\cdot 0 \\ 0 &= a'_{21}\cdot 1 + a'_{22}\cdot 0 + a'_{23}\cdot 0 \\ 0 &= a'_{31}\cdot 1 + a'_{32}\cdot 0 + a'_{33}\cdot 0 \end{aligned}\right\} \tag{4}$$

From this system we find

$$a'_{11} = \lambda_1 \quad a'_{21} = 0, \quad a'_{31} = 0$$

On the basis of the relations

$$\tau_2^* = \lambda_2 \tau_2, \quad \tau_3^* = \lambda_3 \tau_3$$

we find in similar fashion

$$\begin{aligned} a'_{12} &= 0, \quad a'_{22} = \lambda_2, \quad a'_{32} = 0 \\ a'_{13} &= 0, \quad a'_{23} = 0, \quad a'_{33} = \lambda_3 \end{aligned}$$

Thus, the matrix of the transformation is of the form

$$A' = \left\| \begin{matrix} \lambda_1 & 0 & 0 \\ 0 & \lambda_2 & 0 \\ 0 & 0 & \lambda_3 \end{matrix} \right\| \tag{5}$$

The linear transformation is

$$\left.\begin{aligned} y'_1 &= \lambda_1 x'_1 \\ y'_2 &= \qquad \lambda_2 x'_2 \\ y'_3 &= \qquad\qquad \lambda_3 x'_3 \end{aligned}\right\} \tag{6}$$

If $\lambda_1 = \lambda_2 = \lambda_3 = \lambda^*$, then the linear transformation assumes the form

$$\begin{aligned} y'_1 &= \lambda^* x'_1 \\ y'_2 &= \lambda^* x'_2 \\ y'_3 &= \lambda^* x'_3 \end{aligned}$$

This is what is called a *similarity transformation* with a coefficient λ^*. Under this transformation, every vector of the space is an eigenvector with eigenvalue λ^*.

9.13 TRANSFORMING THE MATRIX OF A LINEAR TRANSFORMATION WHEN CHANGING THE BASIS

Let X be an arbitrary vector

$$X = \left\| \begin{matrix} x_1 \\ x_2 \\ x_3 \end{matrix} \right\| = x_1 e_1 + x_2 e_2 + x_3 e_3 \tag{1}$$

given in the basis (e_1, e_2, e_3). The vector X is transformed by the matrix $\boldsymbol{A}$ into the vector $\boldsymbol{Y}$:

$$\boldsymbol{Y} = \begin{Vmatrix} y_1 \\ y_2 \\ y_3 \end{Vmatrix} = y_1\boldsymbol{e}_1 + y_2\boldsymbol{e}_2 + y_3\boldsymbol{e}_3 \tag{2}$$

$$\boldsymbol{Y} = \boldsymbol{AX} \tag{3}$$

In the space under consideration, we introduce a new basis $(\boldsymbol{e}'_1, \boldsymbol{e}'_2, \boldsymbol{e}'_3)$ connected with the old basis by the change-of-basis formulas

$$\left.\begin{aligned} \boldsymbol{e}'_1 &= b_{11}\boldsymbol{e}_1 + b_{21}\boldsymbol{e}_2 + b_{31}\boldsymbol{e}_3 \\ \boldsymbol{e}'_2 &= b_{12}\boldsymbol{e}_1 + b_{22}\boldsymbol{e}_2 + b_{32}\boldsymbol{e}_3 \\ \boldsymbol{e}'_3 &= b_{13}\boldsymbol{e}_1 + b_{23}\boldsymbol{e}_2 + b_{33}\boldsymbol{e}_3 \end{aligned}\right\} \tag{4}$$

Let the vector $\boldsymbol{X}$ be written as follows in the new basis:

$$\boldsymbol{X}' = x'_1\boldsymbol{e}'_1 + x'_2\boldsymbol{e}'_2 + x'_3\boldsymbol{e}'_3 = \begin{Vmatrix} x'_1 \\ x'_2 \\ x'_3 \end{Vmatrix} \tag{5}$$

We can write the equation

$$x_1\boldsymbol{e}_1 + x_2\boldsymbol{e}_2 + x_3\boldsymbol{e}_3 = x'_1\boldsymbol{e}'_1 + x'_2\boldsymbol{e}'_2 + x'_3\boldsymbol{e}'_3 \tag{6}$$

where the expressions (4) are substituted into the right-hand side. Equating the coefficients of the vectors $\boldsymbol{e}_1, \boldsymbol{e}_2, \boldsymbol{e}_3$ on the right and left, we get the equations

$$\left.\begin{aligned} x_1 &= b_{11}x'_1 + b_{12}x'_2 + b_{13}x'_3 \\ x_2 &= b_{21}x'_1 + b_{22}x'_2 + b_{23}x'_3 \\ x_3 &= b_{31}x'_1 + b_{32}x'_2 + b_{33}x'_3 \end{aligned}\right\} \tag{7}$$

or, briefly,

$$\boldsymbol{X} = \boldsymbol{BX}' \tag{8}$$

where

$$\boldsymbol{B} = \begin{Vmatrix} b_{11} & b_{12} & b_{13} \\ b_{21} & b_{22} & b_{23} \\ b_{31} & b_{32} & b_{33} \end{Vmatrix} \tag{9}$$

This is a nonsingular matrix that has an inverse, $\boldsymbol{B}^{-1}$, since system (7) has a definite solution for x'_1, x'_2, x'_3. If we write the vector $\boldsymbol{Y}$ in the new basis

$$\boldsymbol{Y}' = y'_1\boldsymbol{e}'_1 + y'_2\boldsymbol{e}'_2 + y'_3\boldsymbol{e}'_3$$

then quite obviously the equation

$$\boldsymbol{Y} = \boldsymbol{BY}' \tag{10}$$

will hold. Substituting (8) and (10) into (3), we get

$$BY' = ABX' \tag{11}$$

Multiplying both members of the equation by B^{-1}, we obtain

$$Y' = B^{-1}ABX' \tag{12}$$

Hence, the transformation matrix A' in the new basis will be

$$A' = B^{-1}AB \tag{13}$$

Example. Suppose, using the matrix A,

$$A = \begin{Vmatrix} 1 & 1 & 0 \\ 1 & 0 & 1 \\ 0 & 1 & 1 \end{Vmatrix}$$

we transform a vector in the basis (e_1, e_2, e_3). Determine the transformation matrix A' in the basis (e_1', e_2', e_3') if

$$e_1' = e_1 + 2e_2 + e_3$$
$$e_2' = 2e_1 + e_2 + 3e_3$$
$$e_3' = e_1 + e_2 + e_3$$

Solution. Here, matrix B is [see formulas (4) and (9)]

$$B = \begin{Vmatrix} 1 & 2 & 1 \\ 2 & 1 & 1 \\ 1 & 3 & 1 \end{Vmatrix}$$

We find the inverse matrix $[\Delta(B) = 1]$

$$B^{-1} = \begin{Vmatrix} -2 & 1 & 1 \\ -1 & 0 & 1 \\ 5 & -1 & -3 \end{Vmatrix}$$

and then we find

$$B^{-1}A = \begin{Vmatrix} -1 & -1 & 2 \\ -1 & 0 & 1 \\ 4 & 2 & -4 \end{Vmatrix}$$

Finally, by formula (13), we get

$$A' = B^{-1}AB = \begin{Vmatrix} -1 & 3 & 0 \\ 0 & 1 & 0 \\ 4 & -2 & 2 \end{Vmatrix}$$

Let us now prove the following theorem.

Theorem 1. *The characteristic polynomial* [*the left member of* (8), Sec. 9.11] *remains unchanged for any choice of basis under a given linear transformation.*

Proof. We write two matrix equations:

$$A' = B^{-1}AB$$
$$E = B^{-1}EB$$

where $\boldsymbol{A}$ and $\boldsymbol{A}'$ are matrices corresponding to different bases for one and the same linear transformation, $\boldsymbol{B}$ is the change-of-basis matrix from new coordinates to old coordinates, and $\boldsymbol{E}$ is the unit (identity) matrix.

On the basis of the last two equations we get

$$\boldsymbol{A}'-\lambda\boldsymbol{E}=\boldsymbol{B}^{-1}(\boldsymbol{A}-\lambda\boldsymbol{E})\boldsymbol{B}$$

Passing from matrices to determinants and using the rule for multiplication of matrices and determinants, we get

$$\Delta(\boldsymbol{A}'-\lambda\boldsymbol{E})=\Delta(\boldsymbol{B}^{-1}(\boldsymbol{A}-\lambda\boldsymbol{E})\boldsymbol{B})=\Delta(\boldsymbol{B}^{-1})\cdot\Delta(\boldsymbol{A}-\lambda\boldsymbol{E})\,\Delta(\boldsymbol{B})$$

But

$$\Delta(\boldsymbol{B}^{-1})\,\Delta(\boldsymbol{B})=\Delta(\boldsymbol{B}^{-1}\boldsymbol{B})=\Delta(\boldsymbol{E})=1$$

Hence

$$\Delta(\boldsymbol{A}'-\lambda\boldsymbol{E})=\Delta(\boldsymbol{A}-\lambda\boldsymbol{E})$$

On the left and on the right we have the characteristic polynomials of the transformation matrices, which proves the theorem.

9.14 QUADRATIC FORMS AND THEIR TPANSFORMATION

Definition 1. A *quadratic form* in several variables is a homogeneous polynomial of degree two in these variables.

A quadratic form in three variables x_1, x_2, x_3 is of the form

$$F=a_{11}x_1^2+a_{22}x_2^2+a_{33}x_3^2+2a_{12}x_1x_2+2a_{13}x_1x_3+2a_{23}x_2x_3 \tag{1}$$

where a_{ij} are specified numbers and the coefficients 2 are taken so as to simplify subsequent formulas.

Equation (1) can be written thus:

$$\begin{aligned}F=&\,x_1(a_{11}x_1+a_{12}x_2+a_{13}x_3)\\&+x_2(a_{21}x_1+a_{22}x_2+a_{23}x_3)\\&+x_3(a_{31}x_1+a_{32}x_2+a_{33}x_3)\end{aligned} \tag{2}$$

where a_{ij} $(i=1,\ 2,\ 3;\ j=1,\ 2,\ 3)$ are given numbers, and

$$a_{12}=a_{21},\quad a_{13}=a_{31},\quad a_{23}=a_{32} \tag{3}$$

The matrix

$$\boldsymbol{A}=\left\|\begin{matrix}a_{11} & a_{12} & a_{13}\\ a_{21} & a_{22} & a_{23}\\ a_{31} & a_{32} & a_{33}\end{matrix}\right\| \tag{4}$$

is called the *matrix of the quadratic form* (1). The given matrix is symmetric.

We will regard $(x_1,\ x_2,\ x_3)$ as the coordinates of a point of space or the coordinates of a vector in the orthogonal basis $(\boldsymbol{e}_1,\ \boldsymbol{e}_2,\ \boldsymbol{e}_3)$, where $\boldsymbol{e}_1$, $\boldsymbol{e}_2$, $\boldsymbol{e}_3$ are unit vectors.

Let us consider a linear transformation in the basis (e_1, e_2, e_3):

$$\left.\begin{aligned} x_1' &= a_{11}x_1 + a_{12}x_2 + a_{13}x_3 \\ x_2' &= a_{21}x_1 + a_{22}x_2 + a_{23}x_3 \\ x_3' &= a_{31}x_1 + a_{32}x_2 + a_{33}x_3 \end{aligned}\right\} \tag{5}$$

The matrix of this transformation coincides with the matrix of the quadratic form.

Let us now define two vectors

$$X = \begin{Vmatrix} x_1 \\ x_2 \\ x_3 \end{Vmatrix} \tag{6}$$

$$X' = \begin{Vmatrix} x_1' \\ x_2' \\ x_3' \end{Vmatrix} \tag{7}$$

We write transformation (5) as

$$X' = AX \tag{8}$$

Then the quadratic form (2) can be represented as a scalar product of these vectors

$$F = X \cdot AX \tag{9}$$

Let e_1', e_2', e_3' be orthogonal eigenvectors of the transformation (8) corresponding to the eigenvalues $\lambda_1, \lambda_2, \lambda_3$. It may be proved that if the matrix is symmetric, then there exists an orthogonal basis composed of the eigenvectors of matrix A. Let us carry out the transformation (8) in the basis (e_1', e_2', e_3'). Then the transformation matrix in this basis will be diagonal (see Sec. 9.12):

$$\tilde{A} = \begin{Vmatrix} \lambda_1 & 0 & 0 \\ 0 & \lambda_2 & 0 \\ 0 & 0 & \lambda_3 \end{Vmatrix} \tag{10}$$

It may be shown that by applying this transformation to the quadratic form (1), we can reduce the latter to the form

$$F = \lambda_1 \tilde{x}_1^2 + \lambda_2 \tilde{x}_2^2 + \lambda_3 \tilde{x}_3^2 \tag{11}$$

The directions of the eigenvectors e_1', e_2', e_3' are called the *principal directions of the quadratic form.*

9.15 THE RANK OF A MATRIX. THE EXISTENCE OF SOLUTIONS OF A SYSTEM OF LINEAR EQUATIONS

Definition 1. The *minor* of a given matrix $\boldsymbol{A}$ is the determinant composed of elements of the matrix left after striking out certain rows and columns.

Example 1. Suppose we have a matrix

$$\begin{Vmatrix} a_{11} & a_{12} & a_{13} & a_{14} \\ a_{21} & a_{22} & a_{23} & a_{24} \\ a_{31} & a_{32} & a_{33} & a_{34} \end{Vmatrix}$$

Third-order minors of this matrix are obtained by striking out one column and replacing the matrix symbol $\| \ \|$ by the sign $| \ |$ of the determinant. There are four. Second-order minors are obtained by striking out two columns and one row. There are 18. First-order minors number 12.

Definition 2. The *rank* of a matrix $\boldsymbol{A}$ is the highest order of a nonzero minor of $\boldsymbol{A}$.

Example 2. It is easy to verifv that the rank of the matrix

$$\begin{Vmatrix} 1 & -1 & 0 \\ 2 & 0 & 1 \\ 1 & 1 & 1 \end{Vmatrix}$$

is equal to 2.

Example 3. The rank of the matrix

$$\begin{Vmatrix} 1 & 2 & 3 \\ 3 & 6 & 9 \end{Vmatrix}$$

is 1.

If a matrix $\boldsymbol{A}$ is square and of order n, the rank k satisfies the relation $k \leqslant n$. As pointed out above, if $k=n$, the matrix is *nonsingular*, if $k<n$, the matrix is *singular*.

For instance, the matrix

$$\begin{Vmatrix} 0 & 1 & 0 \\ 0 & 0 & 1 \\ 1 & 1 & 0 \end{Vmatrix}$$

is nonsingular since $\Delta(A)=1 \neq 0$; the matrix in Example 2 is singular, since there $n=3$ and $k=2$.

The concept of the rank of a matrix is widely used in the theory of systems of linear equations. The following theorem is valid.

Theorem 1. *Given a system of linear equations:*

$$\left.\begin{aligned} a_{11}x_1 + a_{12}x_2 + a_{13}x_3 = b_1 \\ a_{21}x_1 + a_{22}x_2 + a_{23}x_3 = b_2 \\ a_{31}x_1 + a_{32}x_2 + a_{33}x_3 = b_3 \end{aligned}\right\}. \qquad (1)$$

We introduce the matrix of the system:

$$A = \begin{Vmatrix} a_{11} & a_{12} & a_{13} \\ a_{21} & a_{22} & a_{23} \\ a_{31} & a_{32} & a_{33} \end{Vmatrix} \tag{2}$$

and the augmented matrix

$$B = \begin{Vmatrix} a_{11} & a_{12} & a_{13} & b_1 \\ a_{21} & a_{22} & a_{23} & b_2 \\ a_{31} & a_{32} & a_{33} & b_3 \end{Vmatrix} \tag{3}$$

The system (1) *has solutions if the rank of the matrix* **A** *is equal to the rank of the matrix* **B**. *The system does not have any solutions if the rank of* **A** *is less than the rank of* **B**. *If the rank of* **A** *and the rank of* **B** *is* 3, *then the system has a unique solution. If the rank of the matrices* **A** *and* **B** *is equal to* 2, *then the system has an infinitude of solutions; here, two unknowns are expressed in terms of the third, which has an arbitrary value.*

If the rank of the matrices **A** *and* **B** *is equal to* 1, *then the system has an infinitude of solutions and two unknowns have arbitrary values, the third being expressed in terms of these two.*

The validity of this theorem is readily established on the basis of an analysis (familiar in algebra) of the solutions of a system of equations. This theorem holds true for a system of any number of equations.

9.16 DIFFERENTIATION AND INTEGRATION OF MATRICES

Suppose we have a matrix $\| a_{ij}(t) \|$ where the entries (elements) $a_{ij}(t)$ of the matrix are functions of a certain argument t:

$$\| a_{ij}(t) \| = \begin{Vmatrix} a_{11}(t) & a_{12}(t) & \ldots & a_{1n}(t) \\ a_{21}(t) & a_{22}(t) & \ldots & a_{2n}(t) \\ \ldots & \ldots & \ldots & \ldots \\ a_{m1}(t) & a_{m2}(t) & \ldots & a_{mn}(t) \end{Vmatrix} \tag{1}$$

We can write this more compactly as

$$\| a(t) \| = \| a_{ij}(t) \| \quad (i = 1, 2, \ldots, m; \quad j = 1, 2, \ldots, n) \tag{2}$$

Let the elements of the matrix have derivatives

$$\frac{da_{11}(t)}{dt}, \ \ldots, \ \frac{da_{mn}(t)}{dt}$$

Definition 1. The *derivative of a matrix* $\| a(t) \|$ is a matrix, we denote it by $\frac{d}{dt} \| a(t) \|$, whose entries are derivatives of the elements of the matrix $\| a(t) \|$; that is,

$$\frac{d}{dt}\| a(t) \| = \left\| \begin{matrix} \frac{da_{11}}{dt} & \frac{da_{12}}{dt} & \ldots & \frac{da_{1n}}{dt} \\ \frac{da_{21}}{dt} & \frac{da_{22}}{dt} & \ldots & \frac{da_{2n}}{dt} \\ \ldots & \ldots & \ldots & \ldots \\ \frac{da_{m1}}{dt} & \frac{da_{m2}}{dt} & \ldots & \frac{da_{mn}}{dt} \end{matrix} \right\| \tag{3}$$

Note that this definition of the derivative of a matrix comes quite naturally if to the operations of subtraction of matrices and multiplication by a scalar introduced in Sec. 9.4 we adjoin the operation of passage to a limit:

$$\lim_{\Delta t \to 0} \frac{1}{\Delta t} \{ \| a_{ij}(t+\Delta t) \| - \| a_{ij}(t) \| \}$$
$$= \lim_{\Delta t \to 0} \left\| \frac{a_{ij}(t+\Delta t) - a_{ij}(t)}{\Delta t} \right\| = \left\| \lim_{\Delta t \to 0} \frac{a_{ij}(t+\Delta t) - a_{ij}(t)}{\Delta t} \right\|$$

We can write equation (3) more compactly in the following symbolic form:

$$\frac{d}{dt}\| a(t) \| = \left\| \frac{d}{dt} a_{ij}(t) \right\| \tag{4}$$

or

$$\frac{d}{dt}\| a(t) \| = \left\| \frac{d}{dt} a(t) \right\| \tag{5}$$

In place of the differentiation symbol $\frac{d}{dt}$ the symbol D is often used, and then (5) can be written thus:

$$D \| a \| = \| Da \| \tag{6}$$

Definition 2. The *integral* of the matrix $\| a(t) \|$ is a matrix, which we denote as

$$\int_{t_0}^{t} \| a(z) \| \, dz$$

whose elements are equal to the integrals of the elements of the given matrix:

$$\int_{t_0}^{t} \| a(z) \| \, dz = \left\| \begin{matrix} \int_{t_0}^{t} a_{11}(z)\,dz & \ldots & \int_{t_0}^{t} a_{1n}(z)\,dz \\ \int_{t_0}^{t} a_{21}(z)\,dz & \ldots & \int_{t_0}^{t} a_{2n}(z)\,dz \\ \ldots & \ldots & \ldots \\ \int_{t_0}^{t} a_{m1}(z)\,dz & \ldots & \int_{t_0}^{t} a_{mn}(z)\,dz \end{matrix} \right\| \tag{7}$$

More compacty,

$$\int_{t_0}^{t} \| a(z) \| dz = \left\| \int_{t_0}^{t} a_{ij}(z)\, dz \right\| \tag{8}$$

or

$$\int_{t_0}^{t} \| a(z) \| dz = \left\| \int_{t_0}^{t} a(z)\, dz \right\| \tag{9}$$

The symbol $\int_{t_0}^{t} (\quad)\, dz$ is sometimes denoted by a single letter, say S, and then we can write equation (9), like we did (6), thus:

$$S\|a\| = \|Sa\| \tag{10}$$

9.17 MATRIX NOTATION FOR SYSTEMS OF DIFFERENTIAL EQUATIONS AND SOLUTIONS OF SYSTEMS OF DIFFERENTIAL EQUATIONS WITH CONSTANT COEFFICIENTS

We consider a system of n linear differential equations with n unknown functions $x_1(t)$, $x_2(t)$, ..., $x_n(t)$:

$$\left.\begin{array}{l} \frac{dx_1}{dt} = a_{11}x_1 + a_{12}x_2 + \ldots + a_{1n}x_n \\ \frac{dx_2}{dt} = a_{21}x_1 + a_{22}x_2 + \ldots + a_{2n}x_n \\ \ldots\ldots\ldots\ldots\ldots\ldots\ldots \\ \frac{dx_n}{dt} = a_{n1}x_1 + a_{n2}x_2 + \ldots + a_{nn}x_n \end{array}\right\} \tag{1}$$

The coefficients a_{ij} are constants. We introduce the notation

$$\|x\| = \left\| \begin{array}{c} x_1(t) \\ x_2(t) \\ \vdots \\ x_n(t) \end{array} \right\| \tag{2}$$

This is the solution matrix or the vector solution of system (1). We then define the matrix of derivatives of the solutions:

$$\left\| \frac{dx}{dt} \right\| = \left\| \begin{array}{c} \frac{dx_1}{dt} \\ \frac{dx_2}{dt} \\ \vdots \\ \frac{dx_n}{dt} \end{array} \right\| \tag{3}$$

Let us write down the matrix of coefficients of the system of differential equations:

$$\|a\| = \|a_{ij}\| = \begin{Vmatrix} a_{11} & a_{12} & \dots & a_{1n} \\ a_{21} & a_{22} & \dots & a_{2n} \\ \dots & \dots & \dots & \dots \\ a_{n1} & a_{n2} & \dots & a_{nn} \end{Vmatrix} \tag{4}$$

Using the rule for matrix multiplication (see Sec. 9.4), we can write the system of differential equations (1) in matrix form:

$$\begin{Vmatrix} \frac{dx_1}{dt} \\ \frac{dx_2}{dt} \\ \vdots \\ \frac{dx_n}{dt} \end{Vmatrix} = \begin{Vmatrix} a_{11} & a_{12} & \dots & a_{1n} \\ a_{21} & a_{22} & \dots & a_{2n} \\ \dots & \dots & \dots & \dots \\ a_{n1} & a_{n2} & \dots & a_{nn} \end{Vmatrix} \cdot \begin{Vmatrix} x_1 \\ x_2 \\ \vdots \\ x_n \end{Vmatrix} \tag{5}$$

or, more compactly on the basis of the rule for differentiating matrices

$$\frac{d}{dt}\|x(t)\| = \|a\| \cdot \|x\| \tag{6}$$

This equation can also be written as

$$\frac{d\boldsymbol{x}}{dt} = \boldsymbol{a}\boldsymbol{x} \tag{7}$$

where $\boldsymbol{x}$ is also called the *vector solution*; $\boldsymbol{a}$ is short for the matrix $\|a_{ij}\|$.

Suppose we have

$$\|\alpha\| = \boldsymbol{\alpha} = \begin{Vmatrix} \alpha_1 \\ \alpha_2 \\ \vdots \\ \alpha_n \end{Vmatrix} \tag{8}$$

where α_i are certain scalars.

The set of solutions of a system of differential equations will be sought in the form [see formula (2), Sec. 1.30]

$$\|x\| = e^{kt}\|\alpha\| \tag{9}$$

or

$$\boldsymbol{x} = e^{kt}\boldsymbol{\alpha} \tag{10}$$

Substituting (10) into (7) [or (9) into (6)], via the rule for multiplication of a matrix by a scalar and the rule for differentiating matrices, we get

$$ke^{kt}\boldsymbol{\alpha} = \boldsymbol{a}e^{kt}\boldsymbol{\alpha} \tag{11}$$

whence we obtain

$$k\boldsymbol{\alpha}=\boldsymbol{a}\boldsymbol{\alpha}$$

or

$$\boldsymbol{a}\boldsymbol{\alpha}-k\boldsymbol{\alpha}=0 \tag{12}$$

Recall that in the last equation, $\boldsymbol{a}$ is the matrix (4), k is a scalar, and $\boldsymbol{\alpha}$ is the column matrix (8). The matrix in the left member of (12) can be written as

$$(\boldsymbol{a}-k\boldsymbol{E})\,\boldsymbol{\alpha}=0 \tag{13}$$

where $\boldsymbol{E}$ is an identity (unit) matrix of order n. In expanded form, equation (13) can be rewritten thus:

$$\begin{Vmatrix} a_{11}-k & a_{12} & \dots & a_{1n} \\ a_{21} & a_{22}-k & \dots & a_{2n} \\ \dots & \dots & \dots & \dots \\ a_{n1} & a_{n2} & \dots & a_{nn}-k \end{Vmatrix} \cdot \begin{Vmatrix} \alpha_1 \\ \alpha_2 \\ \vdots \\ \alpha_n \end{Vmatrix} = 0 \tag{14}$$

Equation (12) shows that the vector $\boldsymbol{\alpha}$ can be transformed by the matrix $\boldsymbol{a}$ into a parallel vector $k\boldsymbol{\alpha}$. Hence, the vector $\boldsymbol{\alpha}$ is an eigenvector of the matrix $\boldsymbol{a}$ corresponding to the eigenvalue k (see Sec. 9.11).

In scalar form, equation (12) is written as a system of algebraic equations [see system (3), Sec. 1.30]. The scalar k must be determined from equation (5) of Sec. 1.30, which in matrix form can be written

$$\Delta\,(\boldsymbol{a}-k\boldsymbol{E})=0 \tag{15}$$

That is, the determinant

$$\begin{vmatrix} a_{11}-k & a_{12} & \dots & a_{1n} \\ a_{21} & a_{22}-k & \dots & a_{2n} \\ \dots & \dots & \dots & \dots \\ a_{n1} & a_{n2} & \dots & a_{nn}-k \end{vmatrix} = 0 \tag{16}$$

must be equal to zero.

Let all the roots of equation (16),

$$k_1,\ k_2,\ \dots,\ k_n$$

be distinct. For each value k_i of the system (13) we define a matrix of the values of $\boldsymbol{\alpha}$:

$$\begin{Vmatrix} \alpha_1^{(i)} \\ \alpha_2^{(i)} \\ \vdots \\ \alpha_n^{(i)} \end{Vmatrix}$$

(one of these values is arbitrary). Consequently, in matrix form, the solution of system (1) can be written thus:

$$\begin{Vmatrix} x_1 \\ x_2 \\ \vdots \\ x_n \end{Vmatrix} = \begin{Vmatrix} \alpha_1^{(1)} & \alpha_1^{(2)} & \dots & \alpha_1^{(n)} \\ \alpha_2^{(1)} & \alpha_2^{(2)} & \dots & \alpha_2^{(n)} \\ \alpha_n^{(1)} & \alpha_n^{(2)} & \dots & \alpha_n^{(n)} \end{Vmatrix} \cdot \begin{Vmatrix} C_1 e^{k_1 t} \\ C_2 e^{k_2 t} \\ \vdots \\ C_n e^{k_n t} \end{Vmatrix} \tag{17}$$

where C_i are arbitrary constants, or, briefly,

$$\| x \| = \| \alpha \| \| Ce^{kt} \| \tag{18}$$

In scalar form, the solutions are given by formulas (6), Sec. 1.30.

Example 1. Write down in matrix form the system and the solution of the system of linear differential equations

$$\frac{dx_1}{dt} = 2x_1 + 2x_2$$

$$\frac{dx_2}{dt} = x_1 + 3x_2$$

Solution. We write the matrix of the system

$$A = \begin{Vmatrix} 2 & 2 \\ 1 & 3 \end{Vmatrix}$$

In matrix form, the system of equations is written as [see equation (5)]

$$\begin{Vmatrix} \dfrac{dx_1}{dt} \\ \dfrac{dx_2}{dt} \end{Vmatrix} = \begin{Vmatrix} 2 & 2 \\ 1 & 3 \end{Vmatrix} \cdot \begin{Vmatrix} x_1 \\ x_2 \end{Vmatrix}$$

We now form the characteristic equation (15) and find its roots:

$$\begin{vmatrix} 2-k & 2 \\ 1 & 3-k \end{vmatrix} = 0, \quad \text{i.e.,} \quad k^2 - 5k + 4 = 0$$

hence

$$k_1 = 1, \ k_2 = 4$$

We set up system (14) to determine the values $\alpha_1^{(1)}$, $\alpha_2^{(1)}$ for the root $k_1 = 1$:

$$(2-1)\,\alpha_1^{(1)} + \qquad 2\alpha_2^{(1)} = 0$$

$$\alpha_1^{(1)} + (3-1)\,\alpha_2^{(1)} = 0$$

Setting $\alpha_1^{(1)} = 1$, we get $\alpha_2^{(1)} = -\frac{1}{2}$.

In similar fashion we find $\alpha_1^{(2)}$ and $\alpha_2^{(2)}$ which correspond to the root $k_2 = 4$. We obtain

$$\alpha_1^{(2)} = 1, \qquad \alpha_2^{(2)} = 1$$

Now we can write the solution of the system in matrix form [formula (17)]

$$\begin{Vmatrix} x_1 \\ x_2 \end{Vmatrix} = \begin{Vmatrix} 1 & 1 \\ -\frac{1}{2} & 1 \end{Vmatrix} \cdot \begin{Vmatrix} C_1 e^{t} \\ C_2 e^{4t} \end{Vmatrix}$$

or in the usual form:

$$x_1 = C_1 e^t + C_2 e^{4t}$$

$$x_2 = -\frac{1}{2} C_1 e^t + C_2 e^{4t}$$

Example 2. Write in matrix form the system and the solution of the system of differential equations

$$\frac{dx_1}{dt} = x_1$$

$$\frac{dx_2}{dt} = x_1 + 2x_2$$

$$\frac{dx_3}{dt} = x_1 + x_2 + 3x_3$$

Solution. We write down the matrix of the system:

$$A = \begin{Vmatrix} 1 & 0 & 0 \\ 1 & 2 & 0 \\ 1 & 1 & 3 \end{Vmatrix}$$

Thus, in matrix form, the system of equations is written as [see equation (5)]:

$$\begin{Vmatrix} \frac{dx_1}{dt} \\ \frac{dx_2}{dt} \\ \frac{dx_3}{dt} \end{Vmatrix} = \begin{Vmatrix} 1 & 0 & 1 \\ 1 & 2 & 0 \\ 1 & 1 & 3 \end{Vmatrix} \cdot \begin{Vmatrix} x_1 \\ x_2 \\ x_3 \end{Vmatrix}$$

Let us form the characteristic equation (16) and find its roots:

$$\begin{vmatrix} 1-k & 0 & 0 \\ 1 & 2-k & 0 \\ 1 & 1 & 3-k \end{vmatrix} = 1, \quad \text{that is,} \quad (1-k)(2-k)(3-k) = 0$$

consequently,

$$k_1 = 1, \quad k_2 = 2, \quad k_3 = 3$$

We determine $\alpha_1^{(1)}$, $\alpha_2^{(1)}$, $\alpha_3^{(1)}$, which correspond to the root $k_1 = 1$, from the system of equations (14):

$$\alpha_1^{(1)} + \alpha_2^{(1)} = 0$$

$$\alpha_1^{(1)} + \alpha_2^{(1)} + 2\alpha_3^{(1)} = 0$$

to find

$$\alpha_1^{(1)} = 1, \quad \alpha_2^{(1)} = -1, \quad \alpha_3^{(1)} = 0$$

We determine $\alpha_1^{(2)}$, $\alpha_2^{(2)}$, $\alpha_3^{(2)}$, which correspond to the root $k_2 = 2$, from the system

$$-\alpha_1^{(2)} = 0$$

$$\alpha_1^{(2)} = 0$$

$$\alpha_1^{(2)} + \alpha_2^{(2)} + \alpha_3^{(2)} = 0$$

to find

$$\alpha_1^{(2)} = 0, \quad \alpha_2^{(2)} = 1, \quad \alpha_3^{(2)} = -1$$

We determine $\alpha_1^{(3)}, \alpha_2^{(3)}, \alpha_3^{(3)}$, which correspond to the root $k_3=3$, from the system

$$\begin{aligned} -2\alpha_1^{(3)} \qquad &= 0 \\ \alpha_1^{(3)} - \alpha_2^{(3)} &= 0 \\ \alpha_1^{(3)} + \alpha_2^{(3)} &= 0 \end{aligned}$$

to find

$$\alpha_1^{(3)} = 0, \quad \alpha_2^{(3)} = 0, \quad \alpha_3^{(3)} = 1$$

Now let us write down in matrix form the solution of the system [see formula (17)]:

$$\left\| \begin{matrix} x_1 \\ x_2 \\ x_3 \end{matrix} \right\| = \left\| \begin{matrix} 1 & 0 & 0 \\ -1 & 1 & 0 \\ 0 & -1 & 1 \end{matrix} \right\| \cdot \left\| \begin{matrix} C_1e^t \\ C_2e^{2t} \\ C_3e^{3t} \end{matrix} \right\|$$

or in ordinary form

$$\begin{aligned} x_1 &= C_1e^t \\ x_2 &= -C_1e^t + C_2e^{2t} \\ x_3 &= -C_2e^{2t} + C_3e^{3t} \end{aligned}$$

9.18 MATRIX NOTATION FOR A LINEAR EQUATION OF ORDER n

Suppose we have an nth order linear differential equation with constant coefficients:

$$\frac{d^nx}{dt^n} = a_n\frac{d^{n-1}x}{dt^{n-1}} + a_{n-1}\frac{d^{n-2}x}{dt^{n-2}} + \ldots + a_1x \tag{1}$$

We will see later on that this way of numbering the coefficients is convenient. Set $x = x_1$ and

$$\left.\begin{aligned} \frac{dx_1}{dt} &= x_2 \\ \frac{dx_2}{dt} &= x_3 \\ &\cdots \\ \frac{dx_{n-1}}{dt} &= x_n \\ \frac{dx_n}{dt} &= a_1x_1 + a_2x_2 + \ldots + a_nx_n \end{aligned}\right\} \tag{2}$$

Let us write down the matrix of coefficients of the system:

$$\| a^* \| = \left\| \begin{matrix} 0 & 1 & 0 & 0 & \ldots & 0 \\ 0 & 0 & 1 & 0 & \ldots & 0 \\ . & . & . & . & . & . \\ 0 & 0 & 0 & 0 & \ldots & 1 \\ a_1 & a_2 & a_3 & a_4 & \ldots & a_n \end{matrix} \right\| \tag{3}$$

Then, like formula (5) of Sec. 9.17, the system (2) can be written as follows:

$$\left\| \begin{matrix} \frac{dx_1}{dt} \\ \frac{dx_2}{dt} \\ \vdots \\ \frac{dx_{n-1}}{dt} \\ \frac{dx_n}{dt} \end{matrix} \right\| = \left\| \begin{matrix} 0 & 1 & 0 & \dots & 0 \\ 0 & 0 & 1 & \dots & 0 \\ \dots & \dots & \dots & \dots & \dots \\ 0 & 0 & 0 & \dots & 1 \\ a_1 & a_2 & a_3 & \dots & a_n \end{matrix} \right\| \left\| \begin{matrix} x_1 \\ x_2 \\ \vdots \\ x_{n-1} \\ x_n \end{matrix} \right\| \quad (4)$$

or, briefly.

$$\frac{d}{dt} \| x \| = \| a^* \| \cdot \| x \| \quad (5)$$

The subsequent solution is carried through as in Sec. 9.17, since the matrix equation (5) is a particular case of equation (6), Sec. 9.17.

Example. Write the equation

$$\frac{d^2x}{dt^2} = p\frac{dx}{dt} + qx$$

in matrix form.

Solution. Put $x = x_1$. Then

$$\frac{dx_1}{dt} = x_2$$

$$\frac{dx_2}{dt} = px_2 + qx_1$$

The system of equations looks like this in matrix form:

$$\left\| \begin{matrix} \frac{dx_1}{dt} \\ \frac{dx_2}{dt} \end{matrix} \right\| = \left\| \begin{matrix} 0 & 1 \\ q & p \end{matrix} \right\| \cdot \left\| \begin{matrix} x_1 \\ x_2 \end{matrix} \right\|$$

9.19 SOLVING A SYSTEM OF LINEAR DIFFERENTIAL EQUATIONS WITH VARIABLE COEFFICIENTS BY THE METHOD OF SUCCESSIVE APPROXIMATIONS USING MATRIX NOTATION

Let it be required to find the solution of the system of linear differential equations with variable coefficients

$$\left. \begin{aligned} \frac{dx_1}{dt} &= a_{11}(t)\, x_1 + a_{12}(t)\, x_2 + \dots + a_{1n}(t)\, x_n \\ \frac{dx_2}{dt} &= a_{21}(t)\, x_1 + a_{22}(t)\, x_2 + \dots + a_{2n}(t)\, x_n \\ &\dots\dots\dots\dots\dots\dots\dots\dots \\ \frac{dx_n}{dt} &= a_{n1}(t)\, x_1 + a_{n2}(t)\, x_2 + \dots + a_{nn}(t)\, x_n \end{aligned} \right\} \quad (1)$$

that satisfy the initial conditions

$$x_1 = x_{10},\ x_2 = x_{20},\ \ldots,\ x_n = x_{n0} \quad \text{for} \quad t = t_0 \tag{2}$$

If, besides the matrix of coefficients of the system and the matrix of solution, we introduce the matrix of initial conditions

$$\| x_0 \| = \begin{Vmatrix} x_{10} \\ x_{20} \\ \vdots \\ x_{n0} \end{Vmatrix} \tag{3}$$

then the system of equations (1) with initial conditions (2) can be written thus:

$$\frac{d}{dx} \| x \| = \| a(t) \| \cdot \| x \| \tag{4}$$

for the initial conditions

$$\| x \| = \| x_0 \| \quad \text{for} \quad t = t_0 \tag{5}$$

Here, $\| a(t) \|$ is again the matrix of coefficients of the system.

We will solve the problem by the method of successive approximations.

To get a better grasp of the material that follows let us apply the method of successive approximations first to a single linear equation of the first order (see Sec. 4.26).

It is required to find the solution of the single equation

$$\frac{dx}{dt} = a(t)\, x \tag{6}$$

for the initial conditions

$$x = x_0 \quad \text{for} \quad t = t_0 \tag{7}$$

We will assume that $a(t)$ is a continuous function. As was pointed out in Sec. 4.26, the solution of the differential equation (6) for initial conditions (7) reduces to the solution of the integral equation

$$x = x_0 + \int_{t_0}^{t} a(z)\, x(z)\, dz \tag{8}$$

We will solve this equation by the method of successive approximations:

$$\left.\begin{aligned} x_1 &= x_0 + \int_{t_0}^{t} a(z)\, x_0\, dz \\ x_2 &= x_0 + \int_{t_0}^{t} a(z)\, x_1(z)\, dz \\ &\cdots\cdots\cdots\cdots \\ x_m &= x_0 + \int_{t_0}^{t} a(z)\, x_{m-1}(z)\, dz \\ &\cdots\cdots\cdots\cdots \end{aligned}\right\} \tag{9}$$

To save space we introduce the operator S (the integration operator):

$$S(\ ) = \int_{t_0}^{t} (\ )\, dz \tag{10}$$

Using this operator S, we can write the equations (9) as follows:

$$\begin{aligned} x_1 &= x_0 + S(ax_0) \\ x_2 &= x_0 + S(ax_1) = x_0 + S(a(x_0 + S(ax_0))) \\ x_3 &= x_0 + S(a(x_0 + S(a(x_0 + S(ax_0))))) \\ &\cdots\cdots\cdots\cdots\cdots\cdots \\ x_m &= x_0 + S(a(x_0 + S(a(x_0 + S(a(x_0 + S(a\ldots))))))) \end{aligned}$$

Expanding, we get

$$x_m = x_0 + Sax_0 + Sa\,Sax_0 + Sa\,Sa\,Sax_0 + \ldots + \underbrace{Sa\,Sa\,Sa\ldots Sax_0}_{m \text{ times}}$$

Taking x_0 outside the brackets (x_0 constant), we obtain

$$x_m = [1 + Sa + Sa\,Sa + \ldots + \underbrace{Sa\,Sa\ldots Sa}_{m \text{ times}}]\, x_0 \tag{11}$$

It has been proved (see Sec. 4.26) that if $a(t)$ is a continuous function, then the sequence $\{x_m\}$ converges. The limit of this sequence is a convergent series:

$$x = [1 + Sa + Sa\,Sa + \ldots]\, x_0 \tag{12}$$

Note. If $a(t) = \text{const}$, then formula (12) assumes a simple form. Indeed, by (10) we can write

$$Sa = aS1 = a(t - t_0)$$
$$Sa\,Sa = a^2S(t - t_0) = a^2 \frac{(t - t_0)^2}{2}$$
$$\cdots\cdots\cdots\cdots$$
$$\underbrace{Sa\,Sa \ldots Sa}_{m \text{ times}} = a^m \frac{(t - t_0)^m}{m!}$$

In this case, (12) takes the form

$$x = \left[1 + a\frac{t - t_0}{1} + a^2 \frac{(t - t_0)^2}{2!} + \ldots + a^m \frac{(t - t_0)^m}{m!} + \ldots\right] x_0$$

or

$$x = x_0 e^{a(t - t_0)} \tag{13}$$

The method of solving the single equation (6) that we have just reviewed is carried over in its entirety to the solution of system (1) for the initial conditions (2).

In matrix form, system (1) with initial conditions (2) can be written as

$$\frac{d}{dt} \| x \| = \| a(t) \| \cdot \| x \| \tag{14}$$

for the initial conditions

$$\| x \| = \| x_0 \| \quad \text{for} \quad t = t_0 \tag{15}$$

If we use the rule of matrix multiplication and matrix integration, the solution of system (14), given condition (15), can be reduced to the solution of the matrix integral equation

$$\| x(t) \| = \| x_0 \| + \int_{t_0}^{t} \| a(z) \| \cdot \| x(z) \| \, dz \tag{16}$$

We find the successive approximations

$$\| x_m(t) \| = \| x_0 \| + \int_{t_0}^{t} \| a(z) \| \cdot \| x_{m-1}(z) \| \, dz \tag{17}$$

By successive substitution of the successive approximations under the integral, the solution of the system comes out like this in matrix form:

$$\| x(t) \| = \| x_0 \| + \int_{t_0}^{t} \| a(z_1) \| \Big(\| x_0 \| + \int_{t_0}^{z_1} \| a(z_2) \| \Big(\| x_0 \| + \int_{t_0}^{z_2} \| a(z_3) \| (\ldots) \, dz_3 \Big) dz_2 \Big) dz_1$$

or

$$\| x(t) \| = \| x_0 \| + \int_{t_0}^{t} \| a(z_1) \| \cdot \| x_0 \| \, dz_1 + \int_{t_0}^{t} \| a(z_1) \| \int_{t_0}^{z_1} \| a(z_2) \| \cdot \| x_0 \| \, dz_2 \, dz_1 + \ldots \qquad (18)$$

Using the integration operator S, we can write (18) as

$$\| x(t) \| = [\| E \| + S \| a \| + S \| a \| S \| a \| + \ldots] \, \| x_0 \| \qquad (19)$$

The operator in square brackets can be denoted by a single letter. We denote it by $\mathcal{E}_{\| a(t) \|}^{(t_0 t)}$. Then equation (19) is compactly written as

$$\| x(t) \| = \mathcal{E}_{\| a(t) \|}^{(t_0 t)} \| x_0 \| \qquad (20)$$

It is interesting to note that if the coefficients of system (1) are constants, then, using the rule for taking a common factor of all entries of the matrix outside the matrix symbol,* we can write

$$S \| a \| = \frac{t - t_0}{1} \| a \|$$

$$S \| a \| S \| a \| = \frac{(t - t_0)^2}{2!} \| a \|^2$$

$$S \| a \| S \| a \| S \| a \| = \frac{(t - t_0)^3}{3!} \| a \|^3 \text{ and so on}$$

In the case of constant coefficients, formula (19) assumes the form

$$\| x(t) \| = \Big[\| E \| + \frac{t - t_0}{1} \| a \| + \frac{(t - t_0)^2}{1 \cdot 2} \| a \|^2 + \ldots + \frac{(t - t_0)^m}{m!} \| a \|^m + \ldots \Big] \| x_0 \| \qquad (21)$$

This equation can be symbolized in compact form as

$$\| x(t) \| = e^{(t - t_0) \| a \|} \| x_0 \| \qquad (22)$$

* Here we do not discuss the question of passage to a limit for operations performed on matrices.

Exercises on Chapter 9

1. Find the matrix of the inverse transformation for the linear transformation

$$y_1 = 3x_1 + 2x_2, \quad y_2 = 7x_1 + 5x_2$$

Ans. $\begin{Vmatrix} 5 & -2 \\ -7 & 3 \end{Vmatrix}$.

2. Find the matrix of the inverse transformation

$$y_1 = x_1 - x_2, \quad y_2 = x_1$$

Ans. $\begin{Vmatrix} 0 & 1 \\ -1 & 1 \end{Vmatrix}$.

3. Find the product of the matrices $\begin{Vmatrix} 1 & 2 \\ 3 & 4 \end{Vmatrix} \cdot \begin{Vmatrix} 2 & 0 \\ 3 & -1 \end{Vmatrix}$.

Ans. $\begin{Vmatrix} 8 & -2 \\ 18 & -4 \end{Vmatrix}$.

4. Given the matrices

$$A = \begin{Vmatrix} 1 & 2 & 3 \\ 2 & 0 & 1 \\ 3 & -1 & 1 \end{Vmatrix} \quad \text{and} \quad B = \begin{Vmatrix} 5 & 0 & 7 \\ 1 & 2 & 3 \\ -1 & 0 & 2 \end{Vmatrix}$$

Find the matrices AB and BA.

Ans. $\begin{Vmatrix} 4 & 4 & 19 \\ 9 & 0 & 16 \\ 13 & -2 & 20 \end{Vmatrix}$ and $\begin{Vmatrix} 26 & 3 & 22 \\ 14 & -1 & 8 \\ 5 & -4 & -1 \end{Vmatrix}$.

5. Given: $A = \begin{Vmatrix} 1 & 2 & 3 \\ 5 & 7 & 10 \\ 4 & 3 & 1 \end{Vmatrix}$ and E, a third-order unit matrix. Determine the matrix $A + 2E$.

Ans. $\begin{Vmatrix} 3 & 2 & 3 \\ 5 & 9 & 10 \\ 4 & 3 & 3 \end{Vmatrix}$.

6. Given the matrix $A = \begin{Vmatrix} 1 & 2 \\ 3 & 4 \end{Vmatrix}$. Find the matrix A^2.

Ans. $\begin{Vmatrix} 7 & 10 \\ 15 & 22 \end{Vmatrix}$.

7. Suppose $A = \begin{Vmatrix} 1 & 2 \\ 3 & 4 \end{Vmatrix}$. Find $A^2 + 5A$.

Ans. $\begin{Vmatrix} 12 & 20 \\ 30 & 42 \end{Vmatrix}$.

8. Find the inverse of $A = \begin{Vmatrix} 1 & 1 & 1 \\ 5 & 4 & 3 \\ 10 & 5 & 1 \end{Vmatrix}$.

Ans $\begin{Vmatrix} 11 & -4 & 1 \\ -25 & 9 & -2 \\ 15 & -5 & 1 \end{Vmatrix}$.

9. Write down the solutions of the system of equations

$$x_1+2x_2+x_3=10$$
$$2x_1+x_2+x_3=20$$
$$x_1+3x_2+x_3=30$$

in matrix form and find x_1, x_2, x_3.

Ans. $\begin{Vmatrix} x_1 \\ x_2 \\ x_3 \end{Vmatrix} = \begin{Vmatrix} -2 & 1 & 1 \\ -1 & 0 & 1 \\ 5 & -1 & -3 \end{Vmatrix} \cdot \begin{Vmatrix} 10 \\ 20 \\ 30 \end{Vmatrix}$, $x_1=30$, $x_2=20$, $x_3=-60$.

10. Write down in matrix form the solution of the system of equations

$$x_1+x_2+x_3=3$$
$$5x_1+4x_2+3x_3=11$$
$$10x_1+5x_2+x_3=11.5$$

and find x_1, x_2, x_3.

Ans. $\begin{Vmatrix} x_1 \\ x_2 \\ x_3 \end{Vmatrix} = \begin{Vmatrix} 11 & -4 & 1 \\ -25 & 9 & -2 \\ 15 & -5 & 1 \end{Vmatrix} \cdot \begin{Vmatrix} 3 \\ 11 \\ 11.5 \end{Vmatrix}$, $x_1=0.5$, $x_2=1$, $x_3=1.5$.

11. Find the eigenvalues and the eigenvectors of the matrix

$$\begin{Vmatrix} 1 & 2 & -4 \\ 2 & -2 & -2 \\ -4 & -2 & 1 \end{Vmatrix}$$

Ans. $k_1=6$, $k_2=k_3=-3$; $\tau_1=mi+\frac{1}{2}mj-mk$, τ_2 is any vector satisfying the condition $(\tau_1\tau_2)=0$, and m is an arbitrary scalar.

12. Find the eigenvectors of the matrix

$$\begin{Vmatrix} \cos\alpha & \sin\alpha \\ \sin\alpha & \cos\alpha \end{Vmatrix}$$

Ans. They do not exist.

13. Find the eigenvectors of the matrix $\begin{Vmatrix} 4 & 0 & 0 \\ 0 & 4 & 0 \\ 0 & 0 & 4 \end{Vmatrix}$.

Ans. All the vectors are eigenvectors.

14. Solve the following system of linear differential equations by the matrix method:

$$\frac{dx_1}{dt}+x_2=0$$

$$\frac{dx_2}{dt}+4x_1=0$$

Ans. $x_1=C_1e^{2t}+C_2e^{-2t}$, $x_2=-2(C_1e^{2t}-C_2e^{-2t})$.

APPENDIX

TABLE 1

The values of the function $\Phi(x)=\frac{2}{\sqrt{\pi}}\int_0^x e^{-t^2}dt$

and the reduced Laplace function $\hat{\Phi}(x)=\Phi(\rho x)$

x	$\Phi(x)$	Δ	$\hat{\Phi}(x)$	Δ	x	$\Phi(x)$	Δ	$\hat{\Phi}(x)$	Δ
0.00	0.0000	564	0.0000	269	2.25	0.9985	3	0.8709	83
0.05	0.0564	561	0.0269	269	2.30	0.9988	3	0.8792	79
0.10	0.1125	555	0.0538	268	2.35	0.9991	2	0.8871	74
0.15	0.1680	547	0.0806	267	2.40	0.9993	2	0.8945	71
0.20	0.2227	536	0.1073	266	2.45	0.9995	1	0.9016	66
0.25	0.2763	523	0.1339	265	2.50	0.9996	1	0.9082	64
0.30	0.3286	508	0.1604	262	2.55	0.9997	1	0.9146	59
0.35	0.3794	490	0.1866	261	2.60	0.9998	0	0.9205	56
0.40	0.4284	471	0.2127	258	2.65	0.9998	1	0.9261	53
0.45	0.4755	450	0.2385	256	2.70	0.9999	0	0.9314	50
0.50	0.5205	428	0.2641	252	2.75	0.9999	0	0.9364	46
0.55	0.5633	406	0.2893	250	2.80	0.9999	1	0.9410	44
0.60	0.6039	381	0.3143	246	2.85			0.9454	41
0.65	0.6420	358	0.3389	243	2.90			0.9495	39
0.70	0.6778	334	0.3632	238	2.95			0.9534	36
0.75	0.7112	309	0.3870	235	3.00	1.0000		0.9570	33
0.80	0.7421	286	0.4105	231	3.05			0.9603	32
0.85	0.7707	262	0.4336	226	3.10			0.9635	29
0.90	0.7969	240	0.4562	221	3.15			0.9664	27
0.95	0.8209	218	0.4783	217	3.20			0.9691	25
1.00	0.8427	197	0.5000	212	3.25			0.9716	24
1.05	0.8624	178	0.5212	207	3.30			0.9740	21
1.10	0.8802	159	0.5419	201	3.35			0.9761	21
1.15	0.8961	142	0.5620	197	3.40			0.9782	36
1.20	0.9103	126	0.5817	191	3.50			0.9818	30
1.25	0.9229	111	0.6008	186	3.60			0.9848	26
1.30	0.9340	98	0.6194	181	3.70			0.9874	22
1.35	0.9438	85	0.6375	175	3.80			0.9896	19
1.40	0.9523	74	0.6550	169	3.90			0.9915	15
1.45	0.9597	64	0.6719	164	4.00			0.9930	13
1.50	0.9661	55	0.6883	159	4.10			0.9943	11
1.55	0.9716	47	0.7042	153	4.20			0.9954	9
1.60	0.9736	41	0.7195	147	4.30			0.9963	7
1.65	0.9804	34	0.7342	143	4.40			0.9970	6
1.70	0.9838	29	0.7485	136	4.50			0.9976	5
1.75	0.9867	24	0.7621	132	4.60			0.9981	4
1.80	0.9891	20	0.7753	126	4.70			0.9985	3
1.85	0.9911	17	0.7879	121	4.80			0.9988	3
1.90	0.9928	14	0.8000	116	4.90			0.9991	2
1.95	0.9942	11	0.8116	111	5.00			0.9993	1
2.00	0.9953	10	0.8227	105	5.10			0.9994	2
2.05	0.9963	7	0.8332	102	5.20			0.9996	1
2.10	0.9970	6	0.8434	96	5.30			0.9997	0
2.15	0.9976	5	0.8530	92	5.40			0.9997	
2.20	0.9981	4	0.8622	87					

TABLE 2 The values of the function $f(x)=\frac{1}{\sqrt{2\pi}}e^{-\frac{x^2}{2}}$

x	$f(x)$	x	$f(x)$	x	$f(x)$
0.00	0.3989	1.35	0.1604	2.70	0.0104
0.05	0.3984	1.40	0.1497	2.75	0.0091
0.10	0.3970	1.45	0.1394	2.80	0.0079
0.15	0.3945	1.50	0.1295	2.85	0.0069
0.20	0.3910	1.55	0.1200	2.90	0.0060
0.25	0.3867	1.60	0.1109	2.95	0.0051
0.30	0.3814	1.65	0.1023	3.00	0.0044
0.35	0.3752	1.70	0.0940	3.05	0.0038
0.40	0.3683	1.75	0.0863	3.10	0.0033
0.45	0.3605	1.80	0.0790	3.15	0.0028
0.50	0.3521	1.85	0.0721	3.20	0.0024
0.55	0.3429	1.90	0.0656	3.25	0.0020
0.60	0.3332	1.95	0.0596	3.30	0.0017
0.65	0.3230	2.00	0.0540	3.35	0.0015
0.70	0.3123	2.05	0.0488	3.40	0.0012
0.75	0.3011	2.10	0.0440	3.45	0.0010
0.80	0.2897	2.15	0.0396	3.50	0.0009
0.85	0.2780	2.20	0.0355	3.55	0.0007
0.90	0.2661	2.25	0.0317	3.60	0.0006
0.95	0.2541	2.30	0.0283	3.65	0.0005
1.00	0.2420	2.35	0.0252	3.70	0.0004
1.05	0.2299	2.40	0.0224	3.75	0.0004
1.10	0.2179	2.45	0.0198	3.80	0.0003
1.15	0.2059	2.50	0.0175	3.85	0.0002
1.20	0.1942	2.55	0.0154	3.90	0.0002
1.25	0.1826	2.60	0.0136	3.95	0.0002
1.30	0.1714	2.65	0.0119	4.00	0.0001

TABLE 3 The values of the function $\overline{\Phi(x)}=\frac{1}{\sqrt{2\pi}}\int_0^x e^{-\frac{z^2}{2}}dz$

x	$\overline{\Phi(x)}$	x	$\overline{\Phi(x)}$	x	$\overline{\Phi(x)}$
0.00	0.0000	0.95	0.3289	1.90	0.4713
0.01	0.0040	1.00	0.3413	2.00	0.4772
0.05	0.0199	1.05	0.3531	2.10	0.4821
0.10	0.0398	1.10	0.3643	2.20	0.4861
0.15	0.0596	1.15	0.3749	2.30	0.4893
0.20	0.0793	1.20	0.3849	2.40	0.4918
0.25	0.0987	1.25	0.3944	2.50	0.4938
0.30	0.1179	1.30	0.4032	2.60	0.4953
0.35	0.1368	1.35	0.4115	2.70	0.4965
0.40	0.1554	1.40	0.4192	2.80	0.4974
0.45	0.1736	1.45	0.4265	2.90	0.4981
0.50	0.1915	1.50	0.4332	3.00	0.49865
0.55	0.2088	1.55	0.4394	3.20	0.49931
0.60	0.2257	1.60	0.4452	3.40	0.49966
0.65	0.2422	1.65	0.4505	3.60	0.499841
0.70	0.2580	1.70	0.4554	3.80	0.499927
0.75	0.2734	1.75	0.4599	4.00	0.499968
0.80	0.2881	1.80	0.4641	4.50	0.499997
0.85	0.3023	1.85	0.4678	5.00	0.500000
0.90	0.3159				

INDEX